Lecture Notes in Computer Science   14094

Founding Editors

Gerhard Goos
Juris Hartmanis

Editorial Board Members

The series Lecture Notes in Computer Science (LNCS), including its subseries Lecture Notes in Artificial Intelligence (LNAI) and Lecture Notes in Bioinformatics (LNBI), has established itself as a medium for the publication of new developments in computer science and information technology research, teaching, and education.

LNCS enjoys close cooperation with the computer science R & D community, the series counts many renowned academics among its volume editors and paper authors, and collaborates with prestigious societies. Its mission is to serve this international community by providing an invaluable service, mainly focused on the publication of conference and workshop proceedings and postproceedings. LNCS commenced publication in 1973.

Long Yuan · Shiyu Yang · Ruixuan Li ·
Evangelos Kanoulas · Xiang Zhao
Editors

Web Information Systems and Applications

20th International Conference, WISA 2023
Chengdu, China, September 15–17, 2023
Proceedings

 Springer

Editors
Long Yuan
Nanjing University of Science
and Technology
Nanjing, China

Ruixuan Li (iD)
Huazhong University of Science
and Technology
Wuhan, China

Xiang Zhao
National University of Defense Technology
Changsha, China

Shiyu Yang
Guangzhou University
Guangzhou, China

Evangelos Kanoulas (iD)
University of Amsterdam
Amsterdam, The Netherlands

ISSN 0302-9743   ISSN 1611-3349 (electronic)
Lecture Notes in Computer Science
ISBN 978-981-99-6221-1   ISBN 978-981-99-6222-8 (eBook)
https://doi.org/10.1007/978-981-99-6222-8

This Springer imprint is published by the registered company Springer Nature Singapore Pte Ltd.
The registered company address is: 152 Beach Road, #21-01/04 Gateway East, Singapore 189721, Singapore

Paper in this product is recyclable.

Preface

It is our great pleasure to present the proceedings of the 20th Web Information Systems and Applications Conference (WISA 2023). WISA 2023 was organized by the China Computer Federation Technical Committee on Information Systems (CCF TCIS) and Sichuan University. WISA 2023 provided a premium forum for researchers, professionals, practitioners, and officers closely related to information systems and applications to discuss the theme of intelligent information systems, digital transformation, and information system security, focusing on difficult and critical issues in the metaverse, knowledge graphs, blockchains, and recommendation systems, and the promotion of innovative technology for new application areas of information systems.

WISA 2023 was held in Chengdu, Sichuan, China, during September 15–17, 2023.

This year we received 213 submissions, each of which was assigned to at least three Program Committee (PC) members to review. The peer review process was double blind. The thoughtful discussions on each paper by the PC resulted in the selection of 43 full research papers (an acceptance rate of 20.18%) and 9 short papers. The program of WISA 2023 included keynote speeches and topic-specific invited talks by famous experts in various areas of artificial intelligence and information systems to share their cutting-edge technologies and views about the state of the art in academia and industry. The other events included industrial forums and the CCF TCIS salon.

We are grateful to the general chairs, Leye Yao (Sichuan University), Baowen Xu (Nanjing University), and Ge Yu (Northeastern University), as well as all the PC members and external reviewers who contributed their time and expertise to the paper reviewing process. We would like to thank all the members of the Organizing Committee, and the many volunteers, for their great support in the conference organization. Especially, we would also like to thank the publication chairs, Shiyu Yang (Guangzhou University), Long Yuan (Nanjing University of Science and Technology), Yue Kou (Northeastern University), and Ziqiang Yu (Yantai University) for their efforts on the publication of the conference proceedings. Last but not least, many thanks to all the authors who submitted their papers to the conference.

August 2023

Ruixuan Li
Evangelos Kanoulas
Xiang Zhao

Organization

Steering Committee

Evangelos Kanoulas	University of Amsterdam, Netherlands
Baowen Xu	Nanjing University, China
Ge Yu	Northeastern University, China
Chunxiao Xing	Tsinghua University, China
Ruixuan Li	Huazhong University of Science and Technology, China
Xin Wang	Tianjin University, China
Muhammad Aamir Cheema	Monash University, Australia

General Chairs

Leye Yao	Sichuan University, China
Baowen Xu	Nanjing University, China
Ge Yu	Northeastern University, China

Program Committee Chairs

Ruixuan Li	Huazhong University of Science and Technology, China
Evangelos Kanoulas	University of Amsterdam, The Netherlands
Xiang Zhao	National University of Defense Technology, China

Workshop Co-chairs

Chunxiao Xing	Tsinghua University, China
Xin Wang	Tianjin University, China

Publication Co-chairs

Shiyu Yang   Guangzhou University, China
Long Yuan   Nanjing University of Science and Technology,
                              China
Yue Kou   Northeastern University, China
Ziqiang Yu   Yantai University, China

Publicity Co-chairs

Zhenxing Li   Agile Century, China
Weiwei Ni   Southeast University, China
Junyu Lin   Chinese Academy of Sciences, China

Sponsor Co-chair

Bin Cao   Beijing Small & Medium Enterprises Information
                              Service Co., Ltd., China

Website Co-chairs

Yuhui Zhang   Sichuan University, China
Weixin Zeng   National University of Defense Technology,
                              China

Organizing Committee Co-chair

Shenggen Ju   Sichuan University, China

Program Committee

Baoning Niu   Taiyuan University of Technology, China
Baoyan Song   Liaoning University, China
Bin Li   Yangzhou University, China
Bo Fu   Liaoning Normal University, China
Bohan Li   Nanjing University of Aeronautics and
                              Astronautics, China

Bolin Chen	Northwestern Polytechnical University, China
Buyu Wang	Inner Mongolia Agricultural University, China
Chao Kong	Anhui Polytechnic University, China
Chao Lemen	Renmin University, China
Chen Liu	NCUT, China
Chenchen Sun	TJUT, China
Chengcheng Yu	Shanghai Polytechnic University, China
Cheqing Jin	East China Normal University, China
Chuanqi Tao	NUAA, China
Chunyao Song	Nankai University, China
Dan Yin	Beijing University of Civil Engineering and Architecture, China
Derong Shen	Northeastern University, China
Dian Ouyang	Guangzhou University, China
Dong Li	Liaoning University, China
Erping Zhao	Xizang Minzu University, China
Fan Zhang	Guangzhou University, China
Fang Zhou	East China Normal University, China
Feng Qi	East China Normal University, China
Feng Zhao	Huazhong University of Science and Technology, China
Fengda Zhao	Yanshan University, China
Gansen Zhao	South China Normal University, China
Genggeng Liu	Fuzhou University, China
Guan Yuan	China University of Mining and Technology, China
Guanglai Gao	Inner Mongolia University, China
Guigang Zhang	Institute of Automation, Chinese Academy of Sciences, China
Guojiang Shen	Zhejiang University of Technology, China
Guojun Wang	Guangzhou University, China
Haitao Wang	Zhejiang Lab, China
Haiwei Zhang	Nankai University, China
Haofen Wang	Tongji University, China
Hua Yin	Guangdong University of Finance and Economics, China
Huang Mengxing	Hainan University, China
Hui Li	Guizhou University, China
Jiadong Ren	Yanshan University, China
Jiali Mao	East China Normal University, China
Jianbin Qin	Shenzhen Institute of Computing Sciences, Shenzhen University, China

Jianye Yang	Guangzhou University, China
Jiazhen Xi	Huobi Group, China
Jinbao Wang	Harbin Institute of Technology, China
Jinguo You	Kunming University of Science and Technology, China
Jiping Zheng	Nanjing University of Aeronautics and Astronautics, China
Jun Pang	Wuhan University of Science and Technology, China
Jun Wang	iWudao, China
Junying Chen	South China University of Technology, China
Kai Wang	Shanghai Jiao Tong University, China
Kaiqi Zhang	Harbin Institute of Technology, China
Lan You	Hubei University, China
Lei Xu	Nanjing University, China
Li Jiajia	Shenyang Aerospace University, China
Lin Li	Wuhan University of Technology, China
Lin Wang	Wenge Group, China, China
Lina Chen	Zhejiang Normal University, China
Ling Chen	Yangzhou University, China
Lingyun Song	Northwestern Polytechnical University, China
Linlin Ding	Liaoning University, China
Liping Chen	Tarim University, China
Long Yuan	Nanjing University of Science and Technology, China
Luyi Bai	Northeastern University, China
Lyu Ni	East China Normal University, China
Mei Yu	Tianjin University, China
Meihui Zhang	Beijing Institute of Technology, China
Ming Gao	East China Normal University, China
Minghe Yu	Northeastern University, China
Ningyu Zhang	Zhejiang University, China
Peng Cheng	East China Normal University, China
Qian Zhou	Nanjing University of Posts and Communications, China
Qiaoming Zhu	Soochow University, China
Qingsheng Zhu	Chongqing University, China
Qingzhong Li	Shandong University, China
Qinming He	Zhejiang University, China
Ronghua Li	Beijing Institute of Technology, China
Ruixuan Li	Huazhong University of Science and Technology, China

Shanshan Yao	Shanxi University, China
Shaojie Qiao	Chengdu University of Information Technology, China
Sheng Wang	Wuhan University, China
Shengli Wu	Jiangsu University, China
Shi Lin Huang	Mingyang Digital, China
Shiyu Yang	Guangzhou University, China
Shujuan Jiang	China University of Mining and Technology, China
Shumin Han	Northeastern University, China
Shuo Yu	Dalian University of Technology, China
Shurui Fan	Hebei University of Technology, China
Sijia Zhang	Dalian Ocean University, China
Tianxing Wu	Southeast University, China
Tieke He	Nanjing University, China
Tiezheng Nie	Northeastern Univeristy, China
Wei Li	Harbin Engineering University, China
Wei Song	Wuhan University, China
Wei Wang	ECNU, China
Wei Yu	Wuhan University, China
Weiguang Qu	Nanjing Normal University, China
Weiiyu Guo	Central University of Finance and Economics, China
Weijin Jiang	Hunan University of Commerce, China
Weimin Li	Shanghai University, China
Weiwei Ni	Southeast University, China
Xiang Zhao	National University of Defence Technology, China
Xiangfu Meng	Liaoning Technical University, China
Xiangjie Kong	Zhejiang University of Technology, China
Xiangrui Cai	Nankai University, China
Xiaohua Shi	Shanghai Jiao Tong University, China
Xiaojie Yuan	Nankai Univeristy, China
Xiaoran Yan	Zhejiang Lab, China
Ximing Li	Jilin University, China
Xinbiao Gan	NUDT, China
Xingce Wang	Beijing Normal University, China
Xu Liu	SAP Labs China, China
Xu Lizhen	Southeast University, China
Xueqing Zhao	Xi'an Polytechnic University, China
Xuequn Shang	Northwestern Polytechnical University, China
Xuesong Lu	East China Normal University, China

Yajun Yang	Tianjin University, China
Yanfeng Zhang	Northeastern University, China
Yanhui Ding	Shandong Normal University, China
Yanhui Gu	Nanjing Normal University, China
Yanlong Wen	Nankai University, China
Yanping Chen	Guizhou University, China
Ye Liang	Beijing Foreign Studies University, China
Yi Cai	South China University of Technology, China
Ying Zhang	Nankai Univeristy, China
Yinghua Zhou	USTC, China
Yingxia Shao	BUPT, China
Yong Qi	Xi'an Jiaotong University, China
Yong Tang	South China Normal University, China
Yong Zhang	Tsinghua University, China
Yonggong Ren	Liaoning Normal University, China
Yongquan Dong	Jiangsu Normal University, China
Yu Gu	Northeastern University, China
Yuan Li	North China University of Technology, China
Yuanyuan Zhu	Wuhan University, China
Yue Kou	Northeastern University, China
Yuhua Li	Huazhong University of Science and Technology, China
Yupei Zhang	Northwestern Polytechnical University, China
Yuren Mao	Zhejiang University, China
Zhigang Wang	Ocean University of China, China
Zhiyong Peng	Wuhan University, China
Zhongbin Sun	China University of Mining and Technology, China
Zhongle Xie	Zhejiang University, China
Zhuoming Xu	Hohai University, China
Ziqiang Yu	Yantai University, China
Mingdong Zhu	Henan Institute of Technology, China
Shan Lu	Southeast University, China
Bin Xu	Northeastern University, China

Contents

Data Mining and Knowledge Discovery

Recommender Systems

Natural Language Processing

Security, Privacy and Trust

Blockchain

Parallel and Distributed Systems

Database for Artificial Intelligence

Data Mining and Knowledge Discovery

Research on Long Life Product Prognostics Technology Based on Deep Learning and Statistical Information Fusion

Nan Yang[1]([✉]), Guanghao Ren[1], Ruishi Lin[2], Dongpeng Li[2], and Guigang Zhang[1]

[1] Institute of Automation, Chinese Academy of Sciences, Beijing 100190, China
nan.yang@ia.ac.cn
[2] Beijing Aerospace Automatic Control Institute, Beijing 100070, China

Abstract. The simple use of reliability statistical models to predict the remaining life of products lacks specific information about equipment performance degradation, which may lead to low accuracy in predicting the remaining life of equipment after long-term operation. However, due to the slow or even non changing performance of long-life products in the early stages, using only Deep Learning based state assessment techniques will result in lower accuracy in predicting early remaining life. In order to accurately predict the remaining life of a product throughout its entire lifecycle, this paper proposes a residual life prediction model that integrates reliability and performance information. This method identifies equipment performance degradation indicators in multidimensional time series signals through Deep Learning models, and uses a Discrete Random Damage model to establish the relationship between equipment reliability and operating time. Finally, through Bayesian information fusion technology, the reliability and performance evaluation results are integrated into the remaining life indicator, forming an integrated evaluation method for reliability and performance. Compared with the prediction accuracy of simple ordinary Deep Learning models, this method significantly improves the accuracy of early residual life prediction.

Keywords: Residual Life Prediction · Damage Accumulation Model · LSTM · Bayesian Information Fusion

1 Introduction

1.1 Background Introduction

The key components of long-life cycle products such as aviation, energy, shipbuilding and other fields require a long service life and high reliability, and the corresponding residual service life warning cycle is also required to be longer. For example, the remaining life warning time for ordinary products is 1 Day, while for products with a long life cycle, it may be more than 30 Days. Currently, equipment condition maintenance methods based on machine learning and expert systems have become maintenance assistance methods in fields such as aviation, energy, and computing servers [1–3]. However, for products with a long lifespan, most of them still remain based on reliability theory based on statistical methods. Traditional reliability analysis methods only consider the Mean Time Before Failure (MBTF) as an evaluation indicator, resulting in significant errors in the evaluation of equipment after long-term operation [4–8]. In recent years, Deep Learning models such as LSTM and Bi LSTM have become commonly used methods for predicting the remaining lifespan of equipment in research [9–13]. Although deep learning models can detect degradation trends in complex changes in high-dimensional data, for long-life cycle products, their early life degradation is not reflected in monitoring data, and deep learning models cannot capture degradation information for prediction.

Therefore, this article proposes a new method for predicting the Remaining Useful Life (RUL) of long-life cycle products by combining Deep Learning and Reliability Theory. The method identifies performance degradation indicators in high-dimensional monitoring data based on Long short-term memory network model, and then establishes reliability evaluation model using the Discrete Random Damage Accumulation Model (DRDAM). Finally, using Bayesian information fusion technology, the reliability and performance evaluation results were integrated into the remaining life indicators, forming a comprehensive evaluation method for reliability and performance. Compared with simple deep learning models, this method has higher accuracy in predicting the early remaining life of long-life cycle equipment.

1.2   Method Introduction

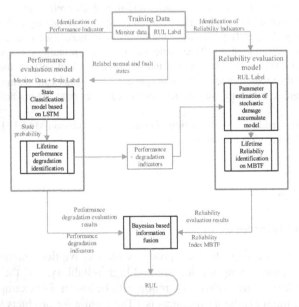

Fig. 1. The flowchart of prognostics based on performance and reliability information fusion

The overall idea of this method is shown in the figure above. This method involves three steps. Firstly, train the LSTM deep learning model using state monitoring data to obtain quantitative indicators of equipment performance degradation at each stage. Integrate high-dimensional monitoring parameters into a single performance indicator through performance degradation indicators. Although this performance indicator is only meaningful before the end of the lifespan of highly reliable products (Fig. 1).

Then, for the health state with unchanged performance degradation indicators, we introduced the DRDAM reliability evaluation model. The Mean Time Before Failure (MBTF) was calculated through DRDAM, and then the running time of the product was subtracted to achieve real-time reliability evaluation of the product.

Finally, based on the Bayesian information fusion method, the reliability and performance evaluation results are combined to obtain the final RUL. Under information fusion, reliability and performance evaluation results play different roles in predicting RUL with different operating times and states, achieving accurate prediction of RUL throughout the entire lifecycle.

2 Reliability Index Prediction Model

2.1 Principle of Damage Accumulation Model

Single impact damage may lead to slight degradation of product performance status. When the impact damage to the product accumulates to a certain extent, the performance of the product will decrease to its fault threshold, and the fault will occur immediately. Record the number of times that impact damage occurs in time $[0, t]$ as N (t). N (t) is a random variable, $\{N(t), t \geq 0\}$ is a point process. Each impact damage causes a certain amount of degradation. Record the cumulative product degradation amount in time $[0, t]$ as $x(t)$. Then the change of degradation amount $x(t)$ with time can be described by the marked point process in the stochastic process.

In engineering practice, degradation prediction $E\{x(t)\}$ is of great significance as it is the main reliability indicator that characterizes the degradation mode of products.

2.2 Discrete Random Damage Accumulation Model

The degradation failure threshold of the product is denoted as l, and the interval time of the i-th impact damage is denoted as $T_i(i = 2, 3...)$. Assuming that $T_i(i = 2, 3...)$ is independent and identically distributed on $H(t)$. Note y_i, $(i = 1, 2, 3...)$ the magnitude of performance degradation caused by the i-th impact damage. Assuming that $y_i(i = 1, 2, 3...)$ is independent and identically distributed on $G(y)$. $N(t)$ represents the number of damage impacts within time, and the cumulative degradation amount at time t can be expressed as:

$$x(t) = \sum_{i=1}^{N(t)} y_i \tag{1}$$

The number of impact damages during the time period $[0, t]$ is $N(t)$, and each impact damage will cause a small amount of degradation. For products with long lifespan and high reliability, the operating time t is generally relatively long, and the number $N(t)$ of impact damages experienced during operation is relatively large. The expected function of the number of impact damages $N(t)$ is $m(t)$, then:

$$m(t) = E[N(t)] = \sum_{n=0}^{\infty} nP\{N(t) = n\} = \sum_{n=0}^{\infty} n[H^{(n)}(t) - H^{(n+1)}(t)] = \sum_{n=1}^{\infty} H^{(n)}(t)$$

$$(2)$$

Represents the average number of impact damages experienced by the internal $[0, t]$, which reflects the law of the impact damage process changing over time. Therefore, when analyzing the cumulative degradation amount within $[0, t]$, the rounding value $m(t)$ can be used instead $N(t)$:

$$x(t) = \sum_{t=0}^{[m(t)]} y_i \tag{3}$$

where $[m(t)]$ represents $m(t)$ rounded. Therefore, the distribution function of the degradation amount $x(t)$ is the $[m(t)]$ convolution of $G(y)$.

2.3 Parameter Estimation

This article constructs an approximate solution for DRDM based on the following assumptions.

Assumption 1: The number of impacts $\{N(t); t \geq 0\}$ is a Poisson process, and its strength is denoted as λ;

Assumption 2: the loss caused by the i-th impact is Δc_i, and it is assumed that $\Delta c_i(i = 1, 2...)$ is independent and identically distributed in the normal distribution $N(\mu_c, \sigma_c^2)$, and its mean and variance are fixed unknown parameters.

If the degradation of device performance within time $[0, t]$ is $\Delta C(t)$ and the performance value at time t is $C(t)$, then

$$\Delta C(t) = C_0 - C(t) \tag{4}$$

The failure threshold of the equipment is $l = 5\% C_0$, and its failure time is T. When the degradation amount reaches the failure threshold, the equipment undergoes degradation failure, and the failure probability at time t is:

$$F(t) = P\{T \leq t\} = P\{\Delta C(t) \geq l\} = 1 - P\{\Delta C(t) < l\} \tag{5}$$

The expected function $E[N(t)] = \lambda t$ of the number of impacts $N(t)$ during time period $[0, t]$ is used to represent the impact pattern of the equipment, that is, it is assumed that the average number of achievable fatigue damages occurring during time period $[0, t]$ is λt, and then the performance degradation of the equipment at time t is $\Delta C(t)$:

$$\Delta C(t) = C_0 - C(t) = \sum_{i=1}^{[\lambda t]} \Delta c_i \tag{6}$$

The mean and variance of the degeneracy $\Delta C(t)$ distribution, respectively:

$$\mu(t) = E[\Delta C(t)] = [\lambda t]E[\Delta c_i] = \mu_c[\lambda t] = \alpha t \tag{7}$$

$$\sigma^2(t) = Var[\Delta C(t)] = [\lambda t]Var[\Delta c_i] = \sigma_c^2[\lambda t] = \beta t \tag{8}$$

Extract m samples from a batch of equipment for impact degradation testing, and regularly measure the performance degradation of all samples. The measurement time is recorded as $t_1 < t_2... < t_n$, and the data format is:

$$\{\Delta c_{ij}; i = 1, 2...m, j = 1, 2,..,n\} \tag{9}$$

$$\hat{\mu}_j = \frac{1}{m}\sum_{i=1}^{m}\Delta c_{ij} \tag{10}$$

$$\hat{\sigma}_j^2 = \frac{1}{m-1}\sum_{i=1}^{m}(\Delta c_{ij} - \hat{\mu}_j)^2 \tag{11}$$

Due to $\mu(t) = \alpha t$ and $\sigma^2(t) = \beta t$, further utilizing the least squares estimates of data $\{(\hat{\mu}_j, \hat{\sigma}_j^2); j = 1, 2, ...n\}$, $\hat{\alpha}$ and $\hat{\beta}$ to estimate α and β, the performance degradation value of the device can be obtained by incorporating the estimated values.

2.4 Definition of Reliability Indicators

MTBF obtained from the DRDAM is a statistical indicator that is only related to the operating time of the equipment and is not related to the specific equipment condition. In order to track the actual operation status and time of electromechanical components and re estimate reliability indicators, this article defines the following formula for calculating reliability indicators based on the characteristics of reliability indicators over time:

$$RX_t = MTBF - TN_t - t \tag{12}$$

Among them, MTBF represents the average failure time of this type of equipment obtained through statistics, and TN_t represents the time it takes for the equipment to run from a fully healthy state to the current time t. The calculation formula is as follows:

$$TN_t = \int_0^\infty td[1 - \Phi\{\frac{DH_{now} - \hat{\alpha}t}{\sqrt{\hat{\beta}t}}\}]$$
$$= \int_0^\infty (\frac{DH_{now}}{2\sqrt{\hat{\beta}t}} - \frac{\hat{\alpha}}{2}\sqrt{\frac{t}{\hat{\beta}}})\phi(\frac{DH_{now} - \hat{\alpha}t}{\sqrt{\hat{\beta}t}})dt \tag{13}$$

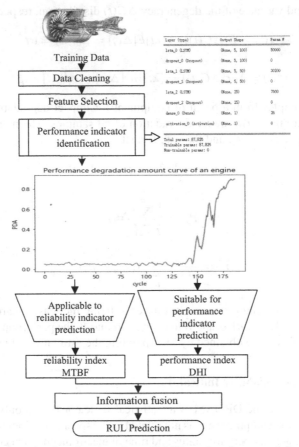

Fig. 2. The prognostics flowchart based on reliability and performance information fusion

3   RUL Prediction Based on Information Fusion

3.1   Performance Degradation Amount Prediction Model

DRDAM assumes that the remaining lifespan or reliability of long-lived devices is solely determined by their usage time. However, performance indicators can more directly reflect the degradation process of equipment life (Fig. 2).

In the PHM08 dataset, each jet engine is associated with 24 dimensional time series data, including 3 control variables and 21 sensor measurements. The dataset provides data from operation to failure, where the first few batches of data for each engine represent sexual health status, and the last few batches indicate that the engine is about to fail or has already failed, with performance degradation approaching or exceeding the threshold. Therefore, we label the first two batches of data for each engine with a random number with a Performance Degradation Amount (PDA) of 0.01–0.05, label the last batch of data for the engine with a random number with a PDA of 0.9–0.91, and label the last batch of data with a PDA of 0.89–0.9. Afterwards, a deep learning model centered on

LSTM was used to train the performance recognition model, as shown in the following figure (Fig. 3).

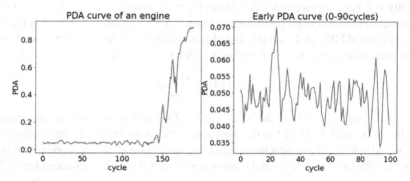

Fig. 3. Lifetime performance degradation amount curve of one engine

From the above experimental results, it can be seen that performance indicators can effectively identify the health status and end-of-life status of equipment. Like most long-life cycle products, their early performance does not change with time, but it rapidly degrades at some point.

The following figure shows the mean and variance variation curves of the performance degradation indicators for 100 engines that we have plotted. From a statistical perspective, their early degradation stage basically conforms to RDAM's assumption (Fig. 4):

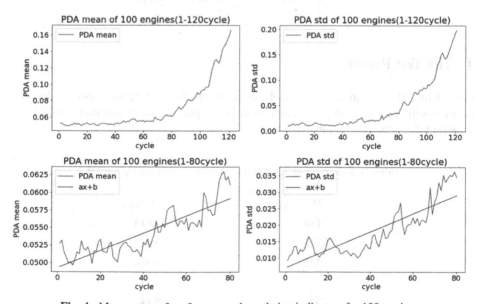

Fig. 4. Mean curve of performance degradation indicators for 100 engines

3.2 Bayes Information Fusion

According to a large number of observations, it can be statistically considered that the reliability and performance index X follows the m-dimensional normal distribution. In this paper, the reliability and performance index X is 2-dimensional, which is the mean time to failure (MTBF) of the reliability index based on the cumulative effect of damage, and the performance index PDA based on the deep learning model, then m = 2:

$$X \sim N_m(\mu(RUL), \sum(RUL)) \tag{14}$$

In the formula, $\mu(RUL)$ is the mean vector of reliability and performance index X under the RUL, and $\sum(RUL)$ is the covariance matrix of reliability and performance index X under RUL. By conducting finite experiments on RUL, the conditional distribution $P(X|RUL)$ of reliability and performance index X can be estimated for all RUL values.

After determining the prior distribution $P(RUL)$ and conditional distribution $P(X|RUL)$, the Bayes formula determines:

$$P(RUL_i|X) = \frac{P(RUL_i)P(X|RUL_i)}{\sum\limits_{j=1}^{k} P(X|RUL_j)P(RUL_j)} \tag{15}$$

The maximum likelihood estimation of RUL is:

$$RUL = E(RUL|X) = \sum\limits_{i}^{k} RUL_i \bullet P(RUL_i|X) \tag{16}$$

4 The Test Result

Predict the RUL of an engine and obtain its reliability and performance evaluation indicators through reliability evaluation models and performance evaluation models. As shown in the table below (Table 1):

Table 1. Reliability and performance indicators of testing engine

CYCLE	$MTBF_{new}$	PDA_{new}
1	150	0.013
2	142	0.012
3	135	0.014
...		...
61	89	0.125

The posterior conditional probability table of residual life $RUL = l$, $l = 1, 2, 3...350$ $l = 1, 2, 3...350$ under the condition of T = 61, reliability RX = 89 and performance index PX = 0.125 can be obtained by bringing the reliability and conditional probability of each residual life corresponding to the performance index at T = 61 into the above formula, as shown below (Table 2):

Table 2. Posterior probability distribution table of RUL of test engine

RUL	$MTBF_{new}$	PDA_{new}	USED TIME	$P(RUL = l \mid RX_{new}, PX_{new})$
1	89	0.125	61	4.1345567e−05
2	89	0.125	61	5.3547985e−05
...		...		...
350	89	0.125	61	7.345527778e−06

According to the posterior conditional probability table of residual life under the reliability and performance indicators of the above test data, the expected value of residual life can be calculated by the following formula through maximum likelihood estimation:

$$\overline{RUL} = E(RUL \mid RX_{new}, PX_{new}) = \sum_{i=1}^{350} i \bullet P(RUL = i \mid RX_{new}, PX_{new}) \qquad (17)$$

By calculation, the expected RUL of this test data is $\overline{RUL} = 86$.

The actual remaining life in this test data is 90, calculated with a fixed value of 350 for the entire life cycle length, and the model's prediction accuracy reaches 98%. Proved the effectiveness of the integrated evaluation method for reliability and performance based on data fusion. The accuracy formula of the remaining life prediction model is defined as follows:

$$accuracy = \left(1 - \frac{abs(real\_rul - predict\_rul)}{total\_life}\right) \times 100\% \qquad (18)$$

Among them: *real_rul* represents the actual RUL, *predict_rul* represents the predicted RUL of the model, while *total_life* represents the entire life cycle of the device.

In order to compare the effectiveness of the method proposed in this article, the LSTM method and Bi-LSTM were used to train a regression prediction model with residual life as the label. The test data of residual life 90, 70, ... 10 were predicted, and the accuracy of the model was evaluated as follows (Table 3):

The above comparison results indicate that compared with simple deep learning models, the method proposed in this paper significantly improves the accuracy of residual life prediction for long life cycle products, with a prediction accuracy of over 90% throughout the entire life cycle.

Table 3. Comparison with LSTM and Bi LSTM methods

ACTUAL RUL	LSTM	BiLSTM	OUR METHOD
90	0.69	0.71	**0.97**
70	0.73	0.88	**0.96**
50	0.95	**0.96**	0.94
30	0.95	**0.96**	0.94
10	**0.98**	**0.98**	0.97
AVERAGE ACC	0.86	0.898	**0.956**

5 Conclusion

This article proposes a fault prediction method based on reliability and performance information fusion. This method utilizes the sensitivity of deep learning models to identify energy degradation indicators in product condition monitoring, and combines the time sensitivity of reliability evaluation models to predict the early decay process of long-life cycle products. Compared with traditional deep learning models, this method significantly improves the accuracy of full life cycle RUL prediction for long-lived products.

References

1. Yao, Q., Wang, J., Zhang, G.: A fault diagnosis expert system based on aircraft parameters. In: Proceedings - 2015 12th Web Information System and Application Conference, WISA 2015, pp. 314–317 (2015).https://doi.org/10.1109/WISA.2015.21
2. Xie, X., Zhang, T., Zhu, Q., Zhang, G.: Design of general aircraft health management system. In: Xing, C., Fu, X., Zhang, Y., Zhang, G., Borjigin, C. (eds.) WISA 2021. LNCS (LNAI and LNB), vol. 12999, pp. 659–667. Springer, Cham (2021). https://doi.org/10.1007/978-3-030-87571-8_57
3. Xu, L., Xu, B., Nie, C.: Testing and fault diagnosis for web application compatibility based on combinatorial method. In: Chen, G., Pan, Y., Guo, M., Lu, J. (eds.) ISPA 2005. LNCS, vol. 3759, pp. 619–626. Springer, Heidelberg (2005). https://doi.org/10.1007/11576259_67
4. Enrico, Z.: Prognostics and health management (PHM): where are we and where do we (need to) go in theory and practice. Reliab. Eng. Syst. Saf. 218(A), 1–16 (2022)
5. Chu, Y., Zhu, Y.: Research on PHM technology framework and its key technologies. In: 2021 IEEE International Conference on Artificial Intelligence and Computer Applications, ICAICA 2021, pp. 952–958 (2021)
6. Yang, H., Miao, X.W.: Prognostics and health management: a review from the perspectives of design, development and decision. Reliab. Eng. Syst. Saf. 217, 1–15 (2021)
7. Heier, H., Mehringskotter, S., Preusche, C.: The use of PHM for a dynamic reliability assessment. In: IEEE Aerospace Conference Proceedings, pp. 1–10 (2018)
8. Compare, M., Bellani, L., Zio, E.: Reliability model of a component equipped with PHM capabilities. Reliab. Eng. Syst. Saf. 168, 4–11 (2017)
9. Khumprom, P., Davila-Frias, A., Grewell, D.: A hybrid evolutionary CNN-LSTM model for prognostics of C-MAPSS aircraft dataset. In: Proceedings - Annual Reliability and Maintainability Symposium (2023)

10. Li, S., Deng, J., Li, Y., Xu, F.: An intermittent fault severity evaluation method for electronic systems based on LSTM network. In: Proceedings - 2022 Prognostics and Health Management Conference, PHM-London, pp. 224–227 (2022)
11. Xu, M., Bai, Y., Qian, P.: Remaining useful life prediction based on improved LSTM hHybrid attention neural network. In: Huang, D.S., Jo, K.H., Jing, J., Premaratne, P., Bevilacqua, V., Hussain, A. (eds.) ICIC 2022. LNCS (LNAI and LNB), vol. 13395, pp. 709–718. Springer, Cham (2022). https://doi.org/10.1007/978-3-031-13832-4_58
12. Rathore, M.S., Harsha, S.P.: Prognostics analysis of rolling bearing based on bi-directional LSTM and attention mechanism. J. Failure Anal. Prevent. **22**(2), 704–723 (2022)
13. Jin, R., Chen, Z., Wu, K., Wu, M., Li, X., Yan, R.: Bi-LSTM-based two-stream network for machine remaining useful life prediction. IEEE Trans. Instrum. Meas. **71**, 1–10 (2022)

A Multi-label Imbalanced Data Classification Method Based on Label Partition Integration

Yuxuan Diao[1,2], Zhongbin Sun[1,2(✉)], and Yong Zhou[1,2]

[1] Mine Digitization Engineering Research Center of Ministry of Education,
Xuzhou 221116, Jiangsu, China
zhongbin@cumt.edu.cn
[2] School of Computer Science and Technology,
China University of Mining and Technology, Xuzhou 221116, Jiangsu, China

Abstract. The problem of multi-label classification is widespread in real life, and its imbalanced characteristics seriously affect classification performance. Currently, resampling methods can be used to solve the problem of imbalanced classification of multi-label data. However, resampling methods ignore the correlation between labels, which may pull in new imbalance while changing the distribution of the original dataset, resulting in a decrease in classification performance instead of an increase. In addition, the resampling ratio needs to be manually set, resulting in significant fluctuations in classification performance. To address this issue, a multi-label imbalanced data classification method ESP based on label partition integration is proposed. ESP divides the dataset into single label datasets and label pair datasets without changing its original distribution, and then learns each dataset to construct multiple binary classification models. Finally, all binary classification models are integrated into a multi-label classification model. The experimental results show that ESP outperforms the five commonly used resampling methods in two common measures: F-Measure and Accuracy.

Keywords: multi-label classification · imbalance · resampling · label partition · model integration

1 Introduction

The problem of multi-label classification widely exists in practical applications, such as biomedical science [3], information security [28] and finance [24]. In recent years, there has been increasing interest in multi-label classification [25] and a number of multi-label learning methods have been proposed [16,30], which can be divided into two different categories, namely problem transformation methods

Supported by the Fundamental Research Funds for the Central Universities under Grant No.2021QN1075.

[4,21] and algorithm adaptation methods [2,11,13]. Both methods have achieved satisfactory classification performance in various multi-label classification tasks.

However, in many practical problems with large number of data, the useful category data is often very limited, only accounting for a small part of the total data. This type of data where the number of samples in a certain category is significantly less than the number of samples in other categories is called imbalanced data, which is prevalent in multi-label data classification [23,27]. For example, in a disease diagnostic problem where the cases of the disease are usually rare as compared to the healthy members of the population, while the main interest of such task is to detect people with diseases. Many traditional classification algorithms usually show poor performance on imbalanced data as they usually ignore the interesting minority class. Thus dealing with the imbalance of multi-label classification tasks is of great significance.

In recent years, a number of different methods have been proposed to address the imbalance data problem in multi-label classification, among which the resampling approaches are the most commonly used techniques due to their independence of the employed classification algorithms [7,9]. However, resampling methods usually change the distribution of the original data and the classification performance of some resampling methods are seriously affected by the sampling ratio as it needs to be manually set.

To address the above issues, the ESP (Ensemble of Single label datasets and label Pair datasets) method is proposed, which does not change the distribution of the original dataset. Specifically, the ESP method first divides the original dataset into single label datasets and label pair datasets and then for each partitioned dataset, a binary classification model is build with a specific classification algorithm. Finally all the binary classification models are integrated into a multi-label data classification model using a specific ensemble rule. In the experiment, six imbalanced datasets from different fields and three popular classification algorithms are employed. The corresponding experimental results demonstrate that ESP is usually better and more stable than five popular resampling methods in terms of two performance metrics, namely Accuracy and F-Measure.

This paper is organized as follows. Section 2 introduces the related work and Sect. 3 presents the ESP method in detail. The experimental results are provided in Sect. 4. Finally, Sect. 5 concludes present study.

2 Related Work

With the increasing interest in multi-label data classification in recent years, a number of resampling methods have been proposed for tackling the class imbalance problem [14,19]. These methods can be divided into oversampling and undersampling methods.

Undersampling methods usually solve the imbalanced data by removing samples associated with the majority class labels. LPRUS [6] is a representative undersampling method based on LP transformation [21]. It randomly removes instances assigned with the most frequent label-set until the number of samples in the multi-label data is reduced to an indicated percentage. However,

LPRUS is usually limited by the label sparseness in the multi-label datasets. Therefore an alternative approach MLRUS [8] is proposed to tackle this limitation. MLRUS is based on the frequency of individual labels, instead of the full label-sets. It resamples the dataset by deleting samples with majority labels. In addition to proposing new methods to solve imbalance, there are many methods that improve on the existing algorithms to solve imbalance problems, such as MLTL [17] which adopts the classic Tomek Link algorithm [20] to address the imbalance. Moreover, MLeNN [9] is built upon the ENN rule [22] to deal with the imbalance. Except for these methods, there are also some other undersampling methods, such as BR-IRUS [18] and ECCRU3 [15].

Oversampling methods generate new samples associated with the minority labels for handling the data imbalance problem. LPROS [6] is the oversampling algorithm corresponding to LPRUS, which balances the dataset by duplicating samples of minority label-sets randomly until the size of the multi-label dataset increases by the prespecified percentage. MLROS [8] is also corresponding to MLRUS, which resamples the dataset by cloning samples with minority labels. SMOTE [10] is a representative oversampling method which solves imbalanced problems by manually synthesizing new samples by searching for nearest neighbor samples of minority samples. However, it can't be used to deal with multi-label data. Therefore an extension of SMOTE for multi-label data named MLSMOTE [5] is proposed. MLSMOTE considers a list of minority labels using the instances in which these labels appear as seeds to generate new instances. In addition, ML-BFO [1] is an oversampling method for duplicating fewer samples based on imbalance minimization and ENN rule.

In summary, both oversampling and undersampling methods operate on the original dataset to address imbalance in which case the corresponding resampling method may introduce new imbalance. The new imbalance may lead to a decrease in classification performance. What's more, some sampling methods may need to set the sampling ratio manually and an unreasonable sampling ratio may not improve the classification performance.

3   ESP Method

3.1   Overview

In order to make full use of the original dataset, the ESP method is proposed and the framework of ESP is represented in Fig. 1. Particularly, the ESP method is mainly divided into three different parts: Label Partition, Binary Classification Model Construction, and Model Integration.

3.2   Label Partition

Considering the co-occurrence of labels in multi-label classification problems, ESP firstly divides multi-label datasets into single label datasets and label pair datasets. A single label dataset is constructed for each label, removing excess

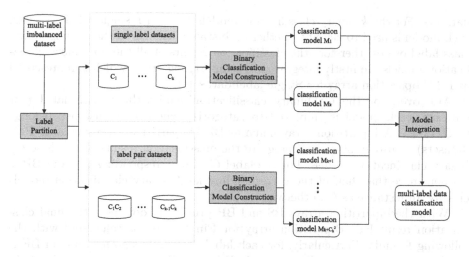

Fig. 1. The framework of ESP.

labels from the instance and retaining only one separate label. For each instance in the multi-label data, if the instance carries a specific label, it is marked as 1. Otherwise it is marked as 0. Thus a multi-label dataset with k labels is partitioned into k different single label datasets.

Label pair datasets are established for a pair of different labels. For a pair of given labels C_i and C_j, the instances only belonging one of them are employed to construct the corresponding label pair dataset as such instances may distinguish the differences between C_i and C_j. By this way, a multi-label dataset with k labels is partitioned into C_k^2 different label pair datasets. Note that the number of instances in different label pair datasets may be discrepant.

In Label Partition of ESP, the distribution of the original multi-label dataset is not altered as no increase or decrease in instances, which is different from the current sampling methods.

3.3   Binary Classification Model Construction

After label partition, the multi-label classification problem is transformed into a number of binary classification problems. Some popular classification algorithms could be used to learn these binary datasets for constructing binary classification models, such as C4.5, SMO and KNN. For a single label dataset, the built model could be used to predict whether an instance belongs to the corresponding label or not. However, for a label pair dataset with labels C_i and C_j, the constructed model is used to categorize an instance into either C_i or C_j.

3.4   Model Integration

When Binary Classification Model Construction is completed, a total of $k + C_k^2$ classification models are obtained from the single label datasets and label pair datasets. For the k binary classification models built with single label datasets, each model is used to classify whether an instance belongs to the corresponding class label or not, therefore the classification results of all these k binary classification models are firstly integrated to a bipartition array, which is represented as BS (**B**ipartition array for **S**ingle label datasets).

Moreover, for the C_k^2 binary classification models built with label pair datasets, each model is employed to categorize whether an instance belongs to C_i or C_j. A bipartition array name as BP (**B**ipartition array for label **P**air datasets) is also employed to integrate the classification results of all these C_k^2 binary classification models. For the label C_i, its corresponding value in BP is 1 when more than half of the corresponding $k-1$ binary classification models classify an instance as C_i, otherwise 0.

With the bipartition arrays BS and BP previously obtained, the final classification result BF (**B**ipartition array for **F**inal result) is calculated with the following formula. Particularly, for each label, its corresponding value in BF is 1 only when the corresponding values in BS and BP are both 1, otherwise 0. To this extend, the single label dataset binary models and label pair dataset binary models are integrated to obtain the final multi-label data classification model.

$$BF = (BS) \oplus (BP)$$

4 Evaluation

4.1 Dataset

In the experiment, six different multi-label imbalanced datasets from various fields are employed, including text, music, audio, biology and image. The statistical information of the dataset is shown in the Table 1, which includes the name of the dataset, the domain of the dataset, number of instances, the number of attributes, the number of labels and MeanIR. Note that MeanIR [19] represents the imbalance degree of the corresponding multi-label datasets. The larger the MeanIR value, the higher the imbalanced degree of the dataset.

Table 1. The statistical information of Dataset.

Dataset	Domain	Instances	Attribute	Labels	MeanIR
Bibtex	text	7395	1836	159	12.4983
Birds	audio	645	260	19	5.4070
CAL500	music	502	68	174	20.5778
Enron	text	1702	1001	53	73.9528
Scene	image	2407	294	6	1.2538
Yeast	biology	2417	103	14	7.1968

From the Table 1, it can be see that all the employed datasets are different with various number of instances and labels. Particularly, the number of dataset instances ranges from 502 to 7396 and the number of labels ranges from 6 to 174. Moreover, the used datasets cover varying degrees of imbalance as the corresponding MeanIR values range from 1.2598 to 73.9529.

4.2   Performance Metrics

Two popular performance metrics are employed in current study, including Accuracy and F-Measure. Suppose $\mathcal{X} = \mathbb{R}^d$ denotes the d-dimensional instance space, and $\mathcal{Y} = \{y_1, y_2, ..., y_k\}$ denotes the label space with k possible class labels. Let $\mathcal{S} = \{(x_i, Y_i)|1 \leq i \leq p\}$ be the test set with p instances. The task of multi-label learning is to learn a function $h : \mathcal{X} \to 2^{\mathcal{Y}}$ from the multi-label training set $\mathcal{D} = \{(x_i, Y_i)|1 \leq i \leq m\}$. For each multi-label example (x_i, Y_i), $x_i \in \mathcal{X}$ is a d-dimensional feature vector $(x_{i1}, x_{i2}, ..., x_{id})^T$ and $Y_i \subseteq \mathcal{Y}$ is the set associated with x_i. For any unseen instance $x \in \mathcal{X}$, the multi-label classifier h(·) predicts $h(x) \subseteq \mathcal{Y}$ as the set of proper labels for x. The example-based classification metrics Accuracy can be defined based on the multi-label classifier h(·):

$$Accuracy = \frac{1}{p} \sum_{i=1}^{p} \frac{|Y_i \cap h(x_i)|}{|Y_i \cup h(x_i)|}$$

Furthermore, F-Measure is an integrated performance metrics of Precision and Recall with balancing factor $\beta \geq 0$. The most common choice is $\beta = 1$ which leads to the harmonic mean of precision and recall.

$$Precision = \frac{1}{p} \sum_{i=1}^{p} \frac{|Y_i \cap h(x_i)|}{|h(x_i)|}$$

$$Recall = \frac{1}{p} \sum_{i=1}^{p} \frac{|Y_i \cap h(x_i)|}{|Y_i|}$$

$$F - Measure = \frac{(1 + \beta^2) \cdot Precision \cdot Recall}{\beta^2 \cdot Precision + Recall}$$

4.3   Experimental Setup

The widely-used ten-fold cross validation is conducted on the six imbalanced datasets with three popular classification algorithms, including SMO [12], C4.5 [14] and KNN [26]. Particularly, the k value of KNN algorithm is set as 3 according to the corresponding study [29]. What's more, to verify the performance of the proposed method ESP, five baselines have been employed for comparison, including MLROS, MLRUS, LPROS, LPRUS and MLTL, which have been introduced in Sect. 2. For comparison, MLROS and MLRUS are selected with two different sampling ratios, respectively 25% and 10%. In the following text, MLROS-25 and MLRUS-25 are used to represent the method using a sampling ratio of 25%, MLROS-10 and MLRUS-10 are used to represent the method using a sampling ratio of 10%.

4.4 Experimental Results

4.4.1   The Results of C4.5

The results of using the C4.5 classification algorithm are shown in Table 2, in which the upper half of the table contains corresponding results for Accuracy, while the lower half presents results for F-Measure. In addition, for a specific dataset, the best performance value in terms of Accuracy and F-Measure is highlighted in bold. Moreover, the row AVG demonstrates the corresponding average performance of ESP and baseline methods, respectively for Accuracy and F-Measure.

Table 2. The results of C4.5 algorithm

Metric	Dataset	MLROS-25	MLRUS-25	MLROS-10	MLRUS-10	LPROS	LPRUS	MLTL	ESP
Accuracy	Bibtex	0.2876	0.2919	0.2993	0.3002	0.2827	0.2925	0.2466	**0.3087**
	Birds	0.5568	0.5738	0.5608	0.5679	0.5408	0.5778	0.5634	**0.5882**
	CAL500	0.2088	0.2143	0.2117	0.2094	0.2067	0.2067	**0.2246**	0.2087
	Enron	0.3971	0.3963	0.4004	0.3931	0.378	0.4055	0.4034	**0.414**
	Scene	0.5455	0.5449	0.5377	0.5371	0.5254	0.5366	0.5394	**0.5515**
	Yeast	0.3997	0.4302	0.4048	0.4386	0.3986	0.4271	0.4343	**0.4511**
	AVG	0.3993	0.4086	0.4025	0.4077	0.3877	0.4077	0.4020	**0.4204**
F-Measure	Bibtex	0.3588	0.3558	0.368	0.3667	0.357	0.3603	0.2972	**0.3758**
	Birds	0.5923	0.6032	0.5895	0.6024	0.5757	0.6128	0.5915	**0.6233**
	CAL500	0.3406	0.3472	0.3437	0.3407	0.3375	**0.3775**	0.3618	0.3401
	Enron	0.5094	0.5043	0.5132	0.5032	0.4888	0.5174	0.5102	**0.5267**
	Scene	**0.5836**	0.5801	0.5728	0.5725	0.5612	0.5766	0.5779	0.5819
	Yeast	0.5234	0.5518	0.5299	0.5617	0.5239	0.551	0.5566	**0.5687**
	AVG	0.4904	0.4847	0.4862	0.4912	0.474	0.4926	0.4825	**0.5028**

From the Table 2, it could be observed that the proposed ESP method obtains the largest number of datasets with best F-Measure values and Accuracy values compared with the seven baseline methods. For the Accuracy metric, except for the CAL500 dataset, the ESP method achieves the best results. For the F-Measure metric, except for the CAL500 dataset and the scene dataset, the ESP method achieves the best results. Moreover, the ESP method outperforms other seven methods in terms of average performance, indicating the effectiveness of ESP with C4.5 as the classification algorithm.

In order to verify the stability of the ESP method, the metrics of each method are ranked on different datasets and we employ the box plot to demonstrate the distribution of corresponding ranking. Figure 2 provides the detailed box plot. Particularly in Fig. 2, MO-25 and MO-10 represent MLROS-25 and MLROS-10, MU-25 and MU-10 represent MLRUS-25 and MLRUS-10. From the Fig. 2(a) and Fig. 2(b), it could be observed that no matter F-Measure or Accuracy, the median of ESP method is 1, which means ESP method can be ranked first in most cases and indicates that ESP is more stable than other methods when using C4.5 as the classification algorithm.

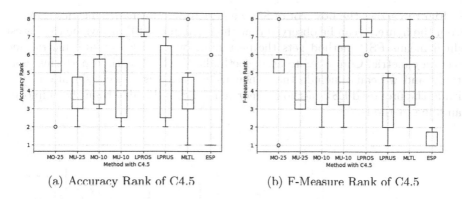

(a) Accuracy Rank of C4.5   (b) F-Measure Rank of C4.5

Fig. 2. The box plot of C4.5 algorithm performance metrics ranking

4.4.2   The Results of SMO

Table 3 shows the results of Accuracy and F-Measure using the SMO classification algorithm. From Table 3, it can be seen that the results of SMO algorithm is different from the results of C4.5 algorithm. For the Accuracy and F-Measure metrics, the ESP method both achieves the best results on the Birds dataset. Though ESP does not obtain the best results on other datasets, it is close to the best results. For example, in the Bibtex dataset, the best method is MLROS-10, which obtains 0.1463 for Accuracy and 0.1839 for F-Measure. ESP obtains 0.1461 and 0.1824 respectively for Accuracy and F-Measure, which are lightly inferior to MLROS-10. However, for the average performance, ESP achieves the best results on both Accuracy and F-Measure, indicating the effectiveness of ESP with SMO as the classification algorithm.

Table 3. The results of SMO algorithm

Metric	Dataset	MLROS-25	MLRUS-25	MLROS-10	MLRUS-10	LPROS	LPRUS	MLTL	ESP
Accuracy	Bibtex	0.1434	0.1341	**0.1463**	0.1413	0.1389	0.1361	0.0827	0.1461
	Birds	0.6049	0.5923	0.6216	0.6325	0.5937	0.616	0.5939	**0.6424**
	CAL500	0.2187	0.221	0.2162	0.2221	0.2199	0.2199	**0.2331**	0.2199
	Enron	0.2878	0.3032	0.2863	0.2989	0.2789	**0.3061**	0.2683	0.2857
	Scene	0.6842	0.6792	**0.6898**	0.6819	0.6652	0.6773	0.686	0.6858
	Yeast	0.4912	0.499	0.4991	0.5078	0.4885	0.5066	**0.5119**	0.5098
	AVG	0.4050	0.4048	0.405	0.4141	0.3975	0.4103	0.396	**0.415**
F-Measure	Bibtex	0.1813	0.1678	**0.1839**	0.176	0.1803	0.1708	0.0992	0.1824
	Birds	0.6438	0.6227	0.6605	0.6652	0.6406	0.6525	0.6238	**0.6782**
	CAL500	0.3488	0.3533	0.3463	0.3541	0.3518	0.3518	**0.3674**	0.3518
	Enron	0.3707	0.3803	0.3689	0.3804	0.3696	**0.3884**	0.3391	0.3673
	Scene	0.6968	0.6907	**0.7025**	0.6939	0.6854	0.6906	0.698	0.6979
	Yeast	0.5886	0.5966	0.597	0.6046	0.5877	0.6048	**0.6084**	0.6062
	AVG	0.4717	0.4686	0.4765	0.479	0.4692	0.4765	0.456	**0.4806**

Figure 3 shows the box plot of SMO algorithm performance metrics ranking. From the figure, it can be observed that the median of ESP always be the lowest which means ESP method gets the best rank on both F-Measure metric and Accuracy metric. Compared to other methods, the ESP method performs more stably and we can conclude that when using SMO as the classification algorithm, the proposed ESP method is more effective and stable than the employed sampling methods.

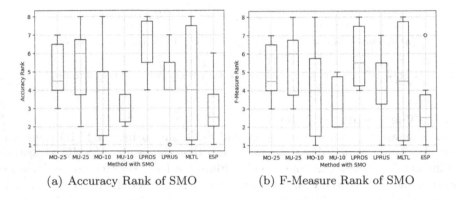

(a) Accuracy Rank of SMO   (b) F-Measure Rank of SMO

Fig. 3. The box plot of SMO algorithm performance metrics ranking

4.4.3   The Results of KNN

Table 4 provides the results of KNN algorithm for the two employed performance metrics. To be specific, ESP achieves the best result on the Bibtex dataset for both metrics. Except for ESP method, it is worth noting that MLTL achieves best results on CAL500 and Enron dataset for two metrics. However, as a whole, the proposed ESP method obtains the best average classification performance once again with Accuracy 0.4464 and F-Measure 0.5266, while MLTL obtains average Accuracy 0.4324 and F-Measure 0.5073. This indicates that when using KNN as the classification algorithm, ESP is usually more effective than the employed sampling methods.

The box plot of KNN algorithm performance metrics ranking is shown in Fig. 4. From the Fig. 4(a), it can be observed that the median of ESP is lowest. Obviously, ESP performs more stably than other seven methods in terms of Accuracy with KNN as the classification algorithm. Figure 4(b) shows the box plot of F-Measure Rank, and it can be seen that ESP method and LPROS achieve the same median rank. However, it is worth noting that LPROS method obtains an exception value, which means that it performs poorly on one employed dataset. On the contrary, there are no exception values for ESP. Therefore the proposed ESP method may be more stable than the employed sampling methods when using KNN as the learning algorithm.

Table 4. The results of KNN algorithm

Metric	Dataset	MLROS-25	MLRUS-25	MLROS-10	MLRUS-10	LPROS	LPRUS	MLTL	ESP
Accuracy	Bibtex	0.3324	0.3288	0.3335	0.3307	0.3295	0.3189	0.2362	**0.3362**
	Birds	0.6301	0.5971	0.6334	0.6112	0.6165	**0.6435**	0.6143	0.6336
	CAL500	0.2046	0.2023	0.204	0.2034	0.2044	0.2044	**0.229**	0.2044
	Enron	0.4001	0.4054	0.4024	0.4057	0.3953	0.3905	**0.4166**	0.405
	Scene	0.5965	0.5969	0.594	**0.6013**	0.5957	0.5972	0.5976	0.5982
	Yeast	0.495	0.498	0.5004	0.4997	**0.5077**	0.5071	0.5005	0.501
	AVG	0.4431	0.4381	0.4446	0.442	0.4415	0.4386	0.4324	**0.4464**
F-Measure	Bibtex	0.4096	0.3972	0.368	0.4018	0.4047	0.393	0.2768	**0.4101**
	Birds	**0.6669**	0.6306	0.5895	0.6454	0.6554	0.6501	0.6456	0.6664
	CAL500	0.3341	0.3311	0.3437	0.3327	0.3343	0.3343	**0.3654**	0.3343
	Enron	0.5116	0.5168	0.5132	0.5168	0.5067	0.5015	**0.5226**	0.5162
	Scene	0.6201	0.6189	0.5728	0.624	**0.626**	0.6229	0.6221	0.6214
	Yeast	0.6066	0.6082	0.5299	0.61	**0.6182**	0.616	0.611	0.6113
	AVG	0.5244	0.5171	0.4862	0.5218	0.5242	0.5196	0.5073	**0.5266**

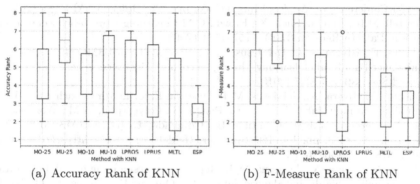

(a) Accuracy Rank of KNN   (b) F-Measure Rank of KNN

Fig. 4. The box plot of KNN algorithm performance metrics ranking

5 Conclusion

In present study, we propose a multi-label imbalanced data classification method ESP which deals with the imbalanced data without changing the distribution of original datasets. To deal with the problem of classification performance degradation caused by existing resampling methods, ESP method firstly divides datasets to single label datasets and label pair datasets, then constructs binary classification models based on different binary datasets, and finally integrates the corresponding binary classification models into a multi-label classification model. The experiments with six different imbalanced datasets and three different classification algorithms demonstrate that ESP is effective and stable, usually outperforming the popular sampling methods in terms of Accuracy and F-Measure.

Although the ESP algorithm has achieved better performance than existing methods, there is still room for further improvement. For example, the partitioned binary datasets in ESP may be imbalanced. In addition, a large number of label pair models need to be constructed in ESP, which consumes a lot of time and memory. Both aforementioned problems may reduce the overall performance of the ESP method and thus our future work will focus on solving these two problems.

References

1. Ai, X., Jian, W., Sheng, V.S., Yao, Y., Cui, Z.: Best first over-sampling for multi-label classification. In: Proceedings of the 24th ACM International on Conference on Information and Knowledge Management, pp. 1803–1806 (2015)
2. Almeida, T.B., Borges, H.B.: An adaptation of the ML-kNN algorithm to predict the number of classes in hierarchical multi-label classification. In: Torra, V., Narukawa, Y., Honda, A., Inoue, S. (eds.) MDAI 2017. LNCS (LNAI), vol. 10571, pp. 77–88. Springer, Cham (2017). https://doi.org/10.1007/978-3-319-67422-3_8
3. Bhattacharya, S., Rajan, V., Shrivastava, H.: ICU mortality prediction: a classification algorithm for imbalanced datasets. In: Proceedings of the Thirty-First AAAI Conference on Artificial Intelligence, pp. 1288–1294. AAAI Press (2017)
4. Boutell, M.R., Luo, J., Shen, X., Brown, C.M.: Learning multi-label scene classification. Pattern Recogn. **37**(9), 1757–1771 (2004)
5. Charte, F., Rivera, A.J., Del Jesus, M.J., Herrera, F.: MLSMOTE: approaching imbalanced multilabel learning through synthetic instance generation. Knowl.-Based Syst. **89**, 385–397 (2015)
6. Charte, F., Rivera, A., del Jesus, M.J., Herrera, F.: A first approach to deal with imbalance in multi-label datasets. In: Pan, J.-S., Polycarpou, M.M., Woźniak, M., de Carvalho, A.C.P.L.F., Quintián, H., Corchado, E. (eds.) HAIS 2013. LNCS (LNAI), vol. 8073, pp. 150–160. Springer, Heidelberg (2013). https://doi.org/10.1007/978-3-642-40846-5_16
7. Charte, F., Rivera, A., del Jesus, M.J., Herrera, F.: Resampling multilabel datasets by decoupling highly imbalanced labels. In: Onieva, E., Santos, I., Osaba, E., Quintián, H., Corchado, E. (eds.) HAIS 2015. LNCS (LNAI), vol. 9121, pp. 489–501. Springer, Cham (2015). https://doi.org/10.1007/978-3-319-19644-2_41
8. Charte, F., Rivera, A.J., del Jesus, M.J., Herrera, F.: Addressing imbalance in multilabel classification: measures and random resampling algorithms. Neurocomputing **163**, 3–16 (2015)
9. Charte, F., Rivera, A.J., del Jesus, M.J., Herrera, F.: MLeNN: a first approach to heuristic multilabel undersampling. In: Corchado, E., Lozano, J.A., Quintián, H., Yin, H. (eds.) IDEAL 2014. LNCS, vol. 8669, pp. 1–9. Springer, Cham (2014). https://doi.org/10.1007/978-3-319-10840-7_1
10. Chawla, N.V., Bowyer, K.W., Hall, L.O., Kegelmeyer, W.P.: SMOTE: synthetic minority over-sampling technique. J. Artif. Intell. Res. **16**(1), 321–357 (2002)
11. Chen, L., Fu, Y., Chen, N., Ye, J., Liu, G.: Rule reduction for EBRB classification based on clustering. In: Xing, C., Fu, X., Zhang, Y., Zhang, G., Borjigin, C. (eds.) WISA 2021. LNCS, vol. 12999, pp. 442–454. Springer, Cham (2021). https://doi.org/10.1007/978-3-030-87571-8_38
12. Chen, P.H., Fan, R.E., Lin, C.J.: A study on SMO-type decomposition methods for support vector machines. IEEE Trans. Neural Netw. **17**(4), 893–908 (2006)
13. Elisseeff, A.E., Weston, J.: A kernel method for multi-labelled classification. In: Proceedings of the 14th International Conference on Neural Information Processing Systems: Natural and Synthetic, pp. 681–687 (2001)

14. Haixiang, G., Yijing, L., Shang, J., Mingyun, G., Yuanyue, H., Bing, G.: Learning from class-imbalanced data: review of methods and applications. Expert Syst. Appl. **73**, 220–239 (2017)
15. Liu, B., Tsoumakas, G.: Making classifier chains resilient to class imbalance. In: Proceedings of The 10th Asian Conference on Machine Learning, ACML 2018, Beijing, China, 14–16 November 2018. Proceedings of Machine Learning Research, vol. 95, pp. 280–295. PMLR (2018)
16. Nguyen, T.T., Nguyen, T.T.T., Luong, A.V., Nguyen, Q.V.H., Liew, A.W.C., Stantic, B.: Multi-label classification via label correlation and first order feature dependance in a data stream. Pattern Recogn. **90**, 35–51 (2019)
17. Pereira, R.M., Costa, Y.M., Silla, C.N., Jr.: MLTL: a multi-label approach for the Tomek Link undersampling algorithm. Neurocomputing **383**, 95–105 (2020)
18. Tahir, M.A., Kittler, J., Yan, F.: Inverse random under sampling for class imbalance problem and its application to multi-label classification. Pattern Recogn. **45**, 3738–3750 (2012)
19. Tarekegn, A.N., Giacobini, M., Michalak, K.: A review of methods for imbalanced multi-label classification. Pattern Recogn. **118**, 107965 (2021)
20. Tomek, I.: Two modifications of CNN. IEEE Trans. Syst. Man Cybern. **SMC-6**, 769–772 (1976)
21. Tsoumakas, G., Vlahavas, I.: Random k-labelsets: an ensemble method for multilabel classification. In: Kok, J.N., Koronacki, J., Mantaras, R.L., Matwin, S., Mladenič, D., Skowron, A. (eds.) ECML 2007. LNCS (LNAI), vol. 4701, pp. 406–417. Springer, Heidelberg (2007). https://doi.org/10.1007/978-3-540-74958-5_38
22. Wilson, D.L.: Asymptotic properties of nearest neighbor rules using edited data. IEEE Trans. Syst. Man Cybern. **SMC-2** (1972)
23. Yu, G., Domeniconi, C., Rangwala, H., Zhang, G., Yu, Z.: Transductive multi-label ensemble classification for protein function prediction. In: ACM SIGKDD International Conference on Knowledge Discovery and Data Mining, pp. 1077–1085 (2012)
24. Zakaryazad, A., Duman, E.: A profit-driven artificial neural network (ANN) with applications to fraud detection and direct marketing. Neurocomputing **175**, 121–131 (2016)
25. Zhang, M., Zhou, Z.: A review on multi-label learning algorithms. IEEE Trans. Knowl. Data Eng. **26**, 1819–1837 (2014)
26. Zhang, M.L., Zhou, Z.H.: A k-nearest neighbor based algorithm for multi-label classification. In: 2005 IEEE International Conference on Granular Computing, vol. 2, pp. 718–721 (2005)
27. Zhang, W.B., Pincus, Z.: Predicting all-cause mortality from basic physiology in the Framingham heart study. Aging Cell **12**, 39–48 (2016)
28. Zhong, W., Raahemi, B., Liu, J.: Classifying peer-to-peer applications using imbalanced concept-adapting very fast decision tree on IP data stream. Peer-to-Peer Netw. Appl. **6**(3), 233–246 (2013)
29. Zhu, X.: Semi-supervised Learning Literature Survey. University of Wisconsin-Madison (2008)
30. Zhu, Y., Kwok, J.T., Zhou, Z.H.: Multi-label learning with global and local label correlation. IEEE Trans. Knowl. Data Eng. **30**, 1081–1094 (2017)

X-ray Prohibited Items Recognition Based on Improved YOLOv5

Wei Li[1] (ID), Xiang Li[1], Wanxin Liu[1]([✉]), Zhihan Liu[1], Jing Jia[2] (ID), and Jiayi Li[3]

[1] College of Computer Science and Technology, Harbin Engineering University, Harbin 150001, China
{wei.li,lzhlzh}@hrbeu.edu.cn, 13019070967@163.com
[2] Faculty of Arts, Design and Architecture, School of Built Environment, The University of New South Wales, Sydney NSW 2052, Australia
jing.jia.1@unsw.edu.au
[3] School of Art and Design, Jiangsu Institute of Technology, Changzhou 213001, China

Abstract. Safety inspections play a crucial role in maintaining social stability and protecting the safety of public life and property. The X-ray security detector is an important scanning device in the security inspection of public transportation and express packages. However, current intelligent detection algorithms face problems such as a limited dataset of prohibited items, the imbalanced distribution of categories, variable target postures, and varying target scales, leading to the occurrence of false positives and missed detections. We propose an improved X-ray prohibited item recognition algorithm, named YOLOv5s-DAB. A deformable convolution module is designed to address the characteristics of different scales and postures of the same prohibited item in different samples. Besides, a multi-scale feature enhancement module SA-ASPP based on attention mechanism is designed, which can handle the problem of overlapping occlusion of multi-scale contraband. Experimental results in the real X-ray prohibited items dataset demonstrate that our model outperforms state-of-the-art methods in terms of detection accuracy.

Keywords: YOLOv5 · Object detection · X-ray security images

1 Introduction

The work on convolutional neural networks in target detection has made great progress and further plays an important role in the item recognition of X-ray images. However, the current algorithms are still inadequate for the detection of prohibited items in X-ray images for two main reasons: (1) the scale of dangerous items is highly variable and the contextual information is unevenly distributed. Conventional convolution cannot flexibly adapt to the real target's perceptual field due to the fixed sampling position, which weakens the feature extraction ability of the network. (2) Items of the same material present similar colors under X-ray fluoroscopy, which can easily cause target confusion when the background is complex and overlapping and obscuring. If these items go through ordinary convolutional layers, they will get similar feature responses, resulting in lower recognition and localization accuracy.

Akcay et al. [1] pioneered the research work on the classification of anomalous objects in X-ray images using deep learning. After that, Akcay et al. [2] further investigated the classification and localization of anomalous objects in X-ray images, and the experimental results proved that the region-based target detection network model is more suitable for the detection of anomalous targets in X-ray images. Tang et al. [3] proposed an X-ray detection algorithm based on feature pyramid detection, which realized effective detection of small targets by constructing a feature pyramid on SSD network. The results show that the detection effect of small dangerous goods is improved by 3.9%. Liang et al. [4] used multi-view images to detect anomalous objects in X-ray images. In the detection phase, if an image proves to be anomalous, the same set of images is considered to contain anomalous objects. Muschi et al. [5] proposed the use of a null-dense convolution module on the target detection algorithm YOLOv4 [6] and an attention-based multiscale feature map to improve the detection accuracy of multi-scale obscured contraband targets, and the average accuracy reached 80.16% on the SIXray [7] dataset. Wang et al. [8] proposed an improved method combining null convolution, moving exponential averaging and positive sample retraining. The method effectively solves the complex background and occlusion problems and improves the detection accuracy of various types of targets. Wang et al. [9] proposed a contraband recognition algorithm based on multilayer attention mechanism, which addresses the shortage of manual feature extraction and the problem of low recognition accuracy in real scenarios for the class-balanced hierarchical excellence (CHR) algorithm in literature [7], and selected ResNet-101 as the base network, combined with CHR and multilayer attention mechanism, and the results showed that compared with the original algorithm detection accuracy is improved by 1.69% compared with the original one. In summary, part of the research on X-ray image detection based on deep learning focuses on solving multi-scale and visual occlusion problems and has some achievements. However, there is still room for optimization in the selection of the underlying network, as well as the rate of missed and false detections.

Based on the above discussion, we adopt YOLOv5s as the basis and designs the detection model YOLOv5s-DAB to handle the issue of insufficient detection of contraband when multi-scale, multi-pose, and overlapping occlusion. Specifically, we propose the deformable convolution module, which enables the network to learn and sample offsets to generate irregular convolution kernels that adapt to different scales or different deformations of objects. We propose the SA-ASPP module, which uses null convolution to replace part of the maximal pooling layer of the original feature extraction module SPPF, reducing the information loss caused by the pooling layer, and at the same time, it can increase the perceptual field and inhibit the interference of the background and noise, etc., guided by the attention module, which leads to more accurate detection. We improve the PANet network of YOLOv5s based on the idea of BiFPN bi-directional cross-scale hopping connectivity, so that the semantic information of the high-level features is weighted and fused with the positional information of the low-level features, which helps the network to focus on the position of small targets. We make the following contributions in this paper.

1. We propose a deformable convolution module (D) to enhance the detection accuracy of contraband with different pose shapes.
2. We propose a multi-scale feature enhancement module (A) based on a permutation attention mechanism to enhance the recognition of targets with complex backgrounds and overlapping occlusions.
3. We propose an improved feature fusion network (B) to improve the recall rate of small targets.
4. We conducted experiments on the SIXray dataset. The experimental results show that the mAP% of the YOLOv5s-DAB model proposed in this paper reached 93.72%, which outperformed other mainstream models.

2   Methodology

As aforementioned, to address the problems of X-ray security image target detection, we design an improved algorithm YOLOv5s-DAB. The overall model is shown in Fig. 1. Compared with YOLOv5s, there are three main improvement parts, which have been marked by red boxes, including (1) introducing DBS to replace CBS to enhance the polymorphic target extraction capability. (2) Replacing SPPF with SA-ASPP. (3) Improving the feature fusion network by BiFPN.

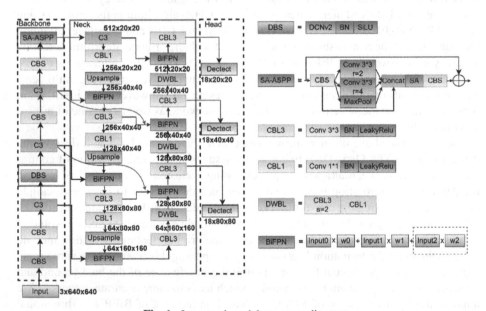

Fig. 1. Improved model structure diagram.

2.1 Deformable Convolution Module

The convolutional kernels in ordinary CNN models have a fixed geometric structure, which makes it difficult to extract object features accurately. Deformable Convolutional DCNv1 (Deformable ConvNets v1) [10] adds an offset variable to the position of each sampling point in the convolutional kernel, thus enabling random sampling around the current position without being limited to the previous regular grid points, as shown in Fig. 2.

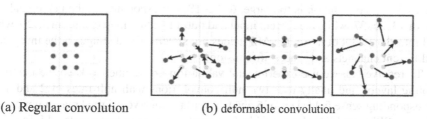

(a) Regular convolution   (b) deformable convolution

Fig. 2. Comparison of regular convolution and deformable convolution sampling.

The output character of DCNv1 is:

$$y(P_0) = \sum_{k=1}^{K} w_k x(p_0 + p_k + \Delta p_k) \tag{1}$$

where the x in Eq. (1) denotes the input feature map, $y(p_0)$ denotes the feature at the position p_0 in the output feature map y, K denotes the number of sampling positions of the convolution kernel, w_k is the weight, p_k indicates the pre-specified offset, and Δp_k represents the learnable offset.

However, since the new position of DCNv1 sampling points will be out of bounds after offset, resulting in some sampling points may be irrelevant parts of the object, so a modulation module is introduced in DCNv2 [11] to learn a change amplitude through modulation Δm, and the weight of the unwanted sampling points can be directly learned to 0 to reduce the influence of irrelevant factors. The output character of DCNv2 is:

$$y(P_0) = \sum_{k=1}^{K} w_k x(p_0 + p_k + \Delta p_k) \cdot \Delta m_k \tag{2}$$

We introduced the above deformable convolution DCNv2 into the YOLOv5s backbone network. The 1×1 deformable convolution does not have the ability to change the perceptual field and is unstable when calculating the offset. Meanwhile, [11] has no description of the effect of different positions of this convolution. Therefore, in this paper, the ordinary 3×3 convolution in the CBS module is replaced by a deformable convolution of the same size to form the DBS module, and the effect of the replacement position is discussed in the subsequent experiments to select the best structure. This module can learn to generate deformable convolution kernels to extract features that adapt to the shape of contraband and improve the detection accuracy.

2.2 Feature Enhancement Based on Attention Mechanism

The SPPF in the YOLOv5s model uses the same size maximum pooling layer to obtain different perceptual field features, which is prone to lose local feature information. For this reason, the maximum pooling layer in SPPF is replaced by the combination of three parallel cavity convolutions with different cavity rates to form the Atrous Spatial Pyramid Pooling (ASPP) module.

The introduction of this module can increase the perceptual field while obtaining multi-scale feature information of the object, but it brings two problems: (1) The void rate of 3 × 3 convolution is too large, 6, 12, 18, corresponding to the perceptual field size of 13, 25, 37, while the current input and output feature map resolution is 20, which will introduce a lot of noise. (2) ASPP does not specifically distinguish the importance of different channels and spaces after fusion.

To resolve (1), we use experimental validation on whether to keep the maximum pooling layers, and finally uses two null convolutions with null rates of 2 and 4 and corresponding sense field sizes of 5 and 9 to replace the two original maximum pooling layers of SPPF, and introduces residual connections to maintain the detail information in the original feature maps, which are constructed as improved ASPP modules.

To address (2), we introduce the replacement attention mechanism (Shuffle Attention, SA) [12], which constitutes the SA-ASPP module, as shown in Fig. 3.

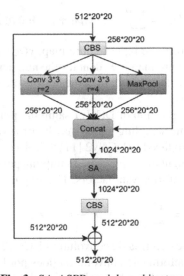

Fig. 3. SA-ASPP module architecture.

SA first splits the input into g groups along the channel dimension to obtain g groups of features, and then each group of features is processed separately with two attention mechanisms.

Among them, the channel attention mechanism embeds global information through global pooling to generate channel features $s \in R^{C/2G \times 1 \times 1}$, where the calculation can be performed by contracting X_{k_1} by $H \times W$:

$$s = F_{gp}(X_{k_1}) = \frac{1}{H \times W} \sum_{i=1}^{H} \sum_{j=1}^{W} X_{k_1}(i,j) \tag{3}$$

Then, the compact feature map is generated by aggregating the information through the Sigmoid activation function, and the final output feature map is calculated as Eq. (4), where $W_1 \in R^{C/2G \times 1 \times 1}$, $b_1 \in R^{C/2G \times 1 \times 1}$ is used for scaling and shifting s:

$$X_{k_1}' = \sigma(W_1 s + b_1) \cdot X_{k_1} \tag{4}$$

The second is the spatial attention mechanism, which is implemented using the GroupNorm, where the mean and variance of each group are calculated and normalized. Then F_C is used to enhance the representation of X_{k2} features. The final output is as Eq. (5), where $W_2 \in R^{C/2G \times 1 \times 1}$, $b_2 \in R^{C/2G \times 1 \times 1}$.

$$X_{k_2}' = \sigma\left(W_2 \cdot GN(X_{k_2}) + b_2\right) \cdot X_{k_2} \tag{5}$$

Finally, the two branches are connected and all sub-features are aggregated together, then the Shuffle operation is used to achieve the flow and fusion of information across groups, thus improving the performance of semantic representation.

2.3 Multi-Scale Feature Fusion Network

According to [13], the PANet structure establishes another bottom-up path on the right side of the FPN structure, making the prediction feature layer contain both semantic information from the higher layers and location information from the lower layers, as a way to improve the prediction accuracy of the model. However, the bidirectional path of PANet does not take into account the variability of contributions from different features, and the feature map loses details in the lateral transfer process, resulting in insufficient detection of small targets.

To achieve efficient bi-directional cross-scale connectivity, the weighted bi-directional feature pyramid network BiFPN removes the nodes in PANet that have only one input edge. Weights are assigned to different inputs using learnable parameters that indicate the degree of contribution of each information in the fusion process. And add lateral connections between the input and output nodes of the middle feature map, this kind of residual structure can well retain the feature information of smaller targets and enhance the feature representation ability.

In this paper, we borrow the idea of BiFPN and make improvements to the feature fusion network PANet, so as to improve the small target detection capability of the model.

Considering the number of parameters, we remove the P7 layer from the above BiFPN network and connect the 40×40 and 80×80 feature maps with the output feature layer of the same size behind, as shown in Fig. 4, in which the red arrows indicate the cross-layer connection, the yellow ones indicate the upsampling operation and the purple ones indicate the downsampling operation.

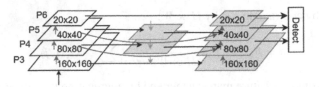

Fig. 4. The improved PANet structure diagram

Since the shallow feature map contains the location information of smaller targets, the lateral connection can directly pass the feature map of larger size in the shallow layer to the prediction layer for weighted fusion and output, which is more helpful for the detection of smaller targets. In addition, the downsampling convolution is replaced by depth-separable convolution, which has a lower number of parameters and lower operation cost compared with the conventional convolution operation. A top-down pathway is designed in layer P6, which is mainly used to transfer semantic information of high-level features, and the bottom-up pathway in layer P3 is retained to transfer location information in low-level features. a pathway is added in layers P4 and P5, respectively, to connect laterally to the output nodes for multi-scale feature fusion. The output of P5 is shown below,

$$P_5^{out} = Conv(\frac{\omega_1' \cdot P_5^{in} + \omega_2' \cdot P_5^{td} + \omega_3' \cdot Resize(P_4^{out})}{\omega_1' + \omega_2' + \omega_3' + \varepsilon}) \tag{6}$$

where P_4^{out} is the feature output from P4, and P_5^{td} is the feature extracted by the fusion of P5 and P6, as shown in Eq. (7).

$$P_5^{td} = Conv(\frac{\omega_1 \cdot P_5^{in} + \omega_2 \cdot Resize(P_6^{in})}{\omega_1 + \omega_2 + \varepsilon}) \tag{7}$$

P_5^{in}, P_6^{in} is the feature output from the P5 and P6 input nodes, *Resize* denotes the upsampling operation, ω indicates the parameters that can be learned and then give importance to different features during the feature fusion process, and ε denotes the bias term.

3 Experiments

After describing our experimental setup (in Sect. 3.1), we compare and analyze the effects of module position and structure on the accuracy of the model in Sect. 3.2. Then, in Sect. 3.3, we conduct ablation experiments to verify the impact of the three improvement methods proposed in this paper on the model performance. In Sect. 3.4, we compare YOLOv5s-DAB with other models to validate the advancement and generality of YOLOv5s-DAB.

3.1 Experimental Setup

Datasets. We used the dataset shown in the table. SIXray is a public dataset containing 1059231 images of the actual scenes of the security check. Among them, 8929 images are positive samples [6]. The SIXray_train is the sum of the training and validation sets divided from the SIXray, which is used in the ablation comparison experiments. The SIXray_test is used to test the algorithm detection accuracy (Table 1).

Table 1. Experimental datasets.

Dataset	Source	Total	Knives	Guns	Wrenches	Pliers	Scissors
SIXray	public	8928	3079	4978	3084	5370	1132
SIXray_train (包含验证集)	SIXray	8035	2768	4526	2750	4839	996
SIXray_test	SIXray	893	311	450	333	531	136

Hyper-Parameter Tuning. We used the official recommended parameter settings of YOLOv5s. The initial learning rate is 0.01, the momentum is 0.937, and the weight decay is 0.0005. We used the trained model on the VOC dataset as a pre-trained model for learning. The training epoch is 300, and the backbone network is frozen for the first 50 epochs to speed up the training, when the batchsize is 16. And the batchsize is 8 when the backbone network is unfrozen. All experiments are implemented on a server with a NVIDIA RTX A5000 24GB GPU.

Metrics. We used the two evaluation metrics including the average accuracy AP and the average accuracy mean mAP. AP is the area enclosed by the P-R curve, and mAP is the mean value of AP for each target category.

In addition, the number of floating point operations FLOPs and number of inferences per second FPS are two processing speed metrics. The smaller the FLOPs, the faster the processing speed.

3.2 Module Validation

Eval-I: Deformable Convolutional Module Position Verification. Using the original YOLOv5s model as a benchmark, the four CBS modules in its backbone network are replaced by the deformable convolutional module DBS in turn.The YOLOv5s-DBS_i in the table indicates the improved algorithm using DBS at different locations separately.

The experimental results show that different degrees of improvement can be obtained in different positions. dBS_2 achieves the highest detection accuracy, with a 92.73% mAP with only 0.06M growth in the number of parameters and 0.38GFLOPs reduction in computation, which is a 1.1% improvement compared to the original model. The data show that the module is better used in the middle feature layer, which is due to the fact that deformable convolution requires both spatial and semantic information, and these two types of information can be expressed in a balanced way in the middle feature layer (Table 2).

Table 2. Deformable convolution module verification experiments.

Models	Size /MB	Parameters /M	Computation/GFLOPs	mAP /%
YOLOv5s	26.99	7.0	8.21	91.60
YOLOv5s-DBS_1	27.02	7.1	7.93	92.29 (+0.7)
YOLOv5s-DBS_2	27.05	7.1	7.83	**92.73 (+1.1)**
YOLOv5s-DBS_3	27.11	7.1	7.78	92.59 (+1.0)
YOLOv5s-DBS_4	27.23	7.1	7.76	91.92 (+0.3)

Eval-II: SA-ASPP Module Structure Validation. The YOLOv5s-DBS_2 model is used as a benchmark to replace the SPPF module at the end of the backbone network. Do comparison experiments on the two structures of the improved ASPP module (without SA) and choose one of them with high accuracy to add different attention mechanisms for comparison. The added attention mechanisms are the feature extraction module guided by SA and SE Based on the first three rows of data, it is obvious that the improvement of ASPP_1 is better than ASPP_2. And the ASPP_1 module guided by different attention mechanisms can all improve the model performance and increase the model detection accuracy by about 0.5 (Table 3).

Table 3. SA_ASPP module validation experiments.

Models	Size /MB	Parameters /M	Computation/GFLOPs	mAP /%
YOLOv5s-DBS_2 -SPPF	27.05	7.1	7.83	92.73
YOLOv5s-DBS_2 -ASPP_1	31.55	8.2	8.31	**92.97 (+0.2)**
YOLOv5s-DBS_2 -ASPP_2	33.81	8.8	8.54	92.85 (+0.1)
YOLOv5s-DBS_2- SE-ASPP_1	33.65	8.7	8.31	93.25 (+0.5)
YOLOv5s-DBS_2- SA-ASPP_1	31.55	8.2	8.31	**93.27 (+0.5)**

3.3 Ablation Study

Eval-III: The Performance of the Three Improvement Methods. In order to verify the impact of D(DBS_2), A(SA-ASPP), and B(BiFPN) on the model performance, ablation experiments are conducted on the SIXray dataset with the original YOLOv5s model as the benchmark The experimental results show that the deformable convolution brings a 1.1% improvement effect. Knife contraband detection accuracy rose the most significantly, with AP improving by nearly 3.6%, and shears and wrenches improving by 1.4% and 1.6%, respectively. The DBS module is more helpful for feature extraction of targets with large size and shape variations. Combined with the SA-ASPP module and the multi-scale feature fusion network, the accuracy of the shears and wrenches improved by 1.7% and 4%, respectively. The final YOLOv5s-DAB algorithm shows an increase in detection accuracy for each class of contraband relative to the unimproved model and a 2.1% improvement in mAP, demonstrating the effectiveness of the YOLOv5s-DAB algorithm for contraband detection in X-ray images (Table 4).

Table 4. Model ablation experiments.

Models	Size /MB	P/M	C/GFLOPs	G /%	K /%	P /%	S /%	W /%	mAP /%
YOLOv5s	26.99	7.0	8.21	99.0	87.2	92.3	92.3	87.2	91.6
YOLOv5s-D	27.05	7.1	7.83	98.7	90.8	93.7	91.6	88.8	92.7 (+1.1)
YOLOv5s-DA	31.55	8.2	8.31	99.4	91.6	93.8	93.0	88.6	93.3 (+1.7)
YOLOv5s-DAB	34.86	9.6	11.6	**99.1**	**90.8**	**93.5**	**94.0**	**91.2**	**93.7 (+2.1)**

3.4 Comparative Experimental Analysis

To further validate the generalization and advancedness of the final model YOLOv5s-DAB we proposed, we compare it with some current mainstream target detection models and the improved X-ray contraband detection model LightRay [14] on the SIXray dataset.

Eval-IV: Quantitative Analysis. As shown in Fig. 5, the loss values of the iterations during the training and validation of the model are recorded. It is obvious that after 50 epochs, the frozen backbone network is unfrozen resulting in a short-lived increase in loss, after which both the training and validation losses gradually decrease and the model gradually converges until the best detection model is obtained.

As shown in Table 5, the highest accuracy of 93.72 is achieved on the SIXray security check dataset, and the mAP of the improved model is nearly 2.1% higher than that before improvement. The experimental results show that our proposed YOLOv5s-DAB model is more advantageous than mainstream target detection models and improved contraband detection models. Although the FPS is slightly lower than the YOLOv5s model, it still meets the needs of practical engineering applications.

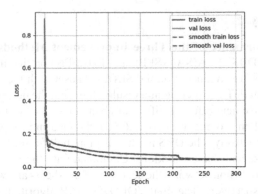

Fig. 5. YOLOv5s-DAB model loss curves.

Table 5. Performance comparison of different detection algorithms.

Model	Size/MB	mAP/%	FPS
SSD	26.28	86.16%	50
CenterNet	32.66	86.26%	43
YOLOv5s	26.99	91.60%	56
YOLOXs	34.21	92.30%	59
LightRay	72.44	82.91%	40
YOLOv5s-DAB	34.86	**93.72%**	53

Eval-V: Qualitative Analysis. To visually evaluate the effectiveness of the algorithm improvement, LightRay, YOLOv5s, and YOLOv5s-DAB are visualized and compared respectively, depicted in Fig. 6. LightRay, YOLOv5s still misses or mis detects some targets with large deformation and serious overlapping occlusion. For example, the wrench with deformation and overlapping background, the wrench with most of the outline obscured, the pliers with confusing and overlapping background, and shears of smaller size and easily overlapped, only the improved model YOLOv5s-DAB we proposed can detect them accurately.

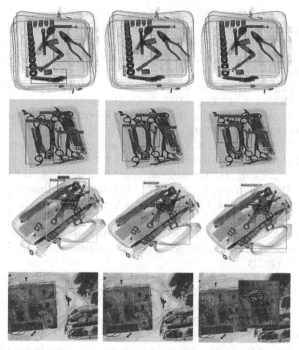

Fig. 6. Algorithm detection effect comparison. Each group from left to right is LightRay, YOLOv5s, YOLOv5s-DAB.

4 Conclusion

We propose an X-ray contraband recognition algorithm named YOLOv5s-DAB to provide reliable recognition results. We design a deformable convolution module to improve the sensitivity of the model to deformed targets. In addition, an attention-based multi-scale feature enhancement module SA-ASPP is designed to suppress noise interference to cope with the overlapping occlusion problem. Experimental results on a real X-ray contraband dataset demonstrate that our model outperforms the state-of-the-art methods in terms of detection accuracy. However, there are still shortcomings to be optimized in this paper, and we will subsequently improve the self-attention module and increase the utility of the model with a lightweight approach.

References

1. Akçay, S., Kundegorski, M.E., Devereux, M., et al.: Transfer learning using convolutional neural networks for object classification within X-ray baggage security imagery. In: 2016 IEEE International Conference on Image Processing (ICIP), pp.1057–1061. IEEE (2016)
2. Akçay, S., Kundegorski, M.E., Willcocks, C.G., et al.: Using deep convolutional neural network architectures for object classification and detection within x-ray baggage security imagery. IEEE Trans. Inf. Forensics Secur. **13**(9), 2203–2215 (2018)

3. Tang, H.Y., Wang, Y., Zhang, X., et al.: Feature pyramid-based algorithm for detecting hazardous materials in X-ray machines. J. Xi'an Post Telecommun. **25**(02), 58–63 (2020)

4. Li, A., Zhang, J., Lv, Y., et al. : Uncertainty-aware joint salient object and camouflaged object detection. In: Proceedings of the IEEE/CVF Conference on Computer Vision and Pattern Recognition 2021, pp. 10071–10081 (2021)

5. Mu, S.Q., Lin, J.J., Wang, H.Q., Wei, X.Z.: X-ray image contraband detection algorithm based on improved YOLOv4. J. Mil. Eng. **42**(12), 2675–2683 (2021)

6. Bochkovskiy, A., Wang, C.Y., Liao, H.: YOLOv4: Optimal Speed and Accuracy of Object Detection. arXiv.2004.10934 (2020)

7. Miao, C., Xie, L., Wan, F., et al.: SIXray: a large-scale security inspection x-ray benchmark for prohibited item discovery in overlapping images. IEEE (2019)

8. Wang, R., Liu, J., Zhang, H.: Research on target detection method based on improved YOLOv5. In: Zhao, X., Yang, S., Wang, X., Li, J. (eds.) Web Information Systems and Applications, WISA 2022, vol. 13579. Springer, Cham (2021). https://doi.org/10.1007/978-3-031-20309-1_40

9. Wang, W., Zhou, Y., Shi, B., He, H., Zhang, J.W.: A multi-layer attention mechanism based algorithm for security screening hazardous materials identification. Adv. Laser Optoelectr. **59**(02), 184–191 (2022)

10. Dai, J., Qi, H., Xiong, Y., et al.: Deformable convolutional networks. In: Proceedings of the IEEE International Conference on Computer Vision, pp.764–773 (2017)

11. Zhu, X., Hu, H., Lin, S., et al.: Deformable convnets v2: more deformable, better results. In:Proceedings of the IEEE/CVF Conference on Computer Vision and Pattern Recognition, pp.9308–9316 (2019)

12. Zhang, Q.L., Yang, Y.B.: Sa-net: shuffle attention for deep convolutional neural networks. In: ICASSP 2021–2021 IEEE International Conference on Acoustics, Speech and Signal Processing (ICASSP), pp. 2235–2239. IEEE (2021)

13. Lin, T.Y., Dollar, P., Girshick, R., et al.: Feature pyramid networks for object detection. IEEE Comput. Soc. (2017)

14. Ren, Y., Zhang, H., Sun, H., et al.: LightRay: lightweight network for prohibited items detection in X-ray images during security inspection. Comput. Electr. Eng. **103**, 108283 (2022)

Temporal Convolution and Multi-Attention Jointly Enhanced Electricity Load Forecasting

Chenchen Sun[1]([✉]), Hongxin Guo[1], Derong Shen[2], Tiezheng Nie[2], and Zhijiang Hou[1]

[1] Tianjin University of Technology, Tianjin, China
suncc_db@163.com, hzj@tjut.edu.cn
[2] Northeastern University, Shenyang, China
{shendr,nietiezheng}@mail.neu.edu.cn

Abstract. Accurate short-term load forecasting (STLF) is useful for power operators to respond to customers' demand and to rationalize generation schedules to reduce renewable energy waste. However, there are still challenges to improve the accuracy of STLF: firstly, modeling the long-term relationships between past observations; secondly, the relationships between variables should be considered in modeling. For this reason, we propose temporal convolution and multi-attention (variable attention and temporal attention) (TC-MA) for electricity load forecasting. Since the electricity load forecast values show different trends influenced by historical loads and covariates, we use variable attention to obtain the dependencies between load values and covariates. The temporal dependence of covariates and loads are extracted separately by temporal convolution, and the temporal attention is then used to assign different weight values to each timestep. We validate the effectiveness of our method using three real datasets. The results show that our model performs excellent results compared to traditional deep learning models.

Keywords: Electricity load forecasting · Temporal convolutional networks · Variable attention · Temporal attention

1 Introduction

Electricity load forecasting is an extremely important task in the energy management industry, which directly affects people's lives and the national economy. Short-term load forecasting (STLF), which forecasts data for the next day or weeks, is the most common form of load forecasting. Accurate short-term load forecasting enables power operators to plan their generation plans to meet electricity demand. It reduces the waste of renewable energy and the cost of electricity generation [1]. According to Bunn and Farmer [2], it is estimated that a 1% reduction in load forecasting error will save electricity operators £10 million per year by 30 years. Therefore, any model that improves the accuracy of electricity load forecasting has the potential to significantly reduce the cost to operators. It can also ensure safe operation of the power system and effective energy management.

Common methods for electricity load forecasting include statistical methods and machine learning methods. Unfortunately, statistical methods [3] fails to effectively

L. Yuan et al. (Eds.): WISA 2023, LNCS 14094, pp. 39–51, 2023.
https://doi.org/10.1007/978-981-99-6222-8_4

model data with nonlinear structures. In contrast, deep learning algorithms play a central role in electricity load forecasting. From advanced feedforward structures with residual connections [1] to convolutional neural networks (CNNs) [8] and recurrent neural networks (RNNs) [10] and their many variants such as short-term memory networks (LSTMs) [11] and GRUs. The nonlinearity and non-smoothness of the input variables increase the difficulty of model fitting, and secondly, the neural network training has overfitting or reaching local minima, which makes it difficult to improve the forecasting accuracy of the model. The paper [12] points out that the hybrid model using multiple models and combining the advantages of different models to extract the features of different aspects of the data are conducive to improving the forecasting accuracy of the model. Therefore, the hybrid architectures of CNNBiLSTM [12] and CNNLSTM [15] are proposed, which aim to capture the temporal dependencies in series by using the recurrent network and perform feature extraction operations by using the convolutional layer.

In summary, improving the accuracy of load forecasting and reducing generation costs are a challenging task. How to efficiently obtain inter-series correlations and how to effectively obtain intra-series time dependencies are still challenges we encounter. In order to solve the above two challenges, we propose an electricity load forecasting method based on temporal convolution and multi-attention. The main contributions of this paper are as follows:

- We propose a multi-attention and temporal convolution-based method for electricity load forecasting. Considering the differences in the correlations of different dimensions of the data, we propose attention of the variables used to extract inter-series correlations, and intra-series correlations are obtained using temporal convolution and temporal attention.
- To solve the first challenge, we use the variable attention to model trends in electricity loads affected by weather and holidays.
- For the second challenge, we use temporal convolution and temporal attention to model temporal dependencies intra-series and to obtain temporal patterns of periodic changes in the series.
- We evaluate our model using three evaluation metrics on three datasets and test the usefulness of the added modules on each. The evaluation metrics show that our model improves the forecasting accuracy compared to the baseline model.

The rest of this paper is organized as follows: In Sect. 2 introduces the research methods related to electricity load forecasting. Section 3 the electricity load forecasting model is proposed. Section 4 verifies the validity of the proposed models through comparative experiments. The conclusion is given in Sect. 5.

2 Related Work

Electricity load itself is a univariate time series, time series forecasting methods can be used for electricity load forecasting. The paper [7] proposed an autoregressive integrated moving average model (ARIMA) time series modeling approach is used for short-term load forecasting for the next 24 h in Iran. Unfortunately, this method has limitations and cannot effectively model the complex nonlinear relationship between load and covariates. Secondly the forecasting accuracy is not high and the computational cost of determining the model parameters is high.

To improve the accuracy of time series forecasting, forecasting methods such as artificial neural networks (ANN) [4], fuzzy logic [6], and support vector machines (SVM) [5] have received increasing attention due to their fairly accurate forecasting results. In this paper, we study the application of deep learning techniques in electricity load forecasting. Convolutional neural networks (CNNs) [8, 9], long short-term memory (LSTMs) [11] and GRUs are commonly used for power load forecasting, which are able to efficiently obtain the local temporal patterns and temporal dependencies of electricity loads. In addition, some hybrid architectures CNNBiLSTM [12] and CNNLSTM [15] are proposed, aiming to reuse the temporal correlation in the captured data of the recurrent network to obtain local time patterns for convolutional electricity loads. Paper [14] proposes that the improved time-series convolutional network (TCN) can not only obtain the local dependency mode of input data, but also make up for the defect that CNNs cannot obtain long dependencies on input data. None of these methods considered the effect of covariates on electricity load, i.e., inter-series correlation. Paper [13] proposes to consider the inter-series correlation of the input data using ConvLSTM and use it for precipitation forecasting, which achieves a high forecasting accuracy. In summary, our goal is to develop a new hybrid load forecasting model to improve the forecasting accuracy.

3 Temporal Convolution and Multi-Attention for Electricity Load Forecasting

3.1 Overview

Here, we propose a method based on temporal convolution and multi-attention for short-term electricity load forecasting. The goal of short-term electricity load forecasting is to find a mapping function that satisfies X to y:

$$f(X) \rightarrow y \tag{1}$$

where $X \in R^{T \times N}$ is the input. The input features include weather features, load features and calendar features.

For any timestep t, we take the observations of the first k timesteps as input to forecasting the load values of the next h timesteps. This is shown in Formula (2):

$$y_t, y_{t+1}, \ldots, y_{t+h-1} = f(x_{t-k}, \ldots, x_{t-1}; \Phi) \tag{2}$$

where $x_{t-k} \in R^{N \times 1}$ is all the input features for timesteps $t - k$, $y_t \in R^{1 \times 1}$ is the forecasted values of the load for timesteps t; Φ is the parameter of the model. We use the data $\{x_{t-k}, \ldots, x_{t-1}\}$ from the sliding window $w(w = k)$ input values to forecast $\{y_t, y_{t+1}, \ldots, y_{t+h-1}\}$.

The general architecture of TC-MA is shown in Fig. 1. The multivariate inputs are first divided into two parts: the electricity load and covariate features. We calculate the

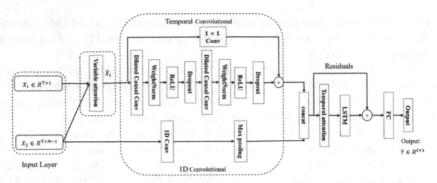

Fig. 1. Architecture of temporal convolution and multi-attention for electricity load forecasting

correlation dependencies X_1 and X_2, and put them in into TCN to extract the temporal dependencies, and use CNN to obtain the local dependencies. We concatenate the outputs of the TCN and CNN and use temporal attention to assign weights to each time period. We use LSTM to extract information from the time periods with larger weights, and the fully connected layer is used to predict the load values.

In our mothed, we use Root Mean Square Error (RMSE) as the loss function in our model, as shown in Formula (15) of Sect. 4.1. Our training goal is to keep the error between y_i and $\widehat{y}_i$ as small as possible.

3.2 Inter-series Modeling with Variable Attention

We use the variable attention to model the correlation between series. From Fig. 2, we can understand that changes in temperature affect human social activities and humans tend to use more electricity whether the temperature is high (July-August) or low (December-January). The different months and temperatures have different effects on the load variables. From Fig. 3, we understand that the load has a clear periodicity during the week. For instance, households use significantly more electricity on weekends than on weekdays. This is due to the difference in electricity consumption caused by the holiday weekend.

According to Fig. 2 and Fig. 3, we can conclude that the load demand varies with temperature during the day. Load demand trends on the same Monday are similar. The load demand is similar or different in different months, days or different moments. Therefore, we need to know the correlation between loads and covariates at each moment. We introduce an attention mechanism to compute the correlation between series, called variable attention. We can calculate the correlation between X_1^t and X_2^t at time t, then we can learn which covariate interacts more with the loading variable at time t.

The basic process of variable attention is:

Step 1: we get to transform the load and covariates into three feature spaces (Q, K and V).

$$Q = X_1 \cdot W^Q, \ K = X_2 \cdot W^K, \ V = X_2 \cdot W^V \tag{3}$$

$$score = softmax(QK^T) \tag{4}$$

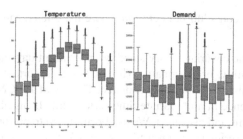

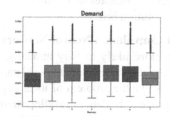

Fig. 2. The relationship of temperature and load   **Fig. 3.** Weekly load pattern changes

$$att(X_1) = score \cdot V \tag{5}$$

where $W^Q \in R^{1 \times 1}$, W^K and $W^V \in R^{(N-1) \times 1}$ are the trainable feature space. $Q \in R^{T \times 1}$ is the query space. $K \in R^{T \times 1}$ is the key space, and $V \in R^{T \times 1}$ is the value space. Variables attention is calculated using features in the Q and K by the operation shown in Formula (4). The attention *score* normalized by softmax is applied to features in the V to calculate the output of variable attention. This step uses Formula (5) to obtain the variable attention, which get X_1 from the representation X_2 of $att(X_1)$.

Step 2: the output is calculated using Formula (6), and then add X_1 and $att(X_1)$ to get $\tilde{X}_1$. $\tilde{X}_1$ contains load and covariate information. So $\tilde{X}_1$ will be fed into the next stage of the model as the final result of the variable attention.

$$\tilde{X}_1 = X_1 + att(X_1) \tag{6}$$

At this point, we effectively model inter-series dependencies.

3.3 Intra-series Temporal Dependence Modeling with Temporal Convolution and Attention

We use TCN to model historical series and CNN to obtain covariate local dependency patterns. Temporal attention mechanism is used to assign attention weights. It is used to resolve inter-series dependencies.

Temporal Convolution Network. We use TCN to obtain the historical time dependence of the electricity load. As the paper [14] says, compared to CNNs and LSTMs, dilation factors are used in the convolutional layers in TCNs to control the receptive field of the convolution. Thus, longer dependencies can be obtained. We hope to capture the long-term dependency to enhance the model prediction accuracy when short term prediction is made.

The TCN architecture is shown in Fig. 1, which consists of 1D-FCN layers with the same input and output lengths and a dilation causal convolution. We use Formula (7) to perform the dilated causal convolution. Given a filter $F = (f_1, f_2, \ldots f_K)$, the operation of the dilated causal convolution with d dilation factor at x_t the input sequence $\tilde{X}_1$ is:

$$h(t) = (\tilde{X}_1 *_d F)(t) = \sum_{k=0}^{K-1} f_k x_{t-dk} \tag{7}$$

where $*_d$ denotes as the dilated causal convolution operation where the dilated causal convolution is d.

1D Convolution Neural Network. The covariates vary steadily on adjacent days and are more likely to vary annually, monthly, or seasonally. This makes it important to obtain the same local temporal pattern than long-term dependence. The use of 1D-CNN can solve the problem of local dependence in multivariate time series while being able to process the data in parallel, saving time and overhead. We use Formula (8) to perform the convolution operation at the input series is: given a filter $F = (f_1, f_2, \ldots f_K)$, the operation of the convolution at x_t the input sequence X_2 is:

$$h(t) = (F * X_2)(t) = \sum_{k=0}^{K-1} f_k x_{t-k} \tag{8}$$

where * denotes the convolution operation.

Temporal Attention. As show in Fig. 3, this Wednesday's electricity load change trend may be the same as last Wednesday's. Then the attention mechanism [17] in the model can give more weight to the Wednesday time period in the input data, and the long-term dependence can be extracted during the training process. We refer to the attention mechanism used for weight assignment applied to the time dimension as temporal attention. We combine the outputs of the TCN and CNN models to obtain the final output of the temporal convolution module, as shown in Formula (9)-(11). Finally, LSTM is used to extract the results of assigning weights to each timestep of time series attention. The use of LSTM after temporal attention avoids information redundancy. Using Formula (12) is noted as the LSTM layer's output.

$$O_{TC} = Concat(O_{tcn}, O_{cnn}) \tag{9}$$

$$\alpha = soft\max(score(O_{TC}, O_{TC}^T)) \tag{10}$$

$$O_{TA} = \alpha \cdot O_{TC} \tag{11}$$

$$O_{lstm} = lstm(O_{TA}) \tag{12}$$

Finally, we use the fully connected layer as the final output layer, and the process of outputting the load value for the next h timesteps is as follows:

$$y_t, y_{t+1}, \ldots, y_{t+h-1} = glue(W \cdot O_{lstm} + b) \tag{13}$$

where W is the weight of the fully connected layer, b denotes the variance of the fully connected layer, and $glue$ is the activation function.

4 Experiment

4.1 Experimental Settings

Datasets. We evaluate the performance of our model using three datasets, each of which is mean-normalized.

TD[1]. The dataset we obtained from the 10th Teddy Cup Data Mining Challenge, named TD, contains 32,126 records from January 1, 2018 to August 31, 2021, and each record contains seven features (Maximum temperature, minimum temperature, wind direction during the day, wind direction at night, weather 1, weather 2, total active power.). We processed the missing data and reworked it to a 1 h sampling frequency.

ISO-NE. We obtained the dataset from the paper [16]. It contains hourly sampling from March 1, 2003 to December 31, 2014. It contains 103776 records sampled hourly from March 1, 2003 to December 31, 2014 and each record contains 7 features (Year, Month, Weekday, Hour, Temperature, Day, demand).

GEF_Area1[2]. We obtain the dataset from the Global Energy Forecasting Competition. We use the historical load and temperature data for "zone_id = 1" as the dataset GEF_Area1. It contains 39,408 records and each record contains 7 features (Year, Month, Weekday, Hour, Temperature, Day, load.)

Evaluation Metrics. In this paper, we evaluate the model using Mean Square Error (MAE), RMSE and R^2-score, which are common evaluation metrics in linear regression tasks. Let y_i and $\widehat{y}_i$ be predicted and ground truth values, n is the total number of timestamps. The evaluation metrics can be computed by Formula (14–16):

$$MAE = \frac{1}{n} \sum_{i=1}^{n} |y_i - \widehat{y}_i| \tag{14}$$

$$RMSE = \sqrt{\frac{1}{n} \sum_{i=1}^{n} (y_i - \widehat{y}_i)^2} \tag{15}$$

$$R^2 = 1 - \frac{\sum_{i=1}^{n} (y_i - \widehat{y}_i)^2}{\sum_{i=1}^{n} (y_i - \bar{y}_i)^2} \tag{16}$$

Baselines. We compared seven baseline models with the following settings:

In CNN, we use a convolutional layer, a Max_pool layer and a fully connected layer. The convolutional layer has 64 filters and the number of kernel size is 2. In our experiments LSTM consists of an LSTM layer (units = 32) and a fully connected layer to predict the load. BiLSTM use a bi-directional LSTM (units = 32) and a fully connected layer for predicting the load. CNNLSTM use a layer of convolution (filter = 64, kernel = 2), Max_pool and LSTM layers (units = 32). CNNBiLSTM use a layer of convolution (filter = 64, kernel = 2), Max_pooling and BiLSTM layers (units = 32). ConvLSTM use ConvLSTM to obtain the spatial dependencies of multivariate inputs. TCN combines dilated convolution and causal convolution of residuals for load prediction. The dilation factor d (d = 1, 2, 4, 8, 16, 32) is used in our experiments.

---

[1] https://www.tipdm.org:10010/#/competition/1481159137780998144/question.

[2] GEFCom2012, https://users.monash.edu.au/~shufan/Competition/index.html.

4.2 Overall Performance

We evaluate our method on three data sets. To reduce the chance and randomness of the experimental results, the average of three experiments with the same experimental setup is taken as the final result. From Table 1 we obtain the following conclusions: Our proposed method outperforms the comparison method on all datasets. The model performance improvement is more obvious on the TD and GEF_Area1 datasets, and the difference is not significant on the ISO-NE dataset. The reason for this is that the load curve on the ISO-NE dataset is smoother, unaffected by noise points, and the load cycle varies regularly. The GEF_Area1 dataset also has a periodical trend, but there is a significant difference in the range of loading values within the same period. Therefore, the GEF_Area1 dataset is more difficult to forecast than the ISO-NE dataset, but easier than the TD dataset with significant noise. The results for the TD and GEF_Area1 datasets are more significantly improved, possibly influenced by the attention of the variables, and the model represents the inter-series correlation well.

Overall, the hybrid models have better forecasting performance than the individual models CNN [8], LSTM, and BiLSTM. Because individual models have limited ability to extract features and can only extract temporal dependencies or local temporal patterns. CNN-LSTM [15] and CNN-BiLSTM [12] hybrid models are able to use CNN and LSTM, BiLSTM to obtain local patterns and temporal dependencies of time series. TCN [14] improves the problem of obtaining long-term dependencies by adding convolutional layers and diffusion factors, and performs significantly better than CNN. Unfortunately, it increases the complexity of the model and makes the model training time increase. In the TD dataset CNN takes only 87 s training time while TCN takes 876 s. ConvLSTM [13] considers sequence-to-sequence dependencies, but the results are not very effective.

Table 1. Overall performance on the three datasets

Metric Methods	TD			GEF_Area1			ISO_NE		
	MAE	RMSE	R^2	MAE	RMSE	R^2	MAE	RMSE	R^2
1D-CNN	0.4604	0.6180	0.1487	0.6275	0.8088	0.3871	0.6675	0.7810	0.3607
LSTM	0.4768	0.6461	0.0692	0.6668	0.8704	0.2900	0.6795	0.8315	0.2748
BiLSTM	0.4727	0.6325	0.1083	0.6677	0.8549	0.3151	0.6820	0.8165	0.3010
TCN	0.2444	0.4580	0.5325	0.4500	0.6128	0.6476	0.2013	0.2785	0.9188
CNN-LSTM	0.2259	0.4504	0.5478	0.4971	0.6701	0.5794	0.1498	0.2128	0.9525
CNN-BiLSTM	0.2459	0.4507	0.5471	0.5022	0.6783	0.5677	0.1505	0.2145	0.9514
ConvLSTM	0.4772	0.6355	0.0995	0.6559	0.8137	0.3566	0.6733	0.7856	0.3531
TC-MA	**0.1782**	**0.4480**	**0.5549**	**0.4227**	**0.5874**	**0.6767**	**0.1356**	**0.2038**	**0.9568**

4.3   Ablation Study

Module Test. We test the effect of variable attention and temporal attention on the model separately. From Fig. 4, we can see that adding both temporal attention and variable attention helps to improve the performance of the model, and presents different gap sizes on different datasets.

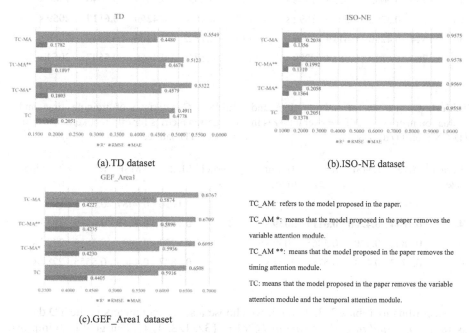

(a).TD dataset   (b).ISO-NE dataset

(c).GEF_Area1 dataset

TC_AM: refers to the model proposed in the paper.

TC_AM *: means that the model proposed in the paper removes the variable attention module.

TC_AM **: means that the model proposed in the paper removes the timing attention module.

TC: means that the model proposed in the paper removes the variable attention module and the temporal attention module.

Fig. 4. Comparison of the effectiveness of different modules in the proposed model on three datasets

In the TD and ISO-NE datasets, we see that removing the temporal and variable attention modules does not change much, and the model performance decreases slightly compared to TC-MA and improves slightly compared to TC. This indicates that both temporal attention and variable attention have improved the model performance, and also indicates that the two types of attention we have added do have their effects. However, the performance of the TC-MA model on the GEF_Area1 dataset is between the results of TC-MA** and TC-MA*, but it is still higher than the performance of TC without both attention modules. These two points show that our two attention modules are effective in improving the model performance.

Hyperparameter Test. We test four types of hyperparameters in the model on each of the three datasets: dilation factor in TCN, number of filters in CNN, kernel size in CNN, and kernel size in TCN.

Table 2. Comparison of training time and evaluation metrics for different dilation factors in TCN on TD dataset

d	MAE	RMSE	R^2	Training time
8	0.1956	0.5085	0.4235	1758 s
16	0.1741	0.5252	0.5670	1892 s
32	**0.1644**	**0.4260**	**0.6119**	**2059 s**
64	0.1784	0.4310	0.5859	2289 s

Table 3. Comparison of training time and evaluation metrics for different number of filters in 1D-CNN on TD dataset

filter	MAE	RMSE	R^2	Training time
64	0.2038	0.4419	0.5646	1504 s
128	**0.1644**	**0.4260**	**0.6119**	**2059 s**
129	0.1822	0.4263	0.5949	2068 s
127	0.1736	0.4466	0.5547	2064 s

Table 4. Comparison of training time and evaluation metrics for different kernel sizes in 1D-CNN on TD dataset

Kernel size	MAE	RMSE	R^2	Training time
3	0.1680	0.4261	0.5946	2070 s
4	**0.1644**	**0.4260**	**0.6119**	**2059 s**
5	0.1862	0.4307	0.5865	2057 s
6	0.1713	0.4347	0.5788	2061 s

Table 5. Comparison of training time and evaluation metrics for different kernel sizes in TCN on TD dataset

Kernel size	MAE	RMSE	R^2	Training time
3	0.1796	0.4487	0.5511	1992 s
4	**0.1644**	**0.4260**	**0.6119**	**2059 s**
5	0.1720	0.4413	0.5655	2126 s
6	0.1837	0.4463	0.5555	2287 s

According to Tables 2, 3, 4 and 5, we choose a set of parameters in the TD dataset as {*d, filter, kernel in CNN, kernel in TCN*} = {32, 128, 4, 4}. Our goal is to improve the accuracy of forecasting, we consider accuracy first and training time second. In the case of small differences in accuracy, the set of parameters that takes less time is given priority. We found that the performance of the model can be effectively improved by adjusting the parameters, and the unexpected finding is that we change the size of the kernel size in the CNN and the training time changes very little compared to the other parameters. This may be because the increase of kernel size in 1D-CNN does not increase the convolutional computation as much as 2D-CNN. Therefore, it leads to less change in training time.

Table 6. Comparison of training time and evaluation metrics for different dilation factors in TCN on GEF_Area1 dataset

d	MAE	RMSE	R^2	Training time
8	0.4271	0.5962	0.6638	1420 s
16	0.4113	0.5634	0.6813	1610 s
32	**0.4030**	**0.5772**	**0.6880**	**1483 s**
64	0.4152	0.5957	0.6675	2113 s

Table 7. Comparison of training time and evaluation metrics for different number of filters in 1D-CNN on GEF_Area1 dataset

filter	MAE	RMSE	R^2	Training time
16	0.4073	0.5894	0.6746	1311 s
32	0.4030	0.5772	0.6880	1483 s
64	0.4002	0.5863	0.6768	1825 s
127	0.4102	0.5878	0.6763	2518 s
128	**0.3966**	**0.5737**	**0.6917**	**2528 s**
129	0.4142	0.5915	0.6719	2551 s

Table 8. Comparison of training time and evaluation metrics for different kernel sizes in 1D-CNN on GEF_Area1 dataset

Kernel size	MAE	RMSE	R^2	Training time
3	0.4192	0.6032	0.6591	2519 s
4	**0.3966**	**0.5737**	**0.6917**	**2528 s**
5	0.4272	0.6072	0.6546	2535 s
6	0.4235	0.6010	0.6617	2769 s

Table 9. Comparison of training time and evaluation metrics for different kernel sizes in TCN on GEF_Area1 dataset

Kernel size	MAE	RMSE	R^2	Training time
2	**0.4000**	**0.5598**	**0.7064**	**2352 s**
3	0.3919	0.5655	0.7005	2444 s
4	0.4072	0.5831	0.6815	2533 s
5	0.3966	0.5737	0.6917	2528 s
6	0.4049	0.5815	0.6833	2811 s

According to Tables 6, 7, 8 and 9, we choose a set of parameters in the GEF_Area1 dataset as {d, *filter, kernel in CNN, kernel in TCN*} = {32, 128, 4, 2}. The method of parameter selection is consistent with the TD dataset, and the variation of its results is consistent with the TD dataset. However, the GEF_Area1 dataset is similar to the TD dataset in that changing the parameters causes large fluctuations in the results. In addition, we choose a set of parameters in the ISO-NE dataset as {d, *filter, kernel in CNN, kernel in TCN*} = {8, 32, 4, 4}.

5 Conclusions

In this paper, we propose an attention and time-series convolution (TC-MA) based electric load forecasting method for daily electric load forecasting. Our model addresses the extraction of inter- series and intra-series dependencies between covariates and load data features. Experiments show that the proposed method achieves lower MAE and RMSE and higher R^2-score compared to common deep learning models on three datasets. it also shows that combining multiple model advantageous hybrid models is beneficial to improve the forecasting accuracy. Therefore, the TC-MA improving the forecasting capacity of STLF problems. Furthermore, the implementation of deep learning in long-term load forecasting will be considered in our following works.

Acknowledgements. This work is supported by the National Natural Science Foundation of China (Grant Nos. 62002262, 62172082, 62072086, 62072084, 71804123).

References

1. Chen, K., Chen, K., Wang, Q., et al.: Short-term load forecasting with deep residual networks. IEEE Trans. Smart Grid **10**(4), 3943–3952 (2018)
2. Bunn, D., Farmer, E.D.: Comparative models for electrical load forecasting (1985)
3. Chakhchoukh, Y., Panciatici, P., Mili, L.: Electric load forecasting based on statistical robust methods. IEEE Trans. Power Syst. **26**(3), 982–991 (2010)
4. Charytoniuk, W., Chen, M.S.: Very short-term load forecasting using artificial neural networks. IEEE Trans. Power Syst. **15**(1), 263–268 (2000)
5. Niu, D.X., Wanq, Q., Li, J.C.: Short term load forecasting model using support vector machine based on artificial neural network. In: 2005 International Conference on Machine Learning and Cybernetics, vol. 7, pp. 4260–4265. IEEE (2005)
6. Yun, Z., Quan, Z., Caixin, S., et al.: RBF neural network and ANFIS-based short-term load forecasting approach in real-time price environment. IEEE Trans. Power Syst. **23**(3), 853–858 (2008)
7. Che, J.X., Wang, J.Z.: Short-term load forecasting using a kernel-based support vector regression combination model. Appl. Energy **132**, 602–609 (2014)
8. Amjady, N.: Short-term hourly load forecasting using time-series modeling with peak load estimation capability. IEEE Trans. Power Syst. **16**(3), 498–505 (2001)
9. Kuo, P.H., Huang, C.J.: A high precision artificial neural networks model for short-term energy load forecasting. Energies **11**(1), 213 (2018)
10. Siddarameshwara, N., Yelamali, A., Byahatti, K.: Electricity short term load forecasting using Elman recurrent neural network. In: 2010 International Conference on Advances in Recent Technologies in Communication and Computing, pp. 351–354 (2010). IEEE
11. Tasarruf, B., Chen, H.Y., et al.: Short-term electricity load forecasting using hybrid prophet-LSTM model optimized by BPNN. Energy Reports **8**, 1678–1686 (2022). ISSN 2352-4847
12. Miao, K., Hua, Q., Shi, H.: Short-term load forecasting based on CNN-BiLSTM with Bayesian optimization and attention mechanism. In: Zhang, Y., Xu, Y., Tian, H. (eds.) Parallel and Distributed Computing, Applications and Technologies. PDCAT 2020. LNCS, vol. 12606, pp. 116–128. Springer, Cham (2021). https://doi.org/10.1007/978-3-030-69244-5_10
13. Yasuno, T., Ishii, A., Amakata, M.: Rain-code fusion: code-to-code ConvLSTM forecasting spatiotemporal precipitation. In: Del Bimbo, A., et al. (eds.) Pattern Recognition. ICPR International Workshops and Challenges. ICPR 2021. LNCS, vol. 12667, pp. 20–34. Springer, Cham (2021). https://doi.org/10.1007/978-3-030-68787-8_2
14. Bai, S., Kolter, J.Z., Koltun, V.: An empirical evaluation of generic convolutional and recurrent networks for sequence modeling. arXiv preprint arXiv:1803.01271 (2018)
15. Liu, Y.F., Yang, Y.H.: Research on short-term power load forecasting based on CNN-LSTM. Sci. Technol. Innov. Appl. **1**, 84–85 (2020)

16. Li, N., Wang, L., Li, X., et al.: An effective deep learning neural network model for short-term load forecasting. Concurr. Comput. Pract. Exp. **32**(7), e5595 (2020)

17. Liang, Y., Wang, H., Zhang, W.: A knowledge-guided method for disease prediction based on attention mechanism. In: Zhao, X., Yang, S., Wang, X., Li, J. (eds.) Web Information Systems and Applications. WISA 2022. LNCS, vol. 13579, pp. 329–340. Springer, Cham (2022). https://doi.org/10.1007/978-3-031-20309-1_29

Jointly Learning Structure-Augmented Semantic Representation and Logical Rules for Knowledge Graph Completion

Jinxing Jiang and Lizhen Xu[✉]

School of Computer Science and Engineering, Southeast University, Nanjing 211189,
Jiangsu, China
{220212039,lzxu}@seu.edu.cn

Abstract. Knowledge Graph Complementation (KGC) aims to predict
the missing triples in incomplete knowledge graphs (KGs). However,
existing approaches rely either on structural features, semantic features
or logical rules. There is not yet a unified way to exploit all three features
mentioned above. To address this problem, this paper proposes a new
KGC framework, SSL, which jointly embeds structure-augmented semantics representation in the natural language description of triples with
their logical rules. First, SSL fine-tunes a pre-trained language model to
capture the semantic information in the natural language description corresponding to the triple. Then, for logical rules corresponding to different
relation types, we augment them to embeddings of entities and relations
to activate stronger representation capabilities. Finally, by jointly training, we can obtain more expressive embeddings for the downstream KGC
task. Our extensive experiments on a variety of KGC benchmarks have
demonstrated the effectivity of our method.

Keywords: Knowledge Graph · Knowledge Representation ·
Knowledge Graph Completion

1 Introduction

In recent years, knowledge graphs (KGs) [6], such as Freebase [3], DBpedia [2],
and Wikidata, have gained increasing popularity for a wide range of applications,
including question answering, natural language processing, computer vision, and
recommendation systems. Knowledge graphs (KGs) are typically represented as
directed graphs, where nodes and edges correspond to entities and relationships, respectively. However, in reality, KGs are often incomplete, with some
relationships between entities missing. To address this challenge, researchers
have focused on knowledge graph completion, which aims to effectively predict
missing links by leveraging known facts expressed in triplets.

The previous KGC models can be broadly classified into three categories:
(1) Structure-based KGE models, such as TransE [4], ComplEx [19], and

© The Author(s), under exclusive license to Springer Nature Singapore Pte Ltd. 2023
L. Yuan et al. (Eds.): WISA 2023, LNCS 14094, pp. 52–63, 2023.
https://doi.org/10.1007/978-981-99-6222-8_5

RotatE [17], which learn embeddings of entities and relations to score the plausibility of triples. (2) Logic-based rules that fully exploit logical rules in the knowledge graph, such as AMIE, DRUM [10], and AnyBurl. (3) Models based on external resources that leverage information beyond the knowledge graph itself, such as entity descriptions, attributes, and other sources. In this paper, we focus on entity description information as it is often readily available.

Despite the success achieved by the three types of methods, we believe they suffer from certain limitations: (1) Structure-based KGE models treat each fact triple independently and ignore other features, such as entity description information and logical rules in the knowledge graph. (2) Rule-based learning models only use structural information and disregard text related to entities and relationships, leading to the loss of prior knowledge contained in these texts. These models' abilities can be further weakened when the knowledge graph is small. (3) Models based on external resources only consider semantic information in the knowledge graph and ignore numerous topological features.

In this paper, we propose SSL, a framework for joint structurally enhanced language-semantic and logical rule representations for knowledge graph completion (KGC). SSL simultaneously combines structural, semantic, and logical information in knowledge graphs. Firstly, SSL embeds the text description corresponding to the triple in the knowledge graph into the vector space using a pre-trained language model. During this process, the forward propagation of the pre-trained language model represents the acquisition of semantic features. Then, SSL optimizes the loss of the structure-based knowledge graph embedding (KGE) model by backpropagation to learn both semantic and structural features. To obtain the logical rule representation in the knowledge graph, we transform hard logic rules into soft logic rules. Instead of requiring large amounts of manpower and time to build and verify hard rules, we use a model based on logical rule learning to obtain specific logical rules for common complex relationship patterns in knowledge graphs, such as symmetry, antisymmetry, and transitivity. We then learn the corresponding soft logical representation constraints for each existing relational schema. The effectiveness and rationality of SSL are proven through a large number of experiments on three widely used public KGC datasets.

To sum up, this paper makes three main contributions:

- We propose a novel knowledge graph completion framework SSL, which can learn joint representations of structural, semantic, and logical rules in the knowledge graph to enhance the expressiveness of embeddings.
- In order to reduce the cost needed to construct hard logic rules, we first propose a model based on logic rule learning to acquire common soft logic rules in the knowledge graph. Then, for each relation pattern, SSL learns the corresponding soft logic representation.
- Experiments on three commonly used knowledge graph complementary datasets demonstrate that the proposed method in this paper outperforms previous methods to establish a new SOTA.

2 Related Work

2.1 Structure-Based KGE Model

Knowledge Graph Embedding (KGE) is a highly valued research direction in representation learning. KGE models can be broadly divided into two categories: translation-based and semantic matching-based.

Translation-Based. Translation-based models, such as TransE [4] and TransR [13], transform entities and relations in KG into low-dimensional vector space. They consider the relation between head and tail entities as a translation operation in the vector space and calculate the distance between head and tail entity vectors to measure the plausibility of a triple. Additionally, translation-based methods can embed entities and relations into different representation spaces, such as complex space and flow space. For example, RotatE [17] models the relation as a rotation from the head entity to the tail entity in the complex vector space. ManifoldE [22]addresses the limitations of TransE [4], which is based on strict head and tail entity vectors and relationship vectors, by approximating all tail entities on a high-dimensional sphere. This approach alleviates the issue of ill-posed algebraic systems and overly strict geometric forms.

Semantic Matching-Based. The model utilizing semantic matching formalizes the Knowledge Graph Completion (KGC) problem into a third-order binary tensor completion problem using tensor decomposition. For a given triple (h, r, t), the score function employed by the semantic matching-based model is $s(\mathbf{h}, \mathbf{r}, \mathbf{t}) = \mathbf{h}^T \mathbf{M}_r \mathbf{t}$, with $\mathbf{h}$ and $\mathbf{t}$ representing the embedding representations of the head and tail entities, respectively, and $\mathbf{M}_r$ representing the relation embedding matrix. RESCAL [16] represents the triples in the Knowledge Graph as a three-dimensional tensor X with size n × n × m, where $\mathbf{X}_{ijk}$ equals 1 if the triple (h, r, t) exists. However, due to its large parameter set, RESCAL [16] is subject to high computational complexity. As a result, DisMult suggested the conversion of the relation matrix $\mathbf{M}_r$ into a diagonal matrix to reduce its constraint. Inspired by DistMult, ComplEx [19] introduced complex vector space into KGE for the first time and used a diagonal matrix to resolve asymmetric relation types.

2.2 Rule-Based Learning Models

The model based on logical rule learning aims to achieve interpretable knowledge graph completion by mining the association of relations and local structures within a knowledge graph. For the first time, KALE jointly learns triples and hard logic rules in a unified model, where triples are represented as atomic formulas and hard logic is represented as complex formulas. However, RUGE [9] argues that creating and verifying hard logic rules requires significant time and resources, making it difficult to apply to large-scale knowledge graphs. Therefore, RUGE [9] proposes to extend KALE [8] with soft rules, which can be mined by atomic tools such as AMIE. Neural LP [24] is the first to propose learning first-order logic rules in an end-to-end differentiable form [25]. It formulates the knowledge graph completion task as a sequence of differentiable operations to model the parameters and structure of logical rules. DRUM [10]introduces an empty relation model based on Neural LP [24], enabling rule learning to be extended to variable-length models.

2.3   Model Based on External Resources

In addition to the structural information of the knowledge graph itself, external resources such as semantic information have been widely used to enhance knowledge graph completion (KGC) systems. These resources include natural language description texts, entity attributes, and entity types. However, obtaining many of these resources can be difficult. In this paper, we focus specifically on the natural language description texts of entities, which are among the most accessible external resources for KGC.

DKRL [23] was the first model to directly utilize entity description texts to learn corresponding entity embeddings. It encoded the semantic information of entity descriptions by using continuous bag-of-words and deep convolutional neural network models, and combined it with TransE to learn entity and relation embeddings. SSP proposed a semantic space projection model based on TransH [21], which jointly learned triplets and textual descriptions to establish interactions between two different features. ATE [1] aimed to learn more precise text representations by first encoding relation mentions and entity descriptions using a BiLSTM, and then introducing a mutual attention mechanism between relation mentions and entity descriptions to learn more accurate text representations of relations and entities. Recently, some text encoding methods, such as KG-BERT [12], have been proposed. These methods use pre-trained language models to encode triplets and output scores for each candidate triplet.

Despite the success of structure-based KGE models, logic rule learning models, and external resource-based models, there is currently no method that can simultaneously leverage structure, logic rules, and external textual descriptions for KGC.

3   Proposed Approach

3.1   Problem Description

A knowledge graph is a type of multi-relational graph that typically consists of a collection of triples in the form of (h, r, t). It can be formally defined as $\mathcal{G} = (\mathcal{E}, \mathcal{R}, \mathcal{T})$, where $\mathcal{T}$ represents the set of all triples and $\mathcal{E}$ and $\mathcal{R}$ respectively represent the sets of all entities and relations.

In this paper, we use boldface to denote vectors. Specifically, we denote the embedding vectors for the head entity h, relation r, and tail entity t as $h \in \mathbb{R}^{d_e}$, $r \in \mathbb{R}^{d_r}$, and $t \in \mathbb{R}^{d_e}$, respectively. Here, d_e and d_r represent the embedding dimensions for entities and relations, respectively.

In KGC, the objective is to infer missing factual triples in an incomplete knowledge graph $\mathcal{G}$. For example, to predict tail entities in the triple $(h, r, ?)$, KGC needs to rank all entities given h and r. Similarly, predicting head entities in $(?, r, t)$ follows a similar approach. In this paper, we adopt the same setting as previous work [17] and add the corresponding reverse triple (t, r^{-1}, h) for each triple (h, r, t), where r^{-1} is the reverse relation of r. With this modification, we only need to focus on predicting tail entities.

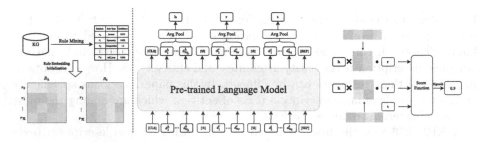

Fig. 1. An overview of our proposed SSL framework.

3.2 Framework Overview

The proposed KGC framework SSL is illustrated in Fig. 1 and comprises three main parts: logical rule embedding, semantic embedding, and structural embedding. SSL initially mines logical rules from the knowledge graph to obtain four predefined types of logical rules: Inverse, Symmetry, Composition, and $Self_loop$. For each type of logical rule, SSL initializes the logical rule representation R_h and R_t for each relation. Additionally, a pre-trained language model is used to encode the textual information of each triple, obtaining the corresponding semantic embedding for each (h, r, t) triple. The semantic embeddings of $\mathbf{h}$, $\mathbf{r}$, and $\mathbf{t}$ are then combined with the logical rule representations to obtain the final fusion vector. Finally, the fusion vector is passed through the Score Function to obtain the corresponding triple score, where the Score Function can be selected from most structural KGE-based score functions. This enables SSL to incorporate structural information into the pre-trained language model during backpropagation for optimization.

3.3 Logical Rule Embedding

Drawing inspiration from previous logic-based models [24] this study employs a triplet-based approach to model logical rules on relationship paths. Specifically, when a triplet (h, r, t) is incomplete, logical rule evidence for completion can be found in the knowledge graph where $r(h, t) \leftarrow r_0(h, e_1) \wedge r_1(e_1, e_2) \wedge \cdots \wedge r_n(e_n, t)$. It is noteworthy that the relationship r is essentially a combination

Table 1. Logical rule forms and soft logical rule constraints. LR stands for Logical Rule and SLR stands for Soft Logical Rule.

LR Type	LR Form	SLR Constraint
Self Loop	$\forall e : r(e, e)$	$\mathbf{L}_r^h \approx \mathbf{L}_r^t$
Symmetry	$\forall (e_1, e_2) : r(e_1, e_2) \Rightarrow r(e_2, e_1)$	$\mathbf{L}_r^h \approx \mathbf{L}_r^t$
Inverse	$\forall (e_1, e_2) : r_1(e_1, e_2) \Rightarrow r_2(e_2, e_1)$	$\mathbf{L}_{r_1}^h \approx \mathbf{L}_{r_2}^t, \mathbf{L}_{r_1}^t \approx \mathbf{L}_{r_2}^h$
Composition	$\forall (e_1, e_2, e_3) : r_1(e_1, e_2) \wedge r_2(e_2, e_3)$ $\Rightarrow r_3(e_1, e_3)$	$\mathbf{L}_{r_1}^t \approx \mathbf{L}_{r_2}^h, \mathbf{L}_{r_2}^t \approx \mathbf{L}_{r_3}^t, \mathbf{L}_{r_1}^h \approx \mathbf{L}_{r_3}^h$

relationship pattern. This research focuses on four fundamental logical rules, namely Inverse, Symmetry, Composition, and Self Loop. For each relationship r, it was discovered that it constitutes a binary relationship with two variables h and t. As such, logical embeddings for each variable are learned for each relationship r, i.e., L_r^h and L_r^t. Table 1 summarizes the four logical rules and their corresponding logical embedding constraints. Importantly, these rules can be mined through the automatic rule mining tool AMIE+ [7] without incurring manual cost. Corresponding constraints are also proposed for each logical rule to limit the relationship embedding. For instance, in the transitive logical rule $(\forall (e_1, e_2, e_3) : r_1(e_1, e_2) \wedge r_2(e_2, e_3) \Rightarrow r_3(e_1, e_3))$, it is believed that the tail entity embedding $\mathbf{L}_{r_1}^t$ of relationship r_1 and the head entity embedding $\mathbf{L}_{r_2}^h$ of relationship r_2 exhibit high similarity, as do the tail entity embedding $\mathbf{L}_{r_2}^t$ of relationship r_2 and the tail entity embedding $\mathbf{L}_{r_3}^t$ of relationship r_3, and the head entity embedding $\mathbf{L}_{r_1}^h$ of relationship r_1 and the head entity embedding $\mathbf{L}_{r_3}^h$ of relationship r_3. This indicates that the logical soft rule constraints presented in Table 1 can be optimized via a similarity scoring function:

$$f_{SL} = \text{Sim}\left(\mathbf{L}_r^t, \mathbf{L}_r^h\right)$$
$$f_{SY} = \text{Sim}\left(\mathbf{L}_r^t, \mathbf{L}_r^h\right)$$
$$f_{IN} = \frac{1}{2}\left(\text{Sim}\left(\mathbf{L}_{r_1}^h, \mathbf{L}_{r_2}^t\right) + \text{Sim}\left(\mathbf{L}_{r_1}^t, \mathbf{L}_{r_2}^h\right)\right) \quad (1)$$
$$f_{TR} = \frac{1}{3}\left(\text{Sim}\left(\mathbf{L}_{r_1}^t, \mathbf{L}_{r_2}^h\right) + \text{Sim}\left(\mathbf{L}_{r_2}^t, \mathbf{L}_{r_3}^t\right) + \text{Sim}\left(\mathbf{L}_{r_1}^h, \mathbf{L}_{r_3}^h\right)\right)$$

In this study, we use the acronyms SL, SY, IN, and TR to represent the four logical rules of Inverse, Symmetry, Composition, and Self Loop, respectively. The similarity calculation function, denoted as $\text{Sim}(\cdot, \cdot)$, is used to measure the similarity between two vectors and can be expressed as follows:

$$\text{Sim}(\mathbf{L}_r^t, \mathbf{L}_r^h) = \exp(-||\mathbf{L}_r^t - \mathbf{L}_r^h||_2) \quad (2)$$

It should be noted that when using the AMIE+ tool to extract logical rules, the length of the logical rules we extract does not exceed two. As for exploring how to use other more complex logical patterns, it is left for future work.

3.4 Semantic Embedding

In KGC, semantic similarity between the head entity h and candidate tail entities, as well as semantic similarity between relation r and candidate tail entities, are crucial for completing the task of predicting the tail entity in a triple $(h, r, ?)$. For example, in the case of the triple $(China, capital, ?)$, the semantic similarity between the head entity China and candidate entity Beijing should be greater than that between China and Nanjing, because Beijing is the correct answer. Similarly, the semantic similarity between the relation "capital" and the candidate entity Beijing should be greater than that between "capital" and Nanjing. Thanks to the advancement of text encoding techniques, pre-trained language models can capture such semantic information through pre-training on large corpora. Fortunately, we can easily obtain the textual descriptions of entities and relations in a triple.

Formally, for a triple (h, r, t), the textual descriptions of the head entity, relation, and tail entity can be represented as $T^h = d_1^h, d_2^h, \cdots, d_h^h$, $T^r = d_1^r, d_2^r, \cdots, d_r^r$, and $T^t = d_1^t, d_2^t, \cdots, d_t^t$, respectively. SSL generates semantic embeddings by forwarding the LM, as shown in Fig. 1. Explicit modeling of dependencies between the head, relation, and tail is required for completing the task in the knowledge graph. For example, connections between the head, relation, and tail aid in predicting the tail in the prediction task. Thus, we concatenate T_h, T_r, and T_t and use the average pooling of the output representation of each token in T_h, T_r, and T_t forwarded through the LM as $\mathbf{h}, \mathbf{r}, \mathbf{t} \in \mathbb{R}^k$, where k is the dimension of the embedding vectors. Specifically, the sequence format input to the LLM is [CLS] Th [\S] Tr [\S] Tt [SEP], where [CLS], [\S], and [SEP] are special tokens. We add an average pooling layer on top of the LM's output layer and take the average pooling of the output representations of each token in T_h to obtain $\mathbf{h}$, as shown in Fig. 1. We obtain $\mathbf{r}$ and $\mathbf{t}$ in the same way.

3.5 Structure Embedding

In the semantic embedding of SSL, structural information cannot be captured during the forward propagation of pre-trained language models. We propose the introduction of structural loss during fine-tuning of pre-trained language models to enhance text representations with structural information. Specifically, for the triple (h, r, t), we minimize the following loss function to achieve this goal.

$$L_{structure} = - \sum_{(h,r,t)\in\mathcal{G}} (\log \mathrm{P}(h|r, t) + \log \mathrm{P}(t|h, r)) \qquad (3)$$

The expressions $\mathrm{P}(h|r, t)$ and $\mathrm{P}(t|h, r)$ represent the probabilities of the correct head and tail entities, respectively, and can be formulated as follows:

$$\mathrm{P}(h \mid r, t) = \frac{\exp(f(\mathbf{h}, \mathbf{r}, \mathbf{t}))}{\sum_{h' \in E - \{h\}} \exp(f(\mathbf{h'}, \mathbf{r}, \mathbf{t}))} \qquad (4)$$

whereby, the function $f(\cdot)$ can be any structure-based KGE scoring function. Table 2 summarizes representative KGE models. h' and t' represent the corrupted head and tail entities, respectively.

Table 2. Score functions of some popular knowledge graph embedding methods. Here, $<x^1, \ldots, x^k> = \sum_i x_i^1 \ldots x_i^k$ denote the generalized dot product, $\bar{x}$ represents the conjugate of complex number x, $\bullet$ presents the Hadamard product.

Method	Score function $f_r(h, t)$
TransE	$-\|h + r - t\|_p$
SimplE	$(< h^{(H)}, r, t^{(T)} > + < t^{(H)}, r^{(inv)}, h^{(T)} >)/2$
ComplEx	$Re(h^T diag(r)\bar{t})$
RotatE	$-\|\mathbf{h} \bullet \mathbf{r} - \mathbf{t}\|^2$

3.6   Optimization

As shown in Table 2, given a triple (h, r, t), when we obtain the corresponding embeddings for the head entity $\mathbf{h}$, the relation $\mathbf{r}$, and the tail entity $\mathbf{t}$, we first calculate the score of the triple as follows:

$$s(h, t) = f_r \left(\boldsymbol{W} \left(\text{concat} \left(\text{x} \times \mathbf{L}_r^h * \text{r}, \text{x} \times \mathbf{L}_r^t * \text{r} \right) \right), \mathbf{t} \right) \tag{5}$$

Similar to the previous method, we use a negative sampling loss function to optimize SSL, where $f_r(\cdot, \cdot)$ is a specific scoring function from Table 2.

The loss function is defined as follows: Similar to the previous method [17], we use a negative sampling loss function to optimize SSL, where $f_r(\cdot, \cdot)$ is a specific scoring function from Table 2. The loss function is defined as follows:

$$\mathcal{L}_{score} = -log\ \sigma(\gamma - s_r(h, t)) - \sum_{i=1}^{n} log\ \sigma(s_r(h, t_i^{'}) - \gamma) \tag{6}$$

We use the sigmoid function σ, where N is the number of negative samples, and γ is a hyperparameter.

Finally, we jointly train the parameters of SSL by optimizing the following overall loss function.

$$\mathcal{L} = \lambda_1 L_{structure} + \lambda_2 L_{score} + \lambda_3 L_R \tag{7}$$

where $L_R = f_{SL} + f_{SY} + f_{IN} + f_{TR}$ and $\lambda_1 + \lambda_2 + \lambda_3 = 1$.

4   Experiments

4.1   Experimental Setup

Datasets. We employed three benchmark Knowledge Graph Completion (KGC) datasets from different knowledge graphs (KGs) to evaluate the performance of our proposed method. These three KGs are Freebase [3], WordNet [15], and UMLS [5]. Freebase is a large-scale KG containing a large number of general knowledge facts. We selected a subset of Freebase, namely FB15K-237 [18],

Table 3. Statistical results of datasets.

Dataset	#Entity	#Relation	#Train	#Dev	#Test
FB15k-237	14,541	237	272,115	17,535	20,466
WN18RR	40,943	11	86,835	3,034	3,134
UMLS	135	46	5,216	652	661

as our test dataset. WordNet provides semantic knowledge of words, and we used a subset of it, namely WN18RR, for testing. UMLS is a semantic network containing medical entities and relationships. The statistical summaries of these datasets are presented in Table 3.

Additionally, for the FB15k-237 dataset, we utilized the entity and relation description text provided by [23]. For the WN18RR dataset, we utilized the entity and relation descriptions provided by WordNet. For the UMLS dataset, we used the entity names themselves as their descriptions.

Evaluation Protocol. KGC (Knowledge Graph Completion) aims to predict the missing entity in incomplete triplets (h, r, t) or $(?, r, t)$ in KGs, and the evaluation of KGC models is based on how well they match the ground truth. In KGC tasks, three evaluation metrics are widely used: Mean rank (MR), mean reciprocal rank (MRR), and Hits@K. For a given incomplete triplet $(h, r, ?)$, all entities in the KG are candidate entities. KGC models need to calculate a score for each candidate entity using a scoring function, rank them in descending order according to the score, and select the top K entities as prediction results. Each incomplete triplet in the test set will obtain the rank of the true entity. MR is the average rank of all correct predictions, but it is easily affected by outliers, so this paper does not report the experimental results of MR. MRR is the average reciprocal rank of all prediction ranks. Hits@K records the percentage of correct predictions that rank in the top K. It should be noted that for a given triplet $(h, r, ?)$, some candidate entities may appear in the training or validation set. These entities will rank higher and affect the ranking of the ground truth entity. Therefore, we need to filter out these entities before calculating the true rank of prediction results, which is the filtered version of evaluation metrics, i.e., filtered MR, filtered MRR, and filtered Hits@K. In this paper, we choose to use the filtered evaluation metrics. Moreover, for the FB15k-237 dataset, we use the entity and relation descriptions provided by [18]. For the WN18RR dataset, we use the entity and relation descriptions provided by WordNet. For the UMLS dataset, we use the entity names as entity descriptions.

Implementation Details. Since our proposed SSL requires the use of pretrained language models, we conducted experiments using two versions of BERT, namely $BERT_{BASE}$ and $BERT_{LARGE}$. Therefore, there are two versions of SSL: SSL-BERT$_{BASE}$ and SSL-BERT$_{LARGE}$. We selected TransE, RotatE, SimplE, ComplEx as the structure-based KGE models used in the scoring function. For

Table 4. Experimental results on three KGC datasets, where bold value represents the best results and underline value represents the second best results. All results are from the original paper, where - denotes results not provided in the original paper.

Method	FB15k-237			WN18RR			UMLS		
	MRR	Hits@1	Hits@10	MRR	Hits@1	Hits@10	MRR	Hits@1	Hits@10
TransE	0.279	19.8	44.1	0.243	4.27	53.2	–	–	98.9
SimplE	0.230	15.0	40.0	0.420	40.0	46.0	–	–	–
ComplEx	0.247	15.8	42.8	0.440	41.0	51.0	–	–	96.7
RotatE	0.337	24.1	**53.3**	0.477	42.8	57.1	0.925	86.3	**99.3**
Neural LP	0.252	18.9	37.5	0.435	37.1	56.6	0.745	62.7	91.8
RUGE	0.169	8.7	34.5	0.231	21.8	43.9	–	–	–
KG-BERT	–	–	42.0	0.216	4.1	52.4	–	–	99.0
stAR	0.348	**25.2**	48.2	0.401	24.3	**70.9**	–	–	99.1
SSL-BERT$_{BASE}$	**0.350**	23.2	48.5	**0.480**	**43.0**	60.6	**0.930**	85.9	99.0
SSL-BERT$_{LARGE}$	**0.351**	23.4	48.5	**0.482**	**43.2**	61.0	**0.931**	86.0	99.0

SSL-BERT$_{BASE}$ and SSL-BERT$_{LARGE}$, the batch size was set to 128, the learning rate was set to 5e-5, and the number of training epochs was set to 20. During training, the embedding dimensions for entities and relations were selected from 256, 512, 1024. The value of γ was selected from 0, 1, 2, 3, 4, 5, 6, 10, 24. In the optimization phase, we optimized each positive triple by generating 15 negative samples by corrupting the head and tail entities. We trained SSL using the AdamW [14] optimizer.

Baselines. To comprehensively evaluate the performance of SSL, we selected three typical types of models, including structure-based KGE models, logic rule learning-based models, and external resource-based models. The structure-based KGE models we selected were TransE [4], SimplE [11], ComplEx [19], and RotatE [17]. The logic rule learning-based models we selected were Neural LP [24] and RUGE [9]. The external resource-based models we selected were KG-BERT [12] and stAR [20].

4.2   Main Results

The experimental results of KGC are presented in Table 4, indicating that SSL outperforms previous methods that rely on only one kind of feature information by combining three different types of feature information. Specifically, on the FB15k-237 dataset, the MRR indicator of SSL increased by 1.4%, 9.9%, and 0.3% compared to the best-performing structure-based KGE model, logic-based learning model, and external resource-based model, respectively. On the WN18RR dataset, SSL showed similar improvements with MRR increasing by 1.4%, 9.9%, and 0.3% compared to the aforementioned models. On the UMLS dataset, the MRR indicator of SSL increased by 1.4%, 9.9%, and 0.3% compared to the best-performing structure-based KGE model, logic-based learning model, and external resource-based model, respectively.

Furthermore, our results showed that $SSL - BERT_{LARGE}$ outperformed $SSL - BERT_{BASE}$, indicating that a better pre-trained language model led to improved SSL performance.

5   Conclusion

This paper proposes a text knowledge graph completion framework, named SSL, which combines logical rules and structural enhancement for knowledge graph completion. Firstly, we extract four basic types of logical rules from different knowledge graphs and learn the soft logical rule embeddings for each relation with corresponding variables. Meanwhile, we capture the semantic embeddings of triplets using pre-trained language models during forward propagation and introduce structural embeddings through adding structural loss embeddings during backpropagation. Finally, we perform KGC using three types of fused vectors. Experimental results demonstrate the effectiveness of SSL on three widely used datasets. We hope that our results will promote further research in this direction.

References

1. An, B., Chen, B., Han, X., Sun, L.: Accurate text-enhanced knowledge graph representation learning. In: Proceedings of the 2018 Conference of the North American Chapter of the Association for Computational Linguistics: Human Language Technologies (Volume 1: Long Papers), pp. 745–755 (2018)
2. Auer, S., Bizer, C., Kobilarov, G., Lehmann, J., Cyganiak, R., Ives, Z.: DBpedia: a nucleus for a web of open data. In: Aberer, K., et al. (eds.) ASWC/ISWC -2007. LNCS, vol. 4825, pp. 722–735. Springer, Heidelberg (2007). https://doi.org/10.1007/978-3-540-76298-0_52
3. Bollacker, K., Evans, C., Paritosh, P., Sturge, T., Taylor, J.: Freebase: a collaboratively created graph database for structuring human knowledge. In: Proceedings of the 2008 ACM SIGMOD International Conference on Management of Data, pp. 1247–1250 (2008)
4. Bordes, A., Usunier, N., Garcia-Duran, A., Weston, J., Yakhnenko, O.: Translating embeddings for modeling multi-relational data. In: Advances in Neural Information Processing Systems, vol. 26 (2013)
5. Dettmers, T., Minervini, P., Stenetorp, P., Riedel, S.: Convolutional 2D knowledge graph embeddings. In: Proceedings of the AAAI Conference on Artificial Intelligence, vol. 32 (2018)
6. Dong, X., et al.: Knowledge vault: a web-scale approach to probabilistic knowledge fusion. In: Proceedings of the 20th ACM SIGKDD International Conference on Knowledge Discovery and Data Mining, pp. 601–610 (2014)
7. Galárraga, L., Teflioudi, C., Hose, K., Suchanek, F.M.: Fast rule mining in ontological knowledge bases with AMIE+. VLDB J. **24**(6), 707–730 (2015)
8. Guo, S., Wang, Q., Wang, L., Wang, B., Guo, L.: Jointly embedding knowledge graphs and logical rules. In: Proceedings of the 2016 Conference on Empirical Methods in Natural Language Processing, pp. 192–202 (2016)
9. Guo, S., Wang, Q., Wang, L., Wang, B., Guo, L.: Knowledge graph embedding with iterative guidance from soft rules. In: Proceedings of the AAAI Conference on Artificial Intelligence, vol. 32 (2018)

10. Hashemi, S., Bahar, R.I., Reda, S.: Drum: a dynamic range unbiased multiplier for approximate applications. In: 2015 IEEE/ACM International Conference on Computer-Aided Design (ICCAD), pp. 418–425. IEEE (2015)
11. Kazemi, S.M., Poole, D.: Simple embedding for link prediction in knowledge graphs. In: Advances in Neural Information Processing Systems, vol. 31 (2018)
12. Kim, B., Hong, T., Ko, Y., Seo, J.: Multi-task learning for knowledge graph completion with pre-trained language models. In: Proceedings of the 28th International Conference on Computational Linguistics, pp. 1737–1743 (2020)
13. Lin, Y., Liu, Z., Sun, M., Liu, Y., Zhu, X.: Learning entity and relation embeddings for knowledge graph completion. In: Proceedings of the AAAI Conference on Artificial Intelligence, vol. 29 (2015)
14. Loshchilov, I., Hutter, F.: Decoupled weight decay regularization. arXiv preprint arXiv:1711.05101 (2017)
15. Miller, G.A.: WordNet: a lexical database for English. Commun. ACM **38**(11), 39–41 (1995)
16. Nickel, M., Tresp, V., Kriegel, H.P., et al.: A three-way model for collective learning on multi-relational data. In: ICML, vol. 11, pp. 3104482–3104584 (2011)
17. Sun, Z., Deng, Z.H., Nie, J.Y., Tang, J.: Rotate: knowledge graph embedding by relational rotation in complex space. arXiv preprint arXiv:1902.10197 (2019)
18. Toutanova, K., Chen, D.: Observed versus latent features for knowledge base and text inference. In: Proceedings of the 3rd Workshop on Continuous Vector Space Models and Their Compositionality, pp. 57–66 (2015)
19. Trouillon, T., Welbl, J., Riedel, S., Gaussier, É., Bouchard, G.: Complex embeddings for simple link prediction. In: International Conference on Machine Learning, pp. 2071–2080. PMLR (2016)
20. Wang, B., Shen, T., Long, G., Zhou, T., Wang, Y., Chang, Y.: Structure-augmented text representation learning for efficient knowledge graph completion. In: Proceedings of the Web Conference 2021, pp. 1737–1748 (2021)
21. Wang, Z., Zhang, J., Feng, J., Chen, Z.: Knowledge graph embedding by translating on hyperplanes. In: Proceedings of the AAAI Conference on Artificial Intelligence, vol. 28 (2014)
22. Xiao, H., Huang, M., Zhu, X.: From one point to a manifold: knowledge graph embedding for precise link prediction. arXiv preprint arXiv:1512.04792 (2015)
23. Xie, R., Liu, Z., Jia, J., Luan, H., Sun, M.: Representation learning of knowledge graphs with entity descriptions. In: Proceedings of the AAAI Conference on Artificial Intelligence, vol. 30 (2016)
24. Yang, F., Yang, Z., Cohen, W.W.: Differentiable learning of logical rules for knowledge base reasoning. In: Advances in Neural Information Processing Systems, vol. 30 (2017)
25. Zhang, Y., et al.: Temporal knowledge graph embedding for link prediction. In: Zhao, X., Yang, S., Wang, X., Li, J. (eds.) WISA 2022. LNCS, vol. 13579, pp. 3–14. Springer, Cham (2022). https://doi.org/10.1007/978-3-031-20309-1_1

Rule-Enhanced Evolutional Dual Graph Convolutional Network for Temporal Knowledge Graph Link Prediction

Huichen Zhai[1], Xiaobo Cao[2], Pengfei Sun[2], Derong Shen[1(✉)], Tiezheng Nie[1], and Yue Kou[1]

[1] Northeastern University, Shenyang 110004, China
shenderong@cse.neu.edu.cn
[2] Beijing System Design Institute of the Electro-Mechanic Engineering, Beijing 100854, China

Abstract. The aim of link prediction over temporal knowledge graphs is to discover new facts by capturing the interdependencies between historical facts. Embedding-based methods learn the interdependencies into low-dimensional vector space while rule-based methods mine logic rules that can precisely infer missing facts. However, most of the embedding-based methods divide the temporal knowledge graph into a snapshot sequence with timestamps which increases the inherent lack of information, and the structural dependency of relations is often overlooked, which is crucial for learning high-quality relation embeddings. To address these challenges, we introduce a novel Rule-enhanced Evolutional Dual Graph Convolutional Network, called RED-GCN, which leverages rule learning to enhance the density of information via inferring and injecting new facts into every snapshot, and an evolutional dual graph convolutional network is employed to capture the structural dependency of relations and the temporal dependency across adjacent snapshots. We conduct experiments on four real-world datasets. The results demonstrate that our model outperforms the baselines, and enhancing information in snapshots is beneficial to learn high-quality embeddings.

Keywords: Temporal knowledge graph · Link prediction · Rule learning · representation learning

1 Introduction

Real world facts are usually time-sensitive [5]. For instance, events that took place at the Beijing Olympics only hold true at 2008. Therefore, temporal knowledge graphs(TKGs) have emerged. In TKGs [4,9,12,13], a fact is represented by a quadruplet (s, r, o, t), where s, o, r and t denote the subject, object, relation and the time at which the fact occurs.

Link prediction, which is an essential task in TKG analysis, aims to learn evolutionary patterns in TKGs to predict new facts [7], and Fig. 1 shows an example.

© The Author(s), under exclusive license to Springer Nature Singapore Pte Ltd. 2023
L. Yuan et al. (Eds.): WISA 2023, LNCS 14094, pp. 64–75, 2023.
https://doi.org/10.1007/978-981-99-6222-8_6

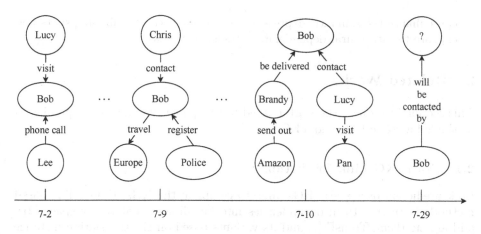

Fig. 1. An example of TKG link prediction. With the facts given from 7-2 to 7-10, the task of TKG link prediction is to predict which entity will contact Bob at 7-29.

Early embedding-based methods [5,6] extend methods [1,20] by introducing the time, but they ignore the temporal dependency between facts. Most of GNN-based methods [7,11] treat the TKG as a sequence of snapshots and capture the temporal dependency using sequential models such as RNNs. However, the density of information reflecting the real world in TKGs is quite low [8] and partitioning a TKG into snapshots further exacerbate the inherent lack of information. Additionally, most GNN-based methods ignore the structural dependency between relations. For example, in Fig. 1, the relation "start war" and "govern" are neighbors connected by the node "Afghanistan", that means these two relations have structural dependency. Capturing the structural dependency between adjacent relations can enhance the semantic expression of relations within snapshots [2]. Rule-based methods are also a prevailing category of TKG link prediction. Most of these methods extract rules from a TKG through preset rule patterns, and infer new facts by calculating the scores of rules. Although rule-based methods can make accurate predictions, they suffer from weak scalability due to the large search space [22].

To address these challenges, we propose the Rule-enhanced Evolutional Dual Graph Convolutional Network(RED-GCN) which leverages the advantages of both GNN-based and rule-based methods, and the contributions include:

- We design a rule-based entity semantic enhancement method to address the lack of information within snapshots and the weak scalability of rule-based methods.
- We design a evolutional dual graph convolutional network to capture the structural dependency between relations and to capture the temporal dependency using a gate component.
- Extensive experiments on four commonly used datasets are conducted. The results show that RED-GCN performs better on TKG link prediction compared to the baselines. The ablation experiments show that RED-GCN

can enhance the semantic expression of sparse entities within snapshots and capture the structural dependency between relations.

2   Related Work

This section introduces the work related to this paper, including link prediction methods for static KGs and TKGs.

2.1   Static KG Link Prediction

Link prediction over static KGs aims to complete the KGs. Embedding-based methods learn structural dependencies into low-dimentional vector space [21], and among them, TransE [1] and its variants based on the translation distance assumption regard the facts as the process of the head entities being translated via the relations to obtain the tail entities, which is simple and efficient; DistMult [20] and its variants based on the semantic matching assumption demonstrate that bilinear mapping functions are better at capturing relation semantics; GNN-based methods such as RGCN [14] and compGCN [19] use a message passing mechanism on the graph structure to aggregate information from the neighbors of target nodes, so that learn structural dependency on KGs more effectively. Another important category of static KG link prediction is rule-based methods. Rule-based methods [3,10] aim to learn deductive and interpretable rules, which are precise and can provide insights for inference results.

Most of these methods do not perform well capturing temporal dependency.

2.2   TKG Link Prediction

There are two settings on the link prediction over TKGs: interpolation and extrapolation [7,11]. Given the facts between timestamp t_0 and t_T, as well as a query $(s, r, ?, t_k)$, the goal of interpolation is to predict which entity should be connected to s via r at t_k, where $t_0 \leq t_k \leq t_T$. Extrapolation is to predict future facts, i.e., $t_k > t_T$. Extrapolation can be applied in various fields such as finance, healthcare, and transportation to predict future trends and risks, and therefore has broad application prospects. Hence, we focus on the extrapolation.

Considering the key to the evolutionary pattern, TA-Dismult [6] and TTransE [5] simply integrate the time into the embedding of relation, and Know-Evolve [17] considers the complex interactions of entities under multi-dimensional relations. However, these models fail to capture the interactions of entities from the neighborhood on the graph structure. RE-NET [7] models local and global interactions of entities by a GCN-based model and a GRU-based model, but it ignores the temporal dependency of relations which is further considered by RE-GCN [11].

These methods still face the challenges caused by partitioning the TKG into a snapshot sequence which further reduces the density of information and less considering the structural dependency of relations in snapshots.

3   Preliminaries

In this section, basic definitions of related terms and tasks are given, and rule learning is introduced.

Temporal Knowledge Graph(TKG). The unit that makes up TKGs is a quadruple in the form of (s, r, o, t). According to the t, a TKG G can be divided into a set of independent snapshots. i.e., $G = \{G_1, G_2, ..., G_t, ...\}$ and $G_t = (\mathcal{E}, \mathcal{R}, \mathcal{T})$, where $\mathcal{E}$, $\mathcal{R}$, and $\mathcal{T}$ indicate the entities, relations and timestamps, respectively. For every quadruple (s, r, o, t) in G_t, where $s, o \in \mathcal{E}$, $r \in \mathcal{R}$ and $t \in \mathcal{T}$, a inverse quadruple (s, r^{-1}, o, t) is also added to the datasets.

TKG Link Prediction. Predicting the missing subject in a query $(s, r, ?, t)$ or the missing object in $(?, r, o, t)$ is called entity prediction. Relation Prediction is aiming at predicting possible relations in a query $(s, ?, o, t)$. We assume that all facts that happen at timestamp t, i.e., G_t, depend on the latest k snapshots(i.e., $\{G_{t-k}, G_{t-k+1}, ..., G_{t-1}\}$), and the structural and temporal dependencies in the historical snapshot sequence have been captured in the embeddings of entities $\mathcal{H}_{t-1} \in \mathbb{R}^{|\mathcal{E}| \times d_{\mathcal{E}}}$ and relations $\mathcal{R}_{t-1} \in \mathbb{R}^{|\mathcal{R}| \times d_{\mathcal{R}}}$ at timestamp $t - 1$, where $d_{\mathcal{E}}$ denotes the dimension of entity embeddings while $d_{\mathcal{R}}$ denotes the dimension of relation embeddings. The entity prediction can be formulated as a ranking problem. Given a query $(s, r, ?, t)$, a snapshot sequence at the latest k timestamps, i.e., $G_{t-k:t-1}$, and the embeddings of entities $\mathcal{H}_{t-1}$ and relations $\mathcal{R}_{t-1}$, the entity prediction is formulated as follows:

$$p(o|G_{t-k:t-1}, s, r) = p(o|\mathcal{H}_{t-1}, \mathcal{R}_{t-1}, s, r), \tag{1}$$

and relation prediction can be similarly formulated as follows:

$$p(r|G_{t-k:t-1}, s, o) = p(r|\mathcal{H}_{t-1}, \mathcal{R}_{t-1}, s, o). \tag{2}$$

Rule Learning. A rule is in the form of

$$head \leftarrow body, \tag{3}$$

where *head* is a triplet (x, r, y) in which x and y are entity variables and r denotes the relation; *body* is a sequence of triplets. A possible example of a rule can be

$$(x, sanction, y) \leftarrow (x, discussWith, z), (z, condemn, y). \tag{4}$$

We can replace the entity variables with entities from a TKG to get a *grounding*. If all of the triples of the *grounding* can be found in the TKG, we call this *grounding* a *support* of the rule. In this paper, the rule learning is to first find rules with the number of *support* greater than 0 and then calculate the score of these rules. Rules with scores higher than a threshold are injected into the neighborhood of sparse entities in the original TKG, thus achieving enhanced semantic representation of sparse entities.

4   Methodology

In this section, we give a detailed introduction to RED-GCN, and the learning objectives. RED-GCN consists of two parts, namely, Rule-based Entity Semantic Enhancement and Evolutional Dual Graph Convolutional Network. These two parts work together to capture the structural and temporal dependencies on the historical snapshot sequence. The evolutional unit at timestamp t is used as an example to illustrate how RED-GCN works, which is shown in Fig. 2.

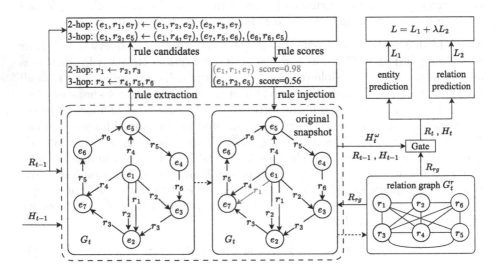

Fig. 2. A Detailed illustration of RED-GCN. The dashed arrows in the diagram represent the transformation of the snapshot, while the solid arrows represent the flow of data. The evolutional unit at t takes the embeddings, i.e., R_{t-1} and H_{t-1}, learned by the unit at $t-1$ as inputs. The outputs of unit at t, which participate the link prediction, are embeddings of entities and relations at t. (Color figure online)

4.1   Rule-Based Entity Semantic Enhancement

The goal of Rule-based Entity Semantic Enhancement is to leverage rule learning to discover hidden facts related to the sparse entities within snapshots, and then inject the hidden facts into the neighborhood of sparse entities to enhance the semantic representation.

Rule Extraction. First, the rule extraction extracts a set of multi-hop rules with *support* greater than 0 from the snapshot. As shown in Fig. 2, a 3-hop rule extracted from G_t may be:

$$(x, r_2, y) \leftarrow (x, r_4, z_1), (z_1, r_5, z_2), (z_2, r_6, y), \tag{5}$$

where r denotes relation, x, y and z are the entity variables. It can be briefly represented as $r_2 \leftarrow r_4, r_5, r_6$, where a relation sequence consists the rule body. After this, rule candidates ϕ are generated by replacing the entity variables in the multi-hop rules with entities from the snapshot. If the resulting rule body can be found as a *grounding* in the snapshot, but the rule head cannot, then this rule head is confirmed as a hidden fact and can be injected into the original snapshot.

Rule Injection. The purpose of rule injection is to determine which hidden facts in ϕ can be injected into the original snapshot. The score of the rule candidate and the sparsity of entities affect this decision. The score is calculated by computing the cosine similarity between the relation in the rule head and the relation sequence in the rule body. The embedding of the relation sequence in the rule body is learned using GRU. The specific formula is as follows:

$$Score_{rule \in \phi} = cosine(h_r^{head}, GRU(R_r^{body})), \tag{6}$$

where $head \in rule$ is the head of a rule candidate, $body \in rule$ is the rule body, h_r^{head} denotes the relation embedding in $head$ and R_r^{body} represents the embedding sequence of relations in $body$.

Paying attention to the sparse entities when learning embeddings can improve the information richness of the snapshot. The sparsity is related to the in-degree and out-degree of the entities. The frequency of an entity appearing in facts can be used to calculate its sparsity, which is represented by the following formula:

$$sp(e) = 1 - \frac{fr(e) - fr_{min}}{fr_{max} - fr_{min}}, \tag{7}$$

where $sp(e)$ is the sparsity of entity $e \in \mathcal{E}$; $fr(e)$ is the frequency of e appearing in facts within the target snapshot. fr_{max} and fr_{min} are the maximum and minimum frequency among all entities. By setting appropriate thresholds for the rule candidate score and entity sparsity, the number of injected facts and the area where they are injected can be effectively controlled.

In the Rule-based Entity Semantic Enhancement, the embeddings of relations is used to compute the score of rule candidates, which establishes a connection between rule learning and embedding learning.

4.2 Evolutional Dual Graph Convolutional Network

The main goal of Evolutional Dual Graph Convolutional Network is to learn embeddings considering the structural dependency of relations. First, a relation graph is constructed based on the original snapshot, then GCN is used to aggregate information on the relation graph to obtain the embeddings of relations. The relation embeddings is then involved in the message passing on original snapshot based on R-GCN to learn embeddings of entities. Finally, a gate component is used to capture the temporal dependency across adjacent snapshots.

Relation Graph. The first step in Evolutional Dual Graph Convolutional Network is to construct a relation graph G_r^t based on the original snapshot G_t. First, all the relations in $\mathcal{R}$ are set as nodes in G_r^t. Then, relation pairs that share the same nodes in G_t are found, and these pairs are linked in G_r^t. By constructing G_r^t, the interactions of relations in the snapshot can be represented.

Message Passing on Relation Graph. based on GCN, the structural dependency of relations are captured into the relation embeddings. The formula is as follows:

$$h_r^{l+1} = f \left(\frac{1}{c_r} \sum_{(r,r_i) \in G_r^t} W_1^l h_{r_i}^l + W_2^l h_r^l \right), \tag{8}$$

where h_r^{l+1} denotes the embedding of lth layer of relation r, and r_i is the i_{th} neighbor of r on G_r^t; W_1 and W_2 are the learnable parameters; c_r denotes the in-degree of r and f is the RRelu activation function. By the message passing on relation graph, the embeddings R_{rg} which capture the interactions of relations at t are learned and taken as an input of the message passing on original snapshot.

Message Passing on Original Snapshot. Considering that TKG is a multi-relational graph and GCN-based representation learning methods have a powerful ability to learn structural dependency, we adopt a multi-layers R-GCN module to capture the structural dependency. Each layer of the module passes information about neighbors and connected relations to the o through a message passing mechanism, i.e.,

$$h_o^{l+1} = f \left(\frac{1}{c_o} \sum_{(s,r) \exists (s,r,o) \in G_t} W_3^l (h_s^l + h_r^l) + W_4^l h_o^l \right), \tag{9}$$

where h_r is learned via message passing on relation graph; h_s^l and h_o^l denote the lth layer embedding of subject entity s and target entity o; W_3^l and W_4^l are learnable parameters; c_o denotes the in-degree of o and it's a constant, as we add inverse triples of facts, c_o actually equals to the degree of o; We choose to apply translational assumption to aggregation and its form is $h_s^l + h_r^l$. f is the RRelu activation function. The output embeddings of entities H_t^ω and the embeddings of relations R_{rg}, are then input into gate component for further learning temporal dependency.

Gate Component. The temporal dependency refers to the correlation between adjacent snapshots. We use a gate component to learn the temporal dependency,

$$H_t \| R_t = U_t \otimes (H_t^\omega \| R_{rg}) + (1 - U_t) \otimes (H_{t-1} \| R_{t-1}), \tag{10}$$

where $\otimes$ and $\|$ denote dot product and concatenation, respectively. U_t denotes the time gate which conducts a nonlinear transformation:

$$U_t = \sigma(W_5(H_{t-1} \| R_{t-1}) + b), \tag{11}$$

where $\sigma(\cdot)$ denotes the sigmoid function; W_5 and b are the weighted matrix and bias of time gate, respectively. Learned by gate component, the final embeddings of snapshot at t, H_t and R_t, are used to predict the future facts at $t + 1$.

4.3 Learning for Link Prediction

Learning Objective. We apply ConvTransE [15] in the score function to keep translational features and calculate the corresponding conditional probabilities of entity prediction and relation prediction in the equations below:

$$p(o|H_t, R_t, s, r) = \sigma(H_t ConvTransE(h_s, h_r)), \tag{12}$$

$$p(r|H_t, R_t, s, o) = \sigma(H_t ConvTransE(h_s, h_o)), \tag{13}$$

where $\sigma(\cdot)$ denotes the sigmoid function and h_s, h_r, h_o are the embeddings of subject entity s, relation r and object entity o in H_t and R_t at timestamp t. ConvTranE can be replaced with other score functions based on translational assumptions.

Parameter Learning. We consider entity prediction and relation prediction as multi-label classification tasks. Entity prediction can be represented as $(s, r, ?, t)$, and similarly relation prediction can be represented as $(s, ?, o, t)$. We also use $y_t^e \in \mathbb{R}^{|\mathcal{E}|}$ and $y_t^r \in \mathbb{R}^{|\mathcal{R}|}$ to represent the label vectors for entity prediction and relation prediction at timestamp t, so the loss functions are as follows:

$$L^e = \sum_{t=1}^{T} \sum_{(s,r,o,t)\in G_t} \sum_{i=0}^{|\mathcal{E}|-1} y_{i,t}^e \log p_i(o|H_{t-1}, R_{t-1}, s, r), \tag{14}$$

$$L^r = \sum_{t=1}^{T} \sum_{(s,r,o,t)\in G_t} \sum_{i=0}^{|\mathcal{E}|-1} y_{i,t}^r \log p_i(r|H_{t-1}, R_{t-1}, s, o), \tag{15}$$

where $y_{i,t}^e$ and $y_{i,t}^r$ denote the ith element of the label vectors, and they equal 1 if the fact appears at timestamp t, otherwise 0; T represents the amount of timestamps. Also we use a multi-task framework to learn the parameters of entity prediction and relation prediction, so the total loss can be expressed as follows:

$$L = L^e + \lambda L^r, \tag{16}$$

where λ denotes the weight value of the tasks.

5 Experiments

5.1 Experimental Setup

Datasets. We use four commonly used datasets, including ICEWS14 [13], ICEWS05-15 [13], ICEWS18 [13] and GDELT [9]. The specific details of these

Table 1. Statistics of the datasets.

Datasets	$|E|$	$|R|$	$|Train|$	$|Valid|$	$|Test|$	Time Inteveral
ICEWS14	6,869	230	74,845	8,514	7,371	24 h
IECWS05-15	10,094	251	368,868	46,302	46,159	24 h
ICEWS18	23,033	256	373,018	45,995	49545	24 h
GDELT	7,691	240	1,734,399	238,765	305,241	15 mins

datasets are shown in Table 1. Following previous methods [7,11,23], we process datasets in a form suitable for extrapolation tasks. The training set, valid set and test set are divided by 80%, 10% and 10% respectively. Moreover, facts in the training set occur earlier than those in the valid set, and the ones in valid set occurs earlier than the ones in test set.

Baselines. To evaluate the performance of our model, we selected a variety of baseline models for comparison study:

- static KG link prediction: TransE [1], DistMult [20], RotatE [16] and ComplEx [18],
- TKG link prediction: TTransE [5], TA-DistMult [6], RE-NET [7] and CyGNet [23].

Static KG link prediction models are selected because these models are representative in dealing with predicting missing facts on KG and they can be compared with RED-GCN on the ability of capturing the temporal dependency. As for TKG link prediction methods, TTransE [5] and TA-DistMult [6] are selected to be compared on the ability of capturing structural dependency. RE-NET [7] and CyGNet [23], designed with extrapolation setting, are more like our model, and the comparison study on these two baselines is to look into whether RED-GCN can perform better while enhancing the semantic expression of snapshots. The results of these baselines are derived from the recent works with the same evaluation metrics as we choose.

Evaluation Metrics. $Hits@1,3,10$ and MRR are employed to measure the prediction results. Noting that since facts evolve in TKG, they may remain valid at the future timestamps, so it's not suitable to remove repeated facts from the ranking list derived from the test set [11].

Hyperparameter. For the Rule-based Entity Semantic Enhancement, the max length of a rule is set at 3, which meas 1-hop, 2-hop and 3-hop rules are extracted from each snapshot; The threshold of rule scores is set to 0.95 while the sparsity threshold of entities is set to 0.90 for ICEWS18 and ICEWS14, and 0.95 for ICEWS05-15 and GDELT. For the Evolutional Dual Graph Convolutional Network, the number of layers in RGCN and the one in GCN are set to 2. The embedding dimension d is set to 200. The drop rate is set to 0.3. Adam is used as the optimizer with the learning rate 0.001. The task weight λ is set to 0.3. The length k of snapshot sequence is set to 2.

5.2   Performance Comparison

Performance Comparison. Table 2 shows the results over four datasets. The results demonstrate that RED-GCN performs better on most of the datasets. All the static KG methods underperform, which shows the importance of modeling temporal dependency; while among the TKG methods, TTransE [5] and TA-Distmult [6] underperform in predicting future facts due to their interpolation setting, and they are also lack of the ability to effectively capture the

Table 2. Results over four datasets.

Datasets	ICEWS14				ICEWS05-15			
Metrics	MRR	Hits@1	Hits@3	Hits@10	MRR	Hits@1	Hits@3	Hits@10
TransE	22.30	7.24	29.41	53.08	17.85	3.13	24.08	48.79
DistMult	20.32	6.13	30.67	47.37	23.43	12.84	27.95	45.70
RotatE	25.71	16.41	29.01	45.16	19.01	10.42	21.35	36.92
ComplEx	22.61	9.88	28.93	47.57	20.26	6.66	26.43	47.31
TTransE	12.86	3.14	15.72	33.65	16.53	5.51	20.77	39.26
TA-DistMult	26.22	16.83	29.72	45.23	27.51	17.57	31.46	47.32
CyGNet	34.90	25.43	39.07	53.45	36.35	25.83	41.51	56.19
RE-NET	35.77	25.99	40.10	54.87	36.86	26.24	41.85	57.60
RED-GCN	**40.03**	**29.16**	**42.23**	**55.85**	**44.39**	**31.57**	**46.71**	**60.44**
Datasets	ICEWS18				GDELT			
Metrics	MRR	Hits@1	Hits@3	Hits@10	MRR	Hits@1	Hits@3	Hits@10
TransE	11.17	0.72	12.91	34.62	4.85	0.31	3.08	14.15
DistMult	13.86	5.61	15.22	31.26	8.61	3.91	8.27	17.04
RotatE	14.53	6.47	15.78	31.86	3.62	0.52	2.26	8.37
ComplEx	15.45	8.04	17.19	30.73	9.84	5.17	9.58	18.23
TTransE	8.44	1.85	8.95	22.38	5.53	0.46	4.97	15.37
TA-DistMult	16.42	8.60	18.13	32.51	10.34	4.44	10.44	21.63
CyGNet	24.80	15.35	28.36	43.53	18.12	11.11	19.22	31.72
RE-NET	26.17	16.43	29.89	44.37	**19.60**	**12.03**	**20.56**	**33.89**
RED-GCN	**26.18**	**17.37**	**31.05**	**46.81**	18.58	10.99	20.07	32.59

structural dependency on TKG; compared with RE-NET [7] and CyGNet [23], RED-GCN performs better, achieving a remarkable improvement on ICEWS14, which means enhancing the information within snapshots and capturing the structural dependency of relations are essential to learn high quality embeddings, and it also demonstrates that rule learning and embedding learning can complement each other, thus enabling the learning of higher quality embeddings for the TKG link prediction.

5.3   Ablation Study

Ablation Study. We conduct an ablation study on the influence of both the Rule-based Entity Semantic Enhancement and the Evolutional Dual Graph Convolutional Network over ICEWS14. Table 3 shows the results, where RED-GCN is our raw model; RED-GCN-NR is constructed without the rule learning; RED-GCN-NRR is constructed by removing both the rule learning and the message passing over the relation graph, i.e., learning without the relation graph. From the results, we can learn that injecting hidden facts into the neighborhood of sparse entities and capturing the structural dependency of relations both enhance the semantic representation in embeddings, and it also shows that rule learning and embedding learning compensate for the disadvantages of each other.

Table 3. Ablation study results on ICEWS14.

Model	MRR	Hits@1	Hits@3	Hits@10
RED-GCN	**40.03**	**29.16**	**42.23**	**55.85**
RED-GCN-NR	39.36	29.03	41.66	55.37
RED-GCN-NRR	38.65	27.91	41.03	54.49

6 Conclusion

In this paper, a novel method RED-GCN utilizing rule learning and embedding learning together to enhance the information in TKGs is proposed, where the rule learning infers and injects hidden facts for better embedding learning for the TKG link prediction, and the rule learning is also guided by embedding learning in the meantime. A large amount of experiments have shown that our model outperforms the baselines on the most of the datasets. Furthermore, we demonstrate the complementarity between rule learning and embedding learning, and that both can play an important role in the TKG link prediction simultaneously.

Acknowledgement. This work was supported by the National Natural Science Foundation of China (62172082, 62072084, 62072086), the Fundamental Research Funds for the central Universities (N2116008).

References

1. Bordes, A., Usunier, N., García-Durán, A., Weston, J., Yakhnenko, O.: Translating embeddings for modeling multi-relational data. In: Advances in Neural Information Processing Systems, vol. 26: 27th Annual Conference on Neural Information Processing Systems 2013, pp. 2787–2795 (2013)
2. Chen, L., Tang, X., Chen, W., Qian, Y., Li, Y., Zhang, Y.: DACHA: a dual graph convolution based temporal knowledge graph representation learning method using historical relation. ACM Trans. Knowl. Discov. Data **16**(3), 46:1–46:18 (2022). https://doi.org/10.1145/3477051
3. Cheng, K., Liu, J., Wang, W., Sun, Y.: Rlogic: recursive logical rule learning from knowledge graphs. In: KDD, pp. 179–189 (2022)
4. Erxleben, F., Günther, M., Krötzsch, M., Mendez, J., Vrandečić, D.: Introducing wikidata to the linked data web. In: Mika, P., et al. (eds.) ISWC 2014. LNCS, vol. 8796, pp. 50–65. Springer, Cham (2014). https://doi.org/10.1007/978-3-319-11964-9_4

5. Han, Z., Chen, P., Ma, Y., Tresp, V.: Dyernie: dynamic evolution of Riemannian manifold embeddings for temporal knowledge graph completion. In: EMNLP, pp. 7301–7316 (2020)
6. Han, Z., Ma, Y., Wang, Y., Günnemann, S., Tresp, V.: Graph Hawkes neural network for forecasting on temporal knowledge graphs. In: Conference on Automated Knowledge Base Construction (2020)
7. Jin, W., Qu, M., Jin, X., Ren, X.: Recurrent event network: autoregressive structure inferenceover temporal knowledge graphs. In: EMNLP, pp. 6669–6683 (2020)
8. Lee, Y., Lee, J., Lee, D., Kim, S.: THOR: self-supervised temporal knowledge graph embedding via three-tower graph convolutional networks. In: ICDM, pp. 1035–1040 (2022)
9. Leetaru, K., Schrodt, P.A.: Gdelt: global data on events, location, and tone, 1979–2012. In: ISA Annual Convention, vol. 2, pp. 1–49. Citeseer (2013)
10. Li, W., Peng, R., Li, Z.: Knowledge graph completion by jointly learning structural features and soft logical rules. IEEE Trans. Knowl. Data Eng. **35**(3), 2724–2735 (2023). https://doi.org/10.1109/TKDE.2021.3108224
11. Li, Z., et al.: Temporal knowledge graph reasoning based on evolutional representation learning. In: SIGIR, pp. 408–417 (2021)
12. Mahdisoltani, F., Biega, J., Suchanek, F.M.: YAGO3: A knowledge base from multilingual wikipedias. In: CIDR (2015)
13. O'Brien, S.P.: Crisis early warning and decision support: contemporary approaches and thoughts on future research. Int. Stud. Rev. **1**, 87–104 (2010)
14. Schlichtkrull, M., Kipf, T.N., Bloem, P., van den Berg, R., Titov, I., Welling, M.: Modeling relational data with graph convolutional networks. In: Gangemi, A., et al. (eds.) ESWC 2018. LNCS, vol. 10843, pp. 593–607. Springer, Cham (2018). https://doi.org/10.1007/978-3-319-93417-4_38
15. Shang, C., Tang, Y., Huang, J., Bi, J., He, X., Zhou, B.: End-to-end structure-aware convolutional networks for knowledge base completion. In: AAAI, pp. 3060–3067 (2019)
16. Sun, Z., Deng, Z., Nie, J., Tang, J.: Rotate: knowledge graph embedding by relational rotation in complex space. In: ICLR (2019)
17. Trivedi, R., Dai, H., Wang, Y., Song, L.: Know-evolve: deep temporal reasoning for dynamic knowledge graphs. In: ICML, vol. 70, pp. 3462–3471. PMLR (2017)
18. Trouillon, T., Welbl, J., Riedel, S., Gaussier, É., Bouchard, G.: Complex embeddings for simple link prediction. In: ICML, vol. 48, pp. 2071–2080. JMLR.org (2016)
19. Vashishth, S., Sanyal, S., Nitin, V., Talukdar, P.P.: Composition-based multi-relational graph convolutional networks. In: ICLR (2020)
20. Yang, B., Yih, W., He, X., Gao, J., Deng, L.: Embedding entities and relations for learning and inference in knowledge bases. In: ICLR (2015)
21. Zhang, J., Shen, D., Nie, T., Kou, Y.: Multi-view based entity frequency-aware graph neural network for temporal knowledge graph link prediction. In: Web Information Systems and Applications - 19th International Conference, WISA, vol. 13579, pp. 102–114 (2022). https://doi.org/10.1007/978-3-031-20309-1_9
22. Zhang, W., et al.: Iteratively learning embeddings and rules for knowledge graph reasoning. In: WWW, pp. 2366–2377 (2019)
23. Zhu, C., Chen, M., Fan, C., Cheng, G., Zhang, Y.: Learning from history: Modeling temporal knowledge graphs with sequential copy-generation networks. In: AAAI, pp. 4732–4740 (2021)

DINE: Dynamic Information Network Embedding for Social Recommendation

Yi Zhang[1], Dan Meng[2], Liping Zhang[1], and Chao Kong[1(✉)]

[1] School of Computer and Information, Anhui Polytechnic University, Wuhu, China
{zhangyi,zhanglp,kongchao}@ahpu.edu.cn
[2] OPPO Research Institute, Shenzhen, China
mengdan@oppo.com

Abstract. The social recommendation aims to integrate social network information to improve the accuracy of traditional recommender systems. Learning embeddings of nodes in networks is one of the core problems of many applications such as recommendation, link prediction, and node classification. Early studies cast the social recommendation task as a vertex ranking problem. Although these methods are effective to some extent, they require assuming social networks and user-item interaction networks as static graphs, whereas real-world information networks evolve over time. In addition, the existing works have primarily focused on modeling users in social networks in general and overlooked the special properties of items. To address these issues, we propose a new method named DINE, short for *Dynamic Information Network Embedding*, to learn the vertex representations for dynamic networks in social recommendation task. We model both users and items simultaneously and integrate the representations in dynamic and static information networks. In addition, the multi-head self-attention mechanism is employed to model the evolution patterns of dynamic information networks from multiple perspectives. We conduct extensive experiments on Ciao and Epinions datasets. Both quantitative results and qualitative analysis verify the effectiveness and rationality of our DINE method.

Keywords: Social recommendation · High-order relations modeling · Attention mechanism · Graph convolutional network

1 Introduction

With the rapid growth of the amount of internet information, recommender system as a prevalent solution to alleviate the problem of information overload becomes an attractive research area over recent years. With the popularization of social media, a huge volume of social network data from users has become available in a variety of domains. Given the social network data, social recommendation aims to leverage social relations among users for generating user-specific

---

Y. Zhang and D. Meng—These authors contribute equally to this work.

L. Yuan et al. (Eds.): WISA 2023, LNCS 14094, pp. 76–87, 2023.
https://doi.org/10.1007/978-981-99-6222-8_7

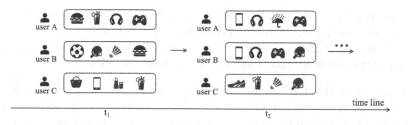

Fig. 1. An illustrative example w.r.t. the possible changes in the interest of users.

items [1,2]. Along with the increasing application of vertex representation, network embedding has caused worldwide concern. To perform predictive analytics and effective social recommendation, it is crucial to first model user preference. Most of the information including user-item interaction information and social information essentially has a graph structure. Network embedding can learn the low-dimensional vector representations for vertices [3,4]. In particular, the edges among vertices in the network embedding space, are captured by proximity or distance between vertex vectors encoding the topological and structural characteristics of networks. Thus the vertex embeddings can be fed into machine learning frameworks to support subsequent various network processing and analysis tasks such as classification [5], clustering [6], visualization [7] and so on (Fig. 1).

To date, existing network embedding methods have primarily focused on static information networks covering homogeneous information networks like social networks which have the same type of vertices [8,9], or heterogeneous information networks like citation networks where vertices (and/or edges) are of different types [10,11]. However real-world networks are evolving over time. For example, in the Twitter network, friendships between users always dynamically change over time, i.e., new friendships are continuously established to the social network while some ones may be broken off. In most of scenarios, network embedding is regarded as graph embedding. They have a similar goal, which is to embed a graph or network into a low-dimensional vector space. While prevalent graph embedding methods like dyngraph2vec and VGRNN can be applied to learn vertex representation for dynamic information networks by a recurrent neural network (RNN), we argue these methods are suboptimal in doing so because there exists long-term dependencies problem and cannot model the dynamic information networks evolution from a different perspective. In addition, existing social recommendation works have primarily focused on modeling users. To some extent, these methods improve the performance of recommender systems. However, the modeling of users is not only affected by the relation between the user and the item but also by the relation of items in real life. In that context, the social recommendation is a challenging problem: 1) graph structures are dynamic and complex. 2) social recommendation researches mainly focus on the modeling of users, ignoring the modeling of items making it insufficient to model users.

To address the limitations of existing methods on social recommendation, we devise a new solution for learning vertex representation in dynamic information networks, namely DINE (short for Dynamic Information Network Embedding). We summarize the main contributions of this work as follows.

- We design a novel dynamic information network representation learning model that utilizes joint GCN and self-attention from the structural and temporal perspective for dynamic information network representation.
- We employ the multi-head self-attention with position embedding and design the relation information aggregation module to capture the relations among items for modeling users and items.
- We illustrate the performance of our model against a comparable baseline on two real datasets. Empirical study results illustrate the effectiveness and rationality of DINE.

The remainder of the paper is organized as follows. We first review related work in Sect. 2. We formulate the problem in Sect. 3, before delving into details of the proposed method in Sect. 4. We perform extensive empirical studies in Sect. 5 and conclude the paper in Sect. 6.

2   Related Work

Social Recommendation. The social recommendation aims at leveraging user correlations implied by social relations to improve the performance of traditional recommender systems. The study of social recommendation has become a hot topic in recent years, and some earlier studies can go back to the 19901990ss [12]. In general, the existing social recommender systems can be categorized into two types according to the basic collaborative filtering (CF) models: memory-based and model-based methods. Following the pioneering work of CF [13], the memory-based methods use either the whole user-item matrix or a sample to generate a prediction, and it typically applies a two-step solution: first obtaining the correlated users $N(u_i)$ for a given user u_i, and then aggregating ratings from $N(u_i)$ for missing ratings [14]. The difference between social recommender systems in this category is employing various approaches to obtain the $N(u_i)$. Generally speaking, memory-based methods have a main drawback: their performance is rather sensitive to the predefined similarity measures for calculating the trust matrix. For model-based methods, they usually assume a model to generate the ratings and apply data mining and machine learning techniques to find patterns from training data. The pioneering work LOCABAL [15] performs a co-factorization in the user-item matrix and the user-user social relation matrix by sharing the same user preference latent factor. Model-based methods indirectly model the propagation of tastes in social networks, which can be used to reduce cold-start users and increase the coverage of items for recommendation.

It is worth pointing out that the above-mentioned methods are designed for extending the basic CF models, for which they only consider rating information and social information. Absolutely, they can improve the performance of recommendation by leveraging the first-order relations, but overlook the study in depth for the utility of high-order relations.

Network Embedding. The pioneering work DeepWalk and node2vec extend the idea of skipgram to model homogeneous information networks. However, they may not be effective to preserve high-order implicit relations of the network. Zhao et al. proposed a heterogeneous social network embedding algorithm for node classification [16]. All types of node representations were learned in a public vector space and performed inference. Pham et al. proposed a deep embedding algorithm for heterogeneous information networks with multiple types of nodes, and the deep embedding representation for multiple types of nodes can well preserve the structure of heterogeneous networks [17]. Zhang et al. proposed a heterogeneous information network embedding algorithm that can preserve the similarity of meta paths, and they utilized a fast dynamic programming approach to compute the proximities based on truncated meta paths [18]. HINchip model the relationship between nodes for each network of coupling heterogeneous networks and calculate the closeness of nodes in different networks by harmonic embedding matrix [19]. However, these methods are not tailored to model temporal evolution in dynamic information networks.

3   Problem Definition

We first introduce the notations used in this paper and then formalize the social recommendation problem to be addressed.

Notations. Let $U = \{u_1, u_2, ..., u_m\}(|U| = m)$ and $I = \{i_1, i_2, ..., i_n\}(|I| = n)$ denote the set of users and items respectively, where m and n denote the number of users and items. A user-item interactive matrix $Y \in \mathbb{R}^{m \times n}$ is defined as the interactive behaviors between users and items. Each element y_{ij} indicates the rating of the user u_i to the item i_j. A user-user adjacency matrix $R \in \mathbb{R}^{m \times m}$ is used to represent the social relations between users. Each element $r_{ii'} = 1$ indicates that user u_i is connected with user $u_{i'}$. A dynamic user-item interaction graph is defined as a sequence of static user-item interaction graph snapshots, $G = \{G^1, \cdots, G^T\}$ where T denotes the number of timesteps. Each user-item interaction graph $G_Y^t = (V, \mathcal{E}^t)$ where V denotes the set of nodes and $\mathcal{E}^t$ denotes the set of links at timestamp t.

Problem Definition. The task of social recommendation aims to map each item $i_j \in I$ to a real value for each user. Formally, the problem can be defined as: given observed user-item interactive matrix Y and user-user adjacency matrix R, we aim to predict whether user u has potential interests in target item v.

4   DINE: Dynamic Information Network Embedding

A good dynamic information network embedding approach needs to well preserve and represent the properties of dynamic information networks for evolution over time. To achieve this aim in a social recommendation scenario, we

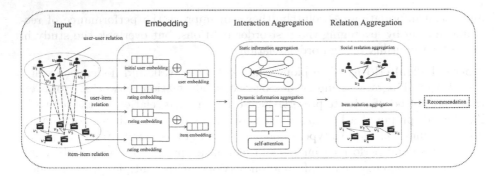

Fig. 2. The overall architecture of the DINE model.

consider constructing a dynamic information network embedding model from two perspectives: constructing item-relation graph, modeling items by capturing the relation among items, and modeling the evolution patterns of dynamic information networks. This section presents our DINE method along this line. We present the overall architecture for the proposed model in Fig. 2.

4.1 Embedding Generation Module

Although previous studies have made great success by focusing on user behaviors, they cannot exploit the relation among items, which are important in recommender systems. In this component, we construct the item relation graph based on the cosine similarity. In addition, due to the randomness of relation construction, explicit relations are not equally important. In this paper, we utilize the ratings in the user-item interaction matrix R to help us model users and items. Inspired by GraphRec [20], we embeds each rating r_{ij} into a D-dimension vector, denoted by $\vec{r}_{ij}$.

Let $\vec{u}_i$ and $\vec{v}_j$ be the initial embeddings of users and items respectively. For each user u_i, the interaction embedding of user u_i and v_j, which integrate the rating information, are calculated as follows:

$$\vec{p}_{ij} = f_{uv}([\vec{r}_{ij} \oplus \vec{v}_j]), \qquad \vec{q}_{ji} = f_{vu}([\vec{r}_{ij} \oplus \vec{u}_i]), \tag{1}$$

where $\vec{p}_{ij}$ denote the interaction embedding from v_j to u_i. In the same way, the interaction embedding from u_i to v_j, denoted by $q_{ji}.\oplus$ denotes the concatenation operation. f_{uv} and f_{vu} denote the two-layer perceptron.

4.2 Interaction Aggregation Module

In this section, we first introduce the major components of the Interaction Aggregation Module. As depicted in Fig. 2, the Interaction Aggregation Module has two components: Static Information Aggregation and Dynamic Information Aggregation.

Static Information Aggregation. In this component, we employ an attention mechanism to aggregate neighbors' context and describe the structural information. In order to make the output vertex embedding to better aggregate the static structural information in a current timestamp t, we introduce an attention mechanism to assign weight to immediate neighbors of a vertex with more flexibility. The structural embeddings of user u_i and item v_j can be defined as a non-linear activation function, as follows:

$$\mathbf{h}_i^S = \sigma \left(\sum_{j \in N_{u_i}^v} \alpha_{ij} \mathbf{W}_0^u p_{ij} + b_0^u \right), \qquad \mathbf{h}_j^S = \sigma \left(\sum_{i \in N_{v_j}^u} \alpha_{ji} \mathbf{W}_0^v q_{ji} + b_0^v \right). \qquad (2)$$

where $\sigma(\cdot)$ is a non-linear activation function. $N_{u_i}^v = \{v \in V : (u_i, v) \in E_u\}$ represents the set of items which user u_i has interacted. $N_{v_j}^u = \{u \in U : (u, v_j) \in E_u\}$ denotes the set of users who have the interactive behavior with item v_j. where $\mathbf{W}_0^u$ and $\mathbf{W}_0^u$ denote the weights of the neural network. b_0^u and b_0^v denote the biases of the neural network. α_{ij} indicates the contribution of interaction embedding p_{ij} to u_i, which can be defined as:

$$\alpha_{ij} = \frac{\exp(e_{ij})}{\sum_{j \in N_{u_i}^v} \exp(e_{ij})},$$

$$e_{ij} = \mathbf{W}_2^\top \cdot \sigma(\mathbf{W}_1 \cdot [u_i \oplus p_{ij}] + b_1) + b_2. \qquad (3)$$

where $\mathbf{W}_1^u$ and $\mathbf{W}_2^u$ denote the weights of the neural network. b_1^u and b_2^v denote the biases of the neural network. In the same way, we can utilize a similar method to get α_{ji}.

Dynamic Information Aggregation. Besides the static structural information aggregation, we need to aggregate the dynamic temporal information to model the evolutionary behavior. Intuitively, different weights to a series of historical representations can be flexibly assigned to capture information about the dynamic evolution of the networks. Based on the learned vertex representations, we can perform tasks such as recommendation, link prediction, and so on.

The dynamic temporal information aggregation module aims to capture evolution information in a dynamic information network. After implementing the embedding generation module, we can get a sequence of user's interaction embedding, denoted by $P(i) = (p_{ij(t)})_{t \in T_i^v}$. In the same way, we can obtain a sequence of the item's interaction embedding, denoted by $Q(j) = (q_{ji(t)})_{t \in T_j^u}$. To reduce the computational cost, we truncate the lengths of long sequences to a fixed value. We aim to obtain a new embedding sequence for users and items through dynamic information modeling. Specifically, the output embedding is mainly calculated by the following formula:

$$h_i^D = \beta_u(P_i \mathbf{W}_v), \qquad h_j^D = \beta_v(Q_j \mathbf{W}_v), \qquad (4)$$

where β_u denotes the attention weight matrix, which calculated by following formula:

$$\beta_u^{ij} = \frac{\exp(e_{ij})}{\sum_{K=1}^{T_i^v} \exp(e_u^{ik})}, \tag{5}$$

$$e_u^{ij} = (\frac{((P_i \mathbf{W}_q)(P_i \mathbf{W}_k)^\top)_{ij}}{\sqrt{F'}} + \mathbf{M}_{ij}). \tag{6}$$

where $\mathbf{W}_q$, $\mathbf{W}_k$, and $\mathbf{W}_v$ denote the linear projection matrices, which can transform the queries, keys, and values to a different space. In the same way, we can obtain the β_v. $\mathbf{M}$ represents a mask matrix to protect the auto-regressive property. The calculation formula is as follows:

$$\mathbf{M}_{ij} = \begin{cases} 0, & i \leq j, \\ -\infty, & otherwise. \end{cases} \tag{7}$$

As you can see from the steps described above, our method can sufficiently capture network evolution. However, the evolution of real-world dynamic information networks is complex. Thus we utilize multi-head attention to model the evolution of networks from multiple different latent perspectives.

After getting the static information representations and the dynamic temporal representation, we can compute the global interaction representation. The formula for calculation is as follows:

$$h_i^I = h_i^S \odot h_i^D, \qquad h_j^I = h_j^S \odot h_j^D. \tag{8}$$

where h_i^I and h_j^I represent the interaction representation of user u_i and item v_j. Similarly, for the user u_i's friends $u_x, x \in N_{u_i}^u$ in social network and the item v_y which is related to item v_j in item relation network, we can utilize the above method to obtain the interaction representation, denoted by h_x^I and h_y^I.

Relation Aggregation Module. Due to the randomness of social relations construction, explicit relations are not always equally important to modeling users. Thus, in this paper, we use attention mechanisms in social networks to characterize different strengths of social relations. Specifically, for a target user u_i, we can utilize the Eqs.(2)–(3) to calculate the social relation representation, denoted by h_i^R. In the same way, we can obtain the item relation representation, denoted by h_j^R.

To better model the user and item, we need to consider both interaction information and related information. Therefore, the final user embedding $\vec{h}_i^u$ of user u_i and the final item embedding $\vec{h}_j^v$ of item v_j are calculated as follows:

$$h_i^u = f^{uu}([h_i^I \odot h_i^R]), \qquad h_j^v = f^{vv}([h_j^I \odot h_j^R]). \tag{9}$$

After getting the final user embedding and item embedding, we compute the predicted rating of the user u_i to the items v_j. The formula for calculation is as follows:

Table 1. Descriptive statistics of datasets.

Dataset	# Users	# Items	# Ratings	# Links
Ciao	49,289	139,738	664,824	487,181
Epinions	2,848	39,586	894,887	35,770

$$\hat{r}_{ij} = f([h_i^u \odot h_j^v]), \tag{10}$$

where $f(\cdot)$ denotes a multi-layer perceptron.

Learning. To enable the learned embeddings to capture the evolution patterns in a dynamic information network, we utilize the mean squared error to train our model:

$$L = \frac{1}{2|K|} \sum_{(u_i, v_j) \in O} (\hat{r}_{ij} - r_{ij}^2), \tag{11}$$

where $O = (u_i, v_j) : r_{ij} \neq 0$ denotes the set with known ratings in the dataset. We adopt the gradient descent for parameter learning. We utilize dropout to prevent overfitting problem.

5   Experiments

To evaluate the performance of our proposed DINE model, we conduct several experiments on two real-world datasets. Through empirical evaluation, we aim to answer the following research questions:

RQ1. How does DINE perform compared with state-of-the-art social recommendation methods?

RQ2. How do the key hyper-parameters affect the performance of DINE?
In what follows, we first introduce the experimental settings and then answer the above research questions in turn.

5.1   Experimental Settings

Datasets. We use two real-world datasets in our experiments. The descriptive statistics about the datasets are shown in Table 1. The Ciao dataset[1] contains the interaction information and the trust relation from the popular product review sites Ciao. The Epinions dataset contains the ratings of users to specific items and the trust relationship between users from Epinions.com website in one month's time.

---

[1] http://www.cse.msu.edu/~tangjili/trust.html.

Evaluation Protocols. For each dataset, 80 % of instances are randomly sampled as the training set, and the remaining 20% of instances are used as the testing set. Following the previous work, we employ the Mean Absolute Error (MAE) and Root Mean Square Error (RMSE) to evaluate the recommendation performance.

Baselines. We compare DINE with the following baselines:

- SocialMF [21]: This is a model-based method for social recommendation, which introuduces the trust propagation to the matrix Factorization(MF).
- DeepSoR [22]: This is a new deep neural network model, which can learn non-linear features from users' social relations.
- NARM [23]: Neural Attentive Recommendation Machine applys attention mechanism to model user's main purpose in the current session. Then using a bi-linear matching scheme to compute recommendation scores.
- STAMP [24]: This is a short-term attention/memory priority model, which can capture complex and intrinsic non-linear features from social relations both the short-term and long-term user interests.
- GraphRec [25]: GraphRec can model graph data in social recommendations coherently;
- DGRec [26]: It is a state-of-the-art deep neural networks method for social recommendation. This method integrates both user features and social structures into convolutional neural networks and attention networks.

Parameter Settings. We implement all baselines and our method on Pytorch. For all methods, the batch size is set as 512 and the dimension of embedding d is set as 24 and 36 respectively. We tune the hyper-parameters by performing a grid search. The learning rate and the regularization coefficient are searched in [0.0001, 0.0005, 0.001, 0.005] and [0.0001, 0.001, 0.01, 0.1] respectively.

5.2 Performance Comparison (RQ1)

Table 2 illustrates the performance of baselines and our DINE, where we have the following key observations:

- GraphRec, DGRec, and DINE outperform the other models significantly. This is due to the factors that: (1) introducing GNN into the recommender systems to aggregate the vertices attribute information and topological structure is helpful to improve the performance; (2) it is insufficient to directly utilize the inner product to optimize user embeddings.
- The tailored social recommendation algorithm DeepSoR is significantly better than SocialMF. This points to the positive effect of modeling high-order relations in the social recommendation process.
- DGRec achieved the best performance among all the baselines. This sheds light on the benefit of employing the information of ratings in the pre-training process.

Table 2. Recommendation performance on Ciao and Epinions

Dataset	RMSE						
	SocialMF	DeepSoR	NARM	STAMP	GraphRec	DGRec	DINE
Ciao	1.05	1.03	1.10	1.07	0.99	0.98	**0.80****
Epinions	1.13	1.09	1.10	1.07	1.05	1.03	**1.01****
Dataset	MAE						
	SocialMF	DeepSoR	NARM	STAMP	GraphRec	DGRec	DINE
Ciao	0.82	0.77	0.81	0.94	0.80	0.75	**0.62****
Epinions	0.88	0.83	0.85	0.88	0.85	0.82	**0.79****

** indicates that the improvements are statistically significant for $p<0.01$ judged by paired t-test.

Table 3. Impact of hyper-parameter on social recommendation.

# heads (h)	Ciao		Epinions	
	MAE	RMSE	MAE	RMSE
1	0.85	1.04	0.99	1.18
2	0.67	0.86	0.91	1.11
4	0.80	0.99	0.88	1.06
8	0.72	0.92	0.81	1.06
16	**0.62**	**0.78**	**0.77**	**1.01**
32	0.77	0.96	0.87	1.07

- Our proposed DINE outperforms the SOTA method. This improvement demonstrates the effectiveness of the self-attention component in the modeling of dynamic and static networks simultaneously.

5.3   Hyper-Parameter Studies (RQ2)

Due to space limitations, we only investigate the impact of the number of heads h in the multi-headed self-attentive mechanism of the social recommendation task. For the sake of fairness, all parameters are kept constant except for the number of heads in the self-attention mechanism. We performed the social recommendation task on the Ciao and Epinions datasets, respectively. The experimental results are shown in Table 3, where the model achieves the best performance when the number of attention heads is 16. In addition, it can be observed that the performance of the remaining models with multiple heads of self-attention mechanism is better than that of the model with one head of self-attentive mechanism. The experimental results show that the multi-head self-attention mechanism can effectively improve the performance of our DINE model. Therefore, modeling dynamic graphs from multiple perspectives using the multi-headed self-attention mechanism is an effective way to improve the performance of social recommendation task.

6   Conclusion

In this paper, we have presented DINE, a novel approach for embedding dynamic information networks in the social recommendation task. It includes user and item embedding generation, user-item interaction aggregation, and relations aggregation modules. Among them, the user-item interaction aggregation is the core step, which models users and items from two perspectives (static structure and dynamic evolution) respectively, considering the impact of item modeling on the recommendation task. Notably, DINE employs a multi-headed self-attention mechanism to achieve multi-perspective modeling of the evolutionary patterns of dynamic information networks. Extensive experiments on several real-world datasets demonstrate the effectiveness and rationality of our DINE method. In the future, we plan to introduce more auxiliary information for social recommendation in dynamic information networks.

Acknowledgment. This work was supported in part by the National Natural Science Foundation of China Youth Fund (No. 61902001) and the Undergraduate Teaching Quality Improvement Project of Anhui Polytechnic University (No. 2022lzyybj02). We would also thank the anonymous reviewers for their detailed comments, which have helped us to improve the quality of this work.

References

1. Jiang, W., Sun, Y.: Social-RippleNet: jointly modeling of ripple net and social information for recommendation. Appl. Intell. **53**(3), 3472–3487 (2023)
2. Saraswathi, K., Mohanraj, V., Suresh, Y., Senthilkumar, J.: Deep learning enabled social media recommendation based on user comments. Comput. Syst. Sci. Eng. **44**(2), 1691–1702 (2023)
3. Mei, G., Ye, S., Liu, S., Pan, L., Li, Q.: Heterogeneous graphlets-guided network embedding via Eulerian-trail-based representation. Inf. Sci. **622**, 1050–1063 (2023)
4. Zhang, Y., et al.: Temporal knowledge graph embedding for link prediction. In: WISA, pp. 3–14 (2022)
5. Shen, X., Dai, Q., Chung, F., Lu, W., Choi, K.: Adversarial deep network embedding for cross-network node classification. In: AAAI, pp. 2991–2999 (2020)
6. Gao, W., Wu, P., Pan, L.: Attribute network embedding method based on joint clustering of representation and network. In: BDCAT, pp. 111–119 (2021)
7. Han, X., Zhao, Y.: Reservoir computing dissection and visualization based on directed network embedding. Neurocomputing **445**, 134–148 (2021)
8. Yu, R., Yang, K., Wang, Z., Zhen, S.: Multimodal interaction aware embedding for location-based social networks. AI Commun. **36**(1), 41–55 (2023)
9. Yan, D., Zhang, Y., Xie, W., Jin, Y., Zhang, Y.: MUSE: multi-faceted attention for signed network embedding. Neurocomputing **519**, 36–43 (2023)
10. Gu, Y., Li, L., Zhang, Y.: Robust android malware detection based on attributed heterogenous graph embedding. In: FCS, pp. 432–446 (2020)

11. Wang, C., Yuan, M., Zhang, R., Peng, K., Liu, L.: Efficient point-of-interest recommendation services with heterogenous hypergraph embedding. IEEE Trans. Serv. Comput. **16**(2), 1132–1143 (2023)
12. Kautz, H.A., Selman, B., Shah, M.A.: Referral web: combining social networks and collaborative filtering. Commun. ACM **40**(3), 63–65 (1997)
13. Koren, Y.: Factorization meets the neighborhood: a multifaceted collaborative filtering model. In: SIGKDD, pp. 426–434 (2008)
14. Dau, A., Salim, N., Rabiu, I.: An adaptive deep learning method for item recommendation system. Knowl. Based Syst. **213**, 106681 (2021)
15. Tang, J., Hu, X., Gao, H., Liu, H.: Exploiting local and global social context for recommendation. In: IJCAI, pp. 2712–2718 (2013)
16. Zhao, X., Jin, Z., Liu, Y., Hu, Y.: Heterogeneous information network embedding for user behavior analysis on social media. Neural Comput. Appl. **34**(7), 5683–5699 (2022)
17. Pham, P., Nguyen, L.T.T., Nguyen, N.T., Kozma, R., Vo, B.: A hierarchical fused fuzzy deep neural network with heterogeneous network embedding for recommendation. Inf. Sci. **620**, 105–124 (2023)
18. Zhang, C., Tang, Z., Yu, B., Xie, Y., Pan, K.: Deep heterogeneous network embedding based on Siamese neural networks. Neurocomputing **388**, 1–11 (2020)
19. Zhao, H., et al.: Hinchip: heterogeneous information network representation with community hierarchy preserving. Knowl. Based Syst. **264**, 110343 (2023)
20. Fan, W., et al.: Graph neural networks for social recommendation. In: WWW, pp. 417–426 (2019)
21. Rendle, S., Freudenthaler, C., Gantner, Z., Schmidt-Thieme, L.: BPR: Bayesian personalized ranking from implicit feedback. In: UAI, pp. 452–461 (2009)
22. He, X., Liao, L., Zhang, H., Nie, L., Hu, X., Chua, T.: Neural collaborative filtering. In: WWW, pp. 173–182 (2017)
23. Guo, G., Zhang, J., Yorke-Smith, N.: Trustsvd: collaborative filtering with both the explicit and implicit influence of user trust and of item ratings. In: AAAI, pp. 123–129 (2015)
24. Wang, X., He, X., Nie, L., Chua, T.: Item silk road: recommending items from information domains to social users. In: SIGIR, pp. 185–194
25. Yu, J., Gao, M., Li, J., Yin, H., Liu, H.: Adaptive implicit friends identification over heterogeneous network for social recommendation. In: CIKM, pp. 357–366 (2018)
26. Qiu, J., Tang, J., Ma, H., Dong, Y., Wang, K., Tang, J.: Deepinf: social influence prediction with deep learning. In: KDD, pp. 2110–2119 (2018)

Empowering Chinese Hypernym-Hyponym Relation Extraction Leveraging Entity Description and Attribute Information

Senyan Zhao[1], ChengZhen Yu[1], Subin Huang[1(✉)], Buyun Wang[2], and Chao Kong[1]

[1] School of Computer and Information, Anhui Polytechnic University, Wuhu, China
zhaosy@mail.ahpu.edu.cn, yuchengzhen@stu.ahpu.edu.cn,
{subinhuang,kongchao}@ahpu.edu.cn
[2] School of Artificial Intelligence, Anhui Polytechnic University, Wuhu, China
ayun@ahpu.edu.cn

Abstract. Hypernym-hyponym relations play a crucial role in various entity-based natural language processing applications. Most previous hypernym-hyponym relation extraction approaches used pattern-based, encyclopedia-based, and clustering-based methods; however, the performance of these approaches in Chinese texts is not ideal. In this paper, we introduce an entity description and attribute information-based approach for extracting Chinese hypernym-hyponym relations. First, we mine entity descriptions and attribute information from a Chinese encyclopedia, followed by applying a filtering strategy to gather candidate entities from plain Chinese texts. Following this, we develop a neural network extraction model that integrates entity descriptions and attribute information to identify hypernym-hyponym relations among the collected Chinese candidate entities. In this model, bidirectional gated recurrent units (Bi-GRU) are utilized to extract Chinese hypernym-hyponym semantic information from entity descriptions, while an attention model captures Chinese hypernym-hyponym semantic information from attribute information. We implement the proposed approach on four real-world Chinese datasets, demonstrating its practicality and superiority to compared approaches concerning evaluation metrics. This paper contributes to enhancing the accuracy of hypernym-hyponym relation extraction in Chinese texts.

Keywords: Hypernym-hyponym relation · Entity description · Attribute information · Attention model

1 Introduction

Hypernym-hyponym relations serve as the foundation for numerous entity-based natural language processing (NLP) applications, including textual entailment [1–3], taxonomy construction [4,5], and question answering [6,7]. In the early stages,

L. Yuan et al. (Eds.): WISA 2023, LNCS 14094, pp. 88–99, 2023.
https://doi.org/10.1007/978-981-99-6222-8_8

some hypernym-hyponym relation repositories, such as WordNet [8], HowNet [9], and CilinE [10], were constructed manually; however, such manual approaches suffer from several limitations (e.g., low coverage, difficulty in keeping it up to date, and manpower and time consumption). To address these issues, effective hypernym-hyponym relation extraction strategies, such as pattern-based [11–13], encyclopedia-based [14–16], and clustering-based approaches [17–19], have been proposed.

Although most of these approaches have been applied to the English corpus and achieved relatively high accuracy, extracting hypernym-hyponym relations for Chinese remains a challenge. Chinese is a low-resource language with complex and flexible grammatical rules and expressions [20,21]. Therefore, it is difficult to directly apply the above approaches to mine hypernym-hyponym relations in Chinese texts.

This study concentrates on extracting hypernym-hyponym relations from Chinese text, with the main contributions outlined below.

- Entity descriptions and attribute information containing abundant Chinese hypernym-hyponym semantic information are collected from a Chinese encyclopedia. This information plays a significant role in enhancing the effectiveness of hypernym-hyponym relation extraction. Furthermore, a filtering strategy is introduced to purify entities in Chinese plain text, facilitating the collection of candidate entities.
- A neural network extraction model combined with entity description and attribute information is designed to extract Chinese hypernym-hyponym relations. In the model, Bi-GRU is employed for capturing hypernym-hyponym semantic information from the entity description, and the attention model is utilized for extracting semantic information from attribute information.

2 Related Works

Hypernym-hyponym relation extraction is a crucial and challenging task that has a long history. This section reviews and summarizes hypernym-hyponym relation extraction from three perspectives: pattern-based, encyclopedia-based, and clustering-based approaches.

The pattern-based approach, initially proposed by Hearst [11], employs syntactic and lexical patterns to match hypernym-hyponym relations. Snow et al. [22] constructed a classifier to extract hypernym-hyponym relationships. In the classifier, parse trees were built to mine the hypernym-hyponym syntactic features. Wu et al. [12] presented a syntactic iterative extraction approach to extract hypernym-hyponym relations from large-scale text corpora using syntactic and lexical patterns. Roller et al. [13] performed a comprehensive experiment on the pattern-based approach, demonstrating that a simple pattern-based method can achieve relatively high performance, offering robust and high-quality hypernym-hyponym relation predictions on large-scale corpora.

Encyclopedias, such as BaiDuBaiKe and Wikipedia, serve as valuable text sources for fine-grained information extraction [23]. Suchanek et al. [24] developed a well-known project called the YAGO from which the hypernym-hyponym relations were extracted by combining the Wikipedia categories and WordNet. Xu et al. [14] constructed a large-scale Chinese knowledge base project named CN-Dbpedia, where the hypernym-hyponym relations were identified based on an end-to-end facts mining method and an existing knowledge base.

Clustering-based approaches extract hypernym-hyponym relations based on distributional hypothesis [25]. Wang et al. [17] proposed an algorithm for clustering hierarchical topics, employing a phrase-centric view for clustering, extracting, and ranking hierarchical topics. Alfarone et al. [18] presented an unsupervised algorithm to automatically extract the hypernym-hyponym relations, employing a clustering-based method to discover hypernym-hyponym relations and exploited a graph-based algorithm to improve the precision of clustered hypernym-hyponym relations. Huang et al. [19] presented an unsupervised approach for learning hypernym-hyponym relations from unstructured text, employing a semantic-clique-based method to cluster hypernym-hyponym relations and utilizing a relation-detection method to promote the precision of the clustered relations.

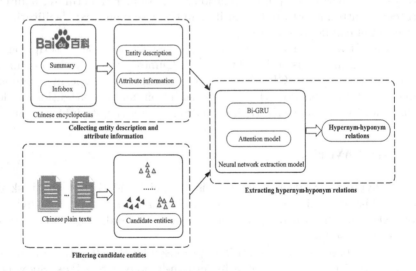

Fig. 1. Overview framework.

3   Approach

As shown in Fig. 1, the proposed approach comprises three components: collecting entity descriptions and attribute information, filtering candidate entities, and extracting hypernym-hyponym relations.

3.1 Collecting Entity Description and Attribute Information

The background knowledge (e.g., entity descriptions and attribute information) of entities is vital for revealing the semantic relations between them. Figure 2 displays the summary (first sentence) and infobox of "Fruit(水果)" and "Banana(香蕉)" (obtained from BaiDuBaiKe[1]). It is a well-established fact that the "Fruit(水果)" is the hypernym of "Banana(香蕉)". From the Figure, the semantic similarities between the summary and infobox of "Fruit(水果)" and "Banana(香蕉)" are apparent. Such similar semantic features can provide background knowledge for Chinese hypernym-hyponym relation extraction. Consequently, this study collects entity descriptions from BaiDuBaiKe summaries and entity attribute information from their respective infoboxes.

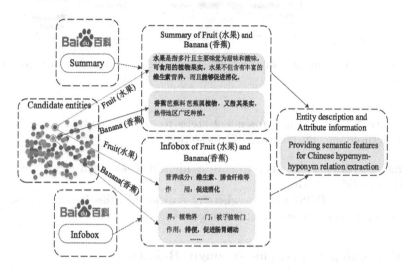

Fig. 2. Examples of entity description and attribute information.

3.2 Filtering Candidate Entities

Chinese semi-structured encyclopedia texts (e.g., encyclopedia titles) and unstructured encyclopedia texts (e.g., encyclopedia article text) encompass numerous entities between which many hypernym-hyponym relations exist. Figure 3 presents examples of Chinese semi-structured and unstructured encyclopedia texts. As evident from the figure, the semi-structured encyclopedia titles "Fruit(水果)" and "Banana(香蕉)" exhibit a hypernym-hyponym relation. In addition, within the unstructured encyclopedia texts, both the entities "Animal (动物)" and "Vertebrate (脊椎动物)" as well as "Weapon (武器)" and "Nuclear warhead (核弹头)" demonstrate hypernym-hyponym relations.

---

[1] https://baike.baidu.com/.

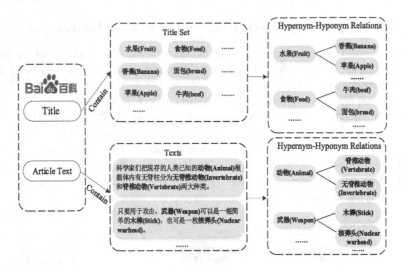

Fig. 3. Examples of Chinese semi-structured encyclopedia texts and unstructured encyclopedia texts.

This study employs the subsequent steps to extract candidate entities from semi-structured and unstructured Chinese texts:

- Named entities parsing whereby entities identified as named entities are added to the candidate entities.
- Part-of-speech (POS) tagging whereby entities identified as nouns or noun phrases are added to candidate entities.

3.3   Extracting Hypernym-Hyponym Relations

In this subsection, a neural network extraction model that combines entity description and attribute information is introduced. As depicted in Fig. 4, the model consists of four components: the entity embedding input layer, entity description feature extraction layer, entity attribute feature extraction layer, and hypernym-hyponym relation classification layer.

- Entity embedding input layer. Given two candidate input entities, e_1 and e_2, the entity embedding input layer links the two entities to the embedding representations, denoted as v_1 and v_1, using the embedding lookup table.
- Entity description feature extraction layer. This layer employs Bi-GRU to capture the hypernym-hyponym semantic features from the descriptions of entities e_1 and e_2 obtained from the summary of BaiDuBaiKe, denoted as

$$d_i = Bi\text{-}GRU(des_i) \quad i \in \{1, 2\} \tag{1}$$

where des_i denotes the description of entity e_i and d_i denotes the output of the entity description feature extraction layer. Subsequently, representations

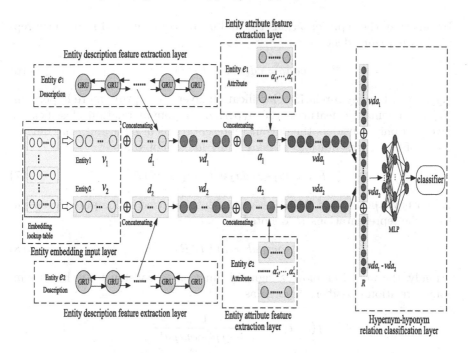

Fig. 4. Architecture of hypernym-hyponym relation extraction model.

v_i and d_i are concatenated into a new representation vd_i, denoted as

$$vd_i = v_i \oplus d_i \quad i \in \{1,2\} \tag{2}$$

where $\oplus$ denotes the concatenation of two representations.

– Entity attribute feature extraction layer. Given the attributes of entities e_1 and e_2 obtained from the infobox of BaiDuBaiKe, the entity attribute feature extraction layer first links the attributes to the embedding representations denoted as $\{attri_1^1, \ldots, attri_1^n\}$ and $\{attri_2^1, \ldots, attri_2^m\}$ after which, it employs an attention model to capture the hypernym-hyponym semantic features from the attributes of entities e_1 and e_2, denoted as

$$H_i^k = atrti_i^k \cdot r_i \quad i \in \{1,2\} \tag{3}$$

$$\alpha_i^k = \frac{exp(H_i^k)}{\sum_{j=1}^{s} exp(H_i^j)} \quad i \in \{1,2\} \quad s \in \{n, m\} \tag{4}$$

where α_i^k denotes an attention weight of the k-th attribute of entity e_i. H_i^k denotes the correlation coefficient of the k-th attribute of entity e_i and r_i (learning to update in model training) denotes a random query matrix of the entity e_i. The output of the entity attribute feature extraction layer is as follows:

$$a_i = \sum_{j=1}^{s} \alpha_i^j \cdot atrti_i^j \quad i \in \{1,2\} \quad s \in \{n, m\} \tag{5}$$

In this case, the representations vd_i and a_i are concatenated into a new representation vda_i, denoted as

$$vda_i = vd_i \oplus a_i \quad i \in \{1, 2\} \tag{6}$$

– Hypernym-hyponym relation classification layer. To capture more hypernym-hyponym semantic features, the hypernym-hyponym relation classification layer not only considers the output representations of vda_1 and vda_2 but the offset of the representations of vda_1 and vda_2, denoted as

$$R = vda_1 \oplus vda_2 \oplus (vda_1 - vda_2) \tag{7}$$

A multilayer perception (MLP) is subsequently utilized to compute the output from the representations R, denoted as

$$output = MLP(R) \tag{8}$$

Finally, the model employs a sigmoid function to construct the hypernym-hyponym relation classifier, denoted as

$$f(e_1, e_2) = \frac{1}{1 + exp(-output)} \tag{9}$$

The logarithmic loss function is exploited to learn the hypernym-hyponym relation classifier, denoted as

$$\mathcal{L} = -[log(f(e_1, e_2) \cdot y_{12} + log(1 - f(e_1, e_2)) \cdot (1 - y_{12}))] \tag{10}$$

where $y_{12} = 1$ if entities e_1 and e_2 possess the hypernym-hyponym relation, and $y_{12} = 0$ if otherwise.

4   Experiments

4.1   Experiment Setup

Dataset. Two corpora, BaiduBaike and SogouCA[2], were used in the study. First, the candidate entities were extracted from Baidu Baike and SogouCA, after which the entity descriptions and attributes of the extracted candidate entities were collected.

To train and evaluate the proposed hypernym-hyponym relation extraction model, four labeled datasets namely FD [20], BK [20], Food [4], and Plant [4] were employed. Detailed statistics of the datasets are listed in Table 1.

Compared Approach. Five approaches were evaluated in the experiments. (1) W-KNN. A K-nearest neighbor (KNN) classifier [26] was utilized to extract Chinese hypernym-hyponym-relations. (2) W-BPNN. A back-propagation neural network (BPNN) classifier [27] was employed for the same purpose. (3) D-SVM.

---

[2] http://www.sogou.com/labs/resource/ca.php.

Table 1. Statistics of the labeled datasets.

Datasets	# hypernym-hyponym-relations	# non-hypernym-hyponym-relations	# entities
FD	1,391	4,294	4,023
BK	3,870	3,582	4,736
Food	6,808	1,760	1,763
Plant	3,031	3,200	1,849

A dynamic weighting neural network [18] and a support vector machine (SVM) [28] classifier were applied to extract Chinese hypernym-hyponym relations. (4) W-Projection. A piecewise linear projection model [29] was exploited to extract the Chinese hypernym-hyponym relations. (5) W-Ours. the proposed neural network extraction model that incorporated entity description and attribute information to extract Chinese hypernym-hyponym relations.

Evaluation Metrics. Three commonly utilized evaluation metrics, namely, the macro average of precision (P), recall (R), and f1 score (F1), were employed to assess the performance of the compared approaches and provide the precision-recall curves of the compared approaches. These were also evaluated according to the mean average precision (MAP), and area under the curve (AUC).

Table 2. Performance comparison for the compared approaches in terms of P, R, and F1.

Approaches	FD			BK			Food			Plant		
	P	R	F1	P	R	F1	P	R	F1	P	R	F1
W-KNN	0.709	0.702	0.701	0.739	0.739	0.739	0.792	0.776	0.778	0.808	0.817	0.812
W-BPNN	0.791	0.789	0.790	0.813	0.819	0.813	0.853	0.848	0.849	0.849	0.867	0.855
D-SVM	0.812	0.807	0.804	0.849	0.843	0.845	0.867	0.864	0.865	0.889	0.887	0.887
W-Projection	0.803	0.799	0.799	0.839	0.847	0.839	0.881	0.867	0.870	0.889	0.885	0.887
W-Ours	**0.859**	**0.850**	**0.851**	**0.874**	**0.881**	**0.875**	**0.923**	**0.906**	**0.910**	**0.930**	**0.928**	**0.929**

4.2 Experimental Results Analysis

Table 2 lists the P, R, and F1 results of the compared approaches. Evidently, the performance of W-Ours surpassed that of the other methods. The W-KNN obtained lower P, R, and F1 results because the prediction results of the KNN were easily affected by noisy and high-dimensional data. The performances of W-BPNN and D-SVM exceeded that of W-KNN, indicating that more sophisticated models can capture additional features to predict classification results in noisy and high-dimensional data. Nevertheless, the P, R, and F1 results of the W-BPNN and D-SVM were lower than those of the W-Ours. For instance, the F1

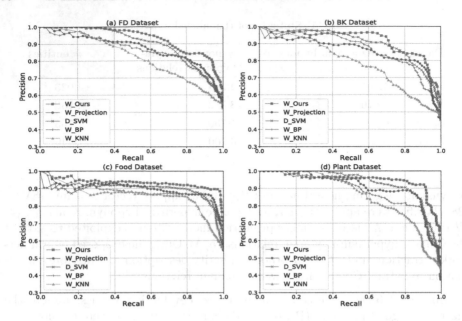

Fig. 5. Comparison of precision-recall curves.

results of W-Ours in FD and Food were 0.851 and 0.910, respectively, while those for W-BPNN were 0.789 and 0.849, and those for D-SVM were 0.804 and 0.865, respectively. W-Projection demonstrated slightly better performance than D-SVM but lagged behind W-Ours. The F1 results of W-Ours in BK and Plant were 0.875 and 0.929, respectively, whereas those for W-Projection were 0.839 and 0.887, improving by 0.036 and 0.042, respectively. These results show that the entity description and attribute information can enhance the performance of the hypernym-hyponym relation extraction.

Figure 5 shows the precision-recall curves of the compared approaches. From the figure, it is evident that the W-KNN obtained lower precision-recall curves in the FD, BK, food, and plant datasets. The precision of W-BPNN and W-Projection outperformed W-KNN throughout the entire recall range but was lower than D-SVM in some instances. For example, the precision of D-SVM was higher than that of W-BPNN and W-Projection in the FD and BK datasets within the recall range [0.0, 0.8]. In comparison to W-BPNN, W-Projection, and D-SVM, W-Ours achieved higher precision-recall curves in most ranges of recall in the FD, BK, Food, and Plant datasets. An exception, here, was that the precisions of W-Ours were marginally lower than those of the other approaches in the BK and Plant datasets with the recall range [0.0, 0.5]. Overall, the precision-recall curves of W-Ours performed better than those of the other approaches in most ranges of recall. These results confirm that the entity description and attribute information are advantageous for the hypernym-hyponym-relation extraction task.

Table 3. Comparison of MAP and AUC.

Approaches	FD		BK		Food		Plant	
	MAP	AUC	MAP	AUC	MAP	AUC	MAP	AUC
W-KNN	0.801	0.799	0.819	0.817	0.846	0.859	0.890	0.889
W-BPNN	0.862	0.864	0.896	0.898	0.892	0.899	0.918	0.924
D-SVM	0.878	0.881	0.918	0.921	0.892	0.906	0.948	0.949
W-Projection	0.872	0.874	0.889	0.899	0.909	0.914	0.931	0.935
W-Ours	**0.928**	**0.926**	**0.933**	**0.938**	**0.932**	**0.945**	**0.958**	**0.965**

Table 3 lists the MAP and AUC results of the compared approaches. The MAP and AUC results of W-BPNN were higher than those of W-KNN, but lower than those of D-SVM and W-Projection. The MAP and AUC results of W-Ours performed better than those of other approaches. For instance, the MAP results of W-Ours in BK and Plant were 0.933 and 0.958, respectively, while those for D-SVM were 0.918 and 0.948, respectively, and for W-Projection are 0.889 and 0.931, respectively, yielding an average improvement of 0.03 and 0.019, respectively. The AUC results of W-Ours in FD and Food were 0.926 and 0.945, respectively, whereas those for D-SVM were 0.881 and 0.906, and 0.874 and 0.914 for W-Projection, resulting in an improvement of by 0.048 and 0.035, respectively.

5 Conclusions

In this paper, an entity description-and attribute information-based approach is proposed for extracting Chinese hypernym-hyponym relations. This approach mines entity description and attribute information from the Chinese encyclopedia and applies a filtering strategy to purify the mined candidate entities. Subsequently, a neural network extraction model that combines entity description and attribute information is employed to extract Chinese hypernym-hyponym relations. The proposed approach is implemented on four real-world Chinese datasets. The experimental results demonstrate its superior performance compared to other approaches in evaluation metrics such as F1 score, MAP, and AUC. In future work, we aim to implement the proposed approach on a larger dataset and further extend it to directly extract Chinese hypernym-hyponym relations from plain text corpora.

Acknowledgement. This research was funded by the University Natural Science Research Projects of Anhui Province (Grant Nos. 2022AH050972, KJ2020A0361, and KJ2021A0516). This research was jointly funded by the Key Technologies R&D Program of the Anhui Province of China (Grant No. 202004a05020013), the Key Project of Collaborative Innovation Fund of Jiujiang District and Anhui Polytechnic University (Grant No. 2021cyxta1), and the Excellent Young Talents Fund Program of Higher Education Institutions of Anhui Province (Grant No. gxyq2020031).

References

1. Yu, C., Zhang, H., Song, Y., Ng, W., Shang, L.: Enriching large-scale eventuality knowledge graph with entailment relations. In: Conference on Automated Knowledge Base Construction, AKBC 2020, Virtual, 22–24 June 2020. https://doi.org/10.24432/C56K5H
2. Yang, G., Jiayu, Y., Dongdong, X., Zelin, G., Hai, H.: Feature-enhanced text-inception model for Chinese long text classification. Sci. Rep. **13**(1), 2087 (2023)
3. Ma, Z., Meng, D., Kong, C., Zhou, L., Li, M., Tao, W.: Coreference resolution with syntax and semantics. In: Web Information Systems and Applications - 19th International Conference, WISA 2022, Dalian, China, 16–18 September 2022, Proceedings, vol. 13579, pp. 181–193 (2022)
4. Huang, S., Luo, X., Huang, J., Wang, H., Gu, S., Guo, Y.: Improving taxonomic relation learning via incorporating relation descriptions into word embeddings. Concurr. Comput. Pract. Exp. **32**(14), e5696 (2020)
5. Bai, H., Wang, T., Sordoni, A., Shi, P.: Better language model with hypernym class prediction. In: Proceedings of the 60th Annual Meeting of the Association for Computational Linguistics (Volume 1: Long Papers), ACL 2022, Dublin, Ireland, 22–27 May 2022, pp. 1352–1362 (2022)
6. Gu, S., Luo, X., Wang, H., Huang, J., Wei, Q., Huang, S.: Improving answer selection with global features. Expert Syst. J. Knowl. Eng. **38**(1), e12603 (2021)
7. Wang, X., Luo, F., Wu, Q., Bao, Z.: How context or knowledge can benefit healthcare question answering? IEEE Trans. Knowl. Data Eng. **35**(1), 575–588 (2023)
8. Miller, G.A.: Wordnet: a lexical database for English. Commun. ACM (CACM) **38**(11), 39–41 (1995)
9. Dong, Z., Dong, Q., Hao, C.: HowNet and its computation of meaning. In: COLING 2010, 23rd International Conference on Computational Linguistics, Demonstrations Volume, 23–27 August 2010, Beijing, China, pp. 53–56 (2010)
10. Che, W., Li, Z., Liu, T.: LTP: a Chinese language technology platform. In: COLING 2010, 23rd International Conference on Computational Linguistics, Demonstrations Volume, 23–27 August 2010, Beijing, China, pp. 13–16 (2010)
11. Hearst, M.A.: Automatic acquisition of hyponyms from large text corpora. In: 14th International Conference on Computational Linguistics, COLING 1992, Nantes, France, 23–28 August, pp. 539–545 (1992)
12. Wu, W., Li, H., Wang, H., Zhu, K.Q.: Probase: a probabilistic taxonomy for text understanding. In: Candan, K.S., Chen, Y., Snodgrass, R.T., Gravano, L., Fuxman, A. (eds.) Proceedings of the ACM SIGMOD International Conference on Management of Data, SIGMOD 2012, Scottsdale, AZ, USA, 20–24 May, pp. 481–492 (2012)
13. Roller, S., Kiela, D., Nickel, M.: Hearst patterns revisited: automatic hypernym detection from large text corpora. In: Proceedings of the 56th Annual Meeting of the Association for Computational Linguistics, ACL 2018, Melbourne, Australia, 15–20 July 2018, Volume 2: Short Papers, pp. 358–363 (2018)
14. Xu, B., et al.: CN-DBpedia: a never-ending Chinese knowledge extraction system. In: Benferhat, S., Tabia, K., Ali, M. (eds.) IEA/AIE 2017. LNCS (LNAI), vol. 10351, pp. 428–438. Springer, Cham (2017). https://doi.org/10.1007/978-3-319-60045-1_44
15. Wu, T., Wang, H., Qi, G., Zhu, J., Ruan, T.: On building and publishing linked open schema from social web sites. J. Web Semant. **51**, 39–50 (2018)

16. Yang, Z., Bi, Y., Wang, L., Cao, D., Li, R., Li, Q.: Development and application of a field knowledge graph and search engine for pavement engineering. Sci. Rep. **12**(1), 7796 (2022)
17. Wang, C., et al.: A phrase mining framework for recursive construction of a topical hierarchy. In: The 19th ACM SIGKDD International Conference on Knowledge Discovery and Data Mining, KDD 2013, Chicago, IL, USA, 11–14 August 2013, pp. 437–445 (2013)
18. Alfarone, D., Davis, J.: Unsupervised learning of an IS-A taxonomy from a limited domain-specific corpus. In: Proceedings of the Twenty-Fourth International Joint Conference on Artificial Intelligence, IJCAI 2015, Buenos Aires, Argentina, 25–31 July 2015, pp. 1434–1441 (2015)
19. Huang, S., Luo, X., Huang, J., Guo, Y., Gu, S.: An unsupervised approach for learning a Chinese IS-A taxonomy from an unstructured corpus. Knowl. Based Syst. **182**, 104861 (2019)
20. Wang, C., Yan, J., Zhou, A., He, X.: Transductive non-linear learning for Chinese hypernym prediction. In: Proceedings of the 55th Annual Meeting of the Association for Computational Linguistics, ACL 2017, Vancouver, Canada, July 30–4 August, Volume 1: Long Papers, pp. 1394–1404 (2017)
21. Huang, S., Xiu, Y., Li, J., Liu, S., Kong, C.: A bilateral context and filtering strategy-based approach to Chinese entity synonym set expansion. Complex Intell. Syst. (2023)
22. Snow, R., Jurafsky, D., Ng, A.Y.: Learning syntactic patterns for automatic hypernym discovery. In: Advances in Neural Information Processing Systems 17 [Neural Information Processing Systems, NIPS 2004, 13–18 December 2004, Vancouver, British Columbia, Canada], pp. 1297–1304 (2004)
23. Hao, M., Li, Z., Zhao, Y., Zheng, K.: Mining high-quality fine-grained type information from Chinese online encyclopedias. In: Hacid, H., Cellary, W., Wang, H., Paik, H.-Y., Zhou, R. (eds.) WISE 2018. LNCS, vol. 11234, pp. 345–360. Springer, Cham (2018). https://doi.org/10.1007/978-3-030-02925-8_25
24. Suchanek, F.M., Kasneci, G., Weikum, G.: YAGO: a large ontology from Wikipedia and wordnet. J. Web Semant. **6**(3), 203–217 (2008)
25. Harris, Z.S.: Distributional structure. Word **10**(2–3), 146–162 (1954)
26. Patrick, E.A., Fischer, F.P.: A generalization of the k-nearest neighbor rule. In: Proceedings of the 1st International Joint Conference on Artificial Intelligence, Washington, DC, USA, 7–9 May, pp. 63–64 (1969)
27. Rumelhart, D.E., Hinton, G.E., Williams, R.J.: Learning internal representation by back-propagation errors. Nature **323**, 533–536 (1986)
28. Cortes, C., Vapnik, V.: Support-vector networks. Mach. Learn. **20**(3), 273–297 (1995)
29. Wang, C., Fan, Y., He, X., Zhou, A.: Predicting hypernym-hyponym relations for Chinese taxonomy learning. Knowl. Inf. Syst. **58**(3), 585–610 (2019)

Knowledge-Concept Diagnosis from fMRIs by Using a Space-Time Embedding Graph Convolutional Network

Ye Lei[1,2], Yupei Zhang[1,2(✉)], Yi Lin[1,2], and Xuequn Shang[1,2]

[1] School of Computer Science, Northwestern Polytechnical University, Xi'an, China
ypzhaang@nwpu.edu.cn
[2] MIIT Big Data Storage and Management Lab, Xi'an, China

Abstract. Diagnosing the contents of learning in brain activities is a long-standing research task in cognitive sciences. The current studies on cognitive diagnosis (CD) in education determine the status of knowledge concept (KC) based on the observed responses to test items. However, the learning process of KC in the brain is left with no touch. This paper proposes to solve the problem of knowledge-concept diagnosis (KCD) from fMRIs by identifying the concepts a student focuses on in learning activities. Using the graph convolutional network (GCN), we introduce the STEGCN approach composed of a spatial GCN for brain-graph structure, a temporal GCN for brain-activity sequence, and a fully connected network for KCD. To evaluate STEGCN, we acquired an fMRI dataset that was collected on five concepts when students were learning a computer course. The experiment results demonstrate that our proposed method yields better performance than traditional models, showing the effectiveness of STEGCN in concept classification. This study contributes to a new fMRI-based route for knowledge-concept diagnosis.

Keywords: Knowledge-Concept Diagnosis · fMRI · Graph Convolution Network · Space-Time Embedding · Educational Data Mining

1 Introduction

Knowledge-Concept Diagnosis (KCD) is one of the core problems of decoding the cognitive process in the learning brain [1]. In education, knowledge diagnosis is usually to predict the knowledge mastery of a student from the observed responses to a set of given test items [2], such as item response theory [3], deep knowledge tracing [4] and score [5], according to the prediction results, help educational institutions or teachers to improve teaching routes, or recommend books

This study was funded in part by the National Natural Science Foundation of China (Nos. 62272392, U1811262, 61802313), the Key Research and Development Program of China (No. 2020AAA0108504), the Higher Research Funding on International Talent cultivation at Northwestern Polytechnical University (No. GJGZZD202202).

L. Yuan et al. (Eds.): WISA 2023, LNCS 14094, pp. 100–111, 2023.
https://doi.org/10.1007/978-981-99-6222-8_9

[6] and courses [7] for students. While these methods all obtain knowledge mastery by using students' external performance, diagnosing students' knowledge process by observing brain activities provides an intrinsically natural understanding.

In neuroscience, brain activities have been measured by various information techniques, e.g., functional Magnetic Resonance Imaging (fMRI), Electroencephalogram (EEG), and Magnetoencephalography (MEG). Compared to others, fMRI provides high spatial and temporal resolution by collecting the blood oxygen level-dependent (BOLD) along the brain activity [8]. Since concept learning is a basic brain process, we in this paper propose to utilize fMRIs for the task of KCD, which is rarely studied as data is hard to acquire.

The current fMRI-based brain studies can be divided into two routes, i.e., analyzing brain images [9] and modeling brain connection networks [10]. The former aims to learn features from image voxel for specific cognitive tasks, e.g., finding out specific brain activity areas corresponding to a cognitive task. Alexander Enge [11] analyzes brain regions using activation likelihood estimation and sccd-based effect size mapping. Several studies [12] have shown that cognitive processes lead to altered patterns of functional connectivity (FC) between different brain regions, making it possible to predict cognitive concepts through the FC status of brain regions.

Generally, the brain's functional connectivity is modeled into an FC graph where the nodes indicate the brain function areas and the edges are the connection statuses between all brain regions. The connection status is artificially set by the correlation during the time series. The common method for creating the FC graph is to compute the similarity between brain areas by Pearson correlation, Spearman correlation or dynamic time warping [13]. The feature of nodes is usually represented by knitting the BOLD signals in each brain region, such as the simple mean of all BOLD of all voxels. Besides, there are many ways to divide many brain regions by using different predefined temples, such as Automated Anatomical Labeling template [14] and Harvard-Oxford template [15].

Recently, with such graph formulation on brain function connectivity, graph learning algorithms have largely been used for brain graph classification, especially for graph convolutional networks (GCN) and its variants [16]. Considering the time-sequential fMRIs, most methods developed to integrate both temporal features and spatial features. Li et al. designed a GCN layer to infer regions of interest (ROIs) that contribute more to the prediction task [17]. Zhang et al. proposed to cascade CNN and LSTM towards fMRI classification [9]. Overall, Spatio-temporal GCNs (STGCN) can simultaneously extract temporal and spatial information from fMRI data.

However, there is a lack of studies on KCD via STGCN models. In this study, we construct the time graph and space graph of the brain FC status, extracting the features of space and time views, respectively, where we set a hyperparameter to balance the two views. The learned features from the views of space and time are finally combined into the fMRI classifier. Our proposed model is shown in Fig. 1. In summary, our study contributes to the followings:

- This study proposes a deep learning approach to diagnosing knowledge concepts by using fMRIs in the learning brain. This method could yield decent results on a dataset collected from a computer course [18]
- Combining fMRI data's time and space features, the proposed model constructs a functional connectivity graph of brain regions from the space and time views, respectively.
- We propose the Space-Time Embedding GCN (STEGCN) for knowledge diagnosis of functional connectivity graphs of brain regions.

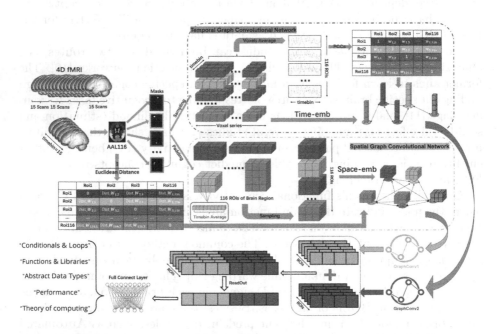

Fig. 1. Space-Time Embedding Graph Convolutional Network

2 Related Work

2.1 Brain Image Analysis

Brain image analysis (BIA) is a long-term research topic in neuroscience and medical imaging. MRI can non-invasively capture the internal structure and BOLD of the brain, helping us understand the physiological changes of the human brain under specific cognitive tasks. Traditional cognitive diagnosis based on MRI focuses on cognitive impairment and related diseases [19], and mostly uses ML and other methods to provide objective analysis [20]. Among them, deep learning methods have excellent performance, especially convolutional neural networks (CNN). As a special multi-layer neural network, CNN can automatically extract features [21], avoiding omissions and misjudgments caused by artificial feature selection.

2.2   fMRI Classification

We usually construct a brain functional connectivity network for brain fMRI, where nodes represent ROI under the template division of a specific brain region, and edges represent the relationship between ROIs. Functional connectivity networks (FCNs) have been widely used to analyze psychiatric problems or specific cognitive diseases. Recently, Spectral Graph Convolutional Network has been proved to be feasible in learning the feature representation of topological graph structure, and has achieved effective results in traffic predictio, action recognition. Studies have shown that the representation of [22] GCN on the brain functional connectivity network is effective.

2.3   Graph Convolutional Network (GCN)

We frames concept classification as a graph classification problem. In this study, we focus on the spectral graph convolution proposed by Kipf and Welling [23]. Suppose a weighted graph is represented by $\mathcal{G} = \{\mathcal{V}, \mathcal{E}, \mathcal{W}\}$, where $\mathcal{V}$, $\mathcal{E}$ and $\mathcal{W}$ are the vertices, edges and weight matrix. Laplacian matrix of the graph is defined as $L = I_N - D^{-\frac{1}{2}} W D^{-\frac{1}{2}}$, where I_N is an identity matrix with size $N \times N$ and D is the diagonal node degree matrix, respectively. L is real symmetric positive semidefinite, it can be factored as $L = U \Lambda U^T$, where $U = [u_0, u_1, \ldots, u_{n-1}] \in \mathbb{R}^{N \times N}$ is the matrix of eigenvectors ordered by eigenvalues, Λ is the diagonal matrix of eigenvalues λ, $\Lambda_{ii} = \lambda_i$. GCN generalizes the spectral convolution operation from grid data to graph data through the Fourier transform of the graph. Suppose $g_\theta(\Lambda) \in \mathbb{R}^{N \times N}$ is a filter function matrix of the eigenvalues Λ with parameter θ, the graph convolution of the input signal $x \in \mathbb{R}^N$ is defined as

$$g_\theta \star x = U g_\theta(\Lambda) U^T x \tag{1}$$

However, in actual calculations, due to the computational complexity of eigen-decomposition, we use Chebyshev polynomials $T_k(x)$ up to Kth as an approximation of $g_\theta(\Lambda)$, i.e.,

$$g_\theta \approx \sum_{i=0}^{K} \theta_i T_i(\tilde{\Lambda}) \tag{2}$$

where $\tilde{\Lambda} = 2\Lambda/\lambda_{max} - I_N$ and θ_i $(i = 1, \ldots, k)$ is Chebyshev coefficient. The Chebyshev polynomials are defined by $T_i(x) = 2x T_{i-1}(x) - T_{i-2}(x)$ with $T_o(x) = 1$ and $T_1(x) = x$ [24]. Let $\tilde{L} = U \tilde{\Lambda} U^T$, then $T_i(\tilde{L}) = U T_i(\tilde{\Lambda}) U^T$, the convolution of a graph signal x can be simplified as

$$g_\theta \star x \approx U \left(\sum_{i=0}^{K} \theta_i T_i\left(\tilde{\Lambda}\right) \right) U^T x = \sum_{i=0}^{K} \theta_i T_i\left(\tilde{L}\right) x \tag{3}$$

3   Materials and Methods

3.1   Data Preprocessing and Graph Prepareing

This study uses the "Think Like an Expert" [18] dataset, which acquired fMRI of 20 students (9 females and 11 males) and 5 experts in a computer class. We only selected data from 20 students to ensure consistency. It contains five categories of computer science concepts: Conditionals & Loops, Functions & Libraries, abstract data type, performance and theory of computing. In order to standardize the data dimension, we uniformly cut the data into NIFTI files with fixed time intervals, each file contains 15 scans, and each scan is a 3D brain neural image. Raw fMRI data needs to be preprocessed to remove noise, artifacts, motion and other possible interference factors, so as to improve the reliability and accuracy of the data. In this study, we used SPM12 for fMRI data preprocessing. The processing pipeline includes: slice temporal layer correction; head motion correction; registration of T1 structural images, T2 functional images and Montreal Neurological Institute (MNI); smoothing; bandpass filtering;

We discard the first ten scans and last time period which is fewer than fifteen scans to ensure a uniform sample size. This procedure yielded 7765 samples with five class sample distributions: 1635(21.05%), 1575(20.29%), 1365(17.58%), 1615(20.79%), and 1575(20.29%).

3.2   Our Proposed Approach

This subsection presents our proposed model, STEGCN, composed of time and space graph convolution networks for knowledge concept recognition. We used the AAL116 template to divide the brain into 116 regions. The graph convolution process can be described mathematically as follows:

$$h_i^{(l+1)} = \sigma(b^l + \sum_{j \in \mathbb{N}(i)} \frac{1}{c_{ji}} h_j^{(l)} W^{(l)}) \tag{4}$$

where $\mathbb{N}(i)$ represents the set of neighbors of node i, c_{ji} is the product of the square root of node degrees, and σ denotes an activation function. In addition, the correlation coefficient serves as the edge weight and participates in the iteration, leading to the following formula:

$$h_i^{(l+1)} = \sigma(b^l + \sum_{j \in \mathbb{N}(i)} \frac{e_{ji}}{c_{ji}} h_j^{(l)} W^{(l)}) \tag{5}$$

The Time-Embedding GCN. To construct the temporal graph G_T, we averaged the BOLD values of voxels in each region during a timebin, we obtain the adjacency matrix W_{Time} from the Pearson Correlation Coefficients(PCCs) between brain regions. If the PCC is less than τ ($\tau = 0.5$), we consider them unconnected and set the w_{ji} to 0, otherwise 1. The edge weights e_{ji}^{Time} are defined as follows:

$$e_{ji}^{Time} = \begin{cases} 0 & x < \tau \\ e_{ji} & x \geq \tau \end{cases} \tag{6}$$

Since the number of voxels in brain regions may be different, We upsample the sequence of voxels to a uniform size, which is defined as the feature of G_T point. We compose t 1D convolution kernels(3×1) into a t-channel filter ($t = timebin$), Compute brain region node embeddings in the temporal dimension using temporal filters, and send them to GCN (Fig. 2).

$$h_i^{Time-embeding} = Time_{embed}(h_i^{Time}) \tag{7}$$

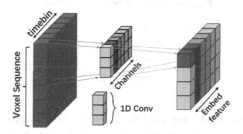

Fig. 2. Time-embeding module

$$h_i^{Time-embeding} = \sigma(b + \sum_{j \in \mathbb{N}(i)} \frac{e_{ji}^{Time}}{c_{ji}} h_j^{Time-embeding} W_{Time}) \tag{8}$$

The Space-Embedding GCN. To construct the spatial graph G_S, we localized the geometric center of each region and calculated the Euclidean distance between brain regions. We obtain the adjacency matrix W_{Space} from the reciprocal of the Euclidean distance between two regions. In space, we believe that there is a universal connection between brain regions, so w_{ji} set to 1, and the spatial graph is a fully connected graph. And the edge weights e_{ji}^{Space} are defined as follows:

$$e_{ji}^{Space} = 1/d_{ji} \tag{9}$$

$$d_{ji} = \sqrt{(x_j - x_i)^2 + (y_j - y_i)^2 + (z_j - z_i)^2} \tag{10}$$

where $x_j, y_j, z_j, x_i, y_i, z_i$ are the coordinates of point j and point i in 3D space of the brain. To further extract Space embeding feature, we average the voxel values of each brain region across the timebin and upsample the voxel space to a uniform size, which is defined as the feature of G_S point. We combine r 3D $3 \times 3 \times 3$ convolution kernels into an r-channel filter ($r = region\ number$).

Compute brain region node embeddings in the Spatial dimension using Spatial filters, and send them to GCN (Fig. 3).

$$h_i^{Space-embeding} = Space_{embed}(h_i^{Space}) \qquad (11)$$

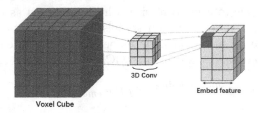

Fig. 3. Space-embeding module

$$h_i^{Space-embeding} = \sigma(b + \sum_{j \in \mathbb{N}(i)} \frac{e_{ji}^{Space}}{c_{ji}} h_j^{Space-embeding} W_{Space}) \qquad (12)$$

The Time-space-embedding GCN. Finally, we concatenate the features of the time and space dimensions, readout the graph to obtain the max feature value of each node, and finally send it to the fully connected layer for classifications.

$$h_i^{FCN} = readout\left(h_i^{Time-embeding},\ h_i^{Space-embeding}\right) \qquad (13)$$

3.3   Model Training and Evaluation

Comparison Models. To show the advantages of the proposed method, we compared STEGCN to Logistic Regression(LR), Random Forest(RF), K-Nearest Neighbor(KNN), Artificial Neural Network(ANN)and Decision Tree(DT).

Model Evaluation. In this study, we employed 5-fold cross-validation to evaluate the model performance due to the sample imbalance problem. The process is as follows:

1. The data set is randomly divided into 5 parts.
2. For each copy, it is selected as the test set in turn, and the remaining 4 copies are used as the training set.
3. Train the model and validate it on the test set. To record the performance of models, we adopt Accuracy (ACC), Sensitivity (SEN), Specificity (SPE), F1-score (F1), Positive Predictive Value (PPV), Matthews Correlation Coefficient (MCC), receiver operating characteristic (ROC) curve and the area under ROC (AUC).
4. Continue selection and training until each part is selected once as a test set.
5. Computes the average performance of the model over 5 validations.

Table 1. Macro-average of models

| Macro-average (5 fold average) | | | | | |
Method	ACC	SEN	SPE	PPV	F1	MCC
DT	70.90±0.26	27.13±0.61	81.80±0.16	27.17±0.65	27.12±0.66	08.95±0.80
KNN	76.15±0.49	40.18±1.15	85.06±0.29	40.36±1.31	39.75±1.26	25.28±1.52
RF	76.82±0.70	41.51±1.75	85.45±0.43	41.91±1.55	40.65±1.77	27.04±2.10
ANN	85.26±1.09	62.62±2.72	90.75±0.68	65.23±2.71	62.36±2.97	54.39±3.13
Logistic	88.07±0.66	69.98±1.62	92.54±0.41	70.02±1.55	69.97±1.60	62.54±2.00
Ours	**91.39±1.73**	**78.95±4.00**	**94.61±1.11**	**78.41±4.22**	**78.36±4.24**	**73.24±5.19**

Table 2. Micro-average of models

| Micro-average (5 fold average) | | | | | |
Method	ACC	SEN	SPE	PPV	F1	MCC
DT	27.25±0.66	27.25±0.66	81.81±0.16	27.25±0.66	27.25±0.66	08.95±0.80
KNN	40.38±1.24	40.38±1.24	85.09±0.31	40.38±1.24	40.38±1.24	25.28±1.52
RF	42.05±1.75	42.05±1.75	85.51±0.43	42.05±1.75	42.05±1.75	27.04±2.10
ANN	63.15±2.73	63.15±2.73	90.78±0.68	63.15±2.73	63.15±2.73	54.39±3.13
Logistic	70.18±1.67	70.18±1.67	92.54±0.41	70.18±1.67	70.18±1.67	62.54±2.00
Ours	**78.48±4.32**	**78.48±4.32**	**94.57±1.08**	**78.48±4.32**	**78.48±4.32**	**73.24±5.19**

4 Experiment Results

4.1 The Entire Evaluation

The table below shows the average accuracy for knowledge diagnosis derived from 5-fold cross-validation. We use the mean and standard deviation of the six indicators to measure the performance of all methods under two macro-average and micro-average calculation methods. Our proposed STEGCN has better performance in all indicators, but the standard deviation is higher than other algorithms. We think this may be related to setting network parameters in the STEGCN process, which is a direction worth optimizing. And other methods perform better than other concepts on the concepts "Conditions & Loops" and "Functions & Libraries" than other concepts. The STEGCN not only performs better on these two concepts but also performs better on the concept "Theory of Computing". It shows that our method extracts more deeply distinguishing features of the concept "Theory of Computing". The comparison of models on each concept proves this point further according to Fig. 5 (Fig. 4 and Tables 1, 2).

4.2 The Evaluations on Per Concept

According to the ROC curve of Fig. 5, all methods have better results in the concept of "Conditions & Loops" and "Functions & Libraries". But our method has

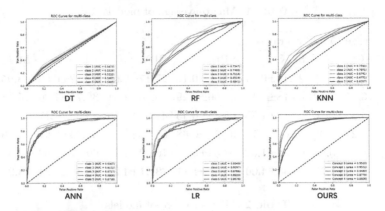

Fig. 4. AUC of knowledge Concepts

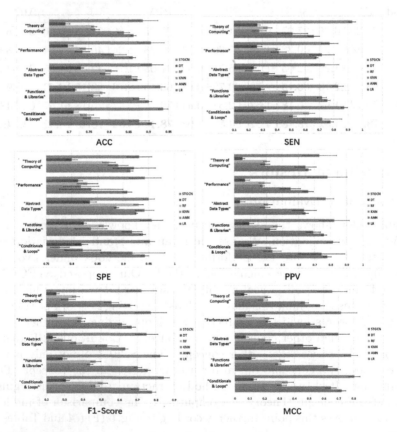

Fig. 5. The performance comparison of each computer course concept on different methods and different indicators. Each graph represents an indicator, the vertical axis in the graph represents each concept, and the horizontal columns of different colors represent different methods

a larger gap in the performance of concepts, and the ROC curves of these two classes of concepts are farther apart than the other classes. In other comparison methods, ANN and LR also show similar situations. This may be due to the unequal level of subordination of concepts: the first two types of concepts are more detailed, while the latter "Theory of Computing", "Abstract Data Type" and "Performance" are more abstract and general, and the specificity reflected in the characteristics is not clear enough. In terms of index comparison, the overall top three are: STEGCN, LR, ANN. However, in the SEN of "Performance", ANN surpassed LR, showing a stronger sensitivity to this concept. "Theory of Computing" and "Abstract Data Type" in the SPE, ANN once again surpassed LR, showing stronger specificity for these two concepts. This shows that the latter three types of concepts also have different conception levels, and ANN will perform better on certain concepts. Therefore, the subordination level division of the concept itself can greatly affect the performance of models. The comprehensive multi-level KCD is also the focus in the future. Being able to identify concepts at different levels at the same time can greatly improve the generality of the KCD models.

4.3 Parameter Discussion

In addition to proving the effectiveness of STEGCN, we also want to know the contribution of the time dimension and space dimension to the KCD task. In the previous experiments, we defaulted that the time dimension and the space dimension are equally important. So we introduce the hyperparameter λ. By λ, we adjust the ratio of space embedding features and time embedding features, and explore the ratio of optimal performance for KCD. When $\lambda = 0$ and $\lambda = 1$, it can be regarded as an ablation experiment with only space dimension and

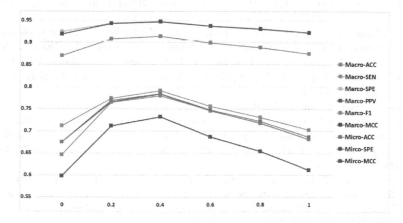

Fig. 6. As the value of λ ranges from 0 to 1, we find that multiple performance metrics of STEGCN reach the highest around 0.4, and the ratio of time and space is 2:3

only time dimension, respectively (Fig. 6).

$$h_i^{FCN} = readout\left(\lambda * h_i^{Time-embeding}, \ (1-\lambda)*h_i^{Space-embeding}\right) \tag{14}$$

5   Conclusions

In this work, we proposed a new model, STEGCN, for KCD. Our model achieves state-of-the-art performance on various metrics, further validating the effectiveness of fMRI and FC for concept classification. Our proposed method has the potential to recognize different concepts in the human brain beyond computer-related concepts. Further research will recognize various concept types by using machine learning models [25–27]. This work contributes to the development of neuroscience and leads to a better understanding of the human brain.

References

1. Zeithamova, D., et al.: Brain mechanisms of concept learning. J. Neurosci. **39**(42), 8259–8266 (2019)
2. Zhang, Y., Dai, H., Yun, Y., Liu, S., Lan, A., Shang, X.: Meta-knowledge dictionary learning on 1-bit response data for student knowledge diagnosis. Knowl.-Based Syst. **205**, 106290 (2020)
3. Pliakos, K., et al.: Integrating machine learning into item response theory for addressing the cold start problem in adaptive learning systems. Comput. Educ. **137**, 91–103 (2019)
4. Piech, C., et al.: Deep knowledge tracing. Adv. Neural Inf. Process. Syst. **28** (2015)
5. Yun, Y., Dai, H., Cao, R., Zhang, Y., Shang, X.: Self-paced graph memory network for student GPA prediction and abnormal student detection. In: Roll, I., McNamara, D., Sosnovsky, S., Luckin, R., Dimitrova, V. (eds.) AIED 2021. LNCS (LNAI), vol. 12749, pp. 417–421. Springer, Cham (2021). https://doi.org/10.1007/978-3-030-78270-2_74
6. Li, S., Xing, X., Liu, Y., Yang, Z., Niu, Y., Jia, Z.: Multi-preference book recommendation method based on graph convolution neural network. In: Zhao, X., Yang, S., Wang, X., Li, J. (eds.) Web Information Systems and Applications. WISA 2022. LNCS, vol. 13579, pp. 521–532. Springer, Cham (2022). https://doi.org/10.1007/978-3-031-20309-1_46
7. Chen, X., Sun, Y., Zhou, T., Wen, Y., Zhang, F., Zeng, Q.: Recommending online course resources based on knowledge graph. In: Zhao, X., Yang, S., Wang, X., Li, J. (eds.) Web Information Systems and Applications. WISA 2022. LNCS, vol. 13579, pp. 581–588. Springer, Cham (2022). https://doi.org/10.1007/978-3-031-20309-1_51
8. Steiner, A.R., Rousseau-Blass, F., Schroeter, A., Hartnack, S., Bettschart-Wolfensberger, R.: Systematic review: anesthetic protocols and management as confounders in rodent blood oxygen level dependent functional magnetic resonance imaging (BOLD fMRI). Animals **11**(1) (2021)
9. Zhang, Y., Xu, Y., An, R., Li, Y., Liu, S., Shang, X.: Markov guided spatio-temporal networks for brain image classification. In: 2022 IEEE International Conference on Bioinformatics and Biomedicine (BIBM), pp. 2035–2041. IEEE (2022)

10. Lynn, C.W., Bassett, D.S.: The physics of brain network structure, function and control. Nat. Rev. Phys. **1**(5), 318–332 (2019)
11. Enge, A., Friederici, A.D., Skeide, M.A.: A meta-analysis of fMRI studies of language comprehension in children. NeuroImage **215**, 116858 (2020)
12. Beaty, R.E., et al.: Robust prediction of individual creative ability from brain functional connectivity. Proc. Natl. Acad. Sci. **115**(5), 1087–1092 (2018)
13. Santana, A.N., Cifre, I., De Santana, C.N., Montoya, P.: Using deep learning and resting-state fMRI to classify chronic pain conditions. Front. Neurosci. **13**, 1313 (2019)
14. Rolls, E.T., Huang, C.C., Lin, C.P., Feng, J., Joliot, M.: Automated anatomical labelling atlas 3. Neuroimage **206**, 116189 (2020)
15. Craddock, R.C., James, G.A., Holtzheimer III, P.E., Hu, X.P., Mayberg, H.S.: A whole brain fMRI atlas generated via spatially constrained spectral clustering. Hum. Brain Mapp. **33**(8), 1914–1928 (2012)
16. Bessadok, A., Mahjoub, M.A., Rekik, I.: Graph neural networks in network neuroscience. IEEE Trans. Pattern Anal. Mach. Intell. (2022)
17. Li, X., et al.: Interpretable brain graph neural network for fMRI analysis. Med. Image Anal. **74**, 102233 (2021)
18. Meshulam, M., et al.: Think like an expert: neural alignment predicts understanding in students taking an introduction to computer science course. bioRxiv, pp. 2020–05 (2020)
19. Feng, W., Liu, G., Zeng, K., Zeng, M., Liu, Y.: A review of methods for classification and recognition of ASD using fMRI data. J. Neurosci. Methods **368**, 109456 (2022)
20. Yin, W., Li, L., Fang-Xiang, W.: Deep learning for brain disorder diagnosis based on fMRI images. Neurocomputing **469**, 332–345 (2022)
21. Qureshi, M.N.I., Jooyoung, O., Lee, B.: 3D-CNN based discrimination of schizophrenia using resting-state fMRI. Artif. Intell. Med. **98**, 10–17 (2019)
22. Zhang, L., Wang, M., Liu, M., Zhang, D.: A survey on deep learning for neuroimaging-based brain disorder analysis. Front. Neurosci. **14**, 779 (2020)
23. Kipf, T.N., Welling, M.: Semi-supervised classification with graph convolutional networks. arXiv preprint arXiv:1609.02907 (2016)
24. Hammond, D.K., Vandergheynst, P., Gribonval, R.: Wavelets on graphs via spectral graph theory. Appl. Comput. Harmon. Anal. **30**(2), 129–150 (2011)
25. Zhang, Y., Xiang, M., Yang, B.: Low-rank preserving embedding. Pattern Recognit. **70**, 112–125 (2017)
26. Zhang, Y., An, R., Liu, S., Cui, J., Shang, X.: Predicting and understanding student learning performance using multi-source sparse attention convolutional neural networks. IEEE Trans. Big Data (2021)
27. Zhang, Y., Liu, S., Qu, X., Shang, X.: Multi-instance discriminative contrastive learning for brain image representation. Neural Comput. Appl. 1–14 (2022)

Interactively Mining Interesting Spatial Co-Location Patterns by Using Fuzzy Ontologies

Jiasheng Yao and Xuguang Bao[✉]

Guangxi Key Laboratory of Trusted Software, Guilin University of Electronic Technology, Guilin 541004, China
bbaaooxx@163.com

Abstract. Spatial data mining refers to the process of extracting implicit and user-interested spatial and non-spatial patterns, general features, rules, and knowledge from spatial databases. As an important branch of spatial data mining, spatial co-location pattern mining is aimed at mining a subset of spatial features that are related, and their instances coexist frequently in geographically adjacent spaces. Numerous efficient methods are able to discover various types of spatial co-location patterns in large datasets. Unfortunately, the problem of identifying patterns that are genuinely interesting to a particular user remains challenging. To address such drawbacks, we propose in this paper an approach that discovers user-preferred spatial co-location patterns according to their specific interest. Generally speaking, the proposed approach filters the discovered frequent co-location patterns using the user's domain knowledge, which is represented by a fuzzy domain ontology. We introduce a generic framework for interactively discovering user-preferred patterns. The user is only asked to express preferences on small sets of patterns, while a filter function is inferred from this feedback by preference learning techniques. We empirically evaluate the accuracy of the algorithm to discover user-preferred patterns by emulating users. Experiments demonstrate that the system is able to accurately identify interesting patterns, accompanying with the 80% of the selected patterns are user-preferred.

Keywords: Spatial data mining · co-location pattern mining · knowledge representation · fuzzy ontology · interactive data exploration

1 Introduction

Knowledge discovery in spatial co-location pattern mining has been defined in Shekhar et al. [1] as the non-trivial process of identifying potentially useful, valid, implicit, and ultimately understandable patterns from spatial data. As an important branch of spatial data mining, spatial co-location pattern mining is aimed at mining a subset of spatial features that are related, and their instances coexist frequently in geographically adjacent spaces. Spatial co-location pattern mining has broad prospects for spatial data owners. Potential value has been found in mining data from environmental science [2], urban planning [3], traffic prediction [4], and so on. For example, spatial co-location patterns can help study the transmission of disease (e.g., Covid-19) as well as reveal environmental factors that may propagate diseases.

© The Author(s), under exclusive license to Springer Nature Singapore Pte Ltd. 2023
L. Yuan et al. (Eds.): WISA 2023, LNCS 14094, pp. 112–124, 2023.
https://doi.org/10.1007/978-981-99-6222-8_10

Typically, spatial co-location pattern mining methods use the frequencies of a set of spatial features participating in a co-location pattern to measure a pattern's prevalence (known as participation index, PI for short) and require a user-specified minimum prevalence threshold, *min_prev*, to filter prevalent co-location patterns [5]. However, determining an appropriate *min_prev* is hard for users. To avoid missing interesting co-location patterns, a very low threshold are often set, resulting in a large number of prevalent co-location patterns of which only a small proportion is interesting to the user. This problem is more aggravated by the downward closure property that holds for the PI measure, whereby all of the 2^l subsets of each l-size prevalent co-location patterns are included in the result set. Consequently, a large amount of prevalent co-location patterns are often generated, however, just a few of these might meet user preferences. User preferences are frequently subjective, and a pattern preferred by one user may not be preferred by another, making objective-oriented PI measures ineffective. Therefore, it is necessary and beneficial to involve user preferences and filter user-preferred co-location patterns.

For example, in the field of environment and vegetation protection, one can mine the prevalent co-location patterns of vegetation distribution data in a region to study the impact of vegetation distribution on the environment. Similarly, an urban air quality researcher will be interested in co-location patterns that include vegetation that regulates the climate and purifies the air. However, a user concerned with soil and water conservation is unlikely to find these co-locations interesting and would prefer to reanalyze the data with their own interests in mind.

Since a user's interest pattern is highly rely on the user's knowledge and preferences, post-mining interesting patterns involving user feedback has been extensively studied in traditional frequent itemset mining. Because there are no transactions or concept- like transactions in spatial data, these mining methods can't be applied directly to co-location pattern mining. Motivated by the interactive mining methods in transactional data mining, several methods of user-preferred co-location patterns mining are proposed. Utilizing the probabilistic model, Wang et al. [6] proposed an approach to discover user-preferred patterns by iteratively involving user feedback and probabilistically quantifying user preferences for co-location patterns. It has difficulty in discovering user-preferred co-location patterns without the assistance of the user's prior knowledge. Utilizing ontology, where ontology was used to estimate the semantic similarity between two co-location patterns, Bao et al. [7] proposed an approach to identify user-preferred patterns by explicitly constructing reasonably precise background knowledge, but its construction suffers from the problem of information loss resulted from the rigid boundaries of crisp relationships, which caused the semantic similarity between two co-location patterns imprecisely.

The conceptual formalism supported by typical ontologies may not be sufficient to represent uncertain information that is commonly found in many application domains. For instance, soybean may be regarded as both food plants and oil bearing crops with different membership degrees, so it is inappropriate to treat all relationships equally as some of them may be more significant than others. To deal with this type of problem, one option is to incorporate fuzzy logic concepts into ontologies so that it can be possible to handle uncertainty on data.

According to the above description, we have made the following new contributions within this paper:

1. We introduce an innovative, interactive method for the efficient discovery of intriguing co-location patterns. This method is designed to facilitate a collaborative process with the user, leveraging their insights and preferences to optimize the pattern discovery.
2. We propose to use fuzzy ontologies to adequately express the user's background knowledge and to calculate the similarity between two co-location patterns.
3. We verified the effectiveness of our algorithm on synthetic datasets and real datasets. Compared with traditional methods, experiments show that our algorithm outperforms traditional methods in the efficient discovery of user-preferred co-location patterns.

2   Related Work

Various spatial data mining experts have focused on applying and extending efficient algorithms for spatial co-location pattern mining for applications in different areas. Shekhar et al. [1] first mentioned the concept of co-location pattern and proposed the participation index (PI) to measure the frequency of a co-location pattern. Tons of methods are based on the PI which satisfies the "downward close" property, and so by using techniques similar to Apriori, it can effectively generate complete and correct prevalent co-location patterns. Huang et al. presented the join-based algorithm [5] which generates the co-location patterns from short to long size. In order to avoid a large number of join operations in the process of generating table instances, the join-less [9] algorithms based on star neighborhood are proposed which use an instance-lookup scheme for identifying co-location patterns. Rest on the join-less framework, CPI-tree algorithm [10] were proposed to further improve the efficiency by making use of the prefix-tree structure, but they consume a lot of storage space. The proposed clique-based approach in [11] avoided identifying row-instances of co-location patterns thus making it much faster than the traditional mining method. The above methods all aimed to improve the mining efficiency or reduce the storage space rather than care about whether the pattern satisfies user preferences.

In order to reduce the number of prevalent co-location patterns, Yoo and Bow [12] proposed a set of compressed co-location patterns, closed co-location patterns, and an efficient method to discover top-k closed co-location patterns. Yoo and Bow [13] then proposed a transactional method for detecting maximum co-location patterns using an efficient filtering process. Bao et al. [14] further removed the redundant patterns from prevalent co-location patterns by utilizing the spatial distribution information of co-location instances. Existing works on condensed representations of prevalent co-locations compute a summary that globally represents the prevalent co-locations. However, such an approach solves interesting pattern discovery from a global perspective which is far from personalization that is needed to meet the pattern discovery demand of a specific user.

Some other studies related to the fuzzy theory in spatial co-location mining are reviewed as follows. Ouyang et al. [15] discussed the spatial co-location pattern mining of fuzzy objects. Lei et al. [16] considered that proximity is a fuzzy concept, fuzzy theory is introduced into co-location pattern mining, and a fuzzy spatial proximity measurement method between instances and a feature were proposed. Fang et al. [17] used the density peaks clustering algorithm to fuzzily partition instance pairs of clusters and mined co-location patterns in a fuzzy cluster approach. Later, Wang et al. [18] found that determining the cluster centers is very hard. They then improved the fuzzy cluster to mine co-location patterns. However, none of the existing work applies fuzzy theory to post-co-location mining.

There exist several interactive frameworks which target personalized co-location pattern discovery by using user feedback. By iteratively incorporating user feedback and probabilistically fine-tuning preferred patterns, Wang et al. [6] proposed an interactive probabilistic post-mining method to discover user-preferred co-location patterns. Zhang et al. [19] apply machine learning methods to interactive spatial co-location patterns mining. Utilizing ontology, where ontology was used to estimate the semantic similarity between two co-location patterns, Bao et al. [8] proposed an approach to identify user-preferred patterns by explicitly constructing reasonably precise background knowledge. Note that all the above approaches are without the user's prior knowledge or are limited in expressing the user's prior knowledge, therefore, these approaches might not be applicable in practice.

3 Preliminary

3.1 Co-Location Pattern Mining

Spatial co-location pattern mining is defined as follows: given a spatial feature type set $F = \{f_1, f_2, \ldots, f_n\}$ and a spatial instances set $O = \{o_1, o_2, \ldots, o_n\}$, where each spatial instance o_i belongs to a spatial feature and can be expressed as a tuple $<$ feature type, instance ID, location $>$. Such that feature type $\in F$, the key to discovering spatial co-location patterns is to find a set of spatial features $cp = \{f_1, f_2, \ldots, f_k\}$ whose instances tend to coexist frequently in geographically adjacent spaces. Let R be a neighbor relationship over pairwise instances. Given two instances $o_i \in O$, $o_i' \in O$, we say they have neighbor relationship if the Euclidean distance between them is no larger than a user-specified distance threshold d. Co-location pattern mining usually uses a support measure, which is called "Participation Index" (PI), to evaluate how frequently the features in a co-location pattern are closely located. Given a co-location pattern $cp = \{f_1, f_2, \ldots, f_k\}$, a participation index $PI(cp)$ is defined as $min_{f_i \in c}\{PR(cp, f_i)\}$, where $PR(cp, f_i)$ is the participation ratio of spatial feature f_i in the pattern cp. The Participation Ratio (PR), $PR(cp, f_i)$, is calculated by $PR(cp, f_i) = |N(cp, f_i)|/|N(f_i)|$, where $|N(cp, f_i)|$ denotes the number of spatial instances of feature f_i that are included in the clique instances of pattern cp (a clique instance of cp is a collection of neighboring instances whose spatial features are included in cp), and $|N(f_i)|$ denotes the number of instances of spatial feature f_i in the spatial data set.

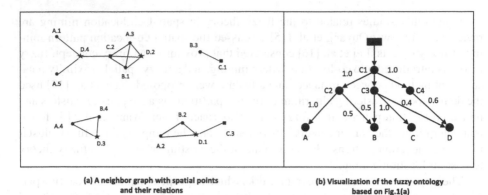

(a) A neighbor graph with spatial points   (b) Visualization of the fuzzy ontology
          and their relations   based on Fig.1(a)

Fig. 1. An example spatial dataset and a constructed fuzzy ontology

Figure 1(a) shows an example of spatial instance distribution with four features, A, B, C and D, where C.1 is represented as the first instance of C. This space contains 5 instances of A, 4 instances of B, 3 instances of C, and 4 instances of D. The connection lines indicate that two spatial instances satisfy the spatial neighbor relationship. For example, B.1 and C.1 are mutual neighbors. {A.4, B.4, D.3} is formed into a clique, and it is a row instance of a 3-size co-location pattern {A, B, D}. If $PI(\{A, B, D\}) \geq min\_prev$, {A, B, D} will be considered as a prevalent co-location pattern.

3.2 Fuzzy Ontologies in Co-Location Pattern Mining

Fuzzy ontology can be seen as extended domain ontology [20]. In this paper, we define fuzzy ontology as the quintuple $O = \{C, P, R, I, X\}$, where C is a set of (fuzzy) concepts (or classes). P is a set of concept properties. R is a set of inter-concept relations between concepts. I. is a set of instances of the concepts. X is not only a set of axioms expressed in a logical language that constrain the meaning of concepts, individuals, relationships and functions, but also a set of fuzzy rules that constrain the fuzzy concepts or fuzzy relations.

Our approach uses fuzzy membership degree in "is-a" relationships between concepts. Every feature in a spatial database is considered as a leaf-concept; generalized concepts are described as the fuzzy ontology concepts that include other ontology concepts in the fuzzy ontology.

Definition 1. A fuzzy relation R is a set of triples set of triples $\{< c_i, c_j, \mu(c_i, c_j) > 0 | c_i \in C, c_j \in C\}$. The $\mu(c_i, c_j)$ is the strength of the relationship between the two concepts. For every $c_i \in C, c_j \in C$, $\mu(c_i, c_j)$ denotes the membership degree of relation R between c_i and c_j.

Definition 2. Given a spatial feature f, $FS(f)$ demonstrates the set of generalized concepts with the degree of membership μ. The set of generalized concepts contains the leaf-concept representing f with the membership degree between f and c_i in a fuzzy ontology.

$$FS(f) = \{< c_i, \mu(f, c_i) > | c_i \in C \wedge \mu(f, c_i) > 0\}$$

For example, Fig. 1(a) shows an example spatial Point of Interest (POI) dataset containing four features (A, B, C and D) with their instances. The solid line between two instances indicates they are neighbors Fig. 1(b) illustrates an example fuzzy ontology constructed over features of a spatial database of Fig. 1(a). From this fuzzy ontology, the generation of B is $FS(B) = \{<C2, 0.5 >, < C3, 0.5 >\}$.

Definition 3. Given a co-location $cp = \{f_1, f_2 \ldots f_n\}(n \geq 1)$, the generalization of cp is defined as

$$FG(cp) = \bigcup_{i=1}^{n} FS(f_i)$$

For example, in Fig. 1(b), given a co-location pattern $cp = \{B, D\}$, the generation of cp is $FG(\{B, D\}) = FS(B) \cup FS(D) = \{< C2, 0.5 >, < C3, 0.5 >\} \cup \{< C3, 0.4 >, < C4, 0.6 >\} = \{< C2, 0.5 >, < C3, 0.5 >, < C4, 0.6 >\}$.

4 Algorithm

The process is similar to the previous works of interactively discovering user-preferred co-location patterns, and the framework of IMCPFO nteractively Mining Co-location Patterns by Using Fuzzy Ontologies) is shown in Fig. 2. Fuzzy ontology is first constructed by domain experts or reused from existing ones since there exist a number of ontologies available in Semantic Web. The IMCPFO takes a set of candidates as the input. In each interaction, several patterns (e.g., 3–10) which selected by sample algorithm are presented to the user as sample co-location patterns. The user provides his/her feedback by indicating whether he/she likes the co-location pattern or not, and the feedback information is used by a fuzzy-ontology-based model to update the candidates, i.e., the model first discovers all similarities of each sample co-location pattern with every candidate and then moves the selected samples and their similarity patterns to the output. Meanwhile, it also removes uninteresting samples from the candidates. The interaction process continues for several rounds until candidate set is empty.

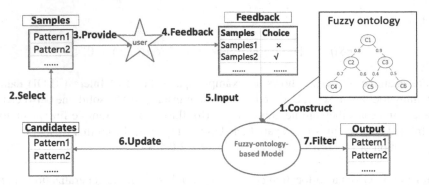

Fig. 2. The framework of interactive post-mining of co-location pattern

4.1 Similarity Between Co-Location Patterns

The fuzzy Jaccard similarity measure [21] is defined as a fuzzy extension of the Jaccard index between two crisp sets by replacing set cardinality with fuzzy set cardinality. Fuzzy ontologies gather similar features and create a generalized concept to include this feature, thus we define similarity between co-location patterns based on fuzzy ontologies. Classical sets are generalized to the fuzzy case. Given two co-location patterns c_1 and c_2, their similarity is calculated as

$$Sim(c_1, c_2) = \frac{|FG(c_1) \cap FG(c_2)|}{|FG(c_1) \cup FG(c_2)|}$$

For example, let $c_1 = \{A, B, C\}, c_2 = \{B, D, E\}$, $FG(c_1) = \{<C2, 0.5>, <C3, 0.5>, <C4, 0.6>\}$, $FG(c_2) = \{<C1, 0.7> <C2, 0.3>, <C3, 0.6>, <C5, 0.8>\}$ $Sim(c_1, c_2) = \frac{|FG(c1) \cap FG(c2)|}{|FG(c1) \cup FG(c2)|} = \frac{|\{<C2,0.3>,<C3,0.5>\}|}{|\{<C1,0.7>,<C2,0.5>,<C3,0.6>,<C4,0.6>,<C5,0.8>\}|} = 1/4 = 0.25$

4.2 Selecting strategy in IMCPFO

In order to select the set of sample co-location patterns efficiently. There should be no redundancy between the selected sample co-location patterns. Redundancy exists between two co-location patterns if there is some similarity in their pattern composition (e.g., spatial features). Because the redundant co-location patterns are similar, providing the user with redundant co-location patterns increases the number of interactions and thus the burden on the user. In the algorithm, we provided that the user-defined set of sample co-location patterns is small in size k.

In IMCPFO, the selecting strategy is to first find the co-location pattern that contains the most information from the candidates. The co-location pattern with the maximum semantic distance from the previous sample pattern is chosen as the second, and so on.

Algorithm 1 shows the pseudo code of the selection process. The candidate co-location pattern having the maximal information is selected as the first sample co-location-pattern (line 2). The following sample co-location patterns will be chosen (lines 5–8): The candidate co-location pattern having maximal semantic distance with the previously chosen sample pattern is chosen as the following sample co-location. The computational complexity of Algorithm 2 is about $O(km)$, where m is the number of co-locations in the initial candidates, and k is the number of co-location patterns in samples.

Algorithm 1 selecting sample patterns

Input:
    PC: A set of prevalent/closed co-location patterns.
    FR: Fuzzy relationship on fuzzy ontology concepts.
    k: The number of sample co-location patterns.
Output:
    SP: A set of sample patterns.
Method:
  1. **Begin**
  2. Pattern p = get_maximal_information (PC, FR)
  3. SP.add(p)
  4. $k=k-1$
  5. **While** $k\, !=0$:
  6.   patterns pd = get_most_different_pattern (PC, FR, p)
  7.   SP.add(pd)
  8. $k=k-1$
  9.   **Return** SP
  10. **End**

4.3 Updating Candidates Using Fuzzy Ontology

The update algorithm is very simple and its pseudo-code is given in Algorithm 2. IMCPFO iterates over each sample co-location pattern sets that is of interest to users and compares the similarity between selected sample co-location and each candidate patterns. If there exists a candidate co-location pattern similar to selected sample co-location, then the pattern is removed from the candidate set (step 2–6) then this pattern is of interest to the user and moves to the DP. Similarly, the same progress will be made for sample co-location pattern sets that are not of interest to users to filter patterns that the user not preferred. The time complexity of the direct update algorithm is $O(km)$, where m is the size of the candidate co-location pattern set.

Algorithm 2 Interactively Mining Co-Location Patterns by Using Fuzzy Ontologies (IMCPFO)

Input:
    PC: A set of prevalent / closed co-location patterns.
    SSP: Sample co-location pattern sets that are of interest to users
    USP: Sample co-location pattern sets that are not of interest to users
    min_simi: Semantic distance threshold
    FR: Fuzzy relationship on fuzzy ontology concepts.
    k: The number of sample co-location patterns.
Output:
    DP: A set of patterns of interest to users.
Method:
  1. **Begin**
  2. **For Each** *ptn* **in** SSP:
  3.   *PC*.remove(*ptn*)
  4.   *DP*.add(*ptn*)
  5.   **For Each** *comptn* **in** PC:
  6.   **If** GetSimi(*ptn, comptn*) > mini_simi **Then** *DP*.add(*comptn*) **and** *PC*.remove(*comptn*)
  7. **For Each** *ptn* **in** USP:
  8.   *PC*.remove(*ptn*)
  9.   **For Each** *comptn* **in** PC:
  10.   **If** GetSimi(*ptn,comptn*) < *mini_simi* **Then** *PC*.remove(*comptn*)
  11. **End**

---

5 Experiments

In this section, we conducted comprehensive experiments to evaluate the proposed approach from multiple perspectives on both real and synthetic data sets. All the algorithms were implemented in Python. All the experiments are performed on a Windows 10 system with a 3.50 GHz CPU and 12.0GB memory.

5.1 Experimental Analysis on the Real Data Set

We validated the performance of IMCPFO on a real Beijing POI (Points of Interests) dataset, which contains 14 spatial features related to tourism. The fuzzy ontology of this dataset was given in Fig. 3. The actual dataset contains 90458 spatial instances with a spatial extent of 324 square kilometers (18*18), and generates a total of 1579 co-location patterns with the distance threshold of 150 m and the minimum participation threshold of 0.3.We empirically evaluate the accuracy of the algorithm to discover user-preferred co-location patterns. The accuracy measure is defined as Accuracy = (H∩M)/(H∪M), where H is the patterns that are preferred by the users, M is expressed as the final result set output by IMCPFO.

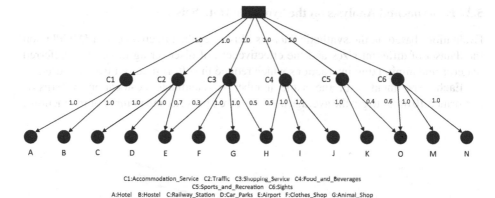

C1:Accommodation_Service C2.Traffic C3.Shopping_Service C4:Food_and_Beverages
C5:Sports_and_Recreation C6:Sights
A:Hotel B:Hostel C:Railway_Station D:Car_Parks E:Airport F:Clothes_Shop G:Animal_Shop
H:Bubble_tea_shop I:Western_Restaurant J:Chinese_Restaurant K:Gymnasium
O:Park M:National_Scenic_spots N:Provincial_Scenic_spots

Fig. 3. Fuzzy onotologies constructed using real datasets

As Fig. 4 shows, we examine the accuracy of the IMCPFO with two algorithms. The first algorithm [6], PICM (Probability-based Interesting Co-location Miner), utilized a simple yet effective probabilistic model. The second algorithm [8], OICM (Ontology-based Interesting Co-location Miner), utilized an ontology-based model. It shows that a larger k causes a higher accuracy because more samples can be fed to the user per iteration, it can be concluded that with k value increases, the accuracy of both algorithms first increases and then has a drop, and this is because a shorter k cannot supply enough information for an effective update within certain rounds while a longer k has more probability to occur the conflicting choice. That is, a particular candidate co-location pattern c with sample co-location pattern c_1 and c_2 are similar, but the user is interested in c_1 and not in c_2. The accuracies estimated with closed co-location patterns are better than those with prevalent co-location patterns because closed co-location patterns are a form of compression of prevalent co-location patterns, which can help them effectively discover preferred patterns. The accuracy of IMCPFO is better than PICM and OICM because IMCPFO uses fuzzy ontology that is more representative of user's background knowledge than conventional ontology.

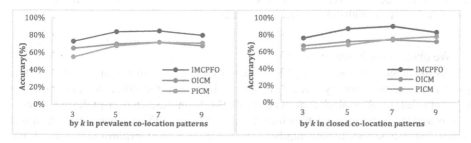

Fig. 4. Comparisons of Accuracy by k value

5.2　Experimental Analysis on the Synthetic Data Sets

Evaluations based on the synthetic datasets illustrated the effectiveness of IMCPFO on the dataset of different sizes and the effectiveness of discovering more user-preferred patterns and minimizing interactions which reduce the effort required from the user.

Each experiment with the same number of candidate co-location patterns is performed 10 times to get the average user-preferred patterns and number of interactions.

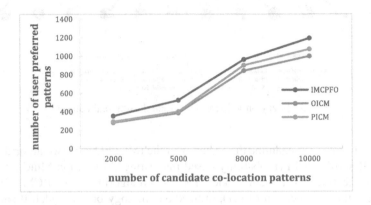

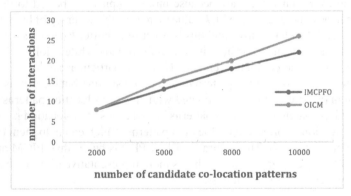

Fig. 5. (a) Evaluations on synthetic data sets with different numbers of patterns. (b) Evaluations on synthetic data sets with different number of patterns.

We observe the following results:

As the number of candidate patterns increases, Fig. 5(a) shows the number of user preferred patterns found by three methods. As it is seen, the IMCPFO finds the most number of all the other algorithms while in PICM the number of user-preferred patterns discovered by the three algorithms increases, but the IMCPFO discovers more than two others. Fig. 5(b) shows that the number of iterations in both algorithms increases. While IMCPFO is less than OICM. This is because fuzzy ontology captures richer semantics than traditional domain knowledge representations, therefore, more user-preferred co-location patterns can be found in each round of interaction.

6 Conclusion

In existing spatial co-location mining approaches, the algorithm or the measure is pre-designed and they generate a great amount of co-location patterns that are most out of the user's interest. In this paper, we proposed an interactive approach to detect user preference co-location patterns using the user's domain knowledge, represented by a fuzzy domain ontology. We use fuzzy ontology to measure the semantic similarity between two co-location patterns and develop an efficient strategy to select patterns for feedback. Then we design a reasonable filter to effectively discover user-preferred co-location patterns.

In future work, we want to apply fuzzy ontology to the co-location mining process in order to obtain fewer and clearer co-location patterns. Moreover, it would be very interesting to have other ways to express user's background knowledge.

Acknowledgement. The work is supported by National Natural Science Foundation of China(No.62066010, No.62006057).

References

1. Shekhar, S., Huang, Y.: Discovering spatial co-location patterns: a summary of results. In: International Symposium on Spatial and Temporal Databases (2001)
2. Li, J., Adilmagambetov, A., Zaiane, O.R., Osornio-Vargas, Á.R., Wine, O.: On discovering co-location patterns in datasets: a case study of pollutants and child cancers. GeoInformatica **20**, 651–692 (2014)
3. Masrur, A., Thakur, G.S., Sparks, K.A., Palumbo, R., Peuquet, D.: Co-location pattern mining of geosocial data to characterize urban functional spaces. IEEE Int. Conf. Big Data (Big Data) **2019**, 4099–4102 (2019)
4. Yao, X., Zhang, Z., Cui, R., Zhao, Y.: Traffic prediction based on multi-graph spatio-temporal convolutional network. In: Web Information System and Application Conference (2021)
5. Huang, Y., Shekhar, S., Xiong, H.: Discovering colocation patterns from spatial data sets: a general approach. IEEE Trans. Knowl. Data Eng. **16**(12), 1472–1485 (2004)
6. Wang, L., Bao, X., Cao, L.: Interactive probabilistic post-mining of user-preferred spatial co-location patterns. In: 2018 IEEE 34th International Conference on Data Engineering (ICDE), pp. 1256–1259 (2018)
7. Bao, X., Gu, T., Chang, L., Xu, Z., Li, L.: Knowledge-based interactive postmining of user-preferred co-location patterns using ontologies. IEEE Trans. Cybern. **52**, 9467–9480 (2022)
8. Bao, X., Wang, L.: Discovering interesting co-location patterns interactively using ontologies. DASFAA Workshops (2017)
9. Yoo, J.S., Shekhar, S., Celik, M.: A join-less approach for co-location pattern mining: a summary of results. In: Fifth IEEE International Conference on Data Mining (ICDM'05), pp. 4 (2005)
10. Wang, L., Bao, Y., Lu, Z.: Efficient discovery of spatial co-location patterns using the iCPI-tree. Open Inform. Syst. J. **3**(1), 69–80 (2009)
11. Bao, X., Wang, L.: A clique-based approach for co-location pattern mining. Inf. Sci. **490**, 244–264 (2019)
12. Yoo, J.S., Bow, M.: Mining top-k closed co-location patterns. In: Proceedings 2011 IEEE International Conference on Spatial Data Mining and Geographical Knowledge Services, pp. 100–105 (2011)

13. Yoo, J.S., Bow, M.: Mining maximal co-located event sets. In: Pacific-Asia Conference on Knowledge Discovery and Data Mining (2011)
14. Bao, X., Lu, J., Gu, T., Chang, L., Wang, L.: NRCP-Miner: towards the discovery of non-redundant co-location patterns. DASFAA (2021)
15. Ouyang, Z., Wang, L., Wu, P.: Spatial co-location pattern discovery from fuzzy objects. Int. J. Artif. Intell. Tools **26**, 1750003 (2017)
16. Lei, L., Wang, L., Wang, X.: Mining spatial co-location patterns by the fuzzy technology. IEEE Int. Conf. Big Knowl. (ICBK) **2019**, 129–136 (2019)
17. Fang, Y., Wang, L., Hu, T.: Spatial co-location pattern mining based on density peaks clustering and fuzzy theory. APWeb/WAIM (2018)
18. Wang, X., Lei, L., Wang, L., Yang, P., Chen, H.: Spatial colocation pattern discovery incorporating fuzzy theory. IEEE Trans. Fuzzy Syst. **30**(6), 2055–2072 (2021)
19. Zhang, Y., Bao, X., Chang, L., Gu, T.: Interactive mining of user-preferred co-location patterns based on SVM. In: International Conference on Intelligent Information Processing, pp. 89–100. Springer, Cham (2022)
20. Ying, W., Ru-bo, Z., Ji-bao, L.: Measuring Concept Similarity between Fuzzy Ontologies (2009)
21. Jaccard, P.: The distribution of the flora in the alpine zone. New Phytol. **11**, 37–50 (1912)

Representation Learning of Multi-layer Living Circle Structure

Haiguang Wang, Junling Liu$^{(\boxtimes)}$, Cheng Peng, and Huanliang Sun

Shenyang Jianzhu University, Shenyang 110168, China
liujl@sjzu.edu.cn

Abstract. Living circle structure reflects the relationship between community residents and surrounding living facilities. Through the study of urban community structure, we can understand the distribution of facilities in the living circle. Most of the existing related work studies the structure of urban communities from the perspective of macro and large-scale, lacking of research on fine-grained and small-scale community. This paper defines the structure of the multi-layer living circle and uses the representation learning method to obtain the structural characteristics of the living circle through the activities of the residents in the surrounding POIs. First, a representation framework of multi-layer living circle structure is proposed. Second, the autoencoder representation learning is used to construct the dynamic activity graphs of the multi-layer living circle and the vector representation of the potential characteristics of the living circle effectively summarizes the multi-layer living circle structure. Finally, an experimental evaluation of the proposed multi-layer living circle structure uses real datasets to verify the validity of the proposed methods in terms of community convenience and community similarity applications.

Keywords: Urban computing · Representation learning · Living circle activity graph · Multi-layer living circle

1 Introduction

Learning the urban community structure can find the relationship between the living people in the living circle and the surrounding facilities, understand the daily life needs of residents and thus create a more livable environment for urban residents [1]. The existing research on urban community structure can be divided into two categories, one is to explore the urban spatial configuration by analyzing the static urban POI distribution [2]. The other is to identify the functions of urban areas and explore the differences between urban areas by analyzing the pattern of urban residents' activities [3]. Most of the work related to urban communities studies focus on the structure of a big living circle rather than a small one.

At present, China takes the quality improvement of facilities in a small living circle of the existing residential area as an important goal of planning and

L. Yuan et al. (Eds.): WISA 2023, LNCS 14094, pp. 125–136, 2023.
https://doi.org/10.1007/978-981-99-6222-8_11

construction [4]. In 2018, the concepts of 5-min, 10-min, 15-min living circle and residential neighborhood are proposed to replace the previous living concepts of residential area and community in the "Urban Residential Planning and Design Standards". These three living circle cover the needs of people in different ranges. Living circle should be a supporting service for residents' daily life and the planning of living circle should take into account the accessibility of important POIs. Therefore, the necessary POIs and their importance in a living circle are the criteria for measuring the quality of the living circle. The study in different ranges of living circle can quantitatively analyze the structure of residents' living circle under different granularity and learn the distribution of residents' living circle facilities.

This paper studies the representation method of multi-layer living circle structure. Through the activity of residents in the multi-layer living circle in the surrounding POIs, the representation learning technology is used to discover the living circle structure.

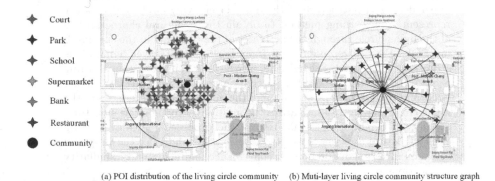

(a) POI distribution of the living circle community   (b) Muti-layer living circle community structure graph

Fig. 1. Examples of different forms of living circle structures

Figure 1 shows two examples of living circle structure. Figure 1(a) shows the distribution of POIs near a living circle in downtown Beijing, with a radius of 1km. The center is the residential area and different types of POIs are distributed around the residential area, such as supermarkets, schools, restaurants and other life services. Figure 1(b) shows the structure of multi-layer living circle and the center is the residential area. Different POIs are distributed in different areas and connected with the center by lines. In addition, different types of POIs in different areas have different levels of importance.

It is challenging to effectively represent multi-layer living circle structure, which mainly includes in the following aspects: Firstly, multi-layer living circle contains a lot of POIs. The selection of appropriate POIs to represent the living circles of different sizes is a challenge. What's more, in the living circle with different sizes, the importance of different types of POIs is different and it is a problem to assign appropriate weights to POIs in living circles with different sizes. Finally, it is a challenge to construct a multi-layer living circle activity graph and embed POIs in different living circles.

In order to solve the above problems, this paper uses different datasets to construct the residents' activity graphs for different living circle. For example, the bicycle trajectory data is used to construct the living circle activity graphs for short travel distance. We assign the weights of POIs according to the visiting frequency of residents, the distance between different living circles and the domain knowledge. In order to construct the multi-layer living circle activity graph, this paper analyzes the living circles of different ranges and the activity pattern of residents in different periods and obtains relatively stable activity patterns to generate dynamic activity graphs. After the construction of activity graphs in different ranges, we design a representation learning framework based on the autoencoder for POI embedding.

The main research work in this paper includes: (1) We define the structure of multi-layer living circle, which is used to describe the necessity and importance of different types of POIs around residential areas. (2) A representation framework of multi-layer living circle structure is proposed for constructing activity graphs of living circles in different ranges and the autoencoder representation learning method is used to embed POIs. (3) The experimental evaluation on real datasets verifies the effectiveness of the proposed method.

2   Problem Definition

This section gives the relevant definitions and formalizes the proposed problems.

Definition 1 Living circle structure. Given a living circle c_k, its structure graph is a star graph denoted by $G^{(k)}(D, E)$, where D is the set of POIs around the living circle c_k, E is the set of relationships between the living circle c_k and D, and any relationship $e(c_k, p) \in E$ denotes p as a necessary POI of the living circle c_k.

Definition 2 Multi-layer living circle structure. The areas $D^{(k)} = \{d_1^{(k)}, d_2^{(k)}, ..., d_n^{(k)}\}$ composed of different living circles are centered on the residential area, where $d_1^{(k)}, d_2^{(k)}, ..., d_n^{(k)}$ are the living circles of different ranges and the relationship among different communities is $d_1^{(k)} < d_2^{(k)} < d_3^{(k)}$.

Definition 3 Living circle graph. Given a living circle, the structure of the living circle is notated as $G^{(k)} = \{G_1^{(k)}, G_2^{(k)}, ..., G_n^{(k)}\}$, where $G_1^{(k)}, G_2^{(k)}, ..., G_n^{(k)}$ are the living circle structure graphs of different scope sizes and $G_1^{(k)}, G_2^{(k)}, ..., G_n^{(k)}$ are formed by the connection between POI p and the living circle c_k within the range of the areas.

Definition 4 Representation learning of multi-layer living circle structure. Given a collection of activity graphs for a set of multi-layer living circles $G_i^{(k)} = \{\{G_{1,1}^{(k)}, G_{1,2}^{(k)}, ..., G_{1,t}^{(k)}, ..., G_{1,n}^{(k)}\}, \{G_{2,1}^{(k)}, G_{2,2}^{(k)}..., G_{2,t}^{(k)}, ..., G_{2,n}^{(k)}\}, \{G_{3,1}^{(k)},$

$G_{3,2}^{(k)}, ..., G_{3,t}^{(k)}, ..., G_{3,n}^{(k)}\}\}$ learns a spatial mapping function: $f(c^{(k)}) : G^{(k)} \rightarrow R^d$, which is used to represent the activity graphs of the living circle c_k as a vector R^d.

This paper will study the multi-layer living circle structure, pay more attention to the importance and distribution of POIs in different scopes and construct the star-type activity graphs between the multi-layer living circle and POIs in different scopes for representation learning.

3   Related Work

3.1   Urban Structure

With the maturity of perception technology and computing environment, urban computing based on urban big data has attracted extensive attention. Urban computing forms a circular process of urban perception, data management, data analysis and service and continuously improves the daily life of urban residents, urban management, operation system and urban environment in an unobtrusive way [1].

In terms of urban community structure, Wang et al. [5] used the skip-gram model to learn the vector representation of urban community structure and realize visualization from urban POI data and a large number of taxi trajectory data. Regarding the identification of cities by urban functional areas, Cai et al. [6] used GPS trajectory data and check-in data to construct a travel pattern graph with spatio-temporal attributes, identified urban functional areas and constructed a detection method combining check-in data and known POI semantic information, which solved the problem of accurate semantic recognition of functional areas to a certain extent. Song et al. [7] provided information for design governance in urban renewal through case studies based on POI data and provided suggestions for governance and comprehensive urban redevelopment planning.

Different from the above research on large-scale community, this paper studies the small living circle structure of residents and constructs the living circle and POI star graphs by representation learning.

3.2   Representation Learning

Representation learning is a set of techniques that transform raw data information into a set of techniques that can be utilized and developed by machine learning, aiming to extract effective low-dimensional features from complex high-dimensional data. According to the development process, it can be divided into word vector representation [2,8], graph representation learning [9], and spatio-temporal graph representation learning [10]. The initial application of representation learning is in the field of natural language processing and computer vision, the most representative is the word2vec model, word2vec uses a layer of neural network to map the one-hot form of word vectors to the distributed form of word vectors [11]. Graph representation learning, also known as graph embedding or network graph embedding, aims to learn a low-dimensional vector to represent vertex information or graph [12]. Graph representation learning algorithms can be divided into probabilistic models, manifold learning methods and reconstruction-based algorithms [13,14]. Spatio-temporal representation learning is the development of graph representation learning in spatio-temporal environment and the time attribute is added to graph representation learning [10].

The living circle structure representation proposed in this paper is a spatio-temporal representation learning, which focuses on the necessity and importance of POIs around residential regions and learns the living circle structure through the activities of residents in different periods of time and different types of POIs.

4   Living Circle Structure Representation Method

This section introduces the representation method of multi-layer living circle structure. Firstly, the proposed framework structure is outlined and then the two key steps of the construction and embedding representation of the multi-layer living circle activity graph are explained in details.

4.1   The Multi-layer Living Circle Community Structure Representation Framework

This section introduces a multi-layer structure graph representation learning framework for living circles, which can capture the dynamic changes of living circle structure generated by residents' activities in different ranges of living areas.

Figure 2 shows the framework structure proposed in this section, which consists of two parts. The first part is the construction of activity graphs by POIs and activity data of residents, where $d_1^{(k)}$, $d_2^{(k)}$, $d_3^{(k)}$ are respectively three different living circles. Their activity graphs $G_1, G_2, ..., G_n$ are constructed by the data of POIs and trajectories of the residents. The second part is the embedding representation of the activity graphs, where the activity graphs are transformed into the structure vectors by an autoencoder.

4.2   Living Circle Activity Graph Construction

The multi-layer living circle activity graph defined in this section is used to reflect the activity regularity of residents and it is necessary to determine which POIs are necessary in the living circle and distinguish the importance of these POIs. At the same time, it is necessary to discover the periodic activity regularity of residents.

Establishment of the Relationship Between Living Circle and POIs. The establishment of the relationship between living circles and POIs requires estimating the likelihood of residents arriving at a POI in bike-sharing trajectory

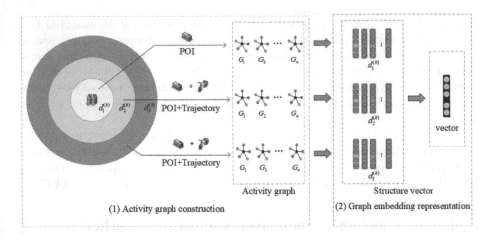

Fig. 2. Representation learning framework for multi-layer living circle

data and quantifying the mobile connectivity between residential areas and POIs. The drop-off point of residents cycling bicycles is often near the POI point, which cannot coincide with the POI point. Meanwhile, residents tend to visit POIs that are close to the drop-off point. Therefore, the method based on probability propagation [15] is used to estimate the probability of residents arriving at POIs, as shown in Eq. 1.

$$P(x) - \frac{\beta_1}{\beta_2} * x * exp(1 - \frac{x}{\beta_1}) \tag{1}$$

In Eq. 1, P is the probability that a user visits a POI and x is the distance between the destination d and the reached POI point p, β_1 and β_2 are two given hyperparameters. β_1 is set as 0.8 which maximizes the access probability, β_2 is set as 10 which maximizes the walking distance from the endpoint to the target POI, and when the distance exceeds the threshold β_2, the probability decreases with exponential long tail. In this paper, only trajectory data from residential areas to POIs and POIs to residential areas are selected to reflect the interaction between residential areas and POIs.

Living Circle Activity Graph Time Division. In order to understand the residents' activity regularity, it is necessary to divide the living circle activity graphs into time periods to obtain relatively stable activity patterns. The data from Monday to Sunday are used to generate 7 dynamic graphs [3]. In this paper, more fine-grained time segments were obtained and different periods of each day were divided. We use the topic model method [10] to divide the segments, where the window is divided by the KL-divergence measure through the sliding window, as shown in Eq. 2.

$$S(\theta_1||\theta_2) = \sum_{i=1}^{|v|} p(s_i||\theta_2) \log \frac{p(s_i||\theta_2)}{p(s_i||\theta_1)} \tag{2}$$

In this paper, θ represents the representative topic in the time segment, which corresponds to the POI category with the first intensity ranking in the time segment, s_i represents a word in the time segment, which corresponds to different POI categories.

The specific process is as follows: firstly, the night period with no trajectory and sparse trajectory data is removed and 1 h is taken as the most fine-grained time segment, then the number of times a certain type of POI is reached by the trajectory in each hour is counted. Next, the intensity of each type of POI in each segment was calculated and the one with the highest intensity was selected as the theme θ. KL-divergence is used to measure the distance between adjacent time segments. The adjacent time segments whose distance is less than a certain threshold are merged to complete the partition of time segments.

POI Weights Generation. In order to distinguish the influence of different POIs on the multi-layer living circle, it is necessary to assign weights to POIs [16]. The larger the weight is, the more important POIs are to the community. In [17], Fu et al. summarize that there are 8 categories of POI related to the convenience of residents and give the importance of each type of POI. In order to reflect the importance degree of different categories of POIs, this section draws on the weight assignment to POI categories. For the sake of generality, the weights are defined uniformly, shopping 0.3, education and school 0.1, catering 0.2, public transportation 0.2, medical 0.06, life services 0.06, sports facilities 0.02 and entertainment 0.02. We use this method to assign the weights of POIs according to living circles in different ranges. For the living circles, three factors are considered in this section, including the frequency of residents' visits, the distance between POI and residential areas and the category, as shown in Eq. 3.

$$w_j^{(k)} = \frac{m_j - m_{min}}{m_{max} - m_{min}} + \frac{dis(p,r)_i - dis_{min}}{dis_{max} - dis_{min}} + w_s^{(k)} \qquad (3)$$

In Eq. 3, $w_j^{(k)}$ is the weight of POI in the living circle, m_j is the number of times residents arrive at POI point p, m_{max} is the maximum value, m_{min} is the minimum value, $dis(p,r)_i$ represents the distance between POI point p and residential area r. dis_{max} and dis_{min} are respectively the maximum distance and minimum distance between POI p and r within the scope of the living circle. $w_s^{(k)}$ is the weight given by the category of the necessary POIs in the living circle.

After obtaining the weights of POIs, a multi-layer living circle activity graph can be constructed. The construction method is applied to the data of different periods to obtain the set of the multi-layer living circle activity graphs $G^{(k)}$.

4.3 Representation of Multi-layer Living Circle Structure

Applying the above construction method, we can obtain the activity graph $G^{(k)}$ of multiple periods, and then the activity graph can be embedded to find the structure formed by the daily activities of residents in different multi-layer living circle ranges. The vectorized multi-layer living circle structure $g_{i,t}^{(k)}$ with different scope sizes contains the POI categories required by different scopes. The connectivity values of different categories of POIs in $g_{i,t}^{(k)}$ are transformed into vectors as the input of the autoencoder respectively.

In this section, the input of the embedded representation is divided into different categories of POIs which contain the vectorized 5-min living circle structure graph $g_{5,t}^{(k)}$, the vectorized 10-min living circle structure graph $g_{10,t}^{(k)}$ and the vectorized 15-min living circle structure graph $g_{15,t}^{(k)}$. The vectorized living circle structure graph $g_{i,t}^{(k)}$ contains the necessary POI types.

In order to make the living circles in different ranges have their unique representation results and characteristics and find the differences of the characteristics in different ranges, the structure graph $g_{i,t}{}^{(k)}$ are divided into vectorization s-minute $g_{s,t}{}^{(k)}$ in different types of POI $p_{s,t}{}^{(k)}$ respectively as input of the encoder, where $\{p_{s-1,t}^{(k)}, p_{s-2,t}^{(k)}, ..., p_{s-n,t}^{(k)}\}$ is the POI category contained in the s-minute living circle. $\{y_{s-1,t}^{(k),1}, y_{s-2,t}^{(k),2}, ..., y_{s-n,t}^{(k),o}\}$ are respectively the latent feature representations in the hidden layer $1, 2, ..., o$ in the encoding step. s-minute living circle structure POI features after representation $\{z_{s-1,t}^{(k)}, z_{s-2,t}^{(k)}, ..., z_{s-n,t}^{(k)}\}$ will form the representation vector of the multi-layer living circle structure $z_t^{(k)}$. Formally, the relationship among these vectors is given by Eq. 4.

$$
\begin{cases}
y_{s-n,t}^{(k),1} = \sigma(w_{s-n,t}^{(k),1} p_{s-n,t}^{(k),1} + b_{s-n,t}^{(k),1}), \forall t \in 1, 2, ..., n, \\
y_{s-n,t}^{(k),r} = \sigma(w_{s-n,t}^{(k),r} y_{s-n,t}^{(k),r-1} + b_{s-n,t}^{(k),r}), \forall r \in 2, 3, ..., m, \\
z_{s-n,t}^{(k)} = \sigma(w_{s-n,t}^{(k),m+1} y_{s-n,t}^{(k),m} + b_{s-n,t}^{(k),m+1})
\end{cases}
\tag{4}
$$

In decoding steps, the input $\{z_{s-1,t}^{(k)}, z_{s-2,t}^{(k)}, ..., z_{s-n,t}^{(k)}\}$ is the structure characteristics of multi-layer living circle, the final output $\{\hat{p}_{s-1,t}^{(k)}, \hat{p}_{s-2,t}^{(k)}, ..., \hat{p}_{s-n,t}^{(k)}\}$ is reconstructed vector. Different living circle of the scope of the hidden layer of potential eigenvectors can be expressed as $\{y_{s-1,t}^{(k),1}, y_{s-2,t}^{(k),2}, ..., y_{s-n,t}^{(k),o}\}$. The relationship among these vector variables is shown in the formula, where s represents the value of the multi-layer living circle time period, w_s and b_s are the weight and bias term, as shown in Eq. 5.

$$
\begin{cases}
\hat{y}_{s-n,t}^{(k),m} = \sigma(\hat{w}_{s-n,t}^{(k),m+1} z_{s-n,t}^{(k)} + \hat{b}_{s-n,t}^{(k),m+1}), \forall t \in 1, 2, ..., n, \\
\hat{y}_{s-n,t}^{(k),r} = \sigma(\hat{w}_{s-n,t}^{(k),r} \hat{y}_{s-n,t}^{(k),r-1} + \hat{b}_{s-n,t}^{(k),r}), \forall r \in 2, 3, ..., m, \\
\hat{p}_{s-n,t}^{(k)} = \sigma(\hat{w}_{s-n,t}^{(k),m+1} \hat{y}_{s,t}^{(k),m} + \hat{b}_{s-n,t}^{(k),1})
\end{cases}
\tag{5}
$$

The autoencoder minimizes the loss of the original characteristics of $\{p_{s-1,t}^{(k)}, p_{s-2,t}^{(k)}, ..., p_{s-n,t}^{(k)}\}$ and reconstructs of feature vector $\{\hat{p}_{s-1,t}^{(k)}, \hat{p}_{s-2,t}^{(k)}, ..., \hat{p}_{s-n,t}^{(k)}\}$. s represents the value of the multi-layer living circle time period, as shown in Eq. 6.

$$
L = \sum_{t \in \{1,2,3,...,n\}} \sum_{s \in \{1,2,...,n\}} \|p_{s-n,t}^{(k)} - \hat{p}_{s-n,t}^{(k)}\|^2
\tag{6}
$$

After representation learning, we have obtained the graph $g_{i,t}^{(k)}$ with multiple time segments and then merged into a living circle activity graph $\hat{g}_{i,t}^{(k)}$. In the input stage, different types of POIs in different ranges are represented separately to the graphs with corresponding living circle ranges and feature representation. After obtaining living circle ranges and feature representation, the POIs characteristics of living circles in different ranges can be compared.

5   Experimental Analysis

5.1   Dataset and Experimental Parameters

We used three real datasets to evaluate the proposed methods. The first dataset is obtained from www.soufun.com, which contains more than 13,000 residents' living circles. The second dataset is downloaded by Tencent Map API interface, including more than 1.8 million POI objects of 19 categories. The third dataset is trajectory data, including Baidu map path query trajectory, taxi trajectory and shared bicycle trajectory data. Among them, the Baidu map path query trajectory data is from the Baidu traffic dataset and the shared bicycle trajectory data is from the data competition community (www.biendata.xyz).

In the process of encoding and decoding, the number of layers for both encoding and decoding of text is set to 3, the number of training rounds of autoencoder is epochs 20 and the number of samples in each training batch-size is 64.

5.2   Convenience Evaluation of Multi-layer Living Circles

This section uses the living circle convenience metric to represent the multi-layer living circle and the 15-min living circle and generate the representation vector. The convenience of living circle refers to the satisfaction and accessibility of residents from living circle to surrounding facilities [18]. We adopt the labeling result of literature [18] as the convenience level label of residential area. The living circle is represented by using the existing representation method and the representation method proposed in this paper to generate the representation vector. Four existing learning ranking algorithms are used to predict the convenience of living circles, and the representation method is evaluated by the prediction results. These methods are as follows: (1) 5-min Living Circle Latent Feature (5-LCLF): In the 5-min living circle, the POI categories and the weights of each category are calculated and the calculated POI weights are embedded. (2) Living Circle Daily Latent Feature (LCDLF): Within the range of 10-min and 15-min living circle, 24 h are divided into multiple periods through a sliding window regardless of the category of POIs. The mobile connectivities among houses and POIs in each community are calculated and embedded. (3) Multi-layer Living Circle Daily Latent Feature (MLCDLF): The 5-min, 10-min and 15-min living circles are represented separately and the resulting methods are learned and combined.

Method 5-LCLF, 10-LCDLF and 15-LCDLF are single-layer living circle representation methods, which are similar with the method in [3]. We consider these three methods as baselines and MLCDLF is the proposed method.

In this section, five learning ranking methods [19] are selected to compare different representations and verify the performance of the proposed representation. The specific learning sorting algorithm includes: (1) Multiple Additive Regression Trees (MART): Enhanced tree models that perform gradient descent in the function space using regression trees. (2) RankBoost(RB): For the enhanced pairwise ranking method, multiple weak rankers are trained and their outputs are combined into the final ranking. (3) ListNet(LN): As a list ranking model, the likelihood of transforming top-k ranking is used as the objective function and neural network and gradient descent are used as the model and algorithm. (4) RankNet(RN): Neural networks are used to model the underlying probabilistic cost function; (5) LambdaMART(LM): A version of the boosted tree based on RankNet improved by LambdaRank, combining MART and LambdaRank.

The learning ranking algorithm is used to compare different representation methods and verify the performance of the proposed multi-layer living circle representation method. We combine the above 4 representation methods and 5 learning sorting algorithms to evaluate the performance. In MART and LambdaMART, the parameter trees are set to 1, the number of leaves is set to 10, the threshold candidates are set to 256, and the learning rate is set to 0.1. In RankBoost, the number of parameter iterations is set to 300, and the threshold candidates are set to 10. In ListNet and RankNet, the parameter learning rate is set to 0.0005, the hidden layers are set to 1, and the hidden nodes per layer are set to 10. The value of NDCG@N is obtained by calculation and NDCG@N is used as the evaluation measure of the ranking results. The higher the NDCG@N value is, the higher the accuracy of the evaluation ranking is.

According to Fig. 3, Fig. 3(a) and Fig. 3(b) shows 5 basic comparison representation methods and DLCF+EF is the best representation method, and our experimental method is also improved on this basis. Figure 3(c)–(f) shows MLCDLF is better than 10-LCDLF and 15-LCDLF. The reason is that MLCDLF divides the range of the same living circle into multiple layers according to the size and learns the potential characteristics of each layer of living circle separately, so MLCDLF can describe the living circle more comprehensively. Method 10-LCDLF, 15-LCDLF and MLCDLF are all better than method 5-LCLF, because 5-LCF only embedds the necessary POIs in the 5-min living circle and these POIs can only meet the basic life needs of residents, so it cannot describe the living circle accurately.

Through experimental results, it can be found that the implicit representation method of multi-layer living circle community has the best performance. The representation framework of multi-layer living circle community can hierarchically describe the living circle structure and learn the potential characteristics of different sizes in a more fine-grained way. The potential characteristics in different ranges makes the convenience evaluation of living circles more accurate.

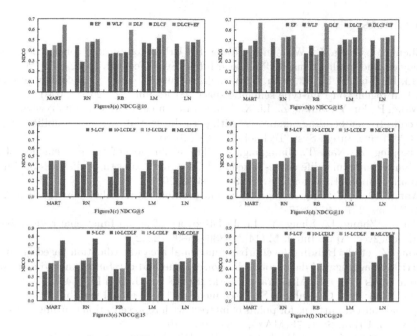

Fig. 3. A comparative study on the convenience in different living circles

6 Conclusion

This paper studies the living circles of urban residents and proposes a framework for learning multi-layer living circles. The framework combines three living circle activity graphs into the structure vectors to describe the embedding representation. The method of propagation probability is used to establish the living circle structure graph, the period of residents' activity is divided by the sliding window and the weights are given to different POIs. Finally, the autoencoder is used to embed the living circle activity graph. In order to evaluate the performance of the proposed framework, it was applied to the convenience evaluation of living circles and the different results of the community convenience ranking by the representation method of multi-layer living circle communities were analyzed. The real datasets were used to conduct experimental evaluation and the effectiveness of the proposed method was verified.

Acknowledgements. This work is supported by the National Natural Science Foundation of China (62073227), the National key R & D Program Foundation, Natural Science Foundation of Liaoning Province (2019-MS-264), the Project of the Educational Department of Liaoning Province (LJZ2021008), Project of China Academic Degree and Graduate Education Association(2020MSA40).

References

1. Zheng, Y., Capra, L., Wolfson, O., Yang, H.: Urban computing: concepts, methodologies, and applications. ACM Trans. Intell. Syst. Technol. **5**(3), 1–55 (2014)
2. Liu, K., Yin, L., Lu, F., Mou, N.: Visualizing and exploring POI configurations of urban regions on POI-type semantic space. Cities **99**(2), 102610 (2020)
3. Wang, P., Fu, Y., Jiawei, Z., Li, X., Lin, D.: Learning urban community structures: a collective embedding perspective with periodic spatial-temporal mobility graphs. ACM Trans. Intell. Syst. Technol. **9**, 1–28 (2018)
4. Xiaoping, Z., Nan, T., Jiangping, W.: The evaluation of the convenience of 15-minute community life circles based on POI data: taking three districts of Wuhan as an example. Huazhong Archit. **38**(6), 4 (2020)
5. Wang, H., Li, Z.: Region representation learning via mobility flow. In: The 2017 ACM, pp. 237–246 (2017)
6. Cai, L., Zhang, L., Liang, Y., Li, J.: Discovery of urban functional regions based on node2vec. Appl. Intell. **52**(14), 16886–16899 (2022)
7. Song, Z., et al.: Building-level urban functional area identification based on multi-attribute aggregated data from cell phones-a method combining multidimensional time series with a SOM neural network. ISPRS Int. J. Geo Inf. **148**(4), 12 (2022)
8. Liu, X., Andris, C., Rahimi, S.: Place niche and its regional variability: measuring spatial context patterns for points of interest with representation learning. Comput. Environ. Urban Syst. **75**(MAY), 146–160 (2019)
9. Ou, M., Peng, C., Jian, P., Zhang, Z., Zhu, W.: Asymmetric transitivity preserving graph embedding. In: ACM SIGKDD International Conference, pp. 1105–1114 (2016)
10. Fu, Y., Wang, P., Du, J., Wu, L., Li, X.: Efficient region embedding with multi-view spatial networks: a perspective of locality-constrained spatial autocorrelations. In: Proceedings of the AAAI Conference on Artificial Intelligence, vol. 33, pp. 906–913 (2019)
11. Turian, J.P., Ratinov, L.A., Bengio, Y.: Word representations: a simple and general method for semi-supervised learning. In: DBLP, pp. 384–394 (2010)
12. Wu, S., et al.: Multi-graph fusion networks for urban region embedding (2022)
13. Lee, H., Grosse, R., Ranganath, R., Ng, A.: Convolutional deep belief networks for scalable unsupervised learning of hierarchical representations, p. 77 (2009)
14. Courville, A., Bergstra, J., Bengio, Y.: Unsupervised models of images by spike-and-slab RBMs, pp. 1145–1152 (2011)
15. Hinton, G.E., Zemel, R.: Autoencoders, minimum description length and Helmholtz free energy. In: Advances in Neural Information Processing Systems, vol. 6, pp. 3–10 (1994)
16. Xu, J., Zhao, Y., Yu, G.: An evaluation and query algorithm for the influence of spatial location based on R k NN. Front. Comput. Sci. **15**, 1–9 (2021)
17. Fu, Y., Xiong, H., Ge, Y., Yao, Z., Zheng, Y., Zhou, Z.H.: Exploiting geographic dependencies for real estate appraisal: a mutual perspective of ranking and clustering. In: Proceedings of the ACM SIGKDD International Conference on Knowledge Discovery and Data Mining, pp. 1047–1056 (2014)
18. Zhang, X., Du, S., Zhang, J.: How do people understand convenience-of-living in cities? A multiscale geographic investigation in Beijing. ISPRS J. Photogrammetry Remote Sens. **148**(FEB), 87–102 (2019)
19. Wang, P., Fu, Y., Zhang, J., Li, X., Li, D.: Learning urban community structures: a collective embedding perspective with periodic spatial-temporal mobility graphs. ACM Trans. Intell. Syst. **9**(6), 63.1–63.28 (2018)

Knowledge Graph Completion with Fused Factual and Commonsense Information

Changsen Liu$^{(\boxtimes)}$, Jiuyang Tang, Weixin Zeng, Jibing Wu, and Hongbin Huang

Laboratory for Big Data and Decision, National University of Defense Technology, ChangSha, China
liuchangsen21@nudt.edu.cn

Abstract. Knowledge graph contains a large number of factual triples, but it is still inevitably incomplete. Previous commonsense-based knowledge graph completion models used concepts to replace entities in triplets to generate high-quality negative sampling and link prediction from the perspective of joint facts and commonsense. However, they did not consider the importance of concepts and their correlation with relationships, resulting in a lot of noise and ample room for improvement. To address this problem, we designed commonsense for knowledge graph completion and filtered the commonsense knowledge based on the analytic hierarchy process. The obtained commonsense can further improve the quality of negative samples and the effectiveness of link prediction. Experimental results on four datasets of the knowledge graph completion (KGC) task show that our method can improve the performance of the original knowledge graph embedding (KGE) model.

Keywords: knowledge graph completion · commonsense · negative sampling

1 Introduction

Knowledge graph is a semantic network that reveals the relationships between entities and can formally describe entities and their relationships in the real world [6]. In knowledge graph, structured knowledge is usually organized as factual triples (head entity, relationship, tail entity) or (h, r, t). In recent years, some knowledge graphs, such as Freebase [1] and DBpedia [8], have been widely applied in the field of artificial intelligence. However, most knowledge graphs are typically sparse and incomplete, lacking a large number of objectively existing factual triples, which has led to research on knowledge graph completion (KGC) tasks.

One mainstream method for knowledge graph completion is based on knowledge graph embedding (KGE). This method can be divided into two stages: first, in the training stage, low-dimensional representations of entities and relationships are learned based on existing triples in the knowledge graph, and this process depends on the quality of negative samples [9]; second, in the inference

L. Yuan et al. (Eds.): WISA 2023, LNCS 14094, pp. 137–148, 2023.
https://doi.org/10.1007/978-981-99-6222-8_12

stage, link prediction is performed by calculating the score of candidate triples using the learned entity and relationship representations, and sorting them to infer missing entities or relationships in the triples.

CAKE [14] was proposed to address the problems of low-quality negative samples and prediction uncertainty in the two stages of knowledge graph completion. However, during the process of generating commonsense knowledge, CAKE directly replaces entities in triples with concepts, lacks judgment on introduced concepts, and results in many noise in the generated commonsense knowledge. In the inference stage, the use of individual form commonsense with noises ignores the relationship between relationships and concepts, resulting in incomplete candidate entities.

To address these issues, we propose a kind of commonsense for knowledge graph completion that does not use commonsense knowledge in the form of triples but instead uses a dictionary of entities or relationships and concepts, which is more suitable for the application scenario of knowledge graphs. To tackle the problem of noise in previously used commonsense knowledge, the study applies the analytic hierarchy process to filter the used commonsense knowledge, resulting in high-quality commonsense knowledge that improves the quality of negative samples in knowledge graph embedding. In the link prediction process, a commonsense-based candidate entity enhancement method is proposed to reduce the sampling space of candidate entities and improve the effectiveness of link prediction.

2   Related Work

KGC Models. There are three main types of knowledge graph completion models: (1) rule-based learning algorithms, such as DRUM [15], and Any-Burl [12], which can automatically mine logical rules from the knowledge graph for inductive link prediction, but suffer from low efficiency due to time-consuming rule search and evaluation; (2) relationship path inference algorithms, such as PRA [11] algorithm, which selects relationship paths under the combination of path constraints and performs maximum likelihood classification, and some RNN models that combine the semantic information of relationship paths through RNN; and (3) knowledge representation-based algorithms, which map compositional triples into different vector spaces, such as TransE [2], ComplEx [20], RotatE [18], and HAKE [25] models, to learn embeddings of entities and relationships, evaluate the credibility of triples, and effectively predict missing triples. Compared to other methods, knowledge graph embedding methods have higher efficiency and better performance in knowledge graph completion.

Negative Sampling. Negative sampling method was initially used to accelerate the training of the Skip-Gram model [13], and later widely applied in areas such as natural language processing and computer vision [7]. There are mainly three types of negative sampling in knowledge graph embedding: (1) Random negative sampling, where most KGE models generate negative triples by randomly replacing entities or relations in the positive triples with those from a uniform distribution [21]; (2) Adversarial negative sampling [5], where KBGAN [4] combines the KGE model with softmax probabilities and selects high-quality negative triples in an adversarial training framework. Self-adversarial sampling [18] performs similarly to KBGAN but uses a self-scoring function without a generator. (3) Commonsense-based negative sampling, where CAKE [14] generates commonsense knowledge from the knowledge graph and designs different sampling strategies for different relationship types to narrow the sampling space, reducing the problem of false negative samples and low-quality negative samples. However, the above negative sampling algorithms all introduce noise or increase the complexity of the model [24], and limit the training of KGE and further restrict the performance of KGC.

Commonsense. In recent years, many studies have attempted to construct a general commonsense graph, such as ConceptNet [17], and ATOMIC [16]. The ConceptNet knowledge base consists of relationship knowledge in the form of triples. Compared with Cyc, ConceptNet uses a non-formalized and more natural language-like description, rather than the formal predicate logic used by Cyc, and is more focused on the relationship between words [23]. ATOMIC is a set of everyday commonsense inference knowledge about events described in natural language and associated with typed if-then relationships [10]. Cai et al. [3] designed a method for generating commonsense knowledge based on concepts and pretrained models, and obtained 500000 commonsense triplets based on ConceptNet. However, these commonsense graphs only contain concepts without links to corresponding entities, making them unsuitable for KGC tasks. CAKE [14] uses concepts instead of entities in triples to build individual and collective forms of commonsense. However, there is a lack of discrimination in generating commonsense triples, resulting in overly abstract concepts and semantically unreasonable commonsense knowledge, leading to many unreasonable commonsense knowledge.

3   Notations and Problem Formalization

Knowledge Graph. A knowledge graph is a structured representation of knowledge using a multi-relationship directed graph structure. Entities and relations are viewed as nodes in the multi-relationship graph and different types of edges. Specifically, knowledge graphs are generally represented as $G = \{E, R, T\}$, where E represents the set of entities, R represents the set of relations. The edge in R connects two entities to form a triple $(h, r, t) \in T$, which represents the directed relationship between the head entity h and the tail entity t with the relation r. $T \subseteq E \times R \times E$ represents the set of triples in the knowledge graph G.

Commonsense. In contrast to using triples to define commonsense knowledge as before, we use concepts to define two types of commonsense: entity-form

$\{(e : c_e)\}$, relation-form $\{(r : C_h)\}$ and $\{(r : C_t)\}$, where $e \in E$ represents an entity, $r \in R$ represents a relation, c_e represents the set of concepts that correspond to the entity, and C_h and C_t respectively represent the set of head concept and tail concept that correspond to the relation r.

KGE Score Function. We define a uniform symbol $E(h, r, t)$ to represent the score function of any KGE model for evaluating the plausibility of a triple (h, r, t). More specifically, the two most typical score function patterns are given as follows:

(1) The **translation**-based score function, such as TransE:

$$E(h, r, t) = ||\mathbf{h} + \mathbf{r} - \mathbf{t}||, \tag{1}$$

where $\mathbf{h}$, $\mathbf{r}$ and $\mathbf{t}$ denote the embeddings of head entity h, relation r and tail entity t, respectively.

(2) The **rotation**-based score function, such as RotatE:

$$E(h, r, t) = ||\mathbf{h} \circ \mathbf{r} - \mathbf{t}||, \tag{2}$$

where $\circ$ indicates the hardmard product.

Link Prediction. Following most of the previous KGC models, we regard link prediction as an entity prediction task. Given a triple query with an entity missing $(h, r, ?)$ or $(?, r, t)$, link prediction takes every entity as a candidate. It calculates the score of each candidate triple by employing the learned KG embeddings and the score function. Then, we rank the candidate entities in light of their scores and output the top n entities as results.

4 Model

The overall framework is shown in the Fig. 1, which mainly consists of three parts: commonsense generation, commonsense-based negative sampling, and candidate entity enhancement.

4.1 Automatic Commonsense Generation

As for the commonsense defined in this article, as long as there are concepts corresponding to entities in the knowledge graph, our method can theoretically automatically generate commonsense from any knowledge graph. Specifically, combining entities with concepts can obtain entity-form commonsense, distinguishing between the two cases of head and tail entities, using concepts to replace entities in triples and removing duplicate concepts can obtain relation-form commonsense.

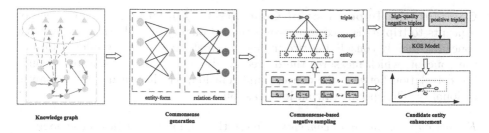

Fig. 1. An overview of the framework

4.2 Commonsense-Aware Negative Sampling

Intuitively, the negative triples satisfying commonsense are more challenging to distinguish from positive triples, contributing to more effective training signals. Therefore, we attempt to assign greater weights to the negative triples satisfying commonsense during sampling.

To reduce the problem of false-negative triples, we adopt the complex relational features defined in TransH [19], including $1-1$, $1-N$, $N-1$, and $N-N$ relations. Here, $\mathbf{1}$ indicates that the entity is unique given a relation and another entity, while $\mathbf{N}$ indicates that there are multiple entities in this case. From these features, when destroying a unique entity, such as the tail entity in an $N-1$ relation, all entities except the destroyed one are negative samples, and entities with the same concept as the destroyed entity are more likely to be high-quality negative samples. Therefore, entity-form commonsense is considered to select negative samples. When destroying a non-unique entity, such as the head entity in an $N-1$ relation, due to the non-uniqueness of the head entity, entities with the same concept as the head entity may be false negative samples. Thus, relation-form commonsense is considered, and the set of candidate concepts is obtained by taking the difference between the relation-form commonsense and entity-form commonsense.

To provide a clearer explanation of the sampling strategy, for a given triplet (h, r, t), we will use the $N-1$ relation as an example to illustrate the negative sampling method.

Select Concept Set Based on Commonsense. Define C to represent the set of selected concepts. For the tail entity, we can use entity-form commonsense to select negative samples.

$$C = c_t, \tag{3}$$

For the head entity, we can use relation-form commonsense to select negative samples.

$$C = C_h - c_h, \tag{4}$$

where $C_h - c_h$ represents the set difference.

Compute the Importance of Concepts to Triplets. The importance of concepts to triplets is used to measure the rationality of replacing entities with concepts, mainly considering two indicators: first, the frequency information of the concept, which measures the degree of abstraction of the concept; second, the correlation between the concept and relation, which measures its semantic rationality. The specific calculation is as follows:

Frequency information of concepts. We use the number of entities corresponding to a concept as basic information.

$$E(c) = \frac{(|E_c|/\tau)^{-\frac{1}{2}}}{\sum_{c' \in C}(|E_{c'}|/\tau)^{-\frac{1}{2}}}, \tag{5}$$

where E_c represents the set of entities corresponding to concept c, $|E_c|$ represents the number of entities in the set. The more entities the concept corresponds to, the lower its importance (more abstract). τ is the temperature control coefficient.

Correlation between the concept and relation. We use the similarity of concepts and relationships in vector space as a measure of correlation.

The correlation between concept c and relation r:

$$r(c, r) = cos(\mathbf{c}, \mathbf{r}). \tag{6}$$

The correlation between concept c and triple (h, r, t):

$$R(c) = r(c, r)\frac{|R_c| - 1}{\sum_{x \in R_c, x \neq r} r(c, x)}. \tag{7}$$

The importance of a concept can be obtained by combining its frequency information and correlation with relations:

$$I(c) = \alpha E(c) + (1 - \alpha)R(c), \tag{8}$$

where α is a hyperparameter used to adjust the ratio between abstractness and rationality, setting to 0.5 in the experiment.

Determine the Sampling Probability of Entities. First, determine the scope of candidate entities. Select all entities corresponding to the concept as candidate entities for negative sampling.

$$S = \{e|e \in c_e, \forall c \in C\}. \tag{9}$$

Then, calculate the sampling probability of the entities. The sampling probability of the entities in the candidate entity set S is calculated using the following formula:

$$w_1(e) = \sum_{c \in C} \frac{I(c)}{|E_c|} \log(\frac{N}{N(e) + 1}). \tag{10}$$

Our method constructs a new sampling space from which negative samples are collected, so it can be combined with other negative sampling methods. In this paper, we incorporate self-adversarial negative sampling during the training process:

$$w_2(h'_j) = 1 - p((h'_j, r, t)|\{(h_i, r_i, t_i)\}) = 1 - \frac{exp\mu E(h'_j, r, t)}{\sum_i exp\mu E(h'_i, r, t)}, \quad (11)$$

$$w_2(t'_j) = p((h, r, t'_j)|\{(h_i, r_i, t_i)\}) = \frac{exp\mu E(h, r, t'_j)}{\sum_i exp\mu E(h, r, t'_i)}, \quad (12)$$

where w and p denote the weight and the probability of the negative triple, respectively. μ is the temperature of sampling motivated by the self adversarial sampling.

4.3   Training the KGE Model

Based on the negative triples obtained, the knowledge graph embedding model is trained to obtain entity embeddings and relation embeddings by increasing the score difference between positive and negative samples. The following loss function is used as the optimization target:

$$L_1 = - \log \sigma(\gamma - E(h, r, t))$$

$$- \sum_{i=1}^{n} \frac{1}{2} [(w_1(h'_i) + w_2(h'_i)) \log \sigma(E(h'_i, r, t) - \gamma) \quad (13)$$

$$+ (w_1(t'_i) + w_2(t'_i)) \log \sigma(E(h, r, t'_i) - \gamma)].$$

In addition, we design a multi-label classification task to optimize concept embeddings:

$$L(e) = -\frac{1}{k} \sum_{i=1}^{k} [y_{e,i} \log p_{e,i} + (1 - y_{e,i}) \log(1 - p_{e,i})], \quad (14)$$

where k represents the total number of corresponding concepts for all entities, $y_{e,i}$ denotes the category label of the entity, and $p_{e,i}$ is the probability of correctly classifying the entity, we calculate it using the cosine similarity between the vectors of the entity and the concepts.

$$y_{e,i} = \begin{cases} 1, & if \quad e \in E_c, \\ 0, & otherwise. \end{cases} \quad (15)$$

$$p_{e,i} = cos(\mathbf{e}, \mathbf{c}_i). \quad (16)$$

Summing the classification loss of head and tail entities together as multi-label classification loss:

$$L_2 = L(h) + L(t). \quad (17)$$

Finally, merging L_1 and L_2 as the loss function for training:

$$L = L_1 + \beta L_2, \tag{18}$$

where β is a hyperparameter used to control the weight of entity classification loss, setting to 0.1 in the experiment.

4.4   Commonsense-Based Candidate Entity Augmentation

Link prediction is generally performed by ranking all entities based on the function scores. By leveraging relation-form commonsense, a more reasonable range can be directly provided for link prediction. Therefore, we propose a commonsense-based method to enhance candidate entities. First, candidate entities are enhanced based on the relation-form commonsense, and entities that conform to the relation-form commonsense are selected as candidate entities because these entities are more likely to be correct triplets. Specifically, taking the query $(h, r, ?)$ as an example, during the candidate entity enhancement stage, the concept set C_t of in the relation-form commonsense $(r : C_t)$ is selected, and entities belonging to the set of concepts C_t can be determined as candidate entities because they meet the commonsense and are more likely to be the correct tail entity from the perspective of commonsense compared to other entities. Then, in the candidate entity ranking stage, each candidate entity obtained from the enhancement stage is scored from the perspective of facts, as shown below:

$$score(e_i) = E(h, r, e_i), \tag{19}$$

where $E(h, r, e_i)$ represents the scoring function for training the KGE model. Subsequently, the predicted results will sort the scores of candidate entities in ascending order and output the entities with higher ranks.

5   Experiments and Results

In this section, we perform extensive experiments of KGC on four widely-used KG datasets containing concepts. First, we introduce the datasets, baseline models, and evaluation protocol. Then, through comparisons with the baseline on multiple datasets, the effectiveness of the proposed method was verified. In addition, further ablation studies were also conducted.

5.1   Experiment Settings

Datasets. The experiments used four real datasets containing ontology concepts, including FB15K [2], FB15K-237 [19], NELL-995 [22], and DBpedia-242. Specifically, DBpedia-242 was extracted from DBpedia [8] and contains a total of 242 concepts. The statistical summary of the datasets is shown in the Table 1. It is worth noting that the entities in FB15K and FB15K-237 always belong to multiple concepts, while each entity in NELL-995 and DBpedia-242 only has one concept.

Table 1. Statistics of the experimental datasets. #Rel,#Ent, #Con represent the number of relations, entities and concepts of each dataset, respectively.

Dataset	#Rel	#Ent	#Con	#Train	#Valid	#Test
FB15K	1,345	14,951	89	483,142	50,000	59,071
FB15K-237	237	14,505	89	272,115	17,535	20,466
NELL-995	200	75,492	270	123,370	15,000	15,838
DBpedia-242	298	99,744	242	592,654	35,851	30,000

Baseline. The model was compared with two state-of-the-art KGE models and their improved methods, including TransE [2], RotatE [18] and CAKE [14]. These baselines were also the basic models combined with our method. It was not necessary to use many baselines because the focus of this work is to observe the impact of applying our method to the original KGE models rather than beating all SOTA models. We provided the baseline results by running the source code of the baselines with the suggested parameters.

Evaluation Metrics. Three commonly used metrics were used to evaluate the performance of link prediction: mean rank (MR), mean reciprocal rank (MRR), and the proportion of correct entities in the top N positions (Hits@N). All metrics were in a filtered setting by removing the candidate triples that already existed in the dataset.

Table 2. Link prediction results on four datasets. **Bold** numbers are the best results for each type of model.

Models	FB15k					FB15k-237				
	MR	MRR	Hits@10	Hits@3	Hits@1	MR	MRR	Hits@10	Hits@3	Hits@1
TransE	50	0.463	0.749	0.578	0.297	195	0.268	0.454	0.298	0.176
CAKE	52	0.518	0.744	0.600	0.390	175	0.301	0.493	0.335	0.206
ours	**49**	**0.573**	**0.765**	**0.646**	**0.460**	**169**	**0.347**	**0.519**	**0.380**	**0.260**
RotatE	39	0.590	0.781	0.686	0.471	204	0.269	0.452	0.298	0.179
CAKE	41	0.610	0.807	0.695	0.489	181	0.318	0.511	0.354	0.223
ours	**38**	**0.651**	**0.818**	**0.727**	**0.544**	184	**0.360**	**0.529**	**0.388**	**0.277**
Models	Dbpedia-242					NELL-995				
	MR	MRR	Hits@10	Hits@3	Hits@1	MR	MRR	Hits@10	Hits@3	Hits@1
TransE	2733	0.242	0.468	0.344	0.100	1081	0.429	0.557	0.477	0.354
CAKE	881	0.330	0.595	0.458	0.160	317	0.533	0.650	0.578	0.461
ours	**505**	**0.391**	**0.653**	**0.490**	**0.199**	**280**	**0.540**	**0.651**	**0.582**	**0.472**
RotatE	1950	0.374	0.582	0.457	0.249	2077	0.460	0.553	0.493	0.403
CAKE	1027	0.423	0.603	0.486	0.320	329	0.546	0.660	0.592	0.474
ours	**664**	**0.486**	**0.649**	**0.515**	**0.344**	**289**	0.552	**0.665**	0.590	**0.480**

5.2 Experimental Results

We report the link prediction results on the four datasets in Table 2, and show the detail results for different models improved by our method respectively. The results indicate that our method can significantly improve the performance of the original model compared to CAKE. This suggests that our method can generate higher quality negative samples and provide more reasonable candidate entity ranges. Compared with the average performance of the two baseline models, our model improved the MRR by 8.7%, 14.2%, 16.7%, and 1.2% on FB15K, FB15K-237, DBpedia-242, and NELL-995, respectively. While on the original model, our method increased the performance by 17.0%, 31.7%, 45.8%, and 22.9% on the four datasets. Among them, the best improvement effect was achieved on the DBpedia-242 dataset, with MRR and Hits@10 improved by 0.062 and 0.052 compared to CAKE, while only 0.007 and 0.003 were improved on the NELL-995 dataset. This is because each entity in these two datasets has only one concept, and commonsense knowledge cannot provide stable guidance for these entities. In contrast, the improvement effect between FB15k and FB15k-237 is relatively small, which indicates that high-quality commonsense knowledge has a more similar effect on negative samples and candidate entities generated for this type of entity. In summary, the results indicate that filtering commonsense knowledge and combining it with the original KGE model can better improve the effectiveness of knowledge graph completion.

5.3 Ablation Study

Table 3. Ablation study of integrating each model into the basic model TransE on FB15K-237 and DBpedia-242.

Models	FB15k-237				
	MR	MRR	Hits@10	Hits@3	Hits@1
our model	169	0.347	0.519	0.380	0.260
w/o fre	174	0.306	0.498	0.340	0.209
w/o cor	169	0.308	0.503	0.342	0.210
w/o enh	170	0.301	0.490	0.334	0.206
Models	DBpedia-242				
	MR	MRR	Hits@10	Hits@3	Hits@1
our model	505	0.391	0.653	0.490	0.199
w/o fre	1415	0.305	0.586	0.446	0.124
w/o cor	1307	0.301	0.569	0.424	0.133
w/o enh	2552	0.239	0.475	0.320	0.104

We validated the effectiveness of each contribution by integrating the entire framework and simplified models, (1) ignoring commonsense frequency information (w/o fre.), (2) ignoring the correlation between concepts and relations (w/o cor.), and (3) ignoring entity enhancement (w/o enh.) into the basic model TransE. The results in Table 3 indicate that our whole model outperforms all ablation models on each dataset. This suggests that adding filtering for commonsense knowledge during negative sampling is meaningful for generating more effective negative samples. Additionally, enhancing candidate entities based on commonsense is beneficial for improving link prediction performance. Overall, each module plays a crucial role in our method.

6   Conclusion

In this paper, we propose a commonsense knowledge representation for knowledge graph completion and design a filtering module for injecting high-quality commonsense knowledge into knowledge graph embedding. Finally, we enhance candidate entities in link prediction by incorporating commonsense knowledge. Experimental results on four datasets demonstrate that our proposed framework outperforms the direct use of concepts to replace entities in triplets with commonsense, showing significant and stable improvements in knowledge graph completion. However, the commonsense proposed in this article is only effective for inductive reasoning. For unknown entities that appear in inductive reasoning, how to effectively utilize commonsense knowledge is a future exploration direction.

References

1. Bollacker, K.D., Evans, C., Paritosh, P.K., Sturge, T., Taylor, J.: Freebase: a collaboratively created graph database for structuring human knowledge. In: SIGMOD Conference, pp. 1247–1250. ACM (2008)
2. Bordes, A., Usunier, N., García-Durán, A., Weston, J., Yakhnenko, O.: Translating embeddings for modeling multi-relational data. In: NIPS, pp. 2787–2795 (2013)
3. Cai, H., Zhao, F., Jin, H.: Commonsense knowledge construction with concept and pretrained model. In: Zhao, X., Yang, S., Wang, X., Li, J. (eds.) WISA 2022. LNCS, vol. 13579, pp. 40–51. Springer, Cham (2022). https://doi.org/10.1007/978-3-031-20309-1_4
4. Cai, L., Wang, W.Y.: KBGAN: adversarial learning for knowledge graph embeddings. In: NAACL-HLT, pp. 1470–1480. Association for Computational Linguistics (2018)
5. Chen, J., et al.: Adversarial caching training: unsupervised inductive network representation learning on large-scale graphs. IEEE Trans. Neural Netw. Learn. Syst. 33(12), 7079–7090 (2022)
6. Ji, S., Pan, S., Cambria, E., Marttinen, P., Yu, P.S.: A survey on knowledge graphs: representation, acquisition, and applications. IEEE Trans. Neural Netw. Learn. Syst. 33(2), 494–514 (2022)
7. Joulin, A., van der Maaten, L., Jabri, A., Vasilache, N.: Learning visual features from large weakly supervised data. In: Leibe, B., Matas, J., Sebe, N., Welling, M. (eds.) ECCV 2016. LNCS, vol. 9911, pp. 67–84. Springer, Cham (2016). https://doi.org/10.1007/978-3-319-46478-7_5

8. Lehmann, J., et al.: DBpedia - a large-scale, multilingual knowledge base extracted from Wikipedia. Semant. Web **6**(2), 167–195 (2015)

9. Li, Z., et al.: Efficient non-sampling knowledge graph embedding. In: WWW, pp. 1727–1736. ACM/IW3C2 (2021)

10. Lin, B.Y., Chen, X., Chen, J., Ren, X.: KagNet: knowledge-aware graph networks for commonsense reasoning. In: EMNLP/IJCNLP (1), pp. 2829–2839. Association for Computational Linguistics (2019)

11. Liu, W., Daruna, A.A., Kira, Z., Chernova, S.: Path ranking with attention to type hierarchies. In: AAAI, pp. 2893–2900. AAAI Press (2020)

12. Meilicke, C., Chekol, M.W., Ruffinelli, D., Stuckenschmidt, H.: Anytime bottom-up rule learning for knowledge graph completion. In: IJCAI, pp. 3137–3143. ijcai.org (2019)

13. Mikolov, T., Chen, K., Corrado, G., Dean, J.: Efficient estimation of word representations in vector space. In: ICLR (Workshop Poster) (2013)

14. Niu, G., Li, B., Zhang, Y., Pu, S.: CAKE: a scalable commonsense-aware framework for multi-view knowledge graph completion. In: ACL (1), pp. 2867–2877. Association for Computational Linguistics (2022)

15. Sadeghian, A., Armandpour, M., Ding, P., Wang, D.Z.: DRUM: end-to-end differentiable rule mining on knowledge graphs. In: NeurIPS, pp. 15321–15331 (2019)

16. Sap, M., et al.: ATOMIC: an atlas of machine commonsense for if-then reasoning. In: AAAI, pp. 3027–3035. AAAI Press (2019)

17. Speer, R., Chin, J., Havasi, C.: ConceptNet 5.5: an open multilingual graph of general knowledge. In: AAAI, pp. 4444–4451. AAAI Press (2017)

18. Sun, Z., Deng, Z., Nie, J., Tang, J.: Rotate: knowledge graph embedding by relational rotation in complex space. In: ICLR (Poster). OpenReview.net (2019)

19. Toutanova, K., Chen, D.: Observed versus latent features for knowledge base and text inference. In: CVSC, pp. 57–66. Association for Computational Linguistics (2015)

20. Trouillon, T., Welbl, J., Riedel, S., Gaussier, É., Bouchard, G.: Complex embeddings for simple link prediction. In: ICML. JMLR Workshop and Conference Proceedings, vol. 48, pp. 2071–2080. JMLR.org (2016)

21. Wang, Z., Zhang, J., Feng, J., Chen, Z.: Knowledge graph embedding by translating on hyperplanes. In: AAAI, pp. 1112–1119. AAAI Press (2014)

22. Xiong, W., Hoang, T., Wang, W.Y.: DeepPath: a reinforcement learning method for knowledge graph reasoning. In: EMNLP, pp. 564–573. Association for Computational Linguistics (2017)

23. Zang, L., Cao, C., Cao, Y., Wu, Y., Cao, C.: A survey of commonsense knowledge acquisition. J. Comput. Sci. Technol. **28**(4), 689–719 (2013)

24. Zhang, Y., Yao, Q., Shao, Y., Chen, L.: NSCaching: simple and efficient negative sampling for knowledge graph embedding. In: ICDE, pp. 614–625. IEEE (2019)

25. Zhang, Z., Cai, J., Zhang, Y., Wang, J.: Learning hierarchy-aware knowledge graph embeddings for link prediction. In: AAAI, pp. 3065–3072. AAAI Press (2020)

Heterogeneous Graphs Embedding Learning with Metapath Instance Contexts

Chengcheng Yu, Lujing Fei, Fangshu Chen$^{(\boxtimes)}$, Lin Chen, and Jiahui Wang

School of Computer and Information Engineering, Institute for Artificial Intelligence,
Shanghai Polytechnic University, Shanghai 201209, China
{ccyu,fschen,chenl,jhwang}@sspu.edu.cn, 20221513018@stu.sspu.edu.cn

Abstract. Embedding learning in heterogeneous graphs consisting of multiple types of nodes and edges has received wide attention recently. Heterogeneous graph embedding is to embed a graph into low-dimensional node representations with the goal of facilitating downstream applications, such as node classification. Existing models using metapaths, either ignore the metapath context information which describes intermediate nodes along a metapath instance, or only using simple aggregation methods to encode metapath contexts, such as mean or linear methods. To address the problems, we propose Metapath Instance Contexts based Graph neural Network (MICGNN). Specifically, MICGNN includes four components: Node transformation, projects the input different types of features into the same feature space. Context aggregation, incorporates the node embeddings along a metapath instance context. Instance aggregation, combines context embeddings obtained from each metapath instance. Semantic aggregation, fuses together the semantic node embeddings derived from different metapaths. Extensive experiments results on two real-world datasets show our model exhibits superior performance in node-related tasks compared to existing baseline models.

Keywords: Metapath Instance Context · Heterogeneous Graph · Graph embedding

1 Introduction

Heterogeneous graph (HetG) are widely used to model various real-world datasets such as citation networks, social networks, and knowledge graphs. A HetG consists of various types of nodes and edges. An example of the HetG is show in Fig. 1. With the rapid development of deep learning, graph neural networks (GNNs) have been proposed, which can effectively combine graph structure and node features to represent nodes as low-dimensional vectors. Nevertheless, most exiting methods focus on homogeneous graph where all nodes and

This work is supported by the National Natural Science Foundation of China (Grant No. 62002216), and the Shanghai Sailing Program (Grant No. 20YF1414400).

L. Yuan et al. (Eds.): WISA 2023, LNCS 14094, pp. 149–161, 2023.
https://doi.org/10.1007/978-981-99-6222-8_13

edges are of a single type, including GCN [1], GraphSAGE [2], GAT [3], and many other variants [4–6]. These techniques are unable to provide better node/graph representation learning for downstream tasks when applied to HetG. Then many methods of GNNs has been expanded to heterogeneous graphs. However, these models based on metapaths, either ignore the metapath context information which describes intermediate nodes along a metapath instance, such as metapath2vec [7], HAN [8], or only using simple aggregation methods to encode metapath contexts, such as MAGNN [9], ConCH [10] and many other variants [12–14]. To tackle this issues, we propose Metapath Instance Contexts based Graph neural Network (MICGNN), which make use of attention mechanism to encode metapath instance contexts.The MICGNN comprises node transformation, context aggregation, instance aggregation and semantic aggregation.

Node Transformation. As a basis for subsequent work, MICGNN projects the input different types of node features into the same feature space using linear transformation, and one type of node share a linear transformation parameter.

Context Aggregation. A metapath instance context is a sequence of nodes along a metapath instance of a target node, which are of different importance to the embedding learning of the target node. Context aggregation is designed to integrate a metapath instance context for getting instance context embedding based on attention mechanism.

Instance Aggregation. One type of metapath may have many metapath instances. The objective of instance aggregation is to combine context embeddings obtained from metapath instances according to Context Aggregation. We use attention mechanism to learn the importance of different metapath instances.

Semantic Aggregation. Different metapaths can represent different semantic relations in a HetG. Semantic aggregation aims to fuse together semantic node embedding derived from different metapaths according to Instance Aggregation, for getting final target node embedding. We use vanilla attention mechanism to learn the importance of different metapaths. The contributions of our work are summarized as follows:

- We propose a novel graph neural network for HetG embedding based on three types of aggregation.
- We design a metapath instance context encoder function based on attention mechanism for embedding metapath instances.
- We perform comprehensive experiments on the IMDb and DBLP datasets for node classification and node clustering. Experiments show that our model exhibits superior performance compared to the baseline models.

2   Related Work

This section gives an overview of related graph representation learning research. We discuss Graph Neural Networks (GNNs) for homogeneous graph and heterogeneous graphs.

Graph Neural Networks. Graph Neural Networks (GNNs) are neural networks that provide a low-dimensional vector representation for graph nodes. Two main types of GNNs exist: spectral-based and spatial-based. Spectral-based GNNs suffer from scalability and generalization issues due to their reliance on the graph's eigenbasis. Spatial-based GNNs, like GraphSAGE and its variants, define convolutions in the graph domain. However, these existing GNNs can't handle well for heterogeneous graphs with diverse types of node and edge. To address this, some methods of GNNs has been expanded to heterogeneous graphs.

Heterogeneous Graph Embedding. HetG embedding learning aims to project nodes into a lower-dimensional vector space. Many Methods have been proposed to learn the embedding of HetG. For instance, metapath2vec is simple but may be biased due to the single metapath used for random walk guidance. HAN transforms a HetG into multiple homogeneous graphs based on metapaths, and uses node-level and semantic-level attention mechanisms to aggregate information from the neighbors of the target node and combine different metapaths. These methods all ignore the metapath context information which describes intermediate nodes along a metapath instance. MAGNN improves HAN for HetG embedding learning, which provides several encoders for distilling information from metapath instances. ConCH utilizes metapath contexts to embed instance for effective classification in heterogeneous information networks. MEOW [11] embedding metapath instances using encoders provided by MAGNN for heterogeneous graph contrastive learning. Although these methods utilizes metapath contexts for heterogeneous graph embedding learning, they just use simple encoder functions for metapath instance embedding learning, such as mean or linear methods. Therefore, we propose MICGNN for heterogeneous graph embedding learning, which makes use of attention mechanism to encode metapath instance contexts to improve performance in node-related tasks.

3 Definition

In this section we formally define various concepts.

Definition 1. *Heterogeneous Graph (HetG)* [9]. *A heterogeneous graph is defined as a graph $\mathcal{G} = (\mathcal{V}, \mathcal{E})$ associated with a node type mapping function $\phi : V \rightarrow \mathcal{A}$ and an edge type mapping function $\psi : \mathcal{E} \rightarrow \mathcal{R}$. $\mathcal{A}$ and $\mathcal{R}$ denote the predefined sets of node types and edge types, respectively, with $|\mathcal{A}| + |\mathcal{R}| > 2$.*

Example. *Figure 1(a) depicts an example, integrating four types of nodes/entities (i.e., author, paper, venue, and term) and three types of edges/relations (i.e., publish, contain, and write). Metapaths encapsulate this complexity.*

Definition 2. *Metapath*. *A metapath P is defined as a path in the form of $A_1 \xrightarrow{R_1} A_2 \xrightarrow{R_2} \cdots \xrightarrow{R_{l-1}} A_l \xrightarrow{R_l} A_{l+1}$ (abbreviated as $A_1 A_2 \ldots A_{l+1}$), which describes a composite relation $R = R_1 \circ R_2 \circ \ldots \circ R_l$ between node types A_1 and A_{l+1}, where $\circ$ denotes the composition operator on relations. The length of P is l.*

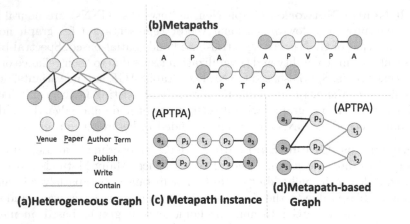

Fig. 1. An exemplary instance of a heterogeneous graph.

Example. *Figure 1(b) showcases three metapaths (i.e., author-paper-author, author-paper-venue-paper-author and author-paper-term-paper-author). $|P|$ refers to the type number of metapaths in a HetG.*

Definition 3. Metapath Instance. *In a heterogeneous graph, for a given metapath P, and the length of P is l. We define a metapath instance p as a node sequence $P(p_1, p_l) = (p_1, p_2, ..., p_i, ..., p_l, p_{l+1})$, where each p_i represents a node and i refer to the position in p. Each node type of the sequence follows the schema defined by P.*

Example. *Figure 1(c) author-paper-term-paper-author, author a_1 is a metapath-based neighbor of author a_2. These three intermediate nodes(p_1, t_1 and p_2) are connected via the metapath instance $a_1 - p_1 - t_1 - p_2 - a_2$.*

Definition 4. Metapath-based Neighbor. *Given a metapath P of a heterogeneous graph, the metapath-based neighbors N_v^P of a node v is defined as the set of nodes that connect with node v via metapath instances of P. A neighbor connected by two different metapath instances is regarded as two different nodes in N_v^P. Note that N_v^P includes v itself if P is symmetric.*

Definition 5. Metapath-based Graph. *Given a metapath P of a heterogeneous graph $\mathcal{G}$, the metapath-based graph $\mathcal{G}_P$ is a graph constructed by all the metapath-P-based neighbor pairs in graph $\mathcal{G}$. Note that $\mathcal{G}^P$ is a homogeneous graph if P is symmetric and starts and ends with the same node type, otherwise it is a heterogeneous graph.*

Example. *Figure 1(d) illustrates the metapath APTPA in a meta-graph. Authors a1 and a3, linked via papers p1, p3, and connected terms t1, t2.*

Definition 6. Heterogeneous Graph Embeddings. *Given a heterogeneous graph $\mathcal{G} = (\mathcal{V}, \mathcal{E})$, with node attribute matrices $X_{A_i} \in \mathbb{R}^{|V_{A_i}| \times d_{A_i}}$ for node types*

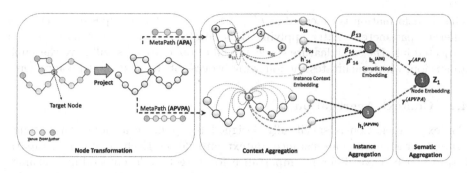

Fig. 2. The overall framework of MICGNN

$A_i \in A$, *heterogeneous graph embedding is the task to learn the d-dimensional node representations* $h_v \in \mathbb{R}^d$ *for all* $v \in \mathcal{V}$ *with* $d \ll |V|$ *that are able to capture rich structural and semantic information involved in* $\mathcal{G}$.

4 Methodology

In this section, we introduce the Metapath Contexts based Graph Neural Network (MICGNN) for heterogeneous graph embedding. As shown in Fig. 2, the overall framework of MICGNN contains four components: node transformation, context aggregation, instance aggregation and semantic aggregation. Beginning with Node transformation, different types of node features are projected into a unified feature space. And then we can obtain multiple metapath-based graphs. The weight of nodes along a metapath for the target node 1 can be learn via attention mechanism, the instance context embedding based on metapath instance can be obtained in the part of context aggregation. The semantic node embedding based on metapath can be got by fusing instance context embedding via attention mechanism in the Instance aggregation part. Finally, The final target node embedding can be learned by integrating semantic node embedding.

The overall forward propagation process is shown in Algorithm 1, lines 1–2 handle Node Transformation, lines 9–10 deal with Context Aggregation, lines 13–14 carry out Instance Aggregation, and lines 19–22 perform Semantic Aggregation.

4.1 Node Transformation

In the context of heterogeneous graphs consisting of various types of nodes. It's necessary to project these diverse node features into a unified feature space, and make sure that different types of node features have the same dimension. Then MICGNN first applies a type specific linear transformation for each type of nodes for projecting input features into the same feature space. Specifically, consider a specific node v of type $A \in \mathcal{A}$, characterized by feature vectors $x_v^A \in \mathbb{R}^{d'}$, and associated with a parameter of weight matrix $M_A \in \mathbb{R}^{d' \times d_A}$ for type A.

This transformation is executed as $h'_v = M_A \cdot x_v^A$. Different types of nodes have different parameters. After applying the operation, each type of node can be projected into the same feature space and all of node features will have the same dimension, which facilitates the following process of the next model component.

4.2   Context Aggregation

Context aggregation is specifically designed to integrate a metapath instance context for generating an instance context embedding. Each node in a metapath instance may have different importance for target node embedding learning. Therefore the weights of nodes along a metapath instance of target node can be learn via attention mechanism, join the weights and fuse the metapath instance context to get each instance context embedding of target node. This process can be represented by the following equations:

$$e_{p_1 p_i}^P = \text{LeakyReLU}(a_{con}^P[h'_{p_1} \parallel h'_{p_i}]), i \in [0, l]$$

$$\alpha_{p_1 p_i}^P = \frac{\exp(e_{p_1 p_i}^P)}{\sum_{j=1}^{|P|} \exp(e_{p_1 p_j}^P)}, \quad h_{P(p_1, p_l)} = \sum_{i=1}^{l} \alpha_{p_1 p_i} \cdot h'_{p_i}. \tag{1}$$

In these equations, p_1 is the target node. $P(p_1, p_l) = (p_1, p_2, ..., p_i, ...p_l)$ is a metapath instance related to p_1. $a_{con}^P \in \mathbb{R}^{2d'}$ is the attention parameter for metapath P, and $\parallel$ denotes the vector concatenation operator. $e_{p_1 p_i}^P$ represents the importance of node p_i for the metapath instance $P(p_1, p_l)$. Then $e_{p_1 p_i}^P$ is normalized across all nodes along the instance using the softmax function. $h_{P(p_1, p_l)}$ is the instance context embedding of the metapath instance $P(p_1, p_l)$, which is calculated by summing up the weighted node embedding along the instance, where the weights are the obtained attention coefficients of $\alpha_{p_1 p_i}^P$. This aggregated representation is subsequently used for instance aggregation.

4.3   Instance Aggregation

Upon the encoding of each metapath instance into instance context embeddings, this section employs a graph attention layer to weight the sum of the instance embeddings of metapath P according to the target node v. Each instance may have different importance for the target node's representation learning. Each instance stemming from the same metapath shares the same attention parameter, while different metapaths have distinct attention parameters. This process can be represented by the following equations:

$$e_{vu}^{P'} = \text{LeakyReLU}(a_{ins}^P \cdot [h'_v \parallel h_{P(v,u)}]),$$

$$\beta_{vu}^P = \frac{\exp(e_{vu}^{P'})}{\sum_{s \in N_v^P} \exp(e_{vs}^{p'})}, \quad h_v^P = \sigma\left(\sum_{u \in N_v^P} \beta_{vu}^P \cdot h_{P(v,u)}\right). \tag{2}$$

In these equations, $a_{ins}^P \in \mathbb{R}^{2d'}$ is the parameterized attention vector for metapath P. e_{vu}^P represents the importance of metapath $P(v, u)$ to the target node v, subsequently normalized across all $u \in N_v^p$ using the softmax function. h_v^P is the semantic node embedding of v which is calculated by summing up the weighted instance context embeddings, where the weights are the obtained attention coefficients of β_{vu}^P.

To address variance caused by the heterogeneity of graphs and enhance the stability of the learning process, we employ a multi-head extension of the self-attention mechanism. This extension involves performing K independent attention mechanisms, followed by concatenating the learned embeddings. Consequently, we obtain the following formulation:

$$h_v^{P'} = \|_{k=1}^K \sigma \left(\sum_{u \in N_{Pv}} [\beta_{vu}^P]_k \cdot h_{P(v,u)} \right). \tag{3}$$

In these equations, $h_v^{P'} \in \mathbb{R}^{d'}$ is the final semantic node embedding, which will be employed subsequently for Semantic aggregation.

4.4 Semantic Aggregation and Training

Now we leverage the attention mechanism to fuse the metapath-specific semantic node vectors of the target node v for obtaining the final node embedding of v. A linear transformation and a nonlinear function are then employed to project the final node embeddings to the desired output dimension.

Given each metapath $P_i \in \mathcal{P}_A$, we transform the metapath-specific node vectors for all nodes of type A, and compute their average as: $s_{P_i} = \frac{1}{|V_A|} \sum_{v \in V_A} \tanh(M_A \cdot h_v^{P_i} + b_A)$, where $M_A \in \mathbb{R}^{d_m \times d'}$ and $b_A \in \mathbb{R}^{d_m}$ are learnable parameters.

$$e_{P_i} = q_A \cdot s_{P_i}, \quad \gamma_{P_i} = \frac{\exp(e_{P_i})}{\sum_{P \in \mathcal{P}_A} \exp(e_P)}$$
$$h_v^{\varphi_A} = \sum_{P \in \varphi_A} \gamma_P \cdot h_v^P, \quad h_v = \sigma(W_o \cdot h_v^{\varphi_A}) \tag{4}$$

In these equations, $q_A \in \mathbb{R}^{d_m}$ is the parameterized attention vector for node type A. γ_{P_i} can be interpreted as the relative importance of metapath P_i for nodes of type A. Once γ_{P_i} is computed for each $P_i \in \mathcal{P}_A$, we perform a weighted sum of all the metapath-specific node vectors of v, where $\sigma(\cdot)$ is an activation function, $W_o \in \mathbb{R}^{d_o \times d'}$ is a weight matrix.

During the training phase, MICGNN employs either semi-supervised learning via cross entropy minimization or unsupervised learning using negative sampling. MICGNN efficiently captures heterogeneous graphs' characteristics, delivering functional node embeddings for diverse downstream tasks.

$$L = -\sum_{v=v_L} \sum_{c=1}^C y_v[C] \cdot \log h_v[c] \tag{5}$$

Algorithm 1. MICGNN Forward Propagation

Input: The heterogeneous graph $\mathcal{G} = (\mathcal{V}, \mathcal{E})$, node types $\mathcal{A} = \{A_1, A_2, \ldots, A_{|\mathcal{A}|}\}$, metapaths $\mathcal{P} = \{P_1, P_2, \ldots, P_{|\mathcal{P}|}\}$, node features $\{x_v, \forall v \in \mathcal{V}\}$, number of attention heads K, number of layers L

Output: The node embeddings $\{z_v, \forall v \in V\}$

1: **for** node type $A \in \mathcal{A}$ **do**
2:     node transformation $h_v^0 \leftarrow M_A \cdot x_v, \forall v \in \mathcal{V}_A$
3: **end for**
4: **for** $l = 1$ to L **do**
5:     **for** node type $A \in \mathcal{A}$ **do**
6:         **for** metapath $P \in \mathcal{P}_A$ **do**
7:             **for** $v \in \mathcal{V}_A$ **do**
8:                 **for** $i = 0$ to l **do**
9:                     Calculate $e_{p_1 p_i}^P = \text{LeakyReLU}(a_{con}^P[h_{p_1}', ||, h_{p_i}'])$,
10:                    Calculate $\alpha_{p_1 p_i}^P = \frac{\exp(e_{p_1 p_i}^P)}{\sum_{j=1}^{|P|} \exp(e_{p_1 p_j}^P)}, h_{P(p_1, p_l)} = \sum_{i=1}^{l} \alpha_{p_1 p_i} \cdot h_{p_i}'$;
11:                **end for**
12:                Calculate $h_{P(v,u)}^l$ for all $u \in N_v^P$
13:                $h_v^P = \|_{k=1}^K \sigma\left(\sum_{u \in N_{P_v}} [\beta_{vu}^P]_k \cdot h_{P(v,u)}\right)$;
14:            **end for**
15:        **end for**
16:        Calculate the weight γ_P for each metapath $P \in \mathcal{P}_A$;
17:        Fuse the embeddings from different metapaths
18:        $[h_v^{P_A}]^l \leftarrow \sum_{P \in \mathcal{P}_A} \gamma_P \cdot [h_v^P]^l, \forall v \in \mathcal{V}_A$;
19:    **end for**
20:    Layer output projection: $[h_v]^l \leftarrow \sigma(W_o^l \cdot [h_v^{P_A}]^l), \forall A \in \mathcal{A}$
21: **end for**
22: $z_v \leftarrow h_v^l, \forall v \in \mathcal{V}$

---

5 Experiments

5.1 Datasets

MICGNN's performance was evaluated on IMDb and DBLP datasets for node classification and clustering tasks. Table 1 summarizes the dataset statistics.

IMDb derived from the Internet Movie Database, comprises 4,278 movies, 2,081 directors, and 5,257 actors. Genres such as Action, Comedy, and Drama are represented. A bag-of-words model is used to capture narrative elements through plot keywords. The dataset is split into 9.35% training, 9.35% validation, and 81.3% testing.

DBLP a computer science bibliography website, includes 4,057 authors, 14,328 papers, 7,723 terms, and 20 venues. It spans areas like Database Management, Data Mining, Artificial Intelligence, and Information Retrieval. Authors' research focuses are represented using a bag-of-words model. The dataset is split into 9.86% training, 9.86% validation, and 80.28% testing.

Table 1. Database Structure and Connectivity Statistics for IMDb and DBLP Datasets.

Database	Node	Edge	Metapath
IMDb	movie (M): 4,278	M-D: 4,278	MDM, MAM
	director (D): 2,081	M-A: 12,828	DMD, AMA
	actor (A): 5,257		DMAMD, AMDMA
DBLP	author (A): 4,057	A-P: 19,645	APA
	paper (P): 14,328	P-T: 85,810	APTPA
	term (T): 7,723	P-V: 14,328	APVPA
	venue (V): 20		

5.2 Baselines

We conducted a comprehensive comparison of MICGNN against several semi-supervised graph embedding models, namely GCN, GAT, HAN, and MAGNN. These models are widely used in graph-based machine learning. We cite MAGNN's study, which evaluated all metapaths for GCN, GAT, HAN and achieved the best performance.

Graph Convolutional Network (GCN) is a homogeneous GNN. This model performs convolutional operations in the graph Fourier domain.

Graph Attention Networks (GAT) is a homogeneous GNN that utilizes an attention mechanism to model relationships on homogeneous graphs, generating node embeddings for graph representation.

Heterogeneous Graph Attention Network (HAN) is a homogeneous GNN leverages attention mechanisms to generate metapath-specific node embeddings from various homogeneous graphs.

Metapath Aggregated Graph Neural Network (MAGNN) is a homogeneous GNN which utilizes intra-metapath aggregation and inter-metapath aggregation to aggregate node information. Here we test all the metapaths for MAGNN and report the best performance. MAGNN_avg refers to MAGNN utilizing the mean metapath instance encoder, MAGNN_rot indicates the usage of the relational rotation encoder.

Implementation Details. General settings for GCN, GAT, HAN, MAGNN and MICGNN: Dropout rate is set to 0.5. Same splits of training, validation, and testing sets are used. Adam optimizer is employed with the learning rate set to 0.005 and the weight decay (L2 penalty) set to 0.001. Models are trained for 100 epochs with early stopping applied, having a patience of 30. For node classification and node clustering, the models are trained in a semi-supervised fashion with a small fraction of nodes labeled as guidance. Embedding dimension of all models is set to 64. Specific settings for GAT, Number of attention heads is set to 8. Specific settings for HAN, MAGNN and MICGNN, Dimension of the attention vector in inter-metapath aggregation is set to 128.

5.3 Performance Comparison

Node Classification. An in-depth evaluation of node classification models was performed using IMDb and DBLP datasets, each encompassing test sets of 3,478 and 3,257 nodes respectively. The learned embeddings of labeled nodes were fed into a linear SVM classifier. To ensure fair comparisons, only test set nodes were included in the SVM as semi-supervised models had already processed the training and validation set nodes. The evaluation outcomes, reflecting Macro-F1 and Micro-F1 scores, are compiled in Table 2.

Table 2. Experiment results (%) on the IMDb and DBLP datasets for the node classification task.

Dataset	Metrics	Train%	GCN	GAT	HAN	MAGNN_avg	MAGNN_rot	MICGNN
IMDb	Macro-F1	20%	52.73	53.64	56.19	57.88	57.37	**59.89**
		40%	53.67	55.50	56.15	58.66	59.32	**60.11**
		60%	54.24	56.46	57.29	59.05	60.12	**61.09**
		80%	54.77	57.43	58.51	59.11	61.20	**61.82**
	Micro-F1	20%	52.80	53.64	56.32	57.91	57.37	**59.04**
		40%	53.76	55.56	57.32	58.78	59.32	**60.26**
		60%	54.23	56.47	58.42	59.15	60.06	**61.03**
		80%	54.63	57.40	59.24	59.27	61.14	**61.78**
DBLP	Macro-F1	20%	88.00	91.05	91.69	92.00	92.22	**92.30**
		40%	89.00	91.24	91.96	92.22	92.52	**92.71**
		60%	89.43	91.42	92.14	92.60	**92.95**	92.83
		80%	89.98	91.73	92.50	92.78	92.97	**93.47**
	Micro-F1	20%	88.51	91.61	92.33	92.54	92.46	**92.68**
		40%	89.22	91.77	92.57	92.71	92.06	**93.21**
		60%	89.57	91.97	92.72	93.07	93.32	**93.36**
		80%	90.33	92.24	93.23	93.34	93.45	**93.94**

MICGNN exhibits a commendable performance, surpassing the standard MAGNN model by approximately 1% in IMDb. This demonstrates the rich information encapsulated in metapath instances, which has been judiciously utilized by the MICGNN. In the DBLP context, while the node classification task seems trivial given the high scores obtained by all models, MICGNN still takes the lead, overtaking the strongest baseline by a small yet significant margin.

Node Clustering. Experiments were conducted on the IMDb and DBLP datasets, focusing on comparing the performance of various models in node clustering tasks. The embeddings of labeled nodes (movies in IMDb and authors in DBLP) produced by each learning model were fed into the K-Means algorithm. For each dataset, the number of clusters was set to match the number of classes - three for IMDb and four for DBLP.

Evaluation metrics employed in this study were the Normalized Mutual Information (NMI) and Adjusted Rand Index (ARI). Given the K-Means algorithm's high dependency on centroid initialization, the algorithm was repeated 10 times for each run of the embedding model, and each embedding model was tested for 10 runs. The comprehensive results averaged over these runs are presented in Table 3.

As can be seen in Table 3, MICGNN consistently outperforms other baseline models in node clustering tasks. It's worth noting, however, that the performance of all models on the IMDb dataset is significantly lower than on the DBLP dataset. This can potentially be attributed to the multi-genre nature of movies in the original IMDb dataset, where the first genre is selected as its class label.

Table 3. Experiment results (%) on the IMDb and DBLP datasets for the node clustering task.

Dataset	Metrics	GCN	GAT	HAN	MAGNN_avg	MAGNN_rot	MICGNN
IMDb	NMI	7.46	7.84	10.79	12.4	14.64	**15.22**
	ARI	7.69	8.87	11.11	14.31	14.52	**15.05**
DBLP	NMI	73.45	70.73	77.49	79.31	**79.50**	79.16
	ARI	77.50	76.04	82.95	84.3	84.29	**84.86**

6 Conclusion

In this paper, we have introduced MICGNN, a novel Graph Neural Network for heterogeneous graph embedding that addresses the limitations of existing methods by leveraging a threefold aggregation strategy. Our system effectively captures the nuances of heterogeneous graphs by integrating node transformation, context aggregation, instance aggregation, and semantic aggregation, thereby providing superior node embeddings useful for various downstream tasks. Our approach utilizes an attention mechanism to encode the metapath instance contexts, further enhancing the meaningfulness of the embeddings. The extensive experiments conducted on IMDb and DBLP datasets underscore MICGNN's superior performance in node classification and clustering tasks as compared to the baseline models. Looking forward, we aim to refine this framework by adapting it for rating prediction tasks in the context of user-item data, facilitated by a heterogeneous knowledge graph. Additionally, we plan to explore the potential of enhancing the attention mechanism within the Context Aggregation component by incorporating multi-head attention, to further improve the performance and adaptability of MICGNN in diverse applications.

Acknowledgements. This work is partially supported by the National Natural Science Foundation of China (Grant No. 62002216), the Shanghai Sailing Program (Grant No. 20YF1414400).

References

1. Kipf, T.N., Welling, M.: Semi-supervised classification with graph convolutional networks. In: 5th International Conference on Learning Representations, ICLR. OpenReview.net (2017). https://doi.org/10.48550/arXiv.1609.02907
2. Hamilton, W.L., Ying, Z., Leskovec, J.: Inductive representation learning on large graphs. In: 31st Conference on Neural Information Processing Systems, pp. 1024–1034. Curran Associates Inc., New York (2017). https://doi.org/10.5555/3294771.3294869
3. Velickovic, P., Cucurull, G., Casanova, A., Romero, A., Liò, P., Bengio, Y.: Graph attention networks. In: 6th International Conference on Learning Representations, ICLR. OpenReview.net (2018). https://doi.org/10.48550/arXiv.1706.02216
4. Yujia, L., Daniel, T., Marc, B., Richard, Z.: Gated graph sequence neural networks. In: 4th International Conference on Learning Representations, ICLR. OpenReview.net (2016). https://doi.org/10.48550/arXiv.1511.05493
5. Jiani, Z., Xingjian, S., Junyuan, X., Hao, M., Irwin, K., Dit-Yan, Y.: GaAN: gated attention networks for learning on large and spatiotemporal graphs. In: 34th Conference on Uncertainty in Artificial Intelligence 2018, Corvallis, Oregon, USA, pp. 339–349. Association for Uncertainty in Artificial Intelligence (2018). https://doi.org/10.48550/arXiv.1803.07294
6. Zhang, J., Shi, X., Zhao, S., King, I.: STAR-GCN: stacked and reconstructed graph convolutional networks for recommender systems. In: Proceedings of the 28th International Joint Conference on Artificial Intelligence (IJCAI 2019), Palo Alto, California, pp. 4264–4270. AAAI Press (2019) . https://doi.org/10.5555/3367471.3367634
7. Dong, Y., Chawla, N.V., Swami, A.: Metapath2Vec: scalable representation learning for heterogeneous networks. In: 23rd SIGKDD Conference on Knowledge Discovery and Data Mining, pp. 135–144. Association for Computing Machinery, New York (2017). https://doi.org/10.1145/3097983.3098036
8. Wang, X., et al.: Heterogeneous graph attention network. In: The Web Conference, pp. 2022–2032. Association for Computing Machinery, New York (2019). https://doi.org/10.1145/3308558.3313562
9. Fu, X., Zhang, J., Meng, Z., King, I.: MAGNN: metapath aggregated graph neural network for heterogeneous graph embedding. In: Proceedings of the Web Conference 2020, pp. 2331–2341. Association for Computing Machinery, New York (2020). https://doi.org/10.1145/3366423.3380297
10. Li, X., Ding, D., Kao, C.M., Sun, Y., Mamoulis, N.: Leveraging metapath contexts for classification in heterogeneous information networks. In: 37th IEEE International Conference on Data Engineering (ICDE 2021), pp. 912–923. Curran Associates Inc., New York (2021). https://doi.org/10.48550/arXiv.2012.10024
11. Jianxiang, Y., Xiang, L.: Heterogeneous graph contrastive learning with metapath contexts and weighted negative samples. In: The 23nd SIAM International Conference on Data Mining (SDM 2023), Philadelphia, PA, USA, pp. 37–45. Society for Industrial and Applied Mathematics (2021). https://doi.org/10.48550/arXiv.2012.10024
12. Guan, W., Jiao, F., Song, X., Wen, H., Yeh, C.H., Chang, X.: Personalized fashion compatibility modeling via metapath-guided heterogeneous graph learning. In: Proceedings of the 45th International ACM SIGIR Conference on Research and Development in Information Retrieval (SIGIR 2022), pp. 482–491. Association for Computing Machinery, New York (2022). https://doi.org/10.1145/3477495.3532038

13. Chen, K., Qiu, D.: Combined metapath based attention network for heteroge-
    nous networks node classification. In: Proceedings of the 3rd International Con-
    ference on Advanced Information Science and System (AISS 2021), Article no. 58.
    Association for Computing Machinery, New York (2022). https://doi.org/10.1145/
    3503047.3503109
14. Li, X., Wang, G., Shen, D., Nie, T., Kou, Y.: Heterogeneous embeddings for rela-
    tional data integration tasks. In: Xing, C., Fu, X., Zhang, Y., Zhang, G., Borjigin,
    C. (eds.) WISA 2021. LNCS, vol. 12999, pp. 680–692. Springer, Cham (2021).
    https://doi.org/10.1007/978-3-030-87571-8_59

Finding Introverted Cores in Bipartite Graphs

Kaiyuan Shu[1], Qi Liang[1], Haicheng Guo[1], Fan Zhang[1(✉)], Kai Wang[2], and Long Yuan[3]

[1] Guangzhou University, Guangzhou 510006, China
kaiyuanshu@e.gzhu.edu.cn, zhangf@gzhu.edu.cn
[2] Shanghai Jiao Tong University, Shanghai 200240, China
[3] Nanjing University of Science and Technology, Nanjing 210094, China

Abstract. In this paper, we propose a novel cohesive subgraph model to find introverted communities in bipartite graphs, named (α, β, p)-core. It is a maximal subgraph in which every vertex in one part has at least α neighbors and at least p fraction of its neighbors in the subgraph, and every vertex in the other part has at least β neighbors. Compared to the (α, β)-core, the additional requirement on the fraction of neighbors in our model ensures that the vertices are introverted in the subgraph regarding their neighbor sets, e.g., for the (α, β)-core of a customer-product network, the shopping interests of the customers inside focus on the products in the subgraph. We propose an $O(m)$ algorithm to compute the (α, β, p)-core with given α, β and p. Besides, we introduce an efficient algorithm to decompose a graph by the (α, β, p)-core. The experiments on real-world data demonstrate that our model is effective and our proposed algorithms are efficient.

Keywords: Cohesive subgraph · Introverted community · Bipartite graph

1 Introduction

Many real-world relationships among various entities are modeled as bipartite graphs, such as collaboration networks [11], gene co-expressions networks [9], customer-product networks [19], and user-page networks [4]. The analysis of bipartite graphs become increasingly important with well applied, makes (α, β)-core a valuable model in mining cohesive subgraphs with applications mainly in [5,7,20,23]. The (α, β)-core of G is a maximal bipartite subgraph G' with two node sets $U' \subseteq U$ and $V' \subseteq V$ and every vertex has at least α and β neighbors. Although it can ensure certain engagement of vertices, the characteristic of vertices would be ignored as vertices have many neighbors outside the subgraph.

Motivated by above, we propose (α, β, p)-core as a novel cohesive subgraph model. It is a maximal subgraph G' with two node sets $U' \subseteq U$ and $V' \subseteq V$, where vertices in U' have at least α neighbors and at least p fraction of its

L. Yuan et al. (Eds.): WISA 2023, LNCS 14094, pp. 162–170, 2023.
https://doi.org/10.1007/978-981-99-6222-8_14

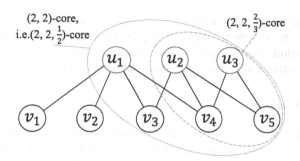

Fig. 1. An Instance of (α, β, p)-core

neighbors, with degree of vertices in V' is at least β in subgraph. The fraction constraint p is to capture introverted users and explore finer granularity in characterizing users accurately with (α, β)-core refined. Figure 1 shows an instance of (α, β, p)-core. The $(2, 2)$-core induced by u_1 to u_3 and v_3 to v_5 also a $(2, 2, \frac{1}{2})$-core. $(2, 2, \frac{2}{3})$-core is induced by u_2, u_3, v_4, v_5, as u_2. u_3 are more introverted and homogeneous while the focus of u_1 is not well covered in the core.

In the following, we summarize the contributions of this paper.

- We propose a novel cohesive subgraph (α, β, p)-core for bipartite graphs. With the extra fraction constraint p, it can customize the degree requirement for every vertex and capture introverted vertices compared to the (α, β)-core.
- We devise an algorithm with $O(m)$ time complexity to compute a (α, β, p)-core subgraph with given α, β and p values. We propose a decomposition algorithm with $O(dmax_U(G) \cdot dmax_V(G) \cdot m)$ time cost to compute (α, β, p)-core for all α, β and p values, where $dmax_U(G)$ (resp. $dmax_V(G)$) is the maximum degree of vertices in set $U(G)$ (resp. $V(G)$), and m is the number of edges in G.
- The experiments on several real-world graphs demonstrate the effectiveness of our (α, β, p)-core model and the efficiency of our algorithms.

2   Related Work

Cohesive subgraph mining is a fundamental problem in graph structure analysis. The fundamental structure of dense subgraph is clique [13], with other new structures like quasi-clique [1], k-plex [17], k-truss [18] and k-core [16]. [3] gives an $O(m)$ algorithm for core decomposition. [22] proposes a finer granularity model and corresponding efficient algorithms. They are all proposed for general graphs. For cohesive subgraphs mining in bipartite graphs, [8] and [10] propose the definition of biclique on bipartite graphs. [24] proposes an efficient algorithm to enumerate all the bicliques. [15] defines a framework of bipartite subgraphs based on the butterfly motif (2,2-biclique) to model the dense regions. [14] studies the problem of butterfly counting. The (α, β)-core was proposed by [2]. [6] and [7] proposed the linear algorithm to compute (α, β)-core. An index-based method computation of (α, β)-core is proposed in [12]. [21] studies the problem of (α, β)-core decomposition. To the best of our knowledge, we are the first to consider the fraction constraint of vertex neighbors in bipartite subgraphs.

3   Preliminaries

The bipartite graph $G = (U, V, E)$ has two node sets U and V, with edges from $E \subseteq U \times V$. $U(G)$ and n_u (resp. $V(G)$ and n_v) are the node set and node number of G, with n as the total number of nodes. $E(G)$ and m are edge set and total edges numbers of G. The degree of $u \in U(G) \cup V(G)$ is $deg(u, G)$, with maximum degree $dmax_U(G)$ (resp. $dmax_V(G)$) in $U(G)$ (resp. $V(G)$). For subgraph G' of G with node sets $U' \subseteq U$ and $V' \subseteq V$, we denote $U(G') = U', V(G') = V'$ and $E(G') = E(G) \cap (U' \times V')$. To simplify, we represent bipartite graphs as graphs.

Definition 1 ((α, β)-core). *Given a graph G and two integers α and β, a subgraph S is the (α, β)-core of G, denote by $C_{\alpha,\beta}(G)$, if (1) $deg(u, S) \geq \alpha$ for every $u \in U(S)$; (2) $deg(v, S) \geq \beta$ for every $v \in V(S)$; (3) S is maximal, i.e., any subgraph S' is not a (α, β)-core if S is a subgraph of S' and $S \neq S'$.*

Obviously, the (α, β)-core of G is unique with given α and β. According to Definition 1, the (α, β)-core of G can be computed by removing vertices which not satisfy the degree constraint with time complexity $O(m)$ in the worst case.

Definition 2 (α/β-core number). *Given an integer α, if a vertex $w \in G$ can be in an (α, β)-core but not in $(\alpha, \beta + 1)$-core, then we say the α-core number of node w is β, denote as $cn_\alpha(w) = \beta$. Correspondingly, the β-core number of w is defined when β is fixed and we denote it by $cn_\beta(w)$.*

From Definition 2, whether a vertex belongs to a (α, β)-core relies on its α(or β)-core number. Since the α-core number and β-core number are similar, we focus on α-core number for the rest. Before introducing the (α, β, p)-core, we define the fraction of a vertex regarding a subgraph S and the graph G.

Definition 3 (fraction). *Given a graph G and a subgraph S of G, the fraction of vertex v in S is defined as the degree of v in S divided by the degree of v in G, i.e., $frac(v, S, G) = deg(v, S)/def(v, G)$.*

Based on the Definition 2 and 3, we define (α, β, p)-core as follows.

Definition 4 ((α, β, p)-core). *Given a graph G, two integers α and β, and a decimal p, a subgraph S is the (α, β, p)-core of G, denoted by $C_{\alpha,\beta,p}(G)$, if every vertex u in $U(G)$ and v in $V(G)$, (1) $deg(u, S) \geq \alpha$ and $deg(v, S) \geq \beta$ (degree constraint); (2) $frac(u, S, G) \geq p$ (fraction constraint) and (3) S is maximal, i.e., any subgraph S' is not a (α, β, p)-core if S is a subgraph of S' and $S \neq S'$.*

Example 1. In Fig. 1, the $(2, 2)$-core have vertices u_1 to u_3 and v_3 to v_5. The 2-core number of vertex u_1 is 2, i.e., $cn_2(u_1) = 2$. As two neighbors of u_1 (i.e., v_1 and v_2) are not in the $(2, 2)$-core, the fraction of u_1 in $(2, 2)$-core is 0.5. The $(2, 2)$-core is a $(2, 2, \frac{1}{2})$-core which also contains $(2, 2, \frac{2}{3})$-core.

Algorithm 1: (α, β, p)-core computation

Input: a graph G, the degree threshold α, β, the fraction threshold p
Output: (α, β, p)-core of G
1 $t[u] \leftarrow max(\alpha, \lceil p \times deg(u, G) \rceil)$ for every vertex $u \in U(G)$;
2 $t[v] \leftarrow \beta$ for every vertex $v \in V(G)$;
3 $G' \leftarrow G$;
4 **while** $\exists u \in U(G') : deg(u, G') < t[u]$ *or* $\exists v \in V(G') : deg(v, G') < t[v]$ **do**
5 $\quad \lfloor$ remove u (or v) and its incident edges from G';
6 **return** G';

Algorithm 2: p-number computation

Input: a (α, β)-core of G named S
Output: the p-core numbers of all vertices in S
1 $S \leftarrow (\alpha, \beta)$-core of G;
2 **while** $U(S)$ *is not empty* **do**
3 $\quad p_{min} \leftarrow min\{deg(u, S)/deg(u, G), \forall u \in U(S)\}$;
4 $\quad$ **while** $\exists u \in U(S) : deg(u, S) < \alpha$ *or* $deg(u, S)/deg(u, G) \leq p_{min}$ **do**
5 $\quad\quad p_{(\alpha,\beta)}[u] \leftarrow p_{min}$;
6 $\quad\quad$ remove u and its incident edges from S;
7 $\quad\quad$ **while** $\exists v \in V(S) : deg(v, S) < \beta$ **do**
8 $\quad\quad\quad p_{(\alpha,\beta)}[v] \leftarrow p_{min}$;
9 $\quad\quad\quad \lfloor$ remove v and its incident edges from S;
10 **return** $p_{(\alpha,\beta)}[w]$;

With any α, β and p, the (α, β, p)-core always be a subgraph of (α, β)-core in G, and unique as it is required to be maximal when α, β and p are fixed. In the subsequent sections, we study the (α, β, p)-core computation in two parts: (1) we compute the (α, β, p)-core of G with specified α, β and p values; and (2) the (α, β, p)-core decomposition retrieves (α, β, p)-core for all possible triples (α, β, p) values where $1 \leq \alpha < dmax_U(G), 1 \leq \beta \leq dmax_V(G)$ and $0 \leq p \leq 1$.

4 (α, β, p)-Core Computation and Decomposition

Algorithm 1 shows the pseudo-code of (α, β, p)-core computation with given α, β, and p, with removing vertices iteratively which not satisfy the constraint until no nodes can be removed. We compute the degree threshold in $U(G)$ as the larger one of α and $\lceil p \times deg(u, G) \rceil$, with degree threshold β of $V(G)$ in Lines 1–2. We check the degree on G in Line 3. If u and v cannot satisfy the degree threshold, we delete them and their incident edges (Line 4). Note that the deletion of a vertex can decrease the degrees of its neighbors, we should delete the vertex and its adjacent edges when its degree is smaller than the degree threshold. We do not need to update the vertex degree which is already under the degree threshold.

Example 2. Figure 1 shows an example graph. For $(2, 2, \frac{2}{3})$-core, we compute the degree threshold for every vertex at first. As u_1, u_2 and u_3 have thresholds $\frac{8}{3}$, 2, and $\frac{4}{3}$, the threshold in $V(G)$ is 2. We delete u_1, v_1, and v_2 and their incident edges since they cannot satisfy the threshold. As the remaining nodes satisfy the degree and fraction constraint, we return the subgraph of u_2, u_3, v_3, v_4, and v_5.

Algorithm 3: (α, β, p)-core decomposition

Input: a graph G
Output: the (α, β, p)-core numbers of all vertices in G
1   **foreach** $\alpha = 1$ to $dmax_U(G)$ **do**
2      $G' \leftarrow G$;
3   **while** $\exists u \in U(G')$ and $deg(u, G') < \alpha$ **do**
4   remove u and its incident edges from G';
5   **while** G' is not empty **do**
6         $\beta_{min} \leftarrow min\{deg(v, G') \mid v \in V(G')\}$;
7   do p-number computation on G' $((\alpha, \beta_{min})$-core$)$;
8   **while** $\exists v \in V(G')$ and $deg(v, G') \leq \beta_{min}$ **do**
9   remove v and its incident edges from G';
10   **while** $\exists u \in U(G')$ and $deg(u, G') < \alpha$ **do**
11   remove u and its incident edges from G';

12   **return** the (α, β, p)-core numbers of all vertices in G;

Although the α (resp. β) ranges from 1 to $dmax_U(G)$ (resp. $dmax_V(G)$), p ranges from 0 to 1 continuously, many (α, β, p)-cores are the same with different α, β or p. Based on this observation, we first define the p-number of a vertex.

Definition 5 (*p-number*). *Given a graph G and degree thresholds α, β, decimal p is the p-number of a vertex u, denoted by $pn(u, \alpha, \beta, p)$, if (1) the (α, β, p)-core contains u and (2) the (α, β, p')-core does not contain u for any $p' > p$.*

Definition 6 (*(α, β, p)-core number*). *Given a graph G, a (α, β, p)-core number of a vertex is a tuple of $\langle \alpha, \beta, p \rangle$, where the value of p is the p-number of the vertex based on the pair of α and β.*

As above mentioned, we compute the (α, β, p)-core in given α, β, and p in $O(m)$ time. However, Algorithm 1 costs too much to return the result with lots of requests of computing (α, β, p)-core in a large graph. So we introduce the (α, β, p)-core decomposition, which aims to find all (α, β, p)-core numbers and make computation more effective with the result of (α, β, p)-core decomposition.

We propose Algorithm 2 to compute the p-number of given (α, β)-core (S for convenience). The fraction of vertices in S may be different. We compute the minimum fraction for all vertices in S first (Line 3). For vertices in $U(S)$ with a degree smaller than α or fraction not larger than p_{min}, we record the p-number of them as p_{min} (Line 5) and remove them with their incident edges from S at Line 6. As removing nodes in $U(S)$ will decrease the degree of the vertex in $V(G)$, we check the degree of all vertices in $V(S)$. If vertices have a smaller degree than β, we record the p-number of them as p_{min} (Line 8) and remove them with their incident edges in S (Line 9). Repeat these processes until S is empty, then we return the p-number of all vertices in given (α, β)-core.

Algorithm 3 is (α, β, p)-core decomposition. We fix α from 1 to $dmax_U(G)$ (Line 1) and delete vertices with degree smaller than α in $U(G)$, ensured the new subgraph is $(\alpha, 1)$-core (denoted G')(Lines 3–4). As the minimum degree value (β_{min}) reveal in Line 6, we do the p-number computation on (α, β_{min})-core and get tuples of (α, β_{min}, p) as (α, β, p)-core numbers. We delete the vertices with degree is β_{min} in $V(G')$ to ensure the degree of vertices in $U(G')$ is more than α (Lines 8–11). We compute a new minimum degree value and repeat the above process until G' is empty. Finally, we will return a (α, β, p)-core number set of each vertex in G. Note that the computation can be also started by fixing β.

Example 3. In Fig. 1, suppose that we begin the decomposition from $\alpha = 2$. The whole graph is a $(2, 1, 1)$-core. When $\beta = 2$, v_1 and v_2 would be deleted as they have the smallest degree, and the subgraph is a $(2, 2)$-core. Node u_1 is the first leaving node as the fraction of u_1 is the smallest, so the (α, β, p)-core number of u_1 is $(2, 2, \frac{1}{2})$. Then v_3 leaves as v_3 does not satisfy the degree threshold β. The u_2 has (α, β, p)-core number $(2, 2, \frac{2}{3})$ since its fraction is the smallest and delete it. After that, the other nodes should be deleted as they do not satisfy the degree constraint, and we finish the decomposition in $\alpha = 2$.

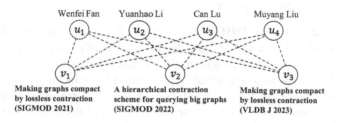

Fig. 2. Case Study on DBLP

Complexity. The time complexity of (α, β, p)-core computation is $O(m)$, with computing degree threshold for each vertex, finding the first vertex to delete after computation, iterative deletion of disqualified vertices while each edge is visited once. The time complexity of p-number computation is $O(m)$. In Algorithm 2, vertex and edge deletion with degree update takes $O(m)$ time as vertices and edges are deleted once. Based on p-number computation, the (α, β, p)-core decomposition costs $O(dmax_U(G) \cdot dmax_V(G) \cdot m)$ time.(Assume $n \ll m$)

5 Experiment

In this section, extensive experiments are conducted to demonstrate the great performance of our model and algorithms.

Datasets. We evaluate the algorithms on five real graphs. The graphs are downloaded from KONECT[1]. The details of these graphs are shown in Table 1.

---

[1] http://konect.cc/networks/.

Table 1. Statistics of Datasets

Dataset	Type	$|U|$	$|V|$	$|E|$	$dmax_U$	$dmax_V$
Writers	Authorship	89k	46k	0.14 m	42	246
Actors	Starring	76k	81k	0.28 m	65	321
YouTube	Membership	94k	30k	0.29 m	1,035	7,591
GitHub	Membership	56k	0.12k	0.44 m	884	3,675
Teams	Membership	0.9 m	34k	1.3 m	17	2,671

Parameters. We conduct experiments under different settings by varying the threshold α, β, p. The default values of α, β, and p are 5, 5, and 0.6, respectively.

Environment. We utilize C++ as the programming language for implementing all algorithms and employ GCC (7.5.0) to compile them with O3 optimization. Our experiments are performed on a computer equipped with an Intel Xeon 2.1GHz CPU and 512G of RAM.

Exp-1: Case Study on DBLP. We extract $(3, 3, 0.15)$-core on DBLP from 2021 to 2023. The chosen subset includes four authors as depicted in Fig. 2. As u_1 to u_4 have 17, 3, 5, and 6 neighbors in $(3, 3, 0.15)$-core, u_1 is not included in the $(3, 3, 0.5)$-core due to p constraint. Li, Lu, and Liu focus on graph contraction as their main research area, except Fan, who focuses on graph partitioning, rule discovery, etc. Thus, with the finer structural constraint derived from p, (α, β, p)-core is capable of mining introverted subgraphs compared to the (α, β)-core.

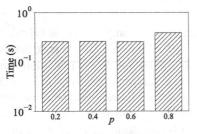

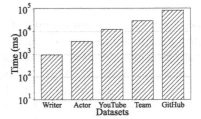

(a) Running Time of Varying p (b) Running Time of Decomposition

Fig. 3. Running Time

Exp-2: Running Time of Varying p. Figure 3(a) shows the computation time of varying p on YouTube with given α and β in 5. The leaving of nodes takes more time as they are smaller than the fraction threshold with p grows. The time-costing deletion proves the remaining graph differ little except $p{=}0.8$. The vertices count in the final subgraph demonstrates our model can find more homogeneous vertices with p growing from 0 to 1. Overall, the fraction p played an important role in finding introverted and homogeneous vertices of (α, β)-core.

Exp-3: Running Time of Decomposition. Figure 3(b) shows the running time of the decomposition in different datasets. We see that the time of decomposition increases as the size of the dataset grows. The result of the decomposition is in an acceptable range which can be evaluated theoretically in Sect. 4.

6   Conclusion

In this paper, we propose a novel cohesive subgraph model in the bipartite graph, (α, β, p)-core, with an algorithm to compute the (α, β, p)-core in given α, β, and p. Besides, we study and design an algorithm in (α, β, p)-core decomposition.

References

1. Abello, J., Resende, M.G.C., Sudarsky, S.: Massive quasi-clique detection. In: Rajsbaum, S. (ed.) LATIN 2002. LNCS, vol. 2286, pp. 598–612. Springer, Heidelberg (2002). https://doi.org/10.1007/3-540-45995-2_51
2. Ahmed, A., Batagelj, V., Fu, X., Hong, S.H., Merrick, D., Mrvar, A.: Visualisation and analysis of the internet movie database. In: 2007 6th International Asia-Pacific Symposium on Visualization, pp. 17–24. IEEE (2007)
3. Batagelj, V., Zaversnik, M.: An o (m) algorithm for cores decomposition of networks. arXiv preprint cs/0310049 (2003)
4. Beutel, A., Xu, W., Guruswami, V., Palow, C., Faloutsos, C.: CopyCatch: stopping group attacks by spotting lockstep behavior in social networks. In: Proceedings of the 22nd International Conference on World Wide Web, pp. 119–130 (2013)
5. Cai, Z., He, Z., Guan, X., Li, Y.: Collective data-sanitization for preventing sensitive information inference attacks in social networks. IEEE Trans. Dependable Secure Comput. **15**(4), 577–590 (2016)
6. Cerinšek, M., Batagelj, V.: Generalized two-mode cores. Soc. Nctw. **42**, 80–87 (2015)
7. Ding, D., Li, H., Huang, Z., Mamoulis, N.: Efficient fault-tolerant group recommendation using alpha-beta-core. In: Proceedings of the 2017 ACM on Conference on Information and Knowledge Management, pp. 2047–2050 (2017)
8. Hochbaum, D.S.: Approximating clique and biclique problems. J. Algorithms **29**(1), 174–200 (1998)
9. Kaytoue, M., Kuznetsov, S.O., Napoli, A., Duplessis, S.: Mining gene expression data with pattern structures in formal concept analysis. Inf. Sci. **181**(10), 1989–2001 (2011)
10. Lehmann, S., Schwartz, M., Hansen, L.K.: Biclique communities. Phys. Rev. E **78**(1), 016108 (2008)
11. Ley, M.: The DBLP computer science bibliography: evolution, research issues, perspectives. In: Laender, A.H.F., Oliveira, A.L. (eds.) SPIRE 2002. LNCS, vol. 2476, pp. 1–10. Springer, Heidelberg (2002). https://doi.org/10.1007/3-540-45735-6_1
12. Liu, B., Yuan, L., Lin, X., Qin, L., Zhang, W., Zhou, J.: Efficient (α, β)-core computation: an index-based approach. In: The World Wide Web Conference, pp. 1130–1141 (2019)
13. Luce, R.D., Perry, A.D.: A method of matrix analysis of group structure. Psychometrika **14**(2), 95–116 (1949)

14. Sanei-Mehri, S.V., Sariyuce, A.E., Tirthapura, S.: Butterfly counting in bipartite networks. In: Proceedings of the 24th ACM SIGKDD International Conference on Knowledge Discovery & Data Mining, pp. 2150–2159 (2018)
15. Sarıyüce, A.E., Pinar, A.: Peeling bipartite networks for dense subgraph discovery. In: Proceedings of the Eleventh ACM International Conference on Web Search and Data Mining, pp. 504–512 (2018)
16. Seidman, S.B.: Network structure and minimum degree. Soc. Netw. **5**(3), 269–287 (1983)
17. Seidman, S.B., Foster, B.L.: A graph-theoretic generalization of the clique concept. J. Math. Sociol. **6**(1), 139–154 (1978)
18. Wang, J., Cheng, J.: Truss decomposition in massive networks. arXiv preprint arXiv:1205.6693 (2012)
19. Wang, J., De Vries, A.P., Reinders, M.J.: Unifying user-based and item-based collaborative filtering approaches by similarity fusion. In: Proceedings of the 29th Annual International ACM SIGIR Conference on Research and Development in Information Retrieval, pp. 501–508 (2006)
20. Xu, K., Williams, R., Hong, S.-H., Liu, Q., Zhang, J.: Semi-bipartite graph visualization for gene ontology networks. In: Eppstein, D., Gansner, E.R. (eds.) GD 2009. LNCS, vol. 5849, pp. 244–255. Springer, Heidelberg (2010). https://doi.org/10.1007/978-3-642-11805-0_24
21. Yu, D., Zhang, L., Luo, Q., Cheng, X., Cai, Z.: Core decomposition and maintenance in bipartite graphs. Tsinghua Sci. Technol. **28**(2), 292–309 (2022)
22. Zhang, C., et al.: Exploring finer granularity within the cores: efficient (k, p)-core computation. In: 2020 IEEE 36th International Conference on Data Engineering (ICDE), pp. 181–192. IEEE (2020)
23. Zhang, H., Chen, Y., Li, X., Zhao, X.: Simplifying knowledge-aware aggregation for knowledge graph collaborative filtering. In: Zhao, X., Yang, S., Wang, X., Li, J. (eds.) WISA 2022. LNCS, vol. 13579, pp. 52–63. Springer, Cham (2022). https://doi.org/10.1007/978-3-031-20309-1_5
24. Zhang, Y., Phillips, C.A., Rogers, G.L., Baker, E.J., Chesler, E.J., Langston, M.A.: On finding bicliques in bipartite graphs: a novel algorithm and its application to the integration of diverse biological data types. BMC Bioinform. **15**, 1–18 (2014)

Recommender Systems

GENE: Global Enhanced Graph Neural Network Embedding for Session-Based Recommendation

Xianlan Sun[1], Dan Meng[2], Xiangyun Gao[1], Liping Zhang[1], and Chao Kong[1(✉)]

[1] School of Computer and Information, Anhui Polytechnic University, Wuhu, China
{XianlanSun,XiangyunGao,LipingZhang,ChaoKong}@ahpu.edu.cn
[2] OPPO Research Institute, Shenzhen, China
mengdan@oppo.com

Abstract. The session-based recommendation aims to generate personalized item suggestions by using short-term anonymous sessions to model user behavior and preferences. Early studies cast the session-based recommendation as a personalized ranking task, and adopt graph neural networks to aggregate information about users and items. Although these methods are effective to some extent, they have primarily focused on adjacent items tightly connected in the session graphs in general and overlooked the global preference representation. In addition, it is difficult to overcome the special properties of popularity bias in the real-world scenario. To address these issues, we propose a new method named GENE, short for *Global Enhanced Graph Neural Network Embedding*, to learn the session graph representations for the downstream session-based recommendation. Our model consists of three components. First, we propose to construct the session graph based on the order in which the items interact in the session with normalization. Second, we employ a graph neural network to obtain the latent vectors of items, then we represent the session graph by attention mechanisms. Third, we explore the session representation fusion for prediction incorporating linear transformation. The three components are integrated in a principled way for deriving a more accurate item list. Both quantitative results and qualitative analysis verify the effectiveness and rationality of our GENE method.

Keywords: Session-based recommendation · Attention mechanism · Representation learning · Popularity bias

1 Introduction

With the popularization of mobile applications, a huge volume of user-generated content (UGC) has become available in a variety of domains [1]. Information overload from massive data has always troubled people. To solve this problem,

---

X. Sun and D. Meng—These authors contribute equally to this work.

© The Author(s), under exclusive license to Springer Nature Singapore Pte Ltd. 2023
L. Yuan et al. (Eds.): WISA 2023, LNCS 14094, pp. 173–184, 2023.
https://doi.org/10.1007/978-981-99-6222-8_15

recommender systems [2] have emerged as the times require. Many of the existing systems provide items of interest for a user based on users' characteristics and historical interactions [3]. However, nowadays, users are more concerned about privacy protection and often choose to anonymously browse websites or log in when making purchases. Additionally, even if users log in, their interaction information may only have short-term records, which makes it difficult for traditional recommender systems to predict accurate ranking lists [4]. To address these issues, researchers have proposed session-based recommender systems, which predict upcoming actions based on a sequence of previously clicked items in the user's current session. Compared with traditional recommender systems, session-based recommender systems [5] can better provide personalized recommendations for anonymous users who are not registered or logged in without personal information such as age, gender, ID, etc.

Recent advances in session-based recommendation have primarily focused on deep learning-based methods [6]. Most of them have employed graph structures to model sessions and they also have employed graph neural networks [7] to propagate information between adjacent items. These methods have improved the performance of recommender systems to some extent. However, they often do not give enough weight to learn the global preference of the session and mainly focus on the local preference of the user's current session. Moreover, existing GNN-based recommendation models [8,9] often suffer from popularity bias in the real-world scenario, they mainly focus on recommending popular items and ignore related long-tail items (less popular or less frequent items). In addition, the candidate items are usually diverse, and users' interests are often varied. Previous work that represents a single user's interests with a fixed-size vector may not fully model users' dynamic preferences, which limits the expressiveness of the recommendation model.

To address the limitations of existing methods on session-based recommendation, we devise a new solution called GENE (Global Enhanced Graph Neural Network Embedding) for the session-based recommendation task. Firstly, in order to alleviate the issue of popularity bias, we normalized the embedding of the item. Secondly, considering that the graph neural network method only aggregates information from adjacent items in the session graph, the self-attention mechanism is introduced to enhance the graph neural network by explicitly focusing on all positions to capture global dependencies. Finally, considering the limitation of over-relying on the last item in the target session, we introduce target-aware attention to calculate the attention scores of all items in the session relative to each target item, in order to understand the contribution of each item to user preferences. We simultaneously considered the issue of popularity bias and the insufficient information captured by graph neural network methods and proposed our session representation model. The main contributions of this work are summarized as follows.

- In order to alleviate the popularity bias problem, we employ item normalization for session-graph representations.

- For session embedding integration, we employ the attention mechanisms to obtain global embedding and target embedding, while considering integrating global embedding, local embedding, and target embedding into a total session embedding to improve the recommendation performance of the current session.
- We perform extensive experiments on several real datasets to illustrate the effectiveness and rationality of GENE.

The remainder of the paper is organized as follows. We first review the related work in Sect. 2. We formulate the problem in Sect. 3, before delving into details of the proposed method in Sect. 4. We perform extensive empirical studies in Sect. 5 and conclude the paper in Sect. 6.

2 Related Work

Our work is related to the following research directions.

2.1 Session-Based Recommendation

Traditional session-based recommendation research can be roughly divided into two categories: item-based neighborhood recommendation methods and sequence recommendation methods based on Markov chains [10]. The former recommends the items with the highest similarity to the currently clicked object. The main idea of the latter is to map the current session to a Markov chain and then infer the user's next action based on the previous session. Rendle et al. [11] proposed a method that combines matrix factorization and Markov chain models to capture sequential behavior between adjacent items. However, most Markov chain-based models only consider the most recent click in the session, so they cannot capture complex high-order sequence patterns.

With the development of deep learning, technology has brought significant performance improvements to session-based recommendations. GRU4Rec [12] is the first to apply RNN to the session-based recommendation. On this basis, PAN [13] captures the short-term and long-term preferences of the target user through the attention network. DGTN [14] uses neighbor information to better predict user preferences. Although the above models have been successful to some extent, they are unable to model complex information between non-adjacent items well.

The development and application of graph neural networks have overcome the limitations of previous methods. Most methods based on graph neural networks first model a session sequence as a session graph [15] or merge multiple session graphs into a global session graph [16], and then use GNN to aggregate the information of related neighbor nodes in the graph. SR-GNN [17] proposes to transform session sequences into session graphs to obtain complex information between items. GC-SAN [18] aims to learn the long-term dependency relationship between items by introducing a self-attention mechanism. GCE-GNN [16] uses information from other sessions as auxiliary information for the recommendation model. It uses graph attention networks to learn item representations at the two levels of the session graph and the global graph. TAGNN [19] further improves session representation by calculating session representation vectors under specified target items through an attention mechanism to adaptively activate different interests of users in a session. Although significant progress has been made in these works, these GNN-based methods have ignored the impact of popularity bias. The work of this article is deeply inspired by these groundbreaking efforts. We comprehensively consider issues related to popularity bias and session embedding to better recommend.

3 Problem Formulation

We first introduce the notations used in this paper and then formalize the session-based recommendation problem to be addressed.

Notations. We assume that V represents the set of items for all sessions, and S represents the set of all past sessions. Each anonymous session $s = [v_{s,1}, v_{s,2}, \ldots, v_{s,n}]$ consists of a series of interactive items in session chronological order, where $v_{s,i} \in V$ represents a click item by the user in the session s. Each session s can be modeled as a directed graph $\mathcal{G}_s = (\mathcal{V}_s, \mathcal{E}_s, \mathbf{A}_s)$, where $\mathcal{V}_s$, $\mathcal{E}_s$, and $\mathbf{A}_s$ represent the node set, edge set, and adjacency matrix respectively. In the session graph, each node represents an item $v_{s,i} \in V$, and each edge $(v_{s,i-1}, v_{s,i}) \in \mathcal{E}_s$ represents a user clicking on item $v_{s,i-1}$ and then clicking on item $v_{s,i}$. Here $\mathbf{A}_s$ is defined as the concatenation of two adjacency matrices $\mathbf{A}_s^{in}$ and $\mathbf{A}_s^{out}$, where $\mathbf{A}_s^{in}$ and $\mathbf{A}_s^{out}$ represent weighted connections of incoming and outgoing edges respectively.

Problem Definition. The goal of session-based recommendation is to predict the next interaction item $v_{s,i+1}$ in session s.

Input: The adjacency matrices $\mathbf{A}_s^{in}$ and $\mathbf{A}_s^{out}$, as well as the normalized item embeddings $\tilde{\mathbf{V}}$.

Output: A ranking list for all candidate items. The top k items in the ranking list will be selected for recommendation.

4 Methodology

To enhance the session recommendation model based on graph neural networks, we consider popularity bias and session embedding issues. In this section, we present the proposed GENE model to address the three major challenges mentioned in Sect. 1 and demonstrate the proposed model framework in Fig. 1.

4.1 Session Graph Construction

Frequent clicks on popular items indicate that popular products have a larger inner product and higher ranking during the final sorting calculation, resulting in more frequent recommendations. For example, the embedded vector $\mathbf{v}_k$ has a larger $\|\mathbf{v}_k\|_2$, resulting in a higher inner product $\mathbf{v}_k^T \mathbf{s} = \|\mathbf{v}_k\|_2 \|\mathbf{s}\|_2 \cos \alpha$ to

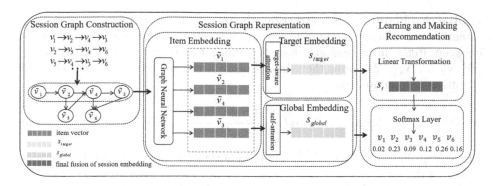

Fig. 1. An overview of GENE model.

ensure the maximum value of **y**. In order to better address the issue of popularity bias, we considered minimizing the impact of embedding norms. We normalize the item embedding as $\tilde{\mathbf{v}}_k = Dropout(Norm(\mathbf{v}_k), p)$, where p is the dropout probability, $Norm(\cdot)$ is the L2 normalization function. Finally, the itemset $\tilde{\mathbf{V}}$ is obtained to enforce the unit norm. After item normalization, complete the graph construction of the current graph based on the session graph in the problem formulation.

4.2 Session Graph Representation

Item Embedding. Graph neural networks are very suitable for session-based recommendations because they can automatically extract the characteristics of session graphs considering node connections. Given the adjacency matrices $\mathbf{A}_s^{in}$ and $\mathbf{A}_s^{out}$, as well as the normalized item embeddings $\tilde{\mathbf{V}}$ as input, for any node in the graph, the current representation of the node and the representation of its adjacent nodes are used to iteratively update the node representation. Specifically, for nodes in the directed session graph, the update rules are:

$$\mathbf{a}_{s,i}^t = \left[\mathbf{A}_{s,i:}^{in}\, \tilde{\mathbf{V}}_s^{t-1}\mathbf{H}_1, \mathbf{A}_{s,i:}^{out}\, \tilde{\mathbf{V}}_s^{t-1}\mathbf{H}_2\right]^T + \mathbf{b}, \tag{1}$$

$$\mathbf{z}_{s,i}^t = \sigma\left(\mathbf{W}_z\mathbf{a}_{s,i}^t + \mathbf{U}_z\tilde{\mathbf{v}}_{s,i}^{t-1}\right), \tag{2}$$

$$\mathbf{r}_{s,i}^t = \sigma\left(\mathbf{W}_r\mathbf{a}_{s,i}^t + \mathbf{U}_r\tilde{\mathbf{v}}_{s,i}^{t-1}\right), \tag{3}$$

$$\hat{\mathbf{v}}_i^t = \tanh\left(\mathbf{W}_o\mathbf{a}_{s,i}^t + \mathbf{U}_o\left(\mathbf{r}_{s,i}^t \odot \tilde{\mathbf{v}}_{s,i}^{t-1}\right)\right), \tag{4}$$

$$\tilde{\mathbf{v}}_i^t = \left(1 - \mathbf{z}_{s,i}^t\right) \odot \tilde{\mathbf{v}}_i^{t-1} + \mathbf{z}_{s,i}^t \odot \hat{\mathbf{v}}_i^t, \tag{5}$$

where $\mathbf{H}_1, \mathbf{H}_2 \in \mathbb{R}^{d\times d}$ control the weight, $\mathbf{W}_z, \mathbf{W}_r, \mathbf{W}_o \in \mathbb{R}^{2d\times d}$ are the weight matrices, $\mathbf{z}_{s,i}^t$ is the reset gate, $\mathbf{r}_{s,i}^t$ is the update gate, t is the training step size, $\sigma(\cdot)$ is the sigmoid function, and $\odot$ is the element-wise multiplication operator.

Global Embedding. Self-attention allows different attention heads to pay attention to information at different positions to capture global preferences between items in a session, after obtaining the vector of all nodes $\tilde{\mathbf{V}} = [\tilde{\mathbf{v}}_1^t, \tilde{\mathbf{v}}_2^t, \ldots, \tilde{\mathbf{v}}_i^t]$. We input them into the self-attention layer, and the ith head attention is computed as:

$$head_i = \text{softmax}(\frac{(\tilde{\mathbf{V}}\mathbf{W}_i^Q)(\tilde{\mathbf{V}}\mathbf{W}_i^K)^T}{\sqrt{d}})(\tilde{\mathbf{V}}\mathbf{W}_i^V), \tag{6}$$

where $\mathbf{W}_i^Q, \mathbf{W}_i^K, \mathbf{W}_i^V \in \mathbb{R}^{d \times (d/h)}$ are parameter matrices of the ith head attention, h is the number of attention heads.

Then, the h heads of self-attention are combined with multi-head attention. The multi-head attention $\mathbf{F}$ is computed as:

$$\mathbf{F} = \text{Concat} (\text{head}_1, \cdots, \text{head}_h) \mathbf{W}^o, \tag{7}$$

where $\mathbf{W}^o \in \mathbb{R}^{d \times d}$ is the projection matrix. On this basis, F is passed through the feedforward network with linear transformation and an activation function, as given below:

$$\mathbf{Y} = \max (0, \mathbf{F}\mathbf{W}_1 + \mathbf{b}_1) \mathbf{W}_2 + \mathbf{b}_2 + \mathbf{F}, \tag{8}$$

where output $\mathbf{Y} = [\mathbf{y}_1, \mathbf{y}_2, \cdots, \mathbf{y}_n] \in \mathbb{R}^{n \times d}$, n is the session length, and d is the item embedding vector dimension. Finally, for each session s, $\mathbf{y}_n$ is the latent vector of the last item, adaptively aggregating all other project information. Therefore, we use the latent vector of the last item as the global embedding vector $\mathbf{s}_{\text{global}} \in \mathbb{R}^d$.

Target Embedding. The target-aware attention is applied to calculate the attention score between all item embedding vectors $\tilde{\mathbf{v}}_i$ and each target item embedding vector $\tilde{\mathbf{v}}_t$ in session s, the model adaptively considers the relevance of the target item. The formula is:

$$\beta_{i,t} = \text{soft max} (e_{i,t}) = \frac{\exp \left(\tilde{\mathbf{v}}_t^T \mathbf{W}\tilde{\mathbf{v}}_i\right)}{\sum_{j=1}^{m} \exp \left(\tilde{\mathbf{v}}_t^T \mathbf{W}\tilde{\mathbf{v}}_j\right)}, \tag{9}$$

for each session s, the target embedding vector is represented by $\mathbf{s}_{\text{target}}^t \in \mathbb{R}^d$,

$$\mathbf{s}_{\text{target}}^t = \sum_{i=1}^{s_n} \beta_{i,t}\tilde{\mathbf{v}}_i. \tag{10}$$

4.3   Learning and Making Recommendation

Since the final behavior of a user is usually determined by the user's last behavior, we simply express the user's short-term preference as a locally embedded $\mathbf{s}_{\text{local}} \in \mathbb{R}^d$ as the representation of the last accessed item $\tilde{\mathbf{v}}_{s,n}$. Finally, we generate

Table 1. The descriptive statistics of datasets.

Datasets	All Clicks	Train Sessions	Test Sessions	Total Items	Avg. Length
Diginetica	982, 961	719, 470	68, 977	43, 097	5.12
Nowplaying	1, 367, 963	825, 304	89, 824	60, 417	7.42

session embedding s_t by performing a linear transformation on the concatenation of local embedding, global embedding, and target embedding:

$$s_t = \mathbf{W}_3 \left[s_{\text{target}}^t \; ; s_{\text{local}} \; ; s_{\text{global}} \right],   \tag{11}$$

where $\mathbf{W_3} \in \mathbb{R}^{d \times 3d}$ is the projection matrix.

After obtaining all item and session embeddings, we calculate the recommended score for each target item:

$$\hat{y} = \frac{\exp \left(\sigma \tilde{\mathbf{v}}_t^T \mathbf{s}_t \right)}{\sum_{i=1}^m \exp \left(\sigma \tilde{\mathbf{v}}_i^T \mathbf{s}_t \right)},   \tag{12}$$

as mentioned in the article [20], the range of cosine similarity $\tilde{\mathbf{v}}_t^T \mathbf{s}_t$ is limited to $[-1, 1]$, indicating that the softmax loss of the training set may reach saturation at high values: scaling factor $\sigma > 1$ is effective in practice, allowing for better convergence (Table 1).

Finally, the model is trained by minimizing the cross-entropy between the ground truth and the prediction:

$$\mathcal{L}(\hat{\mathbf{y}}) = -\sum_{i=1}^m \mathbf{y}_i \log (\hat{\mathbf{y}}_i) + (1 - \mathbf{y}_i) \log (1 - \hat{\mathbf{y}}_i),   \tag{13}$$

where $\mathbf{y}$ denotes the one-hot encoding vector of the ground truth item.

5   Empirical Study

In order to evaluate the proposed model, multiple experiments were conducted on two real-world datasets. Through experimental analysis, we aim to answer the following research questions.

RQ1: How does the performance of the proposed model compare to the state-of-the-art methods?

RQ2: Can the normalized item proposed in this article, as well as the introduction of the self-attention mechanism and target-aware attention, improve the model performance of existing methods?

RQ3: How do key parameters affect the performance of this model?

In what follows, we first introduce the experimental settings and answer the above research questions in turn to demonstrate the rationality of our methods.

5.1   Experimental Settings

Datasets. We conducted experiments on two real-world datasets: Nowplaying[1] and Diginetica[2]. Diginetica is a dataset for CIKM Cup 2016. The Nowplaying dataset contains users' music-listening behaviors. We deleted sessions with a length of less than 2 and items with fewer than 5 occurrences from two datasets. In addition, in order to generate training and test sets, the sessions in the last weeks were used as test sets for Diginetica and Nowplaying. For an existing session, we generate a series of input session sequences and corresponding tags.

Baselines. We compare GENE with the following baselines:

- FPMC [11]: This model combines matrix decomposition and Markov chain to make recommendations, which can capture users' long-term preferences.
- Item-KNN [21]: This model is recommended by calculating the similarity between session items.
- GRU4Rec [12]: This model is the first to apply RNN to the session-based recommendation, modeling user behavior sequences.
- NARM [22]: This model incorporates both RNN and attention mechanism into the session-based recommendation model to capture preferences.
- SR-GNN [17]: This model captures complex transitions between items by modeling session sequences as session graphs.
- GC-SAN [18]: This model utilizes a self-attention mechanism to learn global dependency information between non-adjacent items in a session.
- TAGNN [19]: This model proposes a new target-aware attention graph neural network, which adaptively activates users' interest in different target items.
- NISER+ [20]: This model addresses the prevalence bias in GNN models through embedding normalization and introduces location information.
- Disen-GNN [6]: This model first utilizes disentangling learning technology to transform item embedding into multiple-factor embedding to learn the user's intentions for different factors of each item.

Evaluation Metrics. We use the commonly used metrics in the session-based recommendation, Precision@20 and mean reciprocal rank (MRR@20), to evaluate the performance of all models. Precision@20 represents the recommendation accuracy, and MRR@20 represents the average reciprocal rank of correctly recommended items, where a higher value indicates a better quality of recommendations.

Parameter Settings. To ensure the fairness and impartiality of the experimental results, we optimize parameters using the Adam optimizer. We adjust other hyperparameters according to the random 10% verification set. The initial learning rate of the Adam toolkit is set to 0.001, with a decay of 0.1 every 3 training cycles. The batch size of both datasets is set to 50, the L2 penalty coefficient is set to 10^{-5}.

---

[1] https://dbis.uibk.ac.at/node/263#nowplaying.
[2] http://cikm2016.cs.iupui.edu/cikm-cup.

Table 2. Recommendation Performance on Diginetica and Nowplaying.

Method	Diginetica		Nowplaying	
	Precision@20	MRR@20	Precision@20	MRR@20
FPMC	22.14%	6.66%	7.36%	2.82%
Item-KNN	35.75%	11.57%	15.94%	4.91%
GRU4Rec	29.45%	8.33%	7.92%	4.48%
NARM	49.70%	16.17%	18.59%	6.93%
SR-GNN	50.73%	17.59%	18.87%	7.47%
GC-SAN	51.70%	17.61%	18.85%	7.43%
TAGNN	51.31%	18.03%	19.02%	7.82%
NISER+	53.39%	18.72%	22.35%	8.52%
Disen-GNN	53.79%	18.99%	22.22%	8.22%
GENE	**53.83%**∗∗	**19.21%**∗∗	**22.56%**∗∗	**8.94%**∗∗

∗∗ indicates that the improvements are statistically significant for $p <$ 0.01 judged by paired t-test.

5.2 Performance Comparison (RQ1)

To demonstrate the recommendation performance of our model GENE, we compared it with other state-of-the-art methods. As shown in Table 2, we have the following key observations.

- The performance of GRU4Rec and NARM is superior to the traditional models FPMC. The first two models both use RNN to capture user preferences, indicating that ignoring sequence dependencies is suboptimal. The performance of NARM is superior to GRU4Rec indicating that attention can more accurately capture the main interests and preferences of users.
- The performance of SR-GNN, GC-SAN, TAGNN, NISER+, and Disen-GNN models is superior to NARM and GRU4Rec models, indicating the effectiveness of graph neural networks in recommender systems, which can capture more complex transition information between items. GC-SAN uses a self-attention mechanism on the basis of SR-GNN to more accurately represent global interest preferences. TAGNN obtains target embedding by further considering user interests through target-aware attention on the basis of SR-GNN. NISER+ alleviates popularity bias by normalizing item and session representations based on SR-GNN, which also improves the model performance. Disen-GNN models the potential factors behind the item and analyzes the user's disagreement graph.
- We can see that our method achieved the best performance in terms of Precision and MRR among all methods. These results demonstrate the rationale and effectiveness of GENE in session-based recommendations.

Table 3. Ablation study of key components of GENE

Method	Diginetica		Nowplaying	
	Precision@20	MRR@20	Precision@20	MRR@20
GENE$_{ns}$	53.34%	18.29%	21.63%	8.12%
GENE$_{sa}$	53.45%	18.18%	21.52%	7.46%
GENE$_{ta}$	52.78%	18.25%	20.93%	8.16%
GENE	**53.83%**	**19.21%**	**22.56%**	**8.94%**

5.3 Ablation Study (RQ2)

In this section, keeping all other conditions identical, we design three model variants to analyze the effects on model performance (RQ2): the model GENE$_{ns}$, which removes the normalization step for items, the model GENE$_{sa}$, which removes the self-attention for obtaining global embeddings, and the model GENE$_{ta}$, which removes the target-aware attention for obtaining target embeddings. The results are shown in Table 3. We note that the GENE model performs better than the GENE$_{ns}$ model, indicating that handling popularity bias through normalized embedding is effective for improving performance based on GNN session recommendations; We compared the GENE$_{sa}$ and GENE$_{ta}$ model variants with the GENE model, indicating that the self-attention mechanism can help extract important behavioral information from session data and that the target-aware attention mechanism can better complete correlation recommendations for target items. Overall, the GENE model has achieved optimal performance on both datasets, indicating the necessity and rationality of comprehensively considering normalized embedding and self-attention mechanisms to model global dependencies and target-aware attention for target embedding.

5.4 Hyper-Parameter Study (RQ3)

For the sake of fairness, we keep other parameters unchanged except for the measured parameters. We found that $\sigma = 16.0$ worked best in training. The optimal number of attention heads for Nowplaying is 4 and the optimal number of attention heads for Diginetica is 2 in Table 4. Indicates a slight decrease in performance when the attention head is greater or less than the optimal value. Excessive attention heads make a single attention dimension too small, limiting the ability to express attention.

Table 4. Impact of hyper-parameter on recommendation.

# heads	Diginetica		Nowplaying	
	Precision@20	MRR@20	Precision@20	MRR@20
1	53.62%	18.94%	21.69%	8.22%
2	**53.83%**	**19.21%**	21.75%	8.58%
4	53.63%	18.86%	**22.56%**	**8.94%**
6	53.17%	18.45%	22.42%	8.87%
8	53.23%	18.57%	21.99%	8.27%
10	53.01%	18.26%	21.78%	7.96%
12	53.16%	18.13%	21.49%	7.94%

6   Conclusions

In this paper, we propose a method for learning richer representations in session-based recommendation models. We comprehensively considered popularity bias and session embedding issues. In order to alleviate the model's recommendation of popular items, we normalize item and session graph representations. To enhance session embedding representation, we used target-aware attention and self-attention mechanism to obtain target embedding and global embedding. We observed significant improvements in overall recommendation performance and improved existing state-of-the-art results. In the following work research, it can be considered to obtain higher recommendation accuracy by mining more user behavior information and combining auxiliary information such as knowledge graphs or social networks.

Acknowledgment. This work was supported in part by the National Natural Science Foundation of China Youth Fund (No. 61902001) and the Undergraduate Teaching Quality Improvement Project of Anhui Polytechnic University (No. 2022lzyybj02). We would also thank the anonymous reviewers for their detailed comments, which have helped us to improve the quality of this work. All opinions, findings, conclusions, and recommendations in this paper are those of the authors and do not necessarily reflect the views of the funding agencies.

References

1. Zou, F., Qian, Y., Zhang, Z., Zhu, X., Chang, D.: A data mining approach for analyzing dynamic user needs on UGC platform. In: IEEM 2021, pp. 1067–1071 (2021)
2. Feng, P., Qian, Y., Liu, X., Li, G., Zhao, J.: Robust graph collaborative filtering algorithm based on hierarchical attention. In: Xing, C., Fu, X., Zhang, Y., Zhang, G., Borjigin, C. (eds.) WISA 2021. LNCS, vol. 12999, pp. 625–632. Springer, Cham (2021). https://doi.org/10.1007/978-3-030-87571-8_54

3. Zhang, W., Yan, J., Wang, Z., Wang, J.: Neuro-symbolic interpretable collaborative filtering for attribute-based recommendation. In: WWW 2022, pp. 3229–3238 (2022)
4. Wang, Y., Zhou, Y., Chen, T., Zhang, J., Yang, W., Huang, Z.: Sequence-aware API recommendation based on collaborative filtering. Int. J. Softw. Eng. Knowl. Eng. **32**(8), 1203–1228 (2022)
5. Wang, S., Cao, L., Wang, Y., Sheng, Q.Z., Orgun, M.A., Lian, D.: A survey on session-based recommender systems. ACM Comput. Surv. **54**(7), 154:1–154:38 (2022)
6. Li, A., Zhu, J., Li, Z., Cheng, H.: Transition information enhanced disentangled graph neural networks for session-based recommendation. Expert Syst. Appl. **210**, 118336 (2022)
7. Zhang, S., Huang, T., Wang, D.: Sequence contained heterogeneous graph neural network. In: IJCNN 2021, pp. 1–8 (2021)
8. Liu, C., Li, Y., Lin, H., Zhang, C.: GNNRec: gated graph neural network for session-based social recommendation model. J. Intell. Inf. Syst. **60**(1), 137–156 (2023)
9. Zhang, D., et al.: ApeGNN: node-wise adaptive aggregation in GNNs for recommendation. In: WWW, pp. 759–769 (2023)
10. Zhang, Y., Shi, Z., Zuo, W., Yue, L., Liang, S., Li, X.: Joint personalized Markov chains with social network embedding for cold-start recommendation. Neurocomputing **386**, 208–220 (2020)
11. Rendle, S., Freudenthaler, C., Schmidt-Thieme, L.: Factorizing personalized Markov chains for next-basket recommendation. In: WWW 2010, pp. 811–820 (2010)
12. Hidasi, B., Karatzoglou, A., Baltrunas, L., Tikk, D.: Session-based recommendations with recurrent neural networks. In: ICLR 2016 (2016)
13. Zhu, J., Xu, Y., Zhu, Y.: Modeling long-term and short-term interests with parallel attentions for session-based recommendation. In: Nah, Y., Cui, B., Lee, S.-W., Yu, J.X., Moon, Y.-S., Whang, S.E. (eds.) DASFAA 2020. LNCS, vol. 12114, pp. 654–669. Springer, Cham (2020). https://doi.org/10.1007/978-3-030-59419-0_40
14. Zheng, Y., Liu, S., Li, Z., Wu, S.: DGTN: dual-channel graph transition network for session-based recommendation. In: ICDM 2020, pp. 236–242 (2020)
15. Guo, J., et al.: Learning multi-granularity consecutive user intent unit for session-based recommendation. In: WSDM 2022, pp. 343–352 (2022)
16. Wang, Z., Wei, W., Cong, G., Li, X., Mao, X., Qiu, M.: Global context enhanced graph neural networks for session-based recommendation. In: SIGIR 2020, pp. 169–178 (2020)
17. Wu, S., Tang, Y., Zhu, Y., Wang, L., Xie, X., Tan, T.: Session-based recommendation with graph neural networks. In: AAAI 2019, pp. 346–353 (2019)
18. Xu, C., et al.: Graph contextualized self-attention network for session-based recommendation. In: IJCAI 2019, pp. 3940–3946 (2019)
19. Yu, F., Zhu, Y., Liu, Q., Wu, S., Wang, L., Tan, T.: TAGNN: target attentive graph neural networks for session-based recommendation. In: SIGIR 2020, pp. 1921–1924 (2020)
20. Gupta, P., Garg, D., Malhotra, P., Vig, L., Shroff, G.: NISER: normalized item and session representations with graph neural networks. CoRR abs/1909.04276 (2019)
21. Sarwar, B.M., Karypis, G., Konstan, J.A., Riedl, J.: Item-based collaborative filtering recommendation algorithms. In: WWW 2001, pp. 285–295 (2001)
22. Li, J., Ren, P., Chen, Z., Ren, Z., Lian, T., Ma, J.: Neural attentive session-based recommendation. In: CIKM 2017, pp. 1419–1428 (2017)

Research on Predicting the Impact of Venue Based on Academic Heterogeneous Network

Meifang Fang and Zhijie Ban[✉]

School of Computer Science, Inner Mongolia University, Hohhot 010021, China
banzhijie@imu.edu.cn

Abstract. Academic venues play an important role in the exchange and dissemination of knowledge. The effective prediction of venues' impact plays a positive role in guiding authors to correctly submit manuscripts and evaluating the impact of new papers. Most recent research has utilized relevant features of the author or paper to predict the impact of venues and made some success. In contrast, only a little attempt has been made to simultaneously mine information about papers, authors, and venues based on academic heterogeneous networks for predicting the future impact of venues. The current mainstream methods are based on academic heterogeneous networks. But these don't fully utilize the local network information of nodes, which lead to the neglect of information from strongly correlated neighbor nodes. In addition, the existing studies didn't integrate network topology, textual information, and attribute features effectively. To solve the above problems, we propose a hybrid model of academic heterogeneous network representation learning combined with multivariate random walk, termed as AHRV. The specific content is to mine the heterogeneous local network information of nodes in the academic heterogeneous network, and embed local network structure, text information and attribute features into a unified embedding representation. Then use the graph neural network to aggregate the features and neighbors to get the embedding of nodes. Finally, the embedded vector of the node is combined with the multivariate random walk model to rank the future impact of venues. The experimental results on two real datasets show that the method improved performance.

Keywords: Venue Impact · Academic Heterogeneous Network · Network Representation Learning · Multivariate Random Walk

1 Introduction

In scientific evaluation, quantifying the impact of academic venues is a very practical issue, influencing authors where to submit their research papers and where to search for the latest notable progress in their research fields [1].

Recently, some ranking and predicting venue impact models using random walk on single homogeneous networks have been developed, which begin to consider the network structure [1–3]. However, they ignore different impact of entity. Subsequently, some multivariate random walk models were proposed to sort multiple entities. These

L. Yuan et al. (Eds.): WISA 2023, LNCS 14094, pp. 185–197, 2023.
https://doi.org/10.1007/978-981-99-6222-8_16

models only use global information of the network to recursively sort entity scores, ignoring local network information of nodes and multi-feature fusion. Heterogeneous network local information reflects similarity and strong relevance between nodes [4]. Text features can enrich expressiveness of nodes. For new papers, authors, or venues, their sparse network structure may not accurately reflect their individual impact [5]. Therefore, adding text better reflects complex links in heterogeneous network and approximates the real network.

Nodes in heterogeneous network have different local network structures, attributes and text, making it difficult to represent them in a unified manner. Extensive research on heterogeneous network embedding can solve this problem [6]. Based on this point, this paper proposes a hybrid prediction model based on a combination of heterogeneous network embedding and random walk, called AHRV. Specifically, we design an academic heterogeneous network embedding model fusing multi-feature information. Then, the similarity between nodes is then calculated based on the learning embedding, so that a transfer matrix corresponding to the network can be constructed, and a multivariate random walk algorithm is used to predict the future impact of the venue.

We summarize our main contributions as follows:

- Extracting the complex relationship between academic entities, and using graph neural networks to fuse the heterogeneous sub-network structure, attributes and text features of each entity. Different types of features enrich the structure and content information of academic entities, which increases the expression ability of nodes.
- The academic heterogeneous network embedding model learns a unified representation of different types of nodes. Finally, the node embeddings were then integrated into a multivariate random walk model to simplify the sorting process.

2 Related Work

The initial venue ranking methods are based on PageRank. These methods only use limited information on homogeneous venue citation networks, and while they can give a venue's current impact, it is unable to predict venue's future impact [7, 8].

In recent years, models based on academic heterogeneous network have been used to predict the future impact of papers, authors and venues [9–11], such as univariate random walk. These methods first construct an academic heterogeneous network. They usually divided heterogeneous network into homogeneous networks [11]. These methods ignored the different impact of different types of nodes, thus affecting the predictive capabilities of venue impact. Multivariate random walk algorithms rank multiple entities simultaneously to obtain entities' impact by integrating information to improve prediction accuracy. For example, TAORank [12] considers the role of academic entities, publication time and author order. Wang [11] utilized text information to improve prediction results. The limitation of these methods is that they cannot use rich information (local network information, different types of features).

In recent years, network representation learning models using structural information and relevant features to learn low-dimensional vectors of nodes have been effective in various tasks such as link prediction [13], node classification [6, 14], community detection [15] and recommendation tasks [16]. Heterogeneous network representation learning models such as methpath2vec [17], GATNE [18] and HetGNN [19] have also been proposed. But they are rarely used for predicting academic impact. Only Xiao [5] proposed the ESMR model, which is based on paper and author heterogeneous network embedding captures structural information and text information into a unified embedded representation. The learned embedded representation is integrated into the random walk algorithm to predict the ranking of academic impact. This paper proposes a network representation learning combined with a multivariate random walk model to predict future impact of venue. It obtains local network information and multidimensional features from academic heterogeneous networks to obtain node vector representations, and then ranks venue impact.

3   Our Method

The goal of our proposed model AHRV is to learn the embedding representation of entities in the network by fusing the local network information and multidimensional feature. The embedded vector of node is integrated into the multivariate random walk to predict the future impact of venues. The framework of AHRV is shown in Fig. 1.

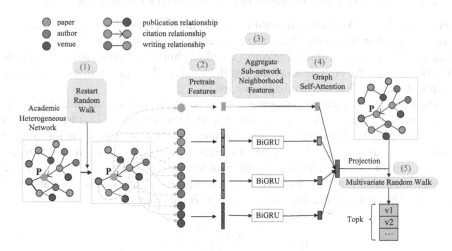

Fig. 1. The framework of AHRV

3.1   Definition

Academic Heterogeneous Network. Given a group of authors $a = \{a_1, a_2, \ldots, a_n\}$, a group of papers $p = \{p_1, p_2, \ldots, p_n\}$, a group of venues $v = \{v_1, v_2, \ldots, v_n\}$. Use E_{pp} is for citation relationship between papers, E_{pa}. is for writing relationship between papers and authors, E_{pv} is for publication relationship between papers and venues. Academic heterogeneous network is defined as $G = \{V, E, R_E\}$. Where $V = \{a, p, v\}$, $E = \{E_{pp}, E_{pa}, E_{pv}\}$, R_E denote different sets of relations. Each node in academic heterogeneous networks contains its own attribute features and text features.

3.2 Academic Heterogeneous Network Embedding

Mining Node Heterogeneous Local Network Information. In the academic hetero-geneous network, each node has a different number of neighbors, and many nodes cannot connect to all types of neighbors, which affects the node's representation abil-ity. In order to make each academic entity not weakened by other weakly related neighbors, this paper uses Restart Random Walk to get the heterogeneous sub-network information of each node to obtain the node's strongly correlated neighbors. The steps are as follows.

(1) We use random walk to sample neighbors for each node. Specifically, starting from node $v \in V$, and then walking to the next neighbor node or return to v with probability p. Since each node has multiple types of neighbors. For example, a paper node has three types of neighbors: paper, author and venue, while author and venue nodes also have three types of neighbors. Therefore, the walking process is restricted to node types to ensure that all types of neighbors are selected. The set of neighbors obtained by sampling is denoted $N(v)$.

(2) The set of sampled neighbors $N(v)$ is grouped by node type and the node type is denoted as t. The top k nodes with the highest sampling frequency are taken as the set of strongly correlated neighbors of node v with type t, denoted as $N_t(v)$.

Pretraining Node's Multidimensional Feature. In order to be able to merge differ-ent feature information in a node, different types of feature information need to be pretrained. Therefore, this paper uses the DeepWalk [20] algorithm to vectorize the sampled sub-network structure. We pretrain text feature with Doc2vec [21] algorithm. As for attribute features of entities, Onehot encoding was applied.

Due to nodes in academic heterogeneous network include unstructured heteroge-neous content (sub-network structure, attribute and text features). That is, we extract the multidimensional feature HC_v from the academic entity. We denote the i-th content feature in HC_v as $x_i \in R^{d_v \times 1}$. (d_v:content feature dimension). Unlike models directly concatenating or linearly transforming different features into a uniform vector, we use a bidirectional recurrent network BiGRU [22] to fuse the different types of features of the nodes. The content embedding of node $v \in V$ is computed as:

$$f_1(v) = \frac{\sum_{i \in HC_v} \left[\underset{GRU}{\leftrightarrow} \{FT_{\theta x}(x_i)\} \right]}{|HC_v|} \tag{1}$$

Where $f_1(v) \in R^{d \times 1}$ (d: content embedding dimension), $FT_{\theta x}$ denotes feature trans-former, fully connected neural network with parameter θx. The GRU is formulated as

$$\begin{aligned}
z_i &= \sigma\left(W_z\left[h_{i-1}, FT_{\theta x}(x_i)\right]\right) \\
r_i &= \sigma\left(W_f\left[h_{i-1}, FT_{\theta x}(x_i)\right]\right) \\
\tilde{h}_i &= \tanh\left(W_c\left[r_i * h_{i-1}, FT_{\theta x}(x_i)\right]\right) \\
h_i &= (1 - z_i) * h_{i-1} + z_i * \tilde{h}_i
\end{aligned} \tag{2}$$

Where $h_i \in R^{(d/2) \times d_f}$ the output hidden state of i-th content, $W_j \in R^{(d/2) \times d_f}$ ($j \in \{z, f, c\}$) are learnable parameters, z_i, r_i are update gate vector, reset gate vector of i-th content feature, respectively.

Node Embedding. In the previous section, nodes in the academic heterogeneous network integrate different types of features to get node content feature vectors. Due to node properties and relationships are supplement each other. In order to represent a node completely, we should first aggregate the content feature vectors of its neighbors of the same type to obtain the paper neighbor feature embedding, the author neighbor feature embedding and the venue neighbor feature embedding. Thus, multiple node content feature vectors of the same type of neighbors are aggregated by BiGRU. The final t type of neighbor feature embedding $f_2^t(v)$:

$$f_2^t(v) = AGG_{v' \in Nt(v)}^t \{f_1(v')\} \tag{3}$$

Where $f_2^t(v) \in R^{d \times 1}$ (d: aggregated content embedding dimension), $f_1(v')$ is the content embedding of v' generated by the module in Subsect. 3.2.2, AGG^t is the t-type neighbor aggregator, which is BiGRU. Thus, we re-formulate $f_2^t(v)$ as follows:

$$f_2^t(v) = \frac{\sum_{v' \in Nt(v)} \left[\overset{\leftrightarrow}{GRU} \{f_1(v')\} \right]}{|Nt(v)|} \tag{4}$$

Due to different neighbor types differently impact node's final representation. We used a graph self-attention mechanism to combine type based neighbor embedding with node content embedding. Thus, the output embedding of v is:

$$\varepsilon_v = \alpha^{v,v} f_1(v) + \sum_{t \in O_V} \alpha^{v,t} f_2^t(v) \tag{5}$$

Where $\varepsilon_v \in R^{d \times 1}$ (d: output embedding dimension), $\alpha^{v,*}$ indicates the importance of different embeddings, $f_1(v)$ is the content embedding of v, $f_2^t(v)$ aggregates multiple types of neighbors. We denote the set of embeddings as $F(v) = \{f_1(v) \cup f_2^t(v)\}$ and reformulate the output embedding of v as:

$$\varepsilon_v = \sum_{f_i \in F(v)} \alpha^{v,i} f_i \tag{6}$$

$$\alpha^{v,i} = \frac{exp\{(u^T[f_i f_1(v)])\}}{\sum_{f_j \in F(v)} exp\{(u^T[f_j f_1(v)])\}} \tag{7}$$

Where $u \in R^{2d \times 1}$ is the attention parameter.

Model Optimization. To perform heterogeneous network representation learning, we defined the following objectives in terms of neural network parameters Θ.

$$o_1 = arg \max_{\Theta} \prod_{v \in V} \prod_{v' \in N_t(v)} \frac{exp\{\varepsilon_{v_c} \cdot \varepsilon_v\}}{\sum_{v_k \in V_t} exp\{\varepsilon_{v_k} \cdot \varepsilon_v\}} \tag{8}$$

Where $N_t(v)$ is the set of t-type context nodes of v such first/second order neighbor the graph or local neighbors in random walks. V_t is the set of t-type nodes in the network, ε_v is the output node embedding.

We use negative sampling technique [19] to optimize the objective function. The negative sample size is set to 1. The impact is less when size > 1. So, the final objective function is:

$$o_2 = \sum_{<v,v_c,v_c'> \in N_t(v)} log\sigma(\varepsilon_{v_c} \cdot \varepsilon_v) + log\sigma(-\varepsilon_{v_c'} \cdot \varepsilon_v) \tag{9}$$

3.3 Predicting the Future Scientific Impact of Venues

Inspired by previous work [5], this section describes predicting venue impact using heterogeneous network embeddings. Different entities with similar potential impact are embedded into the same representation space, bringing them closer. Then based on the learned entities embedding, the cosine similarity is used to measure the similarity between them. For example, the similarity between two papers is measured by

$$sim(p_i, p_j) = \frac{p_i \cdot p_j}{p_i \times p_j} \tag{10}$$

Thus, the transfer matrix in the random walking model can be defined as:

$$M_{pp} = \begin{cases} \lambda \frac{sim(p_i, p_j)}{\sum\limits_{v_k \in Nt(v)} sim(p_i, p_k)} + (1 - \lambda) \frac{\varphi_{pp}^i}{degree(p_i)}, & e_{pp}^{ij} \in E_{pp} \\ 0, & otherwise \end{cases} \tag{11}$$

Where $Nt(v)$ and $degree(p_i)$ denote the set of neighboring nodes and the set of out-degree nodes of p_i, respectively, used λ as adjustable parameters to balance the factors affecting the transfer probability. Similarly, $M_{ap}, M_{pa}, M_{pv}, M_{vp}$ are defined in the same way.

Finally, these transfer matrices are used to perform a random walk over the academic heterogeneous network to calculate the future impact of venues. Each iterative process is defined by the following equation:

$$v^{(t+1)} = \alpha_{vp} M_{vp} v^{(t)} + \beta_{pp} M_{pp} p^{(t)} + \gamma_{pa} M_{pa} a^{(t)} \tag{12}$$

where $a^{(t)}$, $p^{(t)}$ and $v^{(t)}$ are the predicted score vectors at time point t. α_{vp}, β_{pp} and γ_{pa} are the influence weights of other papers, authors and venu on a particular venue. Thus, a score vector can be obtained by iterating over Eq. (12) until convergence.

For example, the vector representation of nodes obtained through embedding in academic heterogeneous networks. The similarity between nodes is calculated according to the node embedding, so as to construct the transfer matrix between networks, and use multivariate random walk to predict the score vector of the site. The score vectors are then sorted in descending order, and the impact ranking of the venues is obtained.

4 Experimental Results

4.1 Datasets

We use two public datasets: ACL and AMiner. We extract papers from 2006 to 2015 year in AMiner and 2006 to 2014 year in ACL. We remove papers missing metadata. Finally, ACL dataset has 13183 papers, 24882 authors and 218 venues. AMiner dataset has 23766 papers, 28646 authors, 18 venues and 124857 citations.

4.2 Experiment Setup

Ground Truth and Evaluation Metrics. Due to lacking criteria, evaluating most works' performance is challenging. We use the modified impact factor [23] as the true standard to evaluate. We split data into training and test sets by time. The training set gets venue ranks before a time point. The test set gets real ranking lists based on modified impact factor. Since most citations occur more than two years after publication, we calculate modified impact factor(mIF) in 2014 year and 2015 year, where mIF for three years. If a venue lacks three years' of data, we use data for a shorter period based on previous work experience [24]. Finally, the results are reported by comparing the similarities of the two rankings.

We adapt the mIF and P@N to evaluate venue ranking results. mIF = C/D, where C is citation number and D is total paper number. P@N = |E ∩ F|/N, where E is method A's topN and F is method B's topN. P@N is used to compare two methods' accuracy. P@N ranges from 0 to 1, the higher the better.

Baselines and Parameter Sensitivity Analysis. To fully validate the performance of our method AHRV, we chose baselines are as follows: (1) PageRank: The basic model of graph-based ranking; (2) MR-Rank: Algorithm [24] for mutually reinforcing ranking of papers and venues based on paper-venue bipartite graph; (3) DeepWalk + Rank learns network structure but ignores node attributes and text information, and apply the multivariate random walking algorithm of Subsect. 3.3 to rank predictions.

In our model, there are two important parameters to learn. The initial value of the dimension d of the node embedding is random. To study the dimension influence, we set vector dimensions ranging from 8 to 256. The top@N venue ranking results of AMiner and ACL datasets are shown in Fig. 2. The results show AHRV's performance varies slightly in different dimensions, and choosing 64 or 128 is reasonable to balance computational complexity and performance. Therefore, the following experiments use 128 dimensions embedding for modeling.

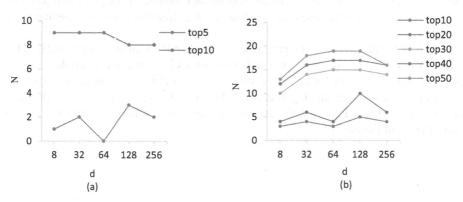

Fig. 2. Analysis of dimensions of vectors. (a) Effect of #Dimension on top@N in Aminer dataset. (b) Effect of #Dimension on top@N in ACL dataset.

Then sampled neighbor sizes from 10 to 40 are analyzed. As shown in Fig. 3, the performance peaks at certain sampled neighbor sizes, while excessively large or small values impact the performance. Therefore, the sampling size is 30. Finally, each node randomly walks 10 times, walk length 30, window size 5, RWR return probability 0.5.

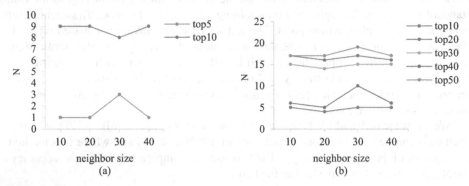

Fig. 3. Analysis of neighbor size. (a) on Effect of neighbor size on top@N in Aminer dataset. (b)Effect of neighbor size on top@N in ACL dataset.

For all baselines parameters, ε is set to 1e-3 for all the algorithms, which is the same as in [24]. τ is set to 0.117, $\alpha = 0.4$ and $\beta = 0.9$ are used in MR-Rank, because such a setting is good for performance prediction. For PageRank, $\alpha = 0.85$ and random jump with a probability of 0.15. For venue ranking algorithm, $\alpha_{vp} = 0.6$, $\beta_{pp} = 0.3$, $\gamma_{pa} = 0.3$.

4.3   Ranking Results of Venue Impact

Tables 1. and 2. shows the Top 10 venues obtained by all algorithms in two datasets. In these tables, the high impact venues are highlighted in black with bold, we can see that the AHRV method is better at identifying high impact venues. Although there is a certain difference in the results between the two datasets, the effectiveness of this method is the best.

In order to fully validate the performance of the AHRV model, we calculated the accuracy index. Due to the different number of venues included in the two datasets, we took different N values on both datasets. We calculated different N values on two datasets based on the real venue ranking lists for 2014 and 2015 year. Finally, we calculated the average accuracy based on the accuracy of the TopN values to verify the overall performance of the model.

Table 1. Top 10 venues in AMiner dataset.

AHRV	DeepWalk + Rank	MR-Rank	PageRank
NAACL	TMM	**CSS**	TMM
WSDM	IJCAI	**ICML**	**ICML**
ACL	WSDM	DASFAA	IJCAI
ECCV	DASFAA	**NAACL**	AL
SIGKDD	ICDM	**VLDB**	CIKM
EMNLP	AL	AL	DASFAA
AL	CIKM	IJCAI	**ACL**
IJCV	ECCV	TMM	**SIGKDD**
IJCAI	**SIGKDD**	ICDM	**VLDB**
ICML	**CSS**	CIKM	**IJCV**

Table 2. Top 10 venues in ACL dataset.

AHRV	DeepWalk + Rank	MR-Rank	PageRank
ACL	**ACL**	wOLP	**HLT@NAACL -Companion**
ACL -Companion	**ACL -Companion**	CoNLL -Companion	CoNLL – Companion
HLT@NAACL – Companion	LREC	wACL -SIGSEM	**ACL**
wASDGML	**HLT@NAACL -Companion**	**HLT@NAACL -Companion**	CoNLL
SemEval	SemEval	**ACL**	HLT@NAACL
WASSA@ACL	HLT@NAACL	LaTeCH	**EACL**
COLING Companion	**COLING**	wCACS	CL
COLING	wSMT	**ACL -Companion**	CoNLL –
EMNLP	CL	BTCLP	Companion
wSMT	CoNLL	SEW	LaTeCH

The accuracy values for different N values as seen in Tables 3., 4., 5. and 6. show that the experimental performance of this paper is overall better than all baseline methods. From the results in the tables, we can see that the accuracy rates vary slightly from year to year, which aptly reflects the dynamic nature of the modified influence factor of venues and demonstrates the effectiveness of AHRV model in identifying high impact venues.

Table 3. Accuracy on the AMiner dataset for 2014 year.

Algorithm	p@3	p@6	p@9	p@12	p@15	Average accuracy
PageRank	0.333	0.167	0.444	0.667	0.8	0.442
MR-Rank	0.333	0.167	0.333	0.583	0.55	0.352
DeepWalk + Rank	0	0.333	0.889	0.833	0.867	0.536
AHRV	0.333	0.5	0.778	0.917	0.867	0.618

Table 4. Accuracy on the AMiner dataset for 2015 year.

Algorithm	p@3	p@6	p@9	p@12	p@15	Average accuracy
PageRank	0.333	0.167	0.444	0.667	0.8	0.448
MR-Rank	0.333	0.333	0.444	0.667	0.533	0.355
DeepWalk + Rank	0	0.333	0.889	0.833	0.867	0.536
AHRV	0.333	0.667	0.778	0.917	0.867	0.638

Table 5. Accuracy on the ACL dataset for 2014 year.

Algorithm	p@10	p@20	p@30	p@40	p@50	Average accuracy
PageRank	0.4	0.4	0.367	0.325	0.34	0.366
MR-Rank	0.4	0.3	0.267	0.2	0.38	0.239
DeepWalk + Rank	0.3	0.3	0.3	0.3	0.34	0.418
AHRV	0.5	0.5	0.5	0.425	0.38	0.461

Table 6. Accuracy on the ACL dataset for 2015 year.

Algorithm	p@10	p@20	p@30	p@40	p@50	Average accuracy
PageRank	0.4	0.45	0.4	0.35	0.36	0.387
MR-Rank	0.3	0.2	0.167	0.15	0.34	0.228
DeepWalk + Rank	0.4	0.35	0.333	0.325	0.36	0.412
AHRV	0.5	0.5	0.5	0.425	0.38	0.449

4.4 Ablation Study

The node embeddings of AHRV capture the features and relationships of academic entities. Do node features affect venue impact? To answer this, we conducted the following
ablation experiments: (a) AHRV-t: Our method AHRV removes text features as a way to

verify the role of text features; (b)AHRV-AV considers only paper attributes; (c)AHRV-PV considers only author attributes; (d)AHRV-PA considers only venue attributes. Results on AMiner and ACL datasets show AHRV has the best performance. The attributes and relationships of academic entities characterize each other, fusing them improves prediction and information representation, enhancing model performance (see Tables 7., 8., 9. and 10.).

Table 7. Accuracy of variant models on the AMiner dataset for 2014 year.

Algorithm	p@3	p@6	p@9	p@12	p@15	Average accuracy
AHRV-T	0	0.333	0.667	0.667	0.667	0.433
AHRV-AV	0	0.5	0.778	0.833	0.867	0.554
AHRV-PV	0.333	0.5	0.778	0.833	0.867	0.576
AHRV-PA	0	0.333	0.778	0.833	0.867	0.529
AHRV	0.333	0.5	0.778	0.917	0.86	0.618

Table 8. Accuracy of variant models on the AMiner dataset for 2015 year.

Algorithm	p@3	p@6	p@9	p@12	p@15	Average accuracy
AHRV-T	0	0.333	0.667	0.75	0.667	0.433
AHRV-AV	0	0.5	0.778	0.833	0.8	0.556
AHRV-PV	0.333	0.5	0.778	0.833	0.785	0.576
AHRV-PA	0	0.5	0.778	0.833	0.867	0.571
AHRV	0.333	0.667	0.778	0.909	0.786	0.638

Table 9. Accuracy of variant models on the ACL dataset for 2014 year.

Algorithm	p@10	p@20	p@30	p@40	p@50	Average accuracy
AHRV-T	0.4	0.3	0.3	0.3	0.38	0.297
AHRV-AV	0.5	0.5	0.5	0.4	0.34	0.434
AHRV-PV	0.5	0.5	0.5	0.4	0.34	0.434
AHRV-PA	0.5	0.5	0.467	0.4	0.34	0.425
AHRV	0.5	0.5	0.5	0.425	0.38	0.461

5 Conclusions

We propose a model to predict venue impact. Our network embedding model learns unified representation of academic entities. The model captures local network information and node features. Integrating the learned vectors into a multivariate random walk

Table 10. Accuracy of variant models on the ACL dataset for 2015 year.

Algorithm	p@10	p@20	p@30	p@40	p@50	Average accuracy
AHRV-T	0.4	0.3	0.3	0.3	0.38	0.307
AHRV-AV	0.5	0.5	0.5	0.4	0.34	0.437
AHRV-PV	0.4	0.5	0.5	0.4	0.34	0.437
AHRV-PA	0.5	0.5	0.467	0.4	0.32	0.425
AHRV	0.5	0.5	0.5	0.425	0.38	0.449

model to obtain the future influence of the venue. Experimental results showed improved performance on both datasets.

Acknowledgements. This work was supported by Natural Science Foundation of Inner Mongolia Autonomous Region (2021MS06015) and Natural Science Foundation of China (61662053).

References

1. An, Z., Shen, Z., Zhou, J., et al.: The science of science: from the perspective of complex systems. Phys. Rep. **714**, 1–73 (2017)
2. Yan, E., Ding, Y.: Measuring scholarly impact in heterogeneous networks. Proc. Am. Soc. Inf. Sci. Technol. **47**(1), 1–7 (2010)
3. Gao, B.J., Kumar, G.: CoRank: simultaneously ranking publication venues and researchers. In: 2019 IEEE International Conference on Big Data, pp. 6055–6057. Los Angeles, CA, USA (2019)
4. Xiao, C., Han, J., Fan, W., Wang, S., Huang, R., Zhang, Y.: Predicting scientific impact via heterogeneous academic network embedding. In: Nayak, A.C., Sharma, A. (eds.) PRICAI 2019. LNCS (LNAI), vol. 11671, pp. 555–568. Springer, Cham (2019). https://doi.org/10.1007/978-3-030-29911-8_43
5. Xiao, C., Sun, L., Han, J., et al.: Heterogeneous academic network embedding based multivariate random walk model for predicting scientific impact. Appl. Intell. **52**(2), 2171–2188 (2022)
6. Jian, T., Meng, Q., Ming, W., et al.: LINE: large-scale information network embedding. In: 24th International Conference on World Wide Web, pp. 1067–1077. Republic and Canton of Geneva, CHE (2015)
7. Bollen, J., Rodriquez, M.: Venue status. Scientometrics **69**(3), 669–687 (2006)
8. Bergstrom, C., West, J., Wiseman, M.: The eigenfactorTM metrics. J. Neurosci **28**(45), 11433–11434 (2008)
9. Wang, Y., Yun, T., Zeng, M.: Ranking scientific articles by exploiting citations, authors, venues, and time information. In: Proceedings of AAAI, pp. 933–939. AAAI (2013)
10. Sayyadi, H., Getoor, L.: Futurerank: ranking scientific articles by predicting their future pagerank. In: Proceedings of SDM, pp. 533–544. Sparks, Nevada, USA DBLP (2009)
11. Wang, S., Xie, S., Zhang, X., Li, Z., Yu, P.S., He, Y.: Coranking the future influence of multiobjects in bibliographic network through mutual reinforcement. ACM Trans. Intell. Syst. Technol. **7**(4), 1–28 (2016)

12. Wu, Z., Lin, W., Liu, P., et al.: Predicting long-term scientific impact based on multi-field feature extraction. In: IEEE Access, pp. 51759–51770. Abramo G, D'Angelo CA (2019)
13. Wang, Z., Chen, C., Li, W.: Predictive network representation learning for link prediction. In: Proceedings of SIGIR, pp. 969–972. Association for Computing Machinery, NY(2017)
14. Hayashi, T., Fujita, H.: Cluster-based zero-shot learning for multivariate data. J. Ambient Intell. Human Comput. **12**(2), 1897–1911 (2021)
15. Cavallari, S., Zheng, V.W., Cai, H., et al.: Learning community embedding with community detection and node embedding on graphs. In: Proceedings of CIKM, pp. 377–386. Association for Computing Machinery, NY, USA (2017)
16. Li, S., Xing, X., Liu, Y., et al.: Multi-preference book recommendation method based on graph convolution neural network. In: Proceedings of WISA '19, pp. 521–532. Springer International Publishing, Dalian (2022)
17. Yu, D., Nitesh, V., Chawla, A.: Metapath2vec: scalable representation learning for heterogeneous networks. In: Proceedings of KDD '17, pp. 135–144. Association for Computing Machinery, NY, USA (2017)
18. Yu, C., Xu, Z., Jian, Z., et al.: Representation learning for attributed multiplex heterogeneous network. In: Proceedings of KDD '19, pp.1358–1368. Anchorage, AK, USA (2019)
19. Chu, Z., Dong, S., Chao, H., et al.: Heterogeneous graph neural network. In: Proceedings of KDD '19. Association for Computing Machinery, pp. 793–803. NY, USA (2019)
20. Bryan, P., Rami, A., Steven, S.: DeepWalk: online learning of social representations. In: Proceedings of KDD '14, pp. 701–710. Association for Computing Machinery, NY, USA (2014)
21. Quoc, L., Tomas, M.: Distributed representations of sentences and documents. In: Proceedings of ICML'14, pp. 1188–1196. JMLR, org (2014)
22. Hoch, S., Jürgen, S.: Long short-term memory. Neural Comput. **9**(8), 1735–1780 (1997)
23. Yan, S., Dongwon, L.: Toward alternative measures for ranking venues: a case of database research community. In: Proceedings of JCDL '07, pp. 235–244. ACM, NY (2007)
24. Zhang, F., Wu, S.: Ranking scientific papers and venues in heterogeneous academic networks by mutual reinforcement. In: Proceedings of JCDL '18, pp. 127–130. Association for Computing Machinery, NY, USA (2018)

Exploiting Item Relationships
with Dual-Channel Attention Networks
for Session-Based Recommendation

Xin Huang[1], Yue Kou[1(✉)], Derong Shen[1], Tiezheng Nie[1], and Dong Li[2]

[1] Northeastern University, Shenyang 110004, China
{kouyue,shenderong,nietiezheng}@cse.neu.edu.cn
[2] Liaoning University, Shenyang 110036, China
dongli@lnu.edu.cn

Abstract. Session-based recommendation (SBR) is the task of recommending the next item for users based on their short-term behavior sequences. Most of the current SBR methods model the transition patterns of items based on graph neural networks (GNNs) because of their ability to capture complex transition patterns. However, GNN-based SBR models neglect the global co-occurrence relationship among items and lack the ability to accurately model user intent due to limited evidence in sessions. In this paper, we propose a new SBR model based on Dual-channel Graph Representation Learning (called DCGRL), which well models user intent by capturing item relationships within and beyond sessions respectively. Specifically, we design a local-level hypergraph attention network to model multi-grained item transition relationships within a session by using sliding windows of different sizes. The experiments demonstrate the effectiveness and the efficiency of our proposed method compared with several state-of-the-art methods in terms of HR@20 and MRR@20.

Keywords: Session-based recommendation · Dual-channel ·
Hypergraph attention network · Multi-head self-attention network

1 Introduction

Session-based recommendation (SBR) is the task of recommending the next item for users based on their short-term behavior sequences. Here two key issues need to be addressed. Most of the current SBR methods model the transition patterns of items based on graph neural networks (GNNs) because of their ability to capture complex transition patterns. However, most models learn the user's current interests by pairs of item transitions, ignoring the complex user intent hidden in a set of items [3]. In addition, GNN-based SBR models neglect the global co-occurrence relationship among items and lack the ability to accurately

This work was supported by the National Natural Science Foundation of China under Grant No. 62072084.

model user intent due to limited evidence in sessions. To overcome the problems of existing SBR methods, we propose a new SBR model based on Dual-channel Graph Representation Learning (called DCGRL), which well models user preference by considering intra- and inter-session relationships among items. We summarise our contributions as follows.

- A new SBR model based on Dual-channel Graph Representation Learning (called DCGRL) is proposed, which includes local-level representation learning and global-level representation learning to capture item relationships within and beyond sessions respectively.
- For local-level representation learning, we design a local-level hypergraph attention network to model multi-grained item transition relationships within a session by using sliding windows of different sizes. For global-level representation learning, we design a global-level multi-head self-attention network to capture global item co-occurrence relationships.
- We conduct extensive experiments on two real-world datasets and the experimental results demonstrate the effectiveness of our proposed model.

2 Related Work

Inspired by the fact that similar users tend to buy similar items, the earliest session-based methods are mostly based on nearest neighbors. Markov-based methods [5,6] treat the recommendation as a sequential optimization problem and solve it to deduct a user's next behavior using the previous one. Hidasi et al. [2] propose the first work called GRU4REC to apply the RNN networks for SBR, which adopts a multi-layer Gated Recurrent Unit (GRU) to model item interaction sequences. On this basis, Tan et al. [8] enhanced the performance of the model above by data augmentation, pre-training, and taking temporal shifts in user behavior into account. Furthermore, STAMP [4] designs another attention component to emphasize the short-term interest in the session. SR-GNN [11] is a seminal GNN-based method that transforms a session into a directed unweighted graph and that utilizes gated GNNs to generate the session representation. GC-SAN [12] improves SR-GNN with the self-attention mechanism to capture long-range dependencies among items. GCE-GNN [10] proposes to enhance session-level representations by using session aware item co-occurrence relations. representation. SHARE [9] use hypergraph attention mechanism to model the user preference for next-item recommendation.

3 Preliminaries

Let $I = \{v_1, v_2, ..., v_n\}$ contain all user-clicked items in the session, where n is the number of items. The initial embedding of the item set is denoted by $X \in \mathbb{R}^{N \times d}$. Given a session $s = [v_1, v_2, \ldots, v_m]$, m is the length of the session s. The goal of our model is to take an anonymous session s and predict the next item $v_{s,m+1}$ that matches the current anonymous user's preference.

Definition 1. Item-Level Hypergraph. We construct a hypergraph $\mathcal{G}^k = (\mathcal{V}^k, \mathcal{E}^k)$ denote the hypergraph constructed from session s. We will connect all the items falling into the specific contextual window. Let $\mathcal{E}_s^W$ represent the collection of all the hyperedges constructed with such a sliding window of size on session s. Then we gather hyperedges based on different sliding windows together to be the set of hyperedges $\mathcal{E}_s$ for session s with $\mathcal{E}_s = \mathcal{E}_s^2 \cup \mathcal{E}_s^3 \cup \ldots \mathcal{E}_s^W$.

Definition 2. Global-Level Graph. Inspired by CE-GNN [10], for each item v_i^p in session s^p, the ϵ-neighbor set of v_i^p denotes a neighbor set of a given item as follows:

$$N_\varepsilon\left(v_i^p\right) = \left\{ v_j^p \mid v_i^p = v_{i'}^p \in S_p \cap S_q; v_j^p \in S_q \right.$$
$$\left. j \in [i' - \varepsilon, i' + \varepsilon]; S_p \neq S_q \right\} \tag{1}$$

We expand the concept of ϵ-neighbor set $N_\varepsilon\left(v_i^p\right)$ by adding items of current session to capture local and global item co-occurence relationship.

$$\mathcal{E}_g = \left\{ e_{ij}^g \mid (v_i, v_j) \mid v_i \in V, v_j \in \mathcal{N}_\varepsilon\left(v_i\right) \cup \mathbf{s} \right\} \tag{2}$$

The global graph is defined as $\mathcal{G}^g = (\mathcal{V}^g, \mathcal{E}^g)$

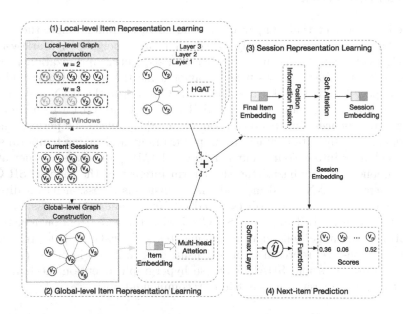

Fig. 1. Overall framework of DCGRL.

4   DCGRL: The Proposed Model

The overall framework of DCGRL is illustrated in Fig. 1.

4.1   Local-Level Item Representation Learning (LIRL)

Node to Hyperedge. Let $\mathbf{m}_{t\sim j}^{(1)}$ denote the information propagating via hyperedge ε_j from node t on the first LIRL layer. We can aggregate the information from each of the nodes connected by hyperedge ε_j with an attention operation to generate the representation $\mathbf{e}_j^{(1)}$ as follows

$$\mathbf{e}_j^{(1)} = \sum_{t\in\mathcal{N}_j} \mathbf{m}_{t\sim j}^{(1)} \quad \mathbf{m}_{t\sim j}^{(1)} = \alpha_{jt}\mathbf{W}_1^{(1)}\mathbf{x}_t^{(0)}$$

$$\alpha_{jt} = \frac{S\left(\hat{\mathbf{W}}_1^{(1)}\mathbf{x}_t^{(0)}, \mathbf{u}^{(1)}\right)}{\sum_{f\in\mathcal{N}_j} S\left(\hat{\mathbf{W}}_1^{(1)}\mathbf{x}_f^{(0)}, \mathbf{u}^{(1)}\right)} \tag{3}$$

where N_j denotes all the nodes connected by hyperedge ε_j and $\mathbf{u}^{(1)}$ represents a trainable node-level context vector for the (1)th LIRL layer. $\mathbf{W}^{(1)}, \hat{\mathbf{W}}^{(1)}$ are the transform matrices and α_{jt} denotes the attention score of node t on hyperedge ε_j. We use $S(\mathbf{a},\mathbf{b}) = \frac{\mathbf{a}^T\mathbf{b}}{\sqrt{D}}$ to calculate the attention scores.

Hyperedge to Node. Let $\mathbf{m}_{t\sim j}^{(1)}$ represent the information (user intent evidence) from hyperedge j to node t. Given the set of hyperedges y^t that are connected to node t, the update embedding for node t is calculated as follows:

$$\mathbf{x}_t^{(1)} = \sum_{j\in\mathcal{Y}_t} \mathbf{m}_{j\to t}^{(1)} \quad \mathbf{m}_{j\to t}^{(1)} = \beta_{tj}\mathbf{W}_2^{(1)}\mathbf{e}_j^{(1)}$$

$$\beta_{tj} = \frac{S\left(\hat{\mathbf{W}}_2\mathbf{e}_j^{(1)}, \mathbf{W}_3^{(1)}\mathbf{x}_t^{(0)}\right)}{\sum_{f\in\mathcal{Y}_t} S\left(\hat{\mathbf{W}}_2^{(1)}\mathbf{e}_f^{(1)}, \mathbf{W}_3^{(1)}\mathbf{x}_t^{(0)}\right)} \tag{4}$$

where $\mathbf{W}_2^{(1)}, \hat{\mathbf{W}}_2^{(1)}, \mathbf{W}_3^{(1)}$ are the trainable matrices, β_{jt} indicates the impact of hyperedges ε_j on node t. The resulting $\mathbf{x}_t^{(1)}$ can be treated as the updated embedding for item x.

With the m layer LIRL encoders, we can obtain the final local itme embedding $\mathbf{X}_l$ which reflect user intent in the session.

4.2   Global-Level Item Representation Learning (GIRL)

The GIRL module aims to capture correlations between any of two items in global graph for item representation by multi-head attention.

$$\mathbf{Q}_i = \mathbf{X}\mathbf{W}_i^Q, \quad \mathbf{K}_i = \mathbf{X}\mathbf{W}_i^K, \quad \mathbf{V}_i = \mathbf{X}\mathbf{W}_i^V,$$

$$\text{MultiHead}(Q, K, V) = \text{Concat}\left(\text{head}_1, \dots, \text{head}_h\right),$$

$$X' = \text{Dropout}(X), \tag{5}$$

$$\mathbf{X}_g^{(1)} = \text{MultiHead}\left(X', X, X\right),$$

where $W_i^Q \in \mathbb{R}^{d \times d_k}, W_i^K \in \mathbb{R}^{d \times d_k}$ and $W_i^V \in \mathbb{R}^{d \times d_k}$ are projection matrices, d_k is the dimension of $head_i$.

We stack m multi-head attention blocks to enhance the model's representation capacity. Each module takes the previous blocks' outputs as the input:

$$\mathbf{X}_g = GIRl(\mathbf{X}_g^{(1)} + \ldots + \mathbf{X}_g^{(m-1)}) \tag{6}$$

where $GIRL()$ denotes the GIRL block, and $\mathbf{X}_g^{(i)}$ is the $i - th$ block's output.

Given item representations of items learning from local and global layer, we fuse them using a linear projection, thus obtaining the final item representation $\tilde{x}$ as follows.

$$\tilde{\mathbf{X}} = [\mathbf{X}_l \| \mathbf{X}_g] \, \mathbf{W}_F \tag{7}$$

4.3   Session Representation Learning

We employ reverse position embedding matrix [7] $\mathbf{P} = [p_1, p_2, \ldots, p_m]$ to reveal the position information for all the items involved in the session as below

$$\tilde{x} = \tanh\left(\mathbf{W_1}\left[\mathbf{x}^* \| p_{m-i+1}\right] + b\right) \tag{8}$$

where $\mathbf{W}_1 \in \mathbb{R}^{2d \times d}$ and $b \in \mathbb{R}^d$ are the trainable parameters and $\|$ denotes the concatenation operation.

We use the last clicked item to represent the user's short-term interests. We then obtain the long-term preferences by employing a soft attention mechanism.

$$\alpha_i = \mathbf{q}^T \sigma \left(\mathbf{W}_2 \tilde{\mathbf{x}}_m + \mathbf{W}_3 \tilde{\mathbf{x}}_i + \mathbf{c}\right) \tag{9}$$

$$\mathbf{s}_g = \sum_{i=1}^{m} \alpha_i \tilde{\mathbf{x}}_i \tag{10}$$

Ultimately, the session embedding representation $\mathbf{s}$ can be generated.

4.4   Next-Item Prediction

We calculate item scores by $\hat{\mathbf{z}} = \mathbf{s}^T \mathbf{X}$. Then, we apply $\hat{\mathbf{y}} = \text{softmax}(\hat{\mathbf{z}})$ to the scores. For each session, we define the loss function as the cross-entropy of the prediction and the ground truth:

$$\mathcal{L}(\hat{\mathbf{y}}) = -\sum_{i=1}^{n} y_i \log(\hat{y}_i) + (1 - y_i) \log(1 - \hat{y}_i) \tag{11}$$

5   Experiments

5.1   Datasets

We conduct experiments on two public datasets: Diginetica and Last.fm. Diginetica is a personalized e-commerce research challenge dataset from CIKM CUP 2016. Last.FM is a music artist recommendation dataset. We kept the top 40,000 most popular artists and treated users' transactions in 8 h as a session.

Table 1. Comparisons of HR@20 and MRR@20 between DCGRL and baselines.

Methods	Diginetica		Last.fm	
	HR@20	MRR@20	HR@20	MRR@20
Pop	0.89	0.28	5.26	1.26
GRU4Rec	29.45	8.22	17.90	5.39
SR-GNN	50.73	17.78	22.33	8.23
GCE-GNN	54.02	19.04	24.39	8.63
SHARE	52.73	18.03	21.79	7.21
DCGRL	55.13	19.29	26.57	10.04

5.2 Baselines and Evaluation Metrics

- **Pop** [1]: This method is a simple benchmark that recommends the most popular (highest ranked) item for users.
- **GRU4Rec** [2]: GRU4Rec utilizes gated recurrent units (GRUs) to capture sequential information and model the short-term intent underlying the current session.
- **SR-GNN** [11]: SR-GNN models explicit dependencies within a session via a graph neural network and then applies a soft attention mechanism to generate session-level embeddings.
- **GCE-GNN** [10]: GCE-GNN is a GNN-based model that constructs a global co-occurrence graph from all sessions to learn global information of items and that integrates global information by considering its similarity to a rough session representation.
- **SHARE** [9]: SHARE proposes a novel session-based recommendation system empowered by hypergraph attention networks.

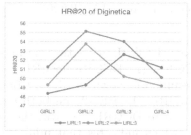
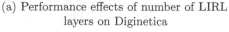
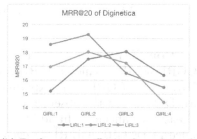

(a) Performance effects of number of LIRL layers on Diginetica

(b) Performance effects of number of GIRL layers on Diginetica

Fig. 2. Parameter sensitivity of the number of LIRL layers and GIRL layers

5.3   Performance Evaluation

Overall Performance. We compare the results of DCGRL with the base-lines, as shown in Table 1. DCGRL consistently outperforms all baseline methods. The above experimental results illustrate the effectiveness of DCGRL. The performance of neural network-based methods is superior to that of the non-personalized approach (Pop). Among these methods, the GNN-based session-based models outperform others due to their ability to dynamically model the complex item transition by constructing graph. In the models based on attention mechanism, DCGRL is better than most models. We believe that the performance improvement of DCGRL verifies the effectiveness of locl-level hypergraph attention network modules to exploit users' multiple granularities intent from user-item interaction.

Effect of Model Hyper-parameters. We studied the number of LIRL layers and GIRL layers in this experiment. We only show the results on the Diginetica dataset on account of limited space. As shown in Fig. 2(a), when the depth of LIRL layers is 2, the model is optimal. With the increase of attention layers, the multi-layer hypergraph neural network would lead to the problem of over-smoothing. As shown in Fig. 2(b), when the depth of GIRL layers is 3, the model is optimal. With the increase of attention layers, the multi-layer hypergraph neural network would lead to the problem of over-smoothing.

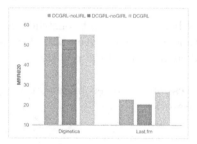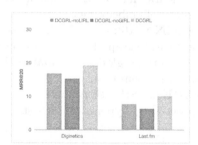

Fig. 3. Comparisons of variants on HR@20 and MRR@20.

Ablation Experiment. Here we compare DCGRL with two variants: (1) DCGRL-noLIRL removes LIRL from DCGRL. (2) DCGRL-noGIRL removes GIRL from DCGRL. The experimental results are shown in Fig. 3. Compared with DCGRL, in the absence of LIRL or GIRL, the HR@20 on both datasets decreases significantly. In addition, compared to DCGRL-noGIRL, the performance of DCGRL-noLIRL is even worse, which indicates that mining global correlation between items through global-level channel is indeed necessary.

6   Conclusions

In this paper, we propose a new Session-based recommendation model based on DCGRL. In this way, our model not only model the intra-session item transition well but also capture inter-session item co-occurence information. We construct two session graphs and then employ two modules to learn the local and global level representations respectively. Then we make predictions based on the session representation.

Acknowledgements. This work was supported by the National Natural Science Foundation of China under Grant Nos. 62072084, 62172082 and 62072086, the Science Research Funds of Liaoning Province of China under Grant No.LJKZ0094, the Natural Science Foundation of Liaoning Province of China under Grant No.2022-MS-171, the Science and Technology Program Major Project of Liaoning Province of China under Grant No.2022JH1/10400009.

References

1. Cremonesi, P., Koren, Y., Turrin, R.: Performance of recommender algorithms on top-n recommendation tasks. In: Proceedings of the Fourth ACM Conference on Recommender Systems, pp. 39–46 (2010)
2. Hidasi, B., Karatzoglou, A., Baltrunas, L., Tikk, D.: Session-based recommendations with recurrent neural networks. arXiv preprint arXiv:1511.06939 (2015)
3. Li, S., Xing, X., Liu, Y., Yang, Z., Niu, Y., Jia, Z.: Multi-preference book recommendation method based on graph convolution neural network. In: Zhao, X., Yang, S., Wang, X., Li, J. (eds.) WISA 2022. LNCS, vol. 13579, pp. 521–532. Springer, Cham (2022). https://doi.org/10.1007/978-3-031-20309-1_46
4. Liu, Q., Zeng, Y., Mokhosi, R., Zhang, H.: STAMP: short-term attention/memory priority model for session-based recommendation. In: Proceedings of the 24th ACM SIGKDD International Conference on Knowledge Discovery & Data Mining, pp. 1831–1839 (2018)
5. Rendle, S., Freudenthaler, C., Schmidt-Thieme, L.: Factorizing personalized markov chains for next-basket recommendation. In: Proceedings of the 19th International Conference on World Wide Web, pp. 811–820 (2010)
6. Shani, G., Heckerman, D., Brafman, R.I., Boutilier, C.: An MDP-based recommender system. J. Mach. Learn. Res. **6**(9) (2005)
7. Sun, F., et al.: BERT4Rec: sequential recommendation with bidirectional encoder representations from transformer. In: Proceedings of the 28th ACM International Conference on Information and Knowledge Management, pp. 1441–1450 (2019)
8. Tan, Y.K., Xu, X., Liu, Y.: Improved recurrent neural networks for session-based recommendations. In: Proceedings of the 1st Workshop on Deep Learning for Recommender Systems, pp. 17–22 (2016)
9. Wang, J., Ding, K., Zhu, Z., Caverlee, J.: Session-based recommendation with hypergraph attention networks. In: Proceedings of the 2021 SIAM International Conference on Data Mining (SDM), pp. 82–90. SIAM (2021)
10. Wang, Z., Wei, W., Cong, G., Li, X.L., Mao, X.L., Qiu, M.: Global context enhanced graph neural networks for session-based recommendation. In: Proceedings of the 43rd International ACM SIGIR Conference on Research and Development in Information Retrieval, pp. 169–178 (2020)
11. Wu, S., Tang, Y., Zhu, Y., Wang, L., Xie, X., Tan, T.: Session-based recommendation with graph neural networks. In: Proceedings of the AAAI Conference on Artificial Intelligence, vol. 33, pp. 346–353 (2019)
12. Xu, C., et al.: Graph contextualized self-attention network for session-based recommendation. In: IJCAI, vol. 19, pp. 3940–3946 (2019)

Interactive Model and Application of Joint Knowledge Base Question Answering and Semantic Matching

Jialing Zeng and Tingwei Chen[✉]

School of Information, Liaoning University, Shenyang, China
twchen@lnu.edu.cn

Abstract. Driven by the new generation of information technologies such as the Internet of Things and mobile communications, smart education is an important strategy for future educational development, and knowledge-based question answering is one of the important methods to help smart education development. But it is only used to answer students' questions. In this regard, on the basis of satisfying question answering, it is important to use questions and corresponding answers to support teaching, implement the educational concept of "student-centered, teacher-led", and carry out relevant research. First, a knowledge-based complex question answering model (KB-CQA) is proposed, and an unsupervised retriever is designed with the help of SBERT and k-dimensional trees to construct a training set that meets the new problems from the original training set, which improves retrieval speed and model accuracy; Second, the information of student and model interaction is matched with classroom teaching content based on BERT semantics, and teachers can adjust the teaching content according to the matching results. Experimental results on the CQA show that the model has achieved certain improvements in indicators such as Macro F1 and Micro F1; the specific results of semantic matching allow teachers to objectively understand the teaching situation and then make informed and targeted changes to the teaching expressions. The goal of improving teaching quality can be achieved through the simultaneous improvement of "students' learning" and "teachers' teaching".

Keywords: Knowledge base · Question answering · BERT · Teaching · Semantic matching

1 Introduction

Nowadays, the development of smarter education driven by technology has become the general trend and is becoming the "direction indicator" of global education reform in the information age. Teaching is a joint activity composed of teachers' teaching and students' learning under the norms of educational purposes, and it is the key to improving the quality of education [17]. Therefore,

© The Author(s), under exclusive license to Springer Nature Singapore Pte Ltd. 2023
L. Yuan et al. (Eds.): WISA 2023, LNCS 14094, pp. 206–217, 2023.
https://doi.org/10.1007/978-981-99-6222-8_18

how to improve teaching with the help of modern technology has become an important topic.

Knowledge base question answering (KBQA) plays a role in many scenarios, such as KBQA in the field of education, medical and financial services [18], etc. It is to query KB by converting natural language questions into logical forms through neural program induction [5], which can be directly executed on KB to produce answers [9]. But there are more and more problems and challenges:

(1) Finding the efficiency of logical forms: the number of logical forms is directly proportional to the difficulty of the problem, resulting in a huge search space and making it difficult to quickly find the correct logical form from a large number of candidate logical forms;
(2) Performance of a single model for all categories of QA: the current QA model is not good enough to answer all types of questions, especially complex questions;
(3) Application of KBQA: usually KBQA is only used for QA in various fields, and the application is not extensible enough.

To solve the above problems, this paper proposes a knowledge-based complex question answering model and then introduces semantic matching, which semantically matches the information interacted between students and the model with classroom teaching content and adjusts the content to improve classroom teaching quality. The main contributions of this paper are as follows:

(1) With the help of kd-tree to divide the search space two-dimensionally and the advantage of SBERT [10] in fast semantic similarity calculation, an unsupervised retriever is designed, and the similar samples retrieved by the retriever are used to quickly learn parameters that meet new questions in order to answer them accurately;
(2) On the CQA with HRED+KV-memNN, NSM-CIPITR and NS-CQA are compared to verify the effectiveness of the model in this paper in all categories of QA;
(3) Combining KBQA with semantic matching not only satisfies "students' learning" but also improves "teachers' teaching".

2   Related Work

At present, KBQA methods are mainly divided into three categories: semantic parsing-based, information retrieval-based, and knowledge embedding-based.

Semantic parsing-based: aims at parsing a natural language utterance into a logic form [1], and queries on the KB to get answers. Liang et al. [7] proposed NSM, which is reinforcement learning that intervenes in probability distributions that generate candidate logical forms but is unaware of the oneness problem. Saha et al. [13] proposed HRED+KVmemNN to predict the answer by classification. Saha et al. proposed CIPITR [12], designed high-order constraints and additional rewards to limit the search space, and trained the model multiple times in order to solve the problem of wholeness. Zhu et al. [21] proposed a Tree2seq model, which maps sentences into the feature space of KB to enhance the accuracy of its mapping. This method can effectively reduce retrieval time, but the error of the method has a great influence on the result. Hua et al. proposed NS-CQA [3], which placed Seq2Seq in reinforcement learning to avoid the problem of difficult model training, but the performance of different problems was quite different.

Information retrieval-based: the entity in question is used to search KB and its related subgraph, and then the answer set is formed. Yih et al. [20] used CNN to solve single-relation and constructed two different matching models, which were used to identify the entities appearing in the question and the similarity between the matching entities and the entities in KB, respectively. Xu et al. [19] introduced description information for entities based on graphs. Although the above methods have achieved good results, they do not take into account the impact of semantic relations between multi-hop entities on the answer.

Knowledge embedding-based: the entities and relations in the KB are embedded into a low-dimensional dense vector semantic space, and specific vector calculations are performed on them. Zhang et al. [4] proposed a method based on embedding. This method takes a question as an input, maps it into KB embedding. Niu et al. [8] introduced the semantic relationship between path and multi-relation question into QA task and proposed PKEEQA based on it. But mapping graphs and questions reduces the interpretability of the method.

Therefore, this paper introduces an unsupervised retriever with SBERT based on the method based on semantic parsing, which makes KB-CQA superior in answering complex questions, including multi-hop questions. Combined with the use of BERT for a more accurate semantic matching method, the information interacting with the model is further mined to achieve the improvement of the teaching quality studied in this paper and can also be used in other fields as required.

3   KB-CQA

The KB-CQA model framework is shown in Fig. 1, whose input is a complex problem q described by natural language, which consists of n characters of the form: $q = (x_1, x_2,..., x_n)$, and the output is the predicted answer a. When faced with a new problem q, we first link entity mentions, type mentions, and relationship patterns involved in the new problem to KB. Secondly, the *top-M* questions Tq that are most similar to the new question are found by the retriever, which is used as the final training sample. Then the parameter θ' of the updated encoder is obtained by training and testing in the KB-CQA model, and the logical form corresponding to the top beam is used as the test question by the parameter θ' and the beam search method. Finally, the predicted answer is obtained by executing this logical form on the KB.

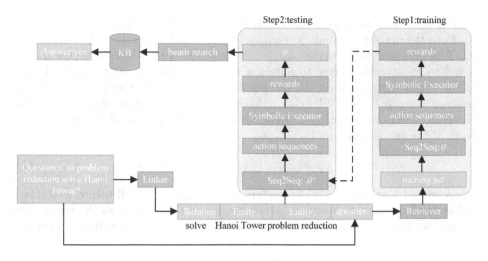

Fig. 1. Framework of KB-CQA model.

3.1   Linker

The linker, for the input problem, first identifies entity mentions and type mentions in the problem using Named Entity Recognition (NER) [6], and in this paper, entity and type mentions are combined into entity mentions for convenience of presentation, after which we need to link the previously identified entity mentions to the entity node in the KB. There may be cases where entity mentions and entity names are highly similar in their literal meanings. Therefore, we consider using a passive entity linker combined with context. Based on the passive entity linker [3], contextual information is added for entity disambiguation, and entity mentions are linked to corresponding entity nodes in the KB. After the above operation, entity mentions and type mentions can be replaced by wildcard characters "<entity>" and "<type>", and after the replacement, the problem is converted into a relational schema. Finally, the resulting relational schema is used as the input of Seq2Seq.

3.2   Seq2Seq

The Seq2Seq model is essentially a network of Encoder-Decoder structure, where the encoder is used to analyze the input sequence and the decoder is used to generate the output sequence. The input of the encoder is the concatenation of questions and question-related information in the form $q_u = (x_1, x_2, ..., x_N)$, and the output is an atomic action sequence in the form $s = (s_1, s_2, ..., s_M)$.

The encoder is a BiLSTM network [15], that takes a question of variable length as input and generates an encoder output vector e_i at each time step i as $(e_i, h_i) = LSTM[\phi_E(x_i), (e_i, h_i - 1)]$. Here (e_i, h_i) is the output and hidden vector of the i-th time step (the dimension setting of e_i and h_i are set as the same), $\phi_E(x_i)$ is the word embedding of x_i.

Our decoder is another attention-based [14] LSTM model that selects output s_m from the output vocabulary V_{output}. The decoder generates a hidden vector g_m from the previous output s_{m-1}. The previous step's hidden vector g_{m-1} is fed to an attention layer to obtain a context vector c_m as a weighted sum of the encoded states using the attention mechanism. The current step's g_m is generated via $g_m = LSTM\{g_{m-1}, [\phi D(s_{m-1}), c_m]\}$, where ϕD is the word embedding of input token s_{m-1}. The decoder state g_m is used to compute the score of the target word $v \in V_{output}$ as,

$$\pi(s_m = v|s <_m, q_u) = softmax(W \cdot g_m + b)_v \tag{1}$$

where w and b are trainable parameters, and $s<_m$ denotes all tokens generated before the time step m. We view all the weights of the encoder as the parameter θ, thus we have the probability that the encoder produces an action sequence s as,

$$\pi(s|q_u; \theta) = \prod_{m=1}^{M} \pi(s_m = v|s <_m, q_u) \tag{2}$$

When adapting the policy to a target question, our encoder outputs action sequences following the distribution computed by Eq. 2. By treating decoding as a stochastic process, the encoder performs random sampling from the probability distribution of action sequences to increase the output sequences' diversity. The encoder and decoder designs are similar to the neural generator and symbolic executor in NS-CQA [3].

After the encoder generates the entire action sequence, the decoder executes the sequence to produce an answer. It compares the predicted answer with the ground-truth answer and outputs a partial reward. If the type of the output answer is different from that of the golden answer, the action sequence that generated this answer will have a reward of 0. In addition, to alleviate the sparse reward problem, the decoder takes the output answer set and the score of the ground-truth answer set as partial rewards, and sends them back to update the parameter of the programmer as the supervision signal.

3.3   Retriever

The retriever retrieves similar samples for each new question and builds a training sample set. It measures the similarity of two questions from two aspects: (1) the number of KB artifacts in the question (2) question semantic similarity. The specific framework is shown in Fig. 2.

The masking mechanism identifies the pattern information of the question, and also combines the attention mechanism, through the attention calculation of the abstract information and the question information, the originally different questions are approximately regarded as questions with the same pattern, which in turn can be regarded as a task. Based on this, a hypothesis is proposed: when two questions have the same number of KB artifacts, the patterns of the questions are similar, so the logical forms corresponding to the two questions are

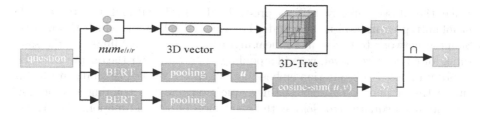

Fig. 2. Framework of the retriever.

also similar in structure. For example, for the two questions "What is Data?" and "What is Information?", after replacing the subject entities "Data" and "Information" with the same wildcards, the question patterns are consistent, and such questions are regarded as the same task, so it can compress a large number of questions in the training set into the same task.

The Number of KB Artifacts. If two questions have the same number of entities, types, and relation artifacts, the structures of their corresponding action sequences are likely to be similar. The entities, types, and relations in the problems are identified by NER and relationship recognition (RR), and their numbers $num_e/num_t/num_r$ are counted, which are expressed as vectors (num_e, num_t, num_r).

According to the above method, the vectors of the three artifacts in the training set and the new question are obtained sequentially. Since the vector is three-dimensional, the training set is constructed into a three-dimensional kd-tree, and top-M questions similar to the new question q are retrieved under the tree as Training set S_1. The kd-tree [2] can choose the axis as a vertical splitting surface arbitrarily. The most common method is to select the axis as the vertical splitting plane in turn, according to the depth of the tree. The tree is balanced, however, it may not produce the best results for every application. Therefore, in this paper, the axis is selected according to the variance. Because the large variance indicates that the data along the direction of the axis is relatively scattered, and data segmentation in this direction has a better resolution.

Semantic Similarity. The higher the semantic similarity between the two questions, the more similar the training set to the new question is retrieved, and the smaller the training set is compared with the original training set, the shorter the time required to obtain the predicted answer. This section mainly uses SBERT to retrieve top-N questions that are similar to the new question in terms of semantic similarity. The specific steps are to use commas and special symbols to obtain the tag list first as the input of the pre-trained BERT and output the tag list of each word, namely the representation of the problem as follows:

$$sentence\ representation = R[CLS] \tag{3}$$

Since the above directly utilizes pre-trained BERT without fine-tuning, the problem representation is inaccurate. For this, BERT is fine-tuned through the Siamese network to derive more meaningful question embeddings. At the same time, for efficient retrieval, we add a pooling operation after the output of BERT to obtain a fixed-size question embedding. Then use the cosine similarity to calculate the similarity between the new question and the training sample, and use the mean square error loss as the regression objective function. Finally, the results are sorted, and the training set S_2 is constructed by selecting samples above the threshold by setting the threshold. The threshold can be adjusted according to the specific test results. We take the intersection of S_1 and S_2, obtained in Sect. 2.3.1, as the final training set S of the KB-CQA model.

4   Adjustment of Teaching Content

Based on KB-CQA, to analyze how far the actual classroom teaching process captures students' error-prone and difficult knowledge points, we analyze the coverage of each sentence in the teaching content by combining semantic matching methods in the form of text semantic quantification. Regarding students' error-prone and difficult-to-understand knowledge points, we start from the information of student and model interaction, first, extract keywords from the information to obtain key knowledge points, and then perform word vector representation. The keyword extraction method used here is the RAKE algorithm [11], which is a very efficient algorithm that actually extracts key phrases and is suitable for the current scene.

Meanwhile, teaching content can be obtained with the help of devices in the smart classroom. To facilitate text similarity matching, it is necessary to convert voice or video into text. Then use each sentence of teaching utterance in the text as the input of BERT to obtain the word vector of each character, and take the average value of all word vectors as the feature vector. Finally, the cosine similarity between the key knowledge points and the text is calculated, and the absolute value of the result is convenient for teachers to compare. If the matching degree is high, it means that the classroom teaching covers a high degree of knowledge points that students are confused about, and this teaching should be strengthened in the future. If the matching degree is low, it means that the coverage degree is not high, and teachers should make targeted adjustments to the teaching contents according to the current problems of students to improve the quality of classroom teaching.

Figure 3 shows the above overall process. At present, the QA model is widely used in various fields. We can use the method in this section to further mine the information on the interaction with the model. For example, in the online consultation, doctors can judge whether to clearly explain the relevant conditions of the disease to the patient based on the information from the patient's multiple consultations.

KB-CQA can answer the above questions, then summarize the questions and answers and extract keywords from them, and the extraction result is "parallel".

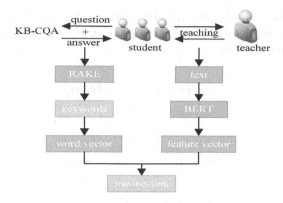

Fig. 3. Method of teaching content adjustment.

"Parallel" is semantically matched with the classroom teaching discourse one by one, and the matching degree is found to be low. The corresponding discourse is "two straight lines are on the same horizontal line". From the results, it can be seen that the teacher simply described the characteristics of "parallel". The teacher can describe it in detail in the next class so that students can have a more thorough understanding of the knowledge point of "parallel".

5 Experiments

5.1 Dataset

The data set used in this model is the CQA data set [13], which is a data set that uses KB Wikidata [16] as the underlying knowledge source for QA. There are only questions and answer, no other information. There are three types of CQA questions. To better study methods, CQA is subdivided into seven types: Simple, Logical Reasoning, Quantitative Reasoning, Comparative Reasoning, Verification (Boolean), Quantitative (Count), and Comparative (Count).

5.2 Evaluation Index and Experiment Settings

KB-CQA uses PyTorch, and model parameters and word vectors are randomly initialized. Set the dimensions of word embedding and LSTM hidden unit to 50 and 128 respectively, and $\eta_1 = 1e-4$, $\eta_2 = 0.1$ in the update of parameter θ.

In the KB-CQA experiment, answers to different types of questions use different evaluation metrics. Boolean value and integer type answers use the accuracy rate to evaluate the QA method, while the answers to the remaining types of questions are divisible entities, using the precision rate to evaluate the correctness of the predicted answers, the recall rate to calculate the proportion of entities in the ground-truth answers of the predicted answers, and F1 to evaluate the effectiveness of the QA method for a certain question, but for certain types of questions, AF1 is required. In addition, Macro F1 and Micro F1 are used to measure the overall F1 of all types of questions answered by the QA method (see formulas 4 and 5),

$$Macro\ AF1 = \frac{1}{M} \sum_{m=1}^{M} AF1(Q_m) \tag{4}$$

$$Micro\ F1 = \frac{\sum_{m=1}^{M} [N_m \cdot AF1(Q_m)]}{\sum_{m=1}^{M} N_m} \tag{5}$$

where M is the number of question types and N_m is the number of samples of a certain question Q_m. Macro F1 is to find the average F1 of all types of questions, but it does not consider the number of samples of each type of question. Micro F1 takes this into account, so overall, Macro F1 and Micro F1 evaluate the QA method from different perspectives for the overall accuracy of all types of questions.

5.3   Baseline Model

KB-CQA uses the following four models for comparative experiments.

HRED+KV-memNN [13] is a memory network that combines HRED and KVmemNN;

NSM [7] uses an EM-like algorithm to find pseudo-golden labels and adds certain probabilities to labels;

CIPITR [12] uses NPI. CIPITR-Sep puts each type of problem into the same model for training; CIPITR-All trains an integrated adaptive model for all types of problems;

NS-CQA [3] uses a weakly supervised learning approach to build end-to-end neural networks. But the problem of integral adaptability has not been solved.

5.4   Experimental Results and Analysis

It can be seen from Table 2 that KB-CQA achieves the best results on both Macro F1 and Micro F1, which are 65.93% and 76.68%. KB-CQA has the highest correct rate in five of the seven types of questions. The answers to these in five types of questions are more complicated than Simple and Logical Reasoning. CIPITR-Sep has the best performance, but the performance of Comparative Reasoning and Comparative (Count) is quite different from KB-CQA. The reason for this is that the correct logical forms corresponding to Simple and Logical Reasoning are relatively simple, while the above two types of questions are difficult. In addition, CIPITR-Sep trains all types of problems separately seven times, and the adaptability is very poor, while KB-CQA can be applied to any new problem. In summary, KB-CQA has the highest quality.

Table 1. Performance comparison of five models.

model	HRED+KV-memNN	NSM	CIPITR-All	CIPITR-Sep	NS-CQA	KB-CQA
Simple	41.40%	88.83%	41.62%	**94.89%**	85.20%	87.52%
Logical Reasoning	37.56%	80.21%	21.31%	**85.33%**	78.23%	80.74%
Quantitative Reasoning	0.89%	36.68%	5.65%	33.27%	44.22%	**44.63%**
Comparative Reasoning	1.63%	59.54%	1.67%	9.60%	59.43%	**61.54%**
Verification (Boolean)	27.28%	58.06%	30.86%	61.39%	84.42%	**84.70%**
Quantitative (Count)	17.80%	58.14%	37.23%	48.40%	61.80%	**61.93%**
Comparative (Count)	9.60%	32.50%	0.36%	0.99%	38.53%	**40.50%**
Macro F1	19.45%	59.12%	19.82%	47.70%	64.55%	**65.93%**
Micro F1	31.18%	74.68%	31.52%	73.31%	75.40%	**76.68%**

To verify the application effect of the method of content adjustment in actual teaching, this study takes the compulsory 1 "Digitalization and Number System Conversion" of high-quality information technology courses in high school as an example.

Table 2. Performance comparison of five models.

keyword	teaching discourse	matching degree
Radix-r	In general, please think about the conversion method between decimal and Radix-r?	0.512
adecimal conversion	To convert octal and hexadecimal to decimal, we can use the summation by full expansion method	0.760
digital tool	To digitize the information in the environment, some digital tools need to be used	0.765

Table 3 above lists the teaching discourses with the highest matching degree. For example, the discourse "To digitize the information in the environment, some digital tools need to be used". describes the knowledge points of "digital tools" in great detail, and the matching degree is 0.765, indicating that this discourse accurately describes the connotation of "digital tools" and covers the knowledge points mentioned by students' questions; however, the discourse "In general, please think about the conversion method between decimal and Radix-r?" only mentions "Radix-r" and does not explain the rest of the knowledge point, and the matching degree is 0.512, which shows that the teacher's explanation of this knowledge point in class has not met the needs of students, and more time should be spent on explaining it.

5.5   Ablation Study

To verify the influence of the retriever on the model, the experiment is carried out by removing the QA model of the retriever, and the results are shown in Table 1. Table 1 show that from the perspective of Micro-F1, the retriever proposed in this paper, the QA model is better than the QA model without the retriever, which verifies that the retriever retrieves samples similar to the new question as training samples instead of the unprocessed original training samples are helpful to improve the accuracy of the model.

Table 3. Analysis of ablation experiment results.

model	Micro AF1
w/o retriever	75.82%
KB-CQA	76.68%

6   Conclusion

This paper starts from the students' questions and always adheres to the "student-centered, teacher-led" idea. On the one hand, it proposes a complex question answering model based on knowledge graph; on the other hand, a method of teaching content adjustment is proposed. The interaction between two aspects, not only improves "students' learning" but also adjusts "teachers' teaching", realizes the goal of helping students solve learning problems in a targeted manner, and truly improves the teaching level. In future research, we can build students' personal KB and consider the differences in students' knowledge and abilities to accurately solve problems for students and reduce the pressure on teachers.

References

1. Berant, J., Liang, P.: Semantic parsing via paraphrasing. In: Proceedings of the 52nd Annual Meeting of the Association for Computational Linguistics (Volume 1: Long Papers), pp. 1415–1425 (2014)
2. Dinh, N.T., Le, T.M., Van, T.T.: An improvement method of KD-tree using k-means and k-NN for semantic-based image retrieval system. In: Rocha, A., Adeli, H., Dzemyda, G., Moreira, F. (eds.) WorldCIST 2022. LNNS, vol. 469, pp. 177–187. Springer, Cham (2022). https://doi.org/10.1007/978-3-031-04819-7_19
3. Hua, Y., Li, Y.F., Qi, G., Wu, W., Zhang, J., Qi, D.: Less is more: data-efficient complex question answering over knowledge bases. J. Web Semant. **65**, 100612 (2020)
4. Huang, X., Zhang, J., Li, D., Li, P.: Knowledge graph embedding based question answering. In: Proceedings of the Twelfth ACM International Conference on Web Search and Data Mining, pp. 105–113 (2019)

5. Jin, H., Li, C., Zhang, J., Hou, L., Li, J., Zhang, P.: XLORE2: large-scale cross-lingual knowledge graph construction and application. Data Intell. 1(1), 77–98 (2019)
6. Li, J., Fei, H., Liu, J., Wu, S., Zhang, M., Teng, C., Ji, D., Li, F.: Unified named entity recognition as word-word relation classification. In: Proceedings of the AAAI Conference on Artificial Intelligence, vol. 36, pp. 10965–10973 (2022)
7. Liang, C., Berant, J., Le, Q., Forbus, K.D., Lao, N.: Neural symbolic machines: learning semantic parsers on freebase with weak supervision. Population 1
8. Niu, G., et al.: Path-enhanced multi-relational question answering with knowledge graph embeddings. arXiv preprint arXiv:2110.15622 (2021)
9. Pasupat, P., Liang, P.: Inferring logical forms from denotations. In: Proceedings of the 54th Annual Meeting of the Association for Computational Linguistics (Volume 1: Long Papers), pp. 23–32 (2016)
10. Reimers, N., Gurevych, I.: Sentence-BERT: sentence embeddings using Siamese BERT-networks. arXiv preprint arXiv:1908.10084 (2019)
11. Rose, S., Engel, D., Cramer, N., Cowley, W.: Automatic keyword extraction from individual documents. Text Min.: Appl. Theory 1–20 (2010)
12. Saha, A., Ansari, G.A., Laddha, A., Sankaranarayanan, K., Chakrabarti, S.: Complex program induction for querying knowledge bases in the absence of gold programs. Trans. Assoc. Comput. Linguist. 7, 185–200 (2019)
13. Saha, A., Pahuja, V., Khapra, M., Sankaranarayanan, K., Chandar, S.: Complex sequential question answering: towards learning to converse over linked question answer pairs with a knowledge graph. In: Proceedings of the AAAI Conference on Artificial Intelligence, vol. 32 (2018)
14. de Santana Correia, A., Colombini, E.L.: Attention, please! A survey of neural attention models in deep learning. Artif. Intell. Rev. 55(8), 6037–6124 (2022)
15. Siami-Namini, S., Tavakoli, N., Namin, A.S.: The performance of LSTM and BiL-STM in forecasting time series. In: 2019 IEEE International Conference on Big Data (Big Data), pp. 3285–3292. IEEE (2019)
16. Vrandečić, D., Krötzsch, M.: WikiData: a free collaborative knowledgebase. Commun. ACM 57(10), 78–85 (2014)
17. Wang, D.: Pedagogy (in Chinese). People's Education Press (2016)
18. Wang, R.: A multi-modal knowledge graph platform based on medical data lake. In: Zhao, X., Yang, S., Wang, X., Li, J. (eds.) WISA 2022. LNCS, vol. 13579, pp. 15–27. Springer, Cham (2022). https://doi.org/10.1007/978-3-031-20309-1_2
19. Xu, Y., Zhu, C., Xu, R., Liu, Y., Zeng, M., Huang, X.: Fusing context into knowledge graph for commonsense question answering. In: Workshop on Commonsense Reasoning and Knowledge Bases (2021)
20. Yih, W.T., He, X., Meek, C.: Semantic parsing for single-relation question answering. In: Proceedings of the 52nd Annual Meeting of the Association for Computational Linguistics (Volume 2: Short Papers), pp. 643–648 (2014)
21. Zhu, S., Cheng, X., Su, S.: Knowledge-based question answering by tree-to-sequence learning. Neurocomputing 372, 64–72 (2020)

Policy-Oriented Object Ranking with High-Dimensional Data: A Case Study of Olympic Host Country or Region Selection

Hengrui Cui(✉)(iD), Jiaxing He(iD), and Weixin Zeng(iD)

Laboratory for Big Data and Decision, National University of Defense Technology, Changsha, China

m18200205761@163.com

Abstract. Policy-oriented object ranking is of great importance in the field of object recommendation and target retrieval, such as e-commerce recommendation system, financial services and selecting hosts of some large-scale sports events. However, the solution of such problem is often accompanied by the processing of massive high-dimensional data. The typical methods may lead to the phenomenon of over-reliance on historical data as well as ignoring the interactions among indicators. This also causes the ranking results to lose part of the timeliness and sensitivity. We designed Entropy-weighted Synthetic Model (ESM) in order to optimize the defects above. We focus on the case of Olympic Host Country or Region Selection to demonstrate the superiority of our method. We first fulfill calculation of synthetic property weights and differential data of each attribute based on time series of the selected synthetic group. Then we evaluate the weight of each indicator, obtain the comprehensive scores and generate the ranking results of the target countries or regions. Last, we implement Placebo Test to eliminate the effects of Synthetic Control Method (SCM) and validate the robustness of the model with the support of Hypothesis Test. From the perspective of comparing with the results of Difference-in-Difference Model (DID) and Propensity Score Matching (PSM), ESM shows better adaptability and predictability and provide conclusion consistent with professional perspective.

Keywords: Policy-oriented Object Ranking · High-dimensional Data · Entropy-weighted Synthetic Model · Placebo Test

1 Introduction

With the rapid development of modern technology, high-dimensional data has become increasingly common in various fields, including finance, healthcare, social media, and scientific research. However, the analysis and processing of high-dimensional data are becoming more challenging due to the following reasons. Firstly, high-dimensional data often contains large amounts of irrelevant or redundant information, which makes it difficult to extract useful features and patterns. Secondly, high-dimensional data requires more complex and time-consuming algorithms for processing and analysis, which may

L. Yuan et al. (Eds.): WISA 2023, LNCS 14094, pp. 218–229, 2023.
https://doi.org/10.1007/978-981-99-6222-8_19

result in computational difficulties and increased costs. Thirdly, high-dimensional data may suffer from the curse of dimensionality, which refers to the phenomenon that the sample size, required for achieving a certain level of statistical significance, increases exponentially with the number of dimensions. Therefore, it is important to develop efficient and effective techniques for handling high-dimensional data, which can help to address various real-world challenges.

Relevant problems frequently appear in the process of international social development, especially for selecting host of some large-scale sports events. In this paper, we focus on the example of Olympic host country or region selection. The growing scale of Olympic events has added social costs to host countries or regions, leading to heavy debts from other countries or regions [1]. The increase in the standards of the Olympic Games has correspondingly promoted the expansion [2] of related facilities and service demand in host countries or regions, which need to match the demand of high-level sports events.

The existing methods for addressing such problems are mainly limited to expert evaluation, paired sample analysis and other approaches. Expert evaluation is constrained by the subjectivity of the evaluation results, while paired sample analysis is limited by the complexity and time-varying indicators of high-dimensional data. In this paper, we propose an Entropy-weighted Synthetic Model. By using the entropy weight method, we provide more objective numerical evaluation results to offer policy recommendations. We also employ the synthetic control method to reduce the complexity of processing data by synthesizing various supporting variables. In the Olympic host country or region selection, we carried out a series of data processing, model solving and comparative experiments to demonstrate the effectiveness and superiority of our function.

2 Preliminary

2.1 Policy-Oriented Object Ranking

Policy-oriented object ranking is to provide support with object recommendation and target retrieval. In solving policy-oriented object ranking with high-dimensional data, we generally regard the variable growth brought about by a policy as a natural experiment. It is up to decision-makers to define two types of scopes. The first type of scope mainly includes the determination of the main variable and secondary variables by relevant experts according to the specific problem. The main variable is used as model observation indicators, and the secondary variables are used as model input variables to support the operation of the model. The second scope consists primarily of synthetic control areas that are regionally delineated in accordance with the policy as well as satisfy the basic standard. The expected output of this method is the score prioritization of natural subjects provided to decision makers.

2.2 Olympic Host Country or Region Selection

The decision of the host country or region of the Olympic Games [3] is a typical policy-oriented object ranking problem. We will take the Olympics host selection as an example to specifically illustrate our methods. We give two types of scopes [4] as input to the model, and expect to get the score ranking of the host country or region list.

For the first scope, we take GDP as the main variable for observation. To determine the secondary variables, we categorize the impact of the Olympics on a country or region into five parts: Regional Consumption, Regional Investment, Foreign Trade, Ecological Environment, and Social Development. We select 11 variables from the 5 main aspects, including GDP per capita, Household final consumption expenditure (Finconexppercap), Gross national expenditure (Totnatexp), International tourism and receipts (Tourevenue), Gross fixed capital formation (GFCF), Foreign direct investment (FDIinflow), Exports of goods and services (EXGS), Imports of goods and services (IMGS), CO_2 emissions kt (lnGHG), CO_2 emissions kg per PPP $ of GDP (CO2emission), Unemployment youth total (Unemployment), GNI (lnGNI) (Table 1).

Table 1. Five categories and eleven indicators.

Main-variable	Sub-variable	Explanation
Per capita GDP	PPP	Per capita GDP of each country or region
Regional consumption	Finconexppercap	The change of local residents' consumption level before and after the Olympic Games
	Totnatexp	
	Tourevenue	
Regional investment	GFCF	Reflect the influence of Olympic Games on regional investment level
	FDIinflow	
Foreign trade	EXGS	Measure the Olympic Games on the host country or region's foreign trade situation change
	IMGS	
Ecological environment	lnGHG	Qualitative assessment of the impact on local air quality and ecological environment
	CO2emission	
Social development	Unemployment	comparing the difference of unemployment rate and the change of resident income level in target countries or regions before and after the Olympic Games
	lnGNI	

For the second scope, we study the countries or regions that have hosted the Olympics in the 21st century, and select the remaining 30 countries or regions with the largest economic sizes in the world according to the world's total GDP [5] ranking as the control group.

3 Related Work

3.1 Difference-in-Difference Analysis

Difference-in-Difference (DID) [6] was originally designed to evaluate the policy-oriented object problem which is a classic means of solving such problems. With the rapid expansion of this approach, DID has gradually expanded in the research of economy and society [7]. In the treatment of selection bias, DID allows for unobservable factors to play a role and allows undetectable factors to influence the individual's decision to intervene. However, DID method still has its own limitations. First, the approach requires more rigorous data. DID is based on panel data, so cross-sectional data and time series data of target objects are essential which brings difficulty to data collection. Second, individual real-time effects are not controlled. When the policy is not implemented, the variables of both the experimental group and the control group change closely with time. The real-time effect results in the inconsistency of the experimental group and the control group before and after the implementation of the policy, resulting in systematic error. Therefore, the application of DID may be problematic.

3.2 Propensity Score Matching

Propensity Score Matching (PSM) is also a classic way to solve the policy-oriented object problem [8]. The basic idea is to pair participants and non-participants with similar observable characteristics to construct a control group similar to the experiment group. The total impact of the project is obtained by calculating the difference in outcomes between each pair of matches. And we weight the scores for all groups. Compared with the exact matching method, PSM does not strictly match individuals one by one according to their observable characteristics, but obtains a propensity score matching score by estimating the probability of each individual participating in the project, and then selects individuals with the same or similar score values to match. Furthermore, PSM assumes [9] that given the observable characteristics of an individual, an individual's participation in the project is random. However, due to the implementation principle of this method, sufficient information must be collected in order to obtain more accurate pairing results. Moreover, the interaction of paired indicators is not taken into consideration mostly, causing possible disorderliness of result.

4 Methodology

4.1 Synthetic Control Method

In order to overcome the complexity and ambiguity brought by high-dimensional data, we comprehensively select synthetic control method as the main function to solve this problem. Abadie and Gardeazabal first proposed synthetic control in their paper [10]. Compared with traditional policy evaluation methods, such as synthetic part of DID, the synthetic control method overcomes the shortcomings to a large extent, and is a non-parametric method that extends the traditional multiplier method. The synthetic control method gives full consideration to the heterogeneity of the treatment group, and constructs a counterfactual linear reference combination according to the additional weight value of the similarity between the target and the economic individual index in the control group. By comparing the gap of the estimated variable in the target area and in the control group, the role of evaluating the implementation of the policy can be demonstrated. On the basis of the above feasible methodological analysis of this issue, we ignore a large number of attributes that are difficult to collect and quantify, and focus on the impact of the Olympics, in this case study, on GDP of the host countries or regions.

Consider the economic growth caused by hosting the Olympics as a policy implementation for the country or region, assuming that we can currently observe economic indicators from $j+1$ countries or regions over a period of T. In the experiment, the hosting country or region is the target country or region that is influenced by the Olympics policy, while the remaining j countries or regions are the control group without being influenced by the Olympics policy.

With T_0 indicating the time when the economic policy for the Olympic Games is implemented, T_0 satisfies the condition that $1 \leq T_0 \leq T$, here T_0 corresponds to the year of the Olympic Games. Using Y_{it}^N to represent the economic growth data of the country or region i at a time t when it has not been affected by the Olympics economic policy, Y_{it}^I represents the economic growth data of a country or region i at a time t when it is affected by the Olympics economic policy.

When the variable $t \in [1, T_0]$, we have $Y_{it}^N = Y_{it}^I$. When $t \in [T_0, T]$, we have $Y_{it}^I = Y_{it}^N - \rho_{i,t}$, where $\rho_{i,t}$ represents the economic growth change of the country or region ranked i at time t under the influence of Olympic policy. There are three scenarios:

- $\rho_{i,t} > 0$: the Olympic Games can promote the economic growth of the target country or region.
- $\rho_{i,t} < 0$: hosting the Olympic Games can inhibit the economic growth of the target country or region.
- $\rho_{i,t} = 0$: the economic growth of the target country or region is not affected by the policy of hosting the Olympic Games.

However, in actual situations, for the country or region $j = 1$ that hosts the Olympics, the data we can obtain is Y_{it}^I, as we can't get access to the data Y_{it}^N if the country or region hadn't hosted the Olympics. For country or region $j = 1, 2, ..., J+1$, we can obtain Y_{it}^N but we don't have Y_{it}^I.

Therefore, our goal is to estimate the counterfactual outcome Y_{it}^N for country or region j. In other words, we want to know what the indicators of country or region j would have been without the intervention of the Olympic Games.

Based on the synthetic control method, the weighted average results of different test units in the control group are aggregated into a composite control. Its weight can be expressed as $W = (w_2, ..., w_{j+1})$, the indicators of the host country or region not affected by the Olympic Games can be expressed as:

$$\hat{Y}_{1t}^N = \sum_{j=2}^{J+1} w_j Y_{jt}^N. \tag{1}$$

We define the optimization problem as follows:

$$\underset{W}{\text{argmin}} \, \|X_1 - X_0 W\| = \left(\sum_{h=1}^{k} v_h \left(X_{h1} - \sum_{j=2}^{J+1} w_j Y_{jt} \right) \right),$$

$$\text{s.t.} \sum_{j=2}^{J+1} w_j = 1, \quad w_j \geq 0$$

(2)

Among the formulas above, v_h represents the weights of different independent features, which need to be determined before solving the optimization problem [11]. The common way to determine the weight value is normalization as the mean of its variable equals 0 and the variance equals 1.

4.2 Entropy Weight Method

Further, in order to provide clear and specific numerical results for decision makers, we comprehensively select the entropy weight method to calculate the relevant weight and comprehensively score the differential results obtained by the synthetic control method.

Entropy weight method is a multi-criteria decision-making technique that is widely used in various fields of research [12]. The method utilizes the concept of entropy to determine the weights of various criteria in decision-making. The basic idea of the entropy weight method is to calculate the entropy of each criterion and then use the entropy values to determine the weight of each criterion. In this way, the method can effectively avoid the subjective biases that may arise from traditional weighting methods. It has been shown to be effective in dealing with complex decision-making problems, especially when there are multiple criteria involved. The method can effectively identify the most significant criteria and assign appropriate weights to each criterion, which can help decision-makers to make more informed and objective decisions. To assess the overall impact of the Olympics on a certain country or region, we assign weights to each indicator. It assumes that the dispersion of indicators is directly proportional to their importance [13].

Accordingly, we use the EWM to calculate the weight of each indicator. We standardize the measured data. The standardized value of the indicator of the sample country or region j is denoted as P_{ij}:

$$P_{ij} = \frac{\tilde{x}_{ij}}{\sum\limits_{j=1}^{m} \tilde{x}_{ij}},$$

(3)

where $i = 1, 2, ..., n; \, j = 1, 2, ..., m$

The entropy value E_i is calculated as:

$$E_i = \frac{\sum\limits_{j=1}^{m} p_{ij} \cdot \ln p_{ij}}{\ln m}.$$

(4)

E_i means that the greater the differentiation degree of indicator i is, the higher weight should be given to the indicator. Therefore, the weight w_i of indicator i is calculated as follows:

$$w_i = \frac{1 - E_i}{\sum\limits_{i=1}^{m} (1 - E_i)}. \tag{5}$$

$$S_j = \sum\limits_{i=1}^{n} w_i \cdot p_{ij}. \tag{6}$$

5   Experiments

5.1   Experiment Settings

Datasets. On the basis of the feasible methodological analysis of selecting Olympics host, we ignore a large number of attributes that are difficult to collect and quantify as we mainly focus on the impact of the Olympics on GDP, serving as estimated indicator, of the host countries or regions. In this case, since objective conditions do not allow us to obtain comparative data on countries or regions affected by having hosted the Olympics, this article uses equilibrium panel data from 2000 to 2016 for 35 economies, including nine host countries or regions of Summer Olympics and Winter Olympics. Australia, Greece, China, the United Kingdom, Brazil, Russia, Canada, Korea and Italy were set up as experiment group. All data are derived from the World Development Indicators published by the World Bank and national official statistics websites. The missing data are supplemented in synthesis way.

Experimental Process. First, we convert the preprocessed dataset to JSON format, construct a transformation matrix and select the countries or regions, support indicators and time series as the three dimensions of the matrix. Second, the transformation matrix is input to the DID model, serving as baseline, and the SCM model to obtain the differential data of the two types. Third, we feed the differential data into the EWM and obtain the comprehensive scores and ranking result of the target countries or regions.

Evaluation. In this experiment, it is difficult to obtain objective evaluation of the final ranking results. We will analyze the ranking results in detail to illustrate the advantages of our approach.

5.2   Ranking Result

The ranking scores are generated as the output of ESM. To make comparison with the baseline, we also generate the ranking scores based on DID and PSM shown as the table below (Table 2):

Table 2. The ranking results of ESM and baselines.

ESM				DID				PSM			
Summer		Winter		Summer		Winter		Summer		Winter	
CR	RS	CR	RS	S	RS	W	RS	S	RS	W	RS
CN	0.54	KR	0.41	UK	0.53	RU	0.41	CN	0.52	KR	0.45
AU	0.42	RU	0.34	CN	0.48	KR	0.36	UK	0.38	RU	0.36
UK	0.35	CA	0.23	AU	0.36	CA	0.23	AU	0.36	CA	0.33
GR	0.12	IT	0.2	GR	0.16	IT	0.18	GR	0.24	IT	0.18
BR	0.1			BR	0.06			BR	0.14		

We initial *country or region* as CR and *Ranking Score* as RS. For simplification, we use abbreviations to denote the countries or regions. For example, CN, AU, UK, GR, BR represent China, Australia, United Kingdom while KR, RU, CA, IT represent Korea, Russia, Canada, Italy.

From the table, we figure out that DID models rely too much on inertia of the historical data while PSM is restricted by the correlation of the indicators. For example, in the summer Olympic target group, the DID model selected the UK while the ESM model selected China as the best choice. The DID model overly focused on the UK's historical high differential economic advantage and did not capture the trend of China's rapid economic growth. PSM also fails to provide a better ranking result due to the correlation of its matching samples which is not expressed in the model. Reflected in the ranking result, the combined scores of the two target countries or regions, such as the United Kingdom and Australia, Russia and Canada, are ranked with little difference, making it difficult to make comprehensive decisions. From the perspective of the current law of economic development, the ranking results given in ESM have better adaptability and predictability.

5.3 Process of Synthetic Control Method

The weights of the control group for the four countries or regions calculated by the Synthetic Control Method are as follows (Table 3):

Table 3. The weights of the control group.

Australia		Greece		Britain		China	
country	weight	country	weight	country	weight	country	weight
Argentina	0.007	Spain	0.254	Holland	0.123	India	0.863
China	0.072	Macao	0.08	USA	0.517	Poland	0.137
Japan	0.04	Poland	0.019	Poland	0.079		
Russia	0.054	Portugal	0.646	Sweden	0.168		
Spain	0.228			Portugal	0.113		
USA	0.443						
Macao	0.022						
Portugal	0.133						

The numbers of each control group are varied as the members within the scope have been filtered for each target country or region. For target country or region, we select members whose attribute values fall within the positive and negative threshold, preset by decision-maker as the control group.

In order to further verify the superiority of synthetic control models in dealing with a class of problems, we further use the PSM and DID to process the data and obtain visualization results. Obviously, the ESM has better performance in the fit of the training set and the stability of its trend.

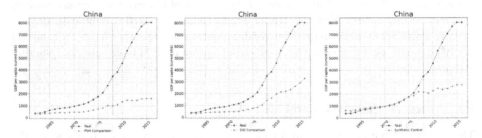

Fig. 1. Comparison between two methods

This comparative experiment, shown in Fig. 1, further verifies that the ESM has better performance in dealing with the problems of inconsistency of original data and complexity of high-dimensional data. Due to significant differences in training results, we simply use Mean Squared Error [14] to illustrate the results shown in Fig. 1:

$$MSE_{PSM} = 860787.63, MSE_{DID} = 748103.79, MSE_{ESM} = 130444.98. \qquad (7)$$

The mean squared error of ESM is significantly smaller than that of the other two methods, indicating its superior fitting performance and better ability to capture the economic development trend of the original target.

5.4 Placebo Test

In order to eliminate the potential influence of ESM on the results, we will further validate the model through placebo test. Addie et al. recommends the use of a placebo test [15] for statistical testing. In order to eliminate the effects of synthetic control methods on the target countries or regions themselves.

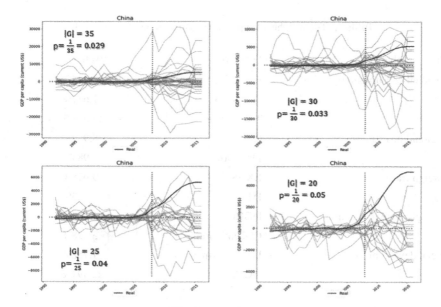

Fig. 2. Placebo test.

In order to eliminate the effects of synthetic control methods on the target countries themselves. We took each country or region $G_{country} \in k_i$, setting China as an example, in turn as a synthetic object, built multiple synthetic groups $G_{country} - k_i$ and plot relevant curves that describes the difference between the real value and the GDP of Synthetic country or region k_i. We gradually reduce, with the reducing step of 5 countries or regions, the number of components, aiming to figure out the robust value of the model.

To demonstrate the robustness of this model, we conduct Hypothesis Testing as follow:

- H_0 : The trend of GDP is effected by Sythetic Control Method.
- H_1 : The trend of GDP is only effected by conducting the Olympic Games.

As is shown in last plot in Fig. 2, the probability that any synthetic curve has a stronger upward trend than the real data is:

$$p \leq \alpha = 0.05. \tag{8}$$

Under these conditions, we reject the original hypothesis so that the Olympics do have a significant impact on the GDP of the host country or region and that the model has good robustness.

5.5 Weights of Secondary Variables

Since the EWM is calculated [16] based on original data. Therefore, differential results of different synthetic models will not affect the weight calculation.

The weights of each indicator and the comprehensive evaluation of the host countries or regions in the past 20 years are as follows (Table 4):

Table 4. The weights of support indicators generated by ESM.

Main-variable	Sub-variable	Weights
Regional consumption	Finconexppercap	2.774%
	Totnatexp	1.621%
	Tourevenue	13.063%
Regional investment	GFCF	28.952%
	FDIinflow	2.98%
Foreign trade	EXGS	9.628%
	IMGS	10.723%
Ecological environment	lnGHG	2.082%
	CO2emission	3.928%
Social development	Unemployment	1.758%
	lnGNI	22.491%

According to the table, there are two major variables as GFCF and lnGNI, three moderate variables, and six minor variables that contribute to GDP as the estimated indicator. The major factors, characterized by their high weight values over 20%, directly affect the economic development. The moderate variables as Tourevenue, EXGS and IMGS weighted around 10% exert a strong influence on economic developments in terms of the flow of urban resources. Conclusions above further suggests the aspects that the government should focus on to seize the opportunity of hosting the Olympics.

6 Conclusion

In this paper, we propose a succinct and time-sensitive solution for policy-oriented object ranking with high-dimensional data. Specifically, we combined two algorithms Synthetic Control Method and Entropy Weight Method for Target objects fitting and comprehensive scores ranking named Entropy-weighted Synthetic Model. For Synthetic Control Method, it effectively converts original data into differential data for each indicator. The outputs are fed to Entropy Weight Method in order to generate the comprehensive scores and the ranking result. We perform experiments on Difference-in-difference model to demonstrate the adaptability and superiority of our proposed method compared with the baseline. Our model significantly outperforms the baseline in the fit of the training set. At the same time, the ranking result generated by our model possesses better predictability. We tend to figure out a better function to integrate and standardize the original data [17] aiming to make the model more scalable, which will be our future work.

References

1. Kang, Y.: Long-term impact of a mega-event on international tourism to the host country or region a conceptual model and the case of the 1988 Seoul Olympics. J. Int. Consum. Market. **6**(3), 205–226 (1994)
2. Mueller, M.: The mega-event syndrome: why so much goes wrong in mega-even planning and what to do about it. J. Am. Plann. Assoc. **81**(1), 6–17 (2015)
3. Baade, R., Baumann, R., Matheson, V.: Assessing the economic impact of college football games on local economies. J. Sports Econ. **9**(6), 628–643 (2008)
4. Zeng, J., Zhou, J.: Variable selection for high-dimensional data model a survey. J. Appl. Stat. Manag. **36**(4), 678–692 (2017). (In Chinese)
5. Bert, R.: World economy. Econ. Outlook **46**(1), 32–34 (2022)
6. Xia, X., Huang, T., Zhang, S.: The impact of intellectual property rights city policy on firm green innovation a quasi-natural experiment based on a staggered did model. Systems **11**(209), 209 (2023)
7. He, T., Chen, W.: Evaluation of sustainable development policy of Sichuan citrus industry in China based on DEA–Malmquist index and did model. Sustainability **15**(4260), 4260 (2023)
8. Wang, X., Hu, S.: Does university-industry collaboration improve the technological innovation performance of manufacturing firms? An empirical study based on propensity score matching method. Technol. Econ. **41**(4), 30–43 (2022). (In Chinese)
9. Liu, J., Zhang, G., Li, C.: Human capital, social capital, and farmers' credit availability in China based on the analysis of the ordered probit and PSM models. Sustainability **12**(4), 1583 (2020)
10. Abadie, A., Gardeazabal, J.: The economic costs of conflict a case study of the Basque country. Am. Econ. Rev. **93**(1), 113–132 (2003)
11. Abadie, A., Diamond, A., Hainmueller, A.: Synthetic control methods for comparative case studies estimating the effect of California's tobacco control program. J. Am. Stat. Assoc. **105**(490), 493–505 (2010)
12. Jin, D., Yang, M., Qin, Z., Peng, J., Ying, S.: A weighting method for feature dimension by semi-supervised learning with entropy. IEEE Trans. Neural Netw. Learn. Syst. **34**(3), 1218–1227 (2023)
13. Fullman, N.: Measuring performance on the healthcare access and quality index for 195 countries and territories and selected subnational locations a systematic analysis from the global burden of disease study 2016. The Lancet **391**(10136), 2236–2271 (2018)
14. Check, A., Nolan, A., Schipper, T.: Forecasting GDP growth using disaggregated GDP revisions. Econ. Bull. **39**(4), 2580–2588 (2019)
15. Abadie, A., Diamond, A., Hainmueller, J.: Comparative politics and the synthetic control method. Am. J. Polit. Sci. **59**(2), 495–510 (2015)
16. Chen, C., Zhang, H.: Evaluation of green development level of Mianyang agriculture, based on the entropy weight method. Sustainability **15**(7589), 7589 (2023)
17. Wang, Y., Gao, S., Li, W., Jiang, T., Yu, S.: Research and application of personalized recommendation based on knowledge graph. In: Xing, C., Fu, X., Zhang, Y., Zhang, G., Borjigin, C. (eds.) Web Information Systems and Applications: 18th International Conference, WISA 2021, Kaifeng, China, September 24–26, 2021, Proceedings, pp. 383–390. Springer International Publishing, Cham (2021). https://doi.org/10.1007/978-3-030-87571-8_33

Natural Language Processing

A Joint Relation Extraction Model Based on Domain N-Gram Adapter and Axial Attention for Military Domain

Zhixiang Yang[1], Zihang Li[2], Ziqing Xu[2(✉)], Zaobin Gan[2], and Wanhua Cao[1]

[1] Wuhan Digital Engineering Research Institute, Wuhan 430074, China
[2] School of Computer Science and Technology,
Huazhong University of Science and Technology, Wuhan 430074, China
{M202273904,kenzie_syu,zgan}@hust.edu.cn

Abstract. Domain-specific relation extraction plays an important role in constructing domain knowledge graph and further analysis. In the field of military intelligence relation extraction, there are challenges such as relation overlapping and exposure bias. Therefore, on the basis of a combination of domain N-gram adapter and axial attention, this paper presents a single-step joint relation extraction model for the field of military text analysis. Considering domain-specific language structures and patterns, the domain-specific N-gram adapter is incorporated into the pre-trained language model to improve the encoding of the proposed model. Furthermore, the axial attention mechanism is applied to capture the dependencies between token pairs and their contexts, so as to enhance the encoding representation ability of the proposed model. After that, entities and relations are jointly extracted by a relation-specific decoding method. The effectiveness of the proposed model is demonstrated through experiments on a military relation extraction dataset with F1-Score 0.6690 and CMeIE with F1-Score 0.6051, which is better than existing joint relation extraction models.

Keywords: military domain · joint relation extraction · domain N-gram adapter · axial attention

1 Introduction

Entity relation extraction is a Natural Language Processing (NLP) task that involves identifying and extracting relationships between entities in a text. The goal of information extraction is to automatically extract structured information from unstructured text data. Entity relation extraction is valuable for military applications such as intelligence gathering, situation awareness, and military planning by identifying key entities and relationships within text data to help decision-making.

While general-purpose relation extraction methods have been extensively studied, domain-specific relation extraction presents several unique challenges

© The Author(s), under exclusive license to Springer Nature Singapore Pte Ltd. 2023
L. Yuan et al. (Eds.): WISA 2023, LNCS 14094, pp. 233–245, 2023.
https://doi.org/10.1007/978-981-99-6222-8_20

that must be addressed to achieve accurate and effective results. Specifically, in the field of military, these challenges include the lack of annotated data, overlapping of relationships within military corpus, and the requirements for military domain-specific knowledge.

There are mainly three complex scenarios that relation extraction models must handle in the field of military intelligence, and they have impact on the accuracy of the models [1]. The scenarios are listed in Fig. 1. An Entity Pair Overlap (EPO) occurs when multiple relation types overlap with an entity pair. Single Entity Overlap (SEO) refers to a situation in which a single entity is involved in multiple relationships with different entities. Subject Object Overlap (SOO) means that nesting exists between a subject and an object.

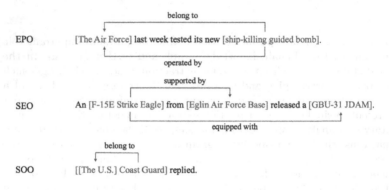

Fig. 1. Examples of EPO, SEO and SOO

Furthermore, despite the fact that traditional pipeline methods are highly scalable, the relation extraction model has proven to be problematic due to error propagation issues [2]. In the training process of joint relation extraction models, ground truth tokens are used for prediction, in the inference process, tokens, generated from the previous step of models, are used for prediction, resulting in an inconsistency between training and inference [3]. Therefore, exposure bias exists in joint relation extraction models.

This paper contributes to the study of relation extraction in the military domain as follows:

- Drawing from proprietary military intelligence and verifiable military news sources, 9 inter-entity relationships have been distilled based on the requirements of intelligence analysis in the military field, and the constraints of entities and relationships have been formally defined.
- To solve the problem of domain-specific knowledge need, relation overlap and exposure bias, a model is proposed in this paper using N-gram adapter, a relation specific tagging strategy and axial attention mechanism, turning joint extraction into a classification problem.

- The meticulous manual annotation of 18,987 corpora has yielded a total of 27,137 triples. Experiments are conducted on the military dataset and another domain-specific dataset to assess the effectiveness of the model.

2   Related Work

As relation extraction has evolved over years, a variety of machine learning approaches have been developed. According to the processing flow of the task, these models can be divided into pipeline methods and joint methods.

(1) Pipeline Method

In pipeline methods, entity and relation extraction are typically divided into two stages. As part of the pipeline methods, named entity recognition (NER) is performed first on the input context, and then relation classification (RC) is conducted between the entity pairs of the NER result.

A support vector machine (SVM) was used by Zhou et al. [4] in order to extract relations among features incorporating lexical, syntactic, and semantic information. Chan and Dan [5] proposed an algorithm for relation extraction that identifies structures first, and then identifies their semantic types within those structures. Zhong and Chen [6] presented a simple end-to-end relation extraction method utilizing two independent encoders to learn the context representations of entities and relations for entity extraction and relation recognition.

Error propagation is a problem associated with pipeline methods. NER errors will adversely affect the performance of the next step of RC. Nevertheless, the relationship between the two tasks is not independent, and pipeline methods ignore this connection and dependency. Additionally, there will be redundancy in the NER step, which will increase computational complexity.

(2) Joint Method

The joint method models the two submodels in a unified manner. This is with the intention of maximizing the potential information between the two tasks and alleviating the problem of error propagation. Input features or internal hidden layer state can be shared between two subtasks. And joint extraction can also be achieved by jointly encoding.

In Miwa and Bansal's work [7], the substructure information of word sequences and dependency trees is captured by stacking bidirectional tree structures on bidirectional Long Short Term Memory Networks(LSTMs). However, the problem of relation overlap remains. Wei et al. [8] proposed CasRel, an annotation method consisting of cascading pointer networks. As a result of the multi-stage decoding process, there are still problems with error propagation and exposure bias, and there will be more redundant calculations as the number of relation categories increases. In the TPLinker [3] proposal by Wang et al., the annotation framework for relation extraction was unified with the linking problem of entity pairs, thus solving the issue of relation overlap. The problem of exposure bias was also addressed using single-stage decoding. In spite of this,

TPLinker's tagging system is redundant and has a low convergence rate. Zhang et al. [9] proposed a multiple-granularity graph method, they then used path reasoning mechanism with attention mechanism to calculate the representation of Entity-Entity edge to extract relation in documents.

A joint method is able to solve the problem of error propagation to a certain extent. Most works focus on proposing novel annotation frameworks or decoding methods. However, the problem of exposure bias still exists. Here, this paper proposes a joint extraction method so as to solve the problem of relation overlap and exposure bias in the military domain.

3   Methodology

To address the problems mentioned above, this paper proposes a single-step decoding joint relation extraction model combining a domain N-gram adapter and axial attention, whose structure is shown in Fig. 2.

The model can be divided into a domain-specific N-gram adapter layer, an attention-based token pair representation layer in combination with a relation specific classifier, and a joint decoding layer.

Here, the definition of the problem is first described. Given a sequence S of length N, the tokens in the sequence are $w_1, w_2, \ldots, w_N$, and the set of C relations $R = \{r_1, r_2, \ldots, r_C\}$. From the sequence S, joint relation extraction aims to extract all entity-relation triples $T = \{(s, r, o) : r \in R\}$, where s, o are slices of the sequence S, representing subjects and objects respectively.

3.1   Domain-Specific N-Gram Adapter

Recent developments in NLP have led to significant improvements in a wide range of tasks due to the development of large, pre-trained language models such as BERT [10] and GPT-3 [11]. These models are trained on massive amounts of text data and can be fine-tuned for specific tasks by adjusting the model's architecture or training data.

However, pre-trained models are usually based on general domain corpus and lack representation ability for specific domains. The inability to generalize has been shown to cause reduced performance when there are domain gaps between pre-training and fine-tuning data [12]. During the pre-training process, there are few corpora related to a specific field of military intelligence. Due to this, the existing pre-trained models are less knowledgeable about the military field, and tokens from military texts are not well represented.

To enhance the adaptability of representing military domain texts, the introduction of domain-specific N-gram adapter [13] is employed as a means. An N-gram adapter based on Transformers is used to enable the model to learn and fuse N-gram representations. Domain-specific language processing performance is enhanced by better representation of larger text granularities.

Figure 3 illustrates the architecture of a pre-trained model with an embedded domain adapter. The right part of the figure shows the architecture of a pre-trained model (RoBERTa [14]), which provides word-level semantic encoding

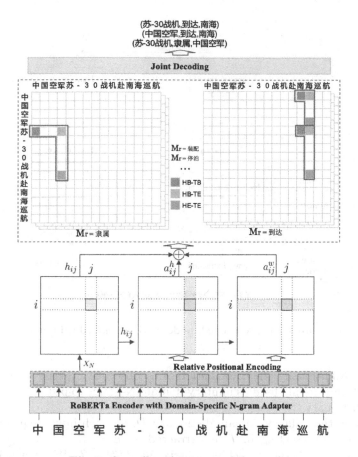

Fig. 2. Overall architecture of the model

but lacks domain knowledge. The left part of the structure is comprised of a domain adapter that extracts N-gram representations from the input text using a pre-built military domain dictionary.

Domain Dictionary and N-Gram Extraction. Pointwise Mutual Information (PMI) is a metric that measures the correlation between two random events. And PMI is widely used in NLP to evaluate relevance between tokens. The military domain dictionary is constructed using the PMI of the training text to extract N-gram information from the specific military domain text.

For any two adjacent tokens (w_i, w_{i+1}) in the sequence S, their PMI is calculated as Eq. 1:

$$PMI(w_i, w_{i+1}) = \log \frac{p(w_i w_{i+1})}{p(w_i) p(w_{i+1})} \tag{1}$$

where $p(w_i)$ represents the probability of occurrence of the token w_i. The N-gram dictionary for $n = 2$ can be obtained by finding all token pairs above a given

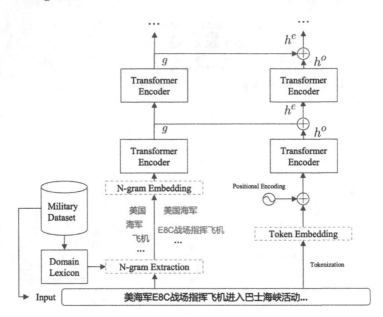

Fig. 3. Structure of the pre-trained model with an embedded domain adapter

threshold. Additionally, the combination of n consecutive tokens with high PMI can be determined and used as military domain N-grams to produce the military domain dictionary.

Based on the military domain dictionary D, for each input sequence S, the substring of S presenting in D is extracted as $E = \{e_1, e_2, \ldots, e_Q : e_i \in D\}$ to construct the N-gram matching matrix $\mathbf{M}^{match} \in R^{N \times Q}$, shown in Fig. 4. $\mathbf{M}^{match}$ records the extracted collection E and the locations of their associated tokens, where $\mathbf{M}^{match}_{i,j} = 1$ denotes $w_i \in e_j$.

Fusing Domain Representation. The backbone network is a pre-trained model based on Transformer encoder architecture with l_1 layers. Every encoder layer contains multiple self-attention sub-layers and feed-forward neural network sub-layers. By stacking multiple encoder layers, the adapter can capture the semantic and contextual information of the input text. For each token w_i in the input text, a hidden state $\mathbf{h}_i^o$ is output and it is used as the input of next layer.

The domain adapter is designed as a Transformer network with l_2 layers in order to better fuse the domain representation. The N-gram information is vectorized through an embedding layer, and then transmitted to the adapter network for encoding to obtain a hidden state sequence $\mathbf{g}$.

Equation 2 and 3 show the process to obtain the hidden state $\mathbf{h}_i^e$ that fuses the pre-trained token representation and domain representation.

$$\mathbf{h}_i^e = \mathbf{h}_i^o + \mathbf{g}_i \tag{2}$$

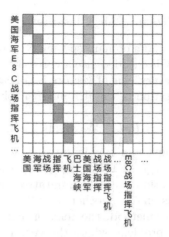

Fig. 4. A sample of N-gram matching matrix

$$\mathbf{h}^o_{i+1} = \mathbf{h}^e_i \qquad (3)$$

Here, $\mathbf{g}_i$ denotes the ith token and the hidden state of the N-gram associated with the token in the matching matrix. The whole fusion process needs to fuse l_2 layers layer by layer from bottom to top.

3.2 Tagging Scheme

The Rel-Spec Horns Tagging scheme [15] is employed, which offers the possibility of future joint decoding while addressing the problem of relation overlap. In the context of C relation types and sequence S of length N, C $N \times N$ matrices are formulated, with each matrix representing a relation. Combining the BIE labels of head and tail entities would require nine types of tags, but the scheme effectively uses three valid tag types to tag head and tail entities and establish their association:

- HB - TB: tags the beginning tokens of both the head and tail entities.
- HB - TE: tags the beginning token of the head entity and ending token of the tail entity.
- HE - TE: tags the ending tokens of both the head and tail entities.

As shown in Fig. 2, HB-TB is tagged on the beginning tokens of the head entity "Su-30" as well as the tail entity "Chinese Air Force", i.e., the position of (Su, Zhong) in the matrix $M_{r=belong}$. Similarly, HB-TE is tagged on the position (Su, Jun), and HE-TE is tagged on the position (Ji, Jun). With the right matrix, this tagging framework can solve the SEO issue, as well as the SOO scenario. By tagging another relation matrix, the EPO issue can be resolved as well.

3.3 Attention-Based Token Pair Classifier

Using a pre-trained model and a domain-specific N-gram adapter, each token w_i in a sequence S of length N is encoded into a vector $\mathbf{h}_i^e$ with context information. To represent token pairs, every pair of vectors is concatenated and input into a linear layer. The representation of the ith and jth token pair $\mathbf{h}_{i,j}$ is calculated from Eq. 4:

$$\mathbf{h}_{i,j} = ReLU(\mathbf{W}_h[\mathbf{h}_i^e; \mathbf{h}_j^e] + \mathbf{b}_h) \tag{4}$$

where $\mathbf{W}_h$ and $\mathbf{b}_h$ are weight and bias learnt during training, $ReLU(\Delta)$ is the ReLU activation function, $i, j \in [1, N]$, the same below. Unlike addition and other operations, concatenation is non-commutative, making it possible to analyze the influence of tokens on one another.

Inspired by the tagging method, the relation extraction problem is transformed into a table-filling problem, where the vector at (i, j) in the table represents the implicit semantic relationship between w_i and w_j. Intuitively, by increasing its representation ability, it is possible to achieve better results in classification. The introduction of positional coding is not difficult to understand since different positions in the table should have varying degrees of impact. The tagging method also reminds us of the attention matrix, and the axial attention mechanism has been chosen to enhance the token pair representation. Using self-attention, it is possible to identify the relationship between two tokens in a sequence, as well as take advantage of parallel processing. In order to represent the order of token pairs, it is necessary to add positional information which is discarded during parallel computing to the token pair representation.

In order to determine both the relationship between the current token and other tokens in the sentence as well as the relationship between other tokens and the current token in the sentence, axial attention [16] is adopted. After calculating vertical and horizontal attention, make residual connections. According to the attention mechanism, the query term $\mathbf{q}_{i,j}$, the key term $\mathbf{k}_{i,j}$, and the value term $\mathbf{v}_{i,j}$ are all linear transformations of the input term $\mathbf{h}_{i,j}$:

$$\begin{cases} \mathbf{q}_{i,j} = \mathbf{W}_Q \mathbf{h}_{i,j} \\ \mathbf{k}_{i,j} = \mathbf{W}_K \mathbf{h}_{i,j} \\ \mathbf{v}_{i,j} = \mathbf{W}_V \mathbf{h}_{i,j} \end{cases} \tag{5}$$

where $\mathbf{W}_Q$, $\mathbf{W}_K$ and $\mathbf{W}_V$ are trainable weight matrices. After this, relative positional encoding is incorporated into axial attention in both directions. This has a higher generalization ability than absolute position encoding and can be adapted to longer input texts [17]. Besides adding position information to the query item of the current token pair representation, the key items of other token pairs on the axis are also incorporated into the position information. After obtaining the attention of each axis, the residual connection is performed as follows:

$$\mathbf{a}_{i,j}^h = \mathbf{h}_{i,j} + \sum_{p=1}^{N} Softmax\left(\mathbf{q}_{i,j}^T\mathbf{k}_{p,j} + \mathbf{q}_{i,j}^T\mathbf{l}_{p-j}^q + \mathbf{k}_{i,j}^T\mathbf{l}_{p-j}^k\right)\left(\mathbf{v}_{p,j} + \mathbf{l}_{p-j}^v\right) \tag{6}$$

$$\mathbf{a}_{i,j}^w = \mathbf{a}_{i,j}^h + \sum_{p=1}^{N} Softmax \left(\mathbf{q}_{i,j}^T \mathbf{k}_{i,p} + \mathbf{q}_{i,j}^T \mathbf{l}_{p-i}^q + \mathbf{k}_{i,p}^T \mathbf{l}_{p-i}^k \right) \left(\mathbf{v}_{i,p} + \mathbf{l}_{p-i}^v \right) \qquad (7)$$

where $\mathbf{a}_{i,j}^h$ and $\mathbf{a}_{i,j}^w$ are axial-attention along the height-axis and width-axis respectively, p is the position of the current token, $p-i$ and $p-j$ stand for relative position, $\mathbf{l}^q$ is the corresponding trainable positional encoding for queries, $\mathbf{l}^k$ is for keys and $\mathbf{l}^v$ is for values.

Next, the token pair representation, which finally incorporates axis attention and positional encoding, is mapped into the tagging matrices in parallel through a matrix $\mathbf{R}$. For each element of the matrix representing a relationship, the score vector $\mathbf{m}$ for each label type is calculated from Eq. 8:

$$\mathbf{m}_{i,j}^c = R^T \mathbf{a}_{i,j}^w, \ 1 \leq c \leq C \qquad (8)$$

where R is a learnable matrix used to compute the score for each tag in parallel. Further, the axial attention mentioned above can also be computed in parallel, which contributes to the model's efficiency.

The final step is to normalize the score vector and predict the tag using SoftMax as Eq. 9, and the training loss is defined by Eq. 10:

$$P\left(y_{i,j}^c\right) = Softmax\left(\mathbf{m}_{i,j}^c\right), \ 1 \leq c \leq C \qquad (9)$$

$$L = -\frac{1}{N \times C \times N} \times \sum_{i=1}^{N} \sum_{c=1}^{C} \sum_{j=1}^{N} \log P\left(y_{i,j}^c = \hat{y}_{i,j}^c\right) \qquad (10)$$

where $\hat{y}_{i,j}^c$ is the true tag of the position (i, j) in the $M_{r=r_C}$ matrix.

3.4 Joint Decoding

The tagging scheme in Sect. 3.3 implements entity relation extraction from the perspective of triplet classification, which allows the extraction of structured entity relation triplet information from unstructured text to be done in only one decoding step. Decoding a single tagging matrix requires finding HB-TB, then HB-TE in the same row, followed by HE-TE in that column. Consequently, the sequence slices at the corresponding positions represent the head and tail entities, and the matrix represents the type of relationship between them.

4   Experiments

In this section, two datasets are described in detail, comparative experiments are conducted using the existing model and the proposed model. To further demonstrate the effectiveness of the model design, ablation experiments are also performed on the models.

The backbone model parameters are initialized using the Chinese RoBERTa-wwm-ext from the HIT Joint Iflytek Laboratory. To compensate for the training gap between the main intervention training parameters and the fasttext word vector trained in this paper, the parameters for the N-gram embedding layer are initialized by the Chinese Wikipedia corpus. By doing so, a relatively fast and smooth training process can be achieved. Training is conducted with a batch size of 4 and a learning rate of 1e−5. NVIDIA RTX 3090 24 GB GPU was used for all experiments in this paper.

4.1 Dataset

Our analysis, informed by confidential military intelligence and credible military news sources, has yielded 9 relationships. 18,987 corpora were meticulously annotated, covering a total of 27,137 triples. The dataset is splited into training, validation and test sets with a ratio of 7:1.5:1.5. Besides, and another public domain-specific dataset CMeIE [19] is adopted to evaluate the effectiveness of the model. There are 44 relation types in CMeIE and it has the training set of 14,339 samples and the validation set of 3,585 samples. The details of the datasets are shown in Table 1.

Table 1. Statistics of datasets

Category	Military			CMeIE		
	Train	Validation	Test	Train	Validation	Test
Normal	8754	1875	1860	5112	1374	1693
EPO	2313	495	557	1257	157	214
SEO	3485	746	738	8794	2211	2789
ALL	13291	2848	2848	14339	3585	4482

4.2 Comparison Models

To assess the effectiveness of the proposed model, a comparative study was conducted with three existing state-of-the-art approaches: CasRel [8], TPLinker [3], GPLinker [18] and OneRel [15]. For each of the methods, the ability to extract relation information from the corpus of military texts was evaluated using the indicators Precision, Recall, and F1-Score.

CasRel: addresses the challenge of relation overlap via a cascading framework. Subsequently, the model utilizes a pointer network consisting of multi-layer relation labels to perform the decoding task.

TPLinker: unifies the annotation framework for relation extraction with the linking problem of entity pairs, effectively resolving the issue of relation overlap and exposure bias.

GPLinker: models relation extraction as a five-tuple extraction task for the relation, beginning and end of head and tail entities. The extraction process is reconstructed from the perspective of probability map.

Table 2. Comparison result of different models on the military dataset

Model	P	R	F1
CasRel	0.6489	0.4822	0.5532
TPLinker	0.6336	0.5941	0.6132
GPLinker	0.6694	0.6259	0.6469
OneRel	0.6766	0.6332	0.6542
Ours	**0.6791**	**0.6593**	**0.6690**

OneRel: optimizes the tagging scheme of TPLinker to reduce the redundancy, and more parallel strategies are added to improve the efficiency of the model.

As shown in Table 2, the proposed model achieved the best performance among all models, with high accuracy. These results suggest that our model can effectively extract entity and relation information from military texts, outperforming existing state-of-the-art methods.

To further verify the effectiveness of the proposed model in domain-specific relation extraction, the Chinese medical dataset, CMeIE was selected to conduct additional comparison experiment. The result in Table 3 illustrates that the proposed model has good adaptation in domain relation extraction.

Table 3. Comparison result of different models on the CMeIE

Model	P	R	F1
CasRel	0.5667	0.4625	0.5093
TPLinker	0.6055	0.4982	0.5466
GPLinker	**0.6523**	0.5174	0.5771
OneRel	0.6269	0.5658	0.5948
Ours	0.6436	**0.5710**	**0.6051**

4.3 Ablation Study

Ablation experiments are conducted to evaluate the effectiveness of each component in the proposed model. Specifically, we compare the performance of the full model with the models that remove the domain N-gram adapter or axial attention mechanism. Based on the ablation experiments results shown in Table 4,

Table 4. Results of ablation experiments on the military dataset

Model	P	R	F1
-N-gram Adapter	0.6819	0.6366	0.6585
-axial attention	**0.7064**	0.6191	0.6599
Ours	0.6791	**0.6593**	**0.6690**

it can be observed that removing the N-gram adapter from the model led to a decrease in recall and F1 score by 2.27% and 1.05% respectively, and a slight increase in precision by 0.28%. Additionally, when removed the axial attention mechanism from the model, we observed an increase in precision (2.73%) but a decrease in recall (4.02%), leading to an overall decrease in F1 score by 0.91%. This could be due to the fact that the components are more robust to handle noisy or ambiguous data. Another possible reason is that the components are increasing the coverage of the model, allowing it to identify more instances of the target relation. This may lead to an increase in the number of true positive relations but also result in the identification of some false positive relations.

The full model achieved a better F1 score, outperforming the model without N-gram adapter or axial attention mechanism. The results show that both components contribute to the performance of the proposed relation extraction model.

Overall, the proposed model achieves state-of-the-art performance on both the military relation extraction dataset and the CMeIE dataset, demonstrating its effectiveness in addressing the challenges in domain relation extraction.

5   Conclusion

In this paper, we collected and annotated a military domain dataset with 9 inter-entity relationships and 18,987 corpora manually, ensuring its accuracy and completeness, and proposed a single-step decoding joint relation extraction model, which combines N-gram adapter and axial attention mechanism, allowing the joint relation extraction problem to be transformed into a form filling classification task. Experimental results indicate that our model achieves better results than other models in domain-specific dataset, as well as being more adaptable and context-aware. But, the proposed model can only handle sentence-level texts and not for long texts. In future work, we will further abstract the model into a unified solution to address the document-level relation extraction problem.

References

1. Zheng, S., Wang, F., Bao, H., et al.: Joint extraction of entities and relations based on a novel tagging scheme. In: Proceedings of the 55th Annual Meeting of the Association for Computational Linguistics, vol. 1 (2017)
2. Li, Q., Ji, H.: Incremental joint extraction of entity mentions and relations. In: Proceedings of the 52nd Annual Meeting of the Association for Computational Linguistics, vol. 1, pp. 402–412 (2014)
3. Wang, Y., Yu, B., Zhang, Y., et al.: TPLinker: single-stage joint extraction of entities and relations through token pair linking. In: Proceedings of the 28th International Conference on Computational Linguistics, pp. 1572–1582 (2020)
4. Zhou, G., Su, J., Zhang, J., Zhang, M.: Exploring various knowledge in relation extraction. In: Proceedings of the 43rd Annual Meeting of the Association for Computational Linguistics, pp. 427–434 (2005)

5. Chan, Y., Roth, D.: Exploiting syntactico-semantic structures for relation extraction. In: Proceedings of the 49th Annual Meeting of the Association for Computational Linguistics: Human Language Technologies, pp. 551–560 (2011)
6. Zhong, Z., Chen, D.: A frustratingly easy approach for entity and relation extraction. In: Proceedings of the Conference of the North American Chapter of the Association for Computational Linguistics: Human Language Technologies, pp. 50–61 (2021)
7. Miwa, M., Bansal, M.: End-to-end relation extraction using LSTMs on sequences and tree structures. In: Proceedings Of the 54th Annual Meeting of the Association for Computational Linguistics, vol. 1 (2016)
8. Wei, Z., Su, J., Wang, Y., et al.: A novel cascade binary tagging framework for relational triple extraction. In: Proceedings of the 58th Annual Meeting of the Association for Computational Linguistics, pp. 1476–1488 (2020)
9. Zhang, J., Liu, M., Xu, L.: Multiple-granularity graph for document-level relation extraction. In: Zhao, X., Yang, S., Wang, X., Li, J. (eds.) WISA 2022. LNCS, vol. 13579, pp. 126–134. Springer, Cham (2022). https://doi.org/10.1007/978-3-031-20309-1_11
10. Kenton, J., Toutanova, L.: Bert: Pre-training of deep bidirectional transformers for language understanding. In: Proceedings of NaacL-HLT, vol. 1, p. 2 (2019)
11. Brown, T., Mann, B., Ryder, N., et al.: Language models are few-shot learners. Adv. Neural. Inf. Process. Syst. **33**, 1877–1901 (2020)
12. Beltagy, I., Lo, K., Cohan, A.: SciBERT: a pretrained language model for scientific text. In: Proceedings of the 2019 Conference on Empirical Methods in Natural Language Processing and the 9th International Joint Conference on Natural Language Processing, pp. 3615–3620 (2019)
13. Diao, S., Xu, R., Su, H., et al.: Taming pre-trained language models with n-gram representations for low-resource domain adaptation. In: Proceedings of the 59th Annual Meeting of the Association for Computational Linguistics and the 11th International Joint Conference on Natural Language Processing, vol. 1, pp. 3336–3349 (2021)
14. Liu, Y., Ott, M., Goyal, N., et al.: A robustly optimized bert pretraining approach. ArXiv Preprint ArXiv:1907.11692 (2019)
15. Shang, Y., Huang, H., Mao, X.: OneRel: joint entity and relation extraction with one module in one step. In: Proceedings of the AAAI Conference on Artificial Intelligence, vol. 36, pp. 11285–11293 (2022)
16. Wang, H., Zhu, Y., Green, B., Adam, H., Yuille, A., Chen, L.-C.: Axial-DeepLab: stand-alone axial-attention for panoptic segmentation. In: Vedaldi, A., Bischof, H., Brox, T., Frahm, J.-M. (eds.) ECCV 2020, Part IV. LNCS, vol. 12349, pp. 108–126. Springer, Cham (2020). https://doi.org/10.1007/978-3-030-58548-8_7
17. Shaw, P., Uszkoreit, J., Vaswani, A.: Self-attention with relative position representations. In: Proceedings of NaacL-HLT, pp. 464–468 (2018)
18. Su, J.: GPLinker: joint entity relation extraction based on GlobalPointer (2022). https://kexue.fm/archives/8888
19. Guan, T., Zan, H., Zhou, X., Xu, H., Zhang, K.: CMeIE: construction and evaluation of Chinese medical information extraction dataset. In: Zhu, X., Zhang, M., Hong, Yu., He, R. (eds.) NLPCC 2020, Part I. LNCS (LNAI), vol. 12430, pp. 270–282. Springer, Cham (2020). https://doi.org/10.1007/978-3-030-60450-9_22

TCM Function Multi-classification Approach Using Deep Learning Models

Quanying Ren[1], Keqian Li[2], Dongshen Yang[1], Yan Zhu[3],
Keyu Yao[3], and Xiangfu Meng[1(✉)]

[1] College of Electronic and Information, Engineering Liaoning Technical University,
Huludao, China
marxi@126.com
[2] School of Medical Information, Changchun University of Chinese Medicine,
Changchun, China
[3] Information Institute of Traditional Chinese Medicine,
Chinese Academy of Traditional Chinese Medicine, Beijing, China

Abstract. Traditional Chinese Medicine prescriptions are regarded as
an important resource that brings together the treatment experience
and wisdom of doctors of all ages. Effectively sorting out and excavating
them, especially combining with their functional indications and medication rules, can improve clinical efficacy and new drug research and
development. Because the classification system of prescriptions is not
completely unified, the data of Chinese patent medicines, national standard formulae and the seventh editions of Chinese formula textbooks are
manually integrated, and a quadrat data set containing 21 efficacy classifications is constructed. Since on the text description of prescription
information (prescription name, composition, indications and efficacy)
in the data set, a variety of deep learning text classification models are
used to automatically judge prescription classification, so as to establish an efficient and accurate classification model, and finally construct
a prescription efficacy classification data set. The experimental results
show that the pre-trained Bert-CNN model has the best effect, with the
accuracy rate of 77.87%, and the weighted accuracy rate, weighted recall
rate and weighted F1 value of 79.46%, 77.87% and 77.44%, respectively.
This study provides a useful reference for further realizing the automatic
information processing of ancient formulas.

Keywords: Text classification · Deep learning · Prescription efficacy
classification · Bert · CNN

1 Introduction

Through thousands of years of clinical practice of traditional Chinese medicine,
a large number of formulae have been selected and accumulated. By the end

Supported by the National Natural Science Foundation of China (82174534, 61772249),
and the Fundamental Research Funds for the Central Public Welfare Research Institutes (ZZ160311).

of the late Qing Dynasty, there were more than 100,000 ancient formulae. In addition, the new formulae developed and created, the self-made formulae of hospitals, and the Chinese patent medicine formulas on the market, they were even more "vast". These formulas embody the therapeutic experience and wisdom of ancient and modern physicians. If they can be effectively sorted out and excavated, especially their functions and indications, they will provide important support for improving clinical efficacy and developing new drugs.

In order to sort out and study the formulae of past dynasties, colleges and research institutes around the country have established various prescription databases and analysis systems [1,2], which still face problems such as insufficient standardization of data and difficulty in efficient retrieval and analysis [3]. Among all kinds of information contained in formulae, in addition to the structure and standardization of drug composition information, more attention has been paid and better solutions have been achieved [4,5]. However, there are few studies on information extraction and standardization in terms of the information of the opposite side (including indications and efficacy). The main reason is that its description covers a long history, and there are many phenomena such as polysemy, synonymy, ambiguity, cross meaning. In addition, the classification system of formula is not completely unified, and there is a lack of high-quality and large-scale training corpus. At present, there are few researches on information extraction and standardization based on natural language processing technology of deep learning model.

In recent years, with the extensive application of depth model in the medical field and the release and improvement of the national standards related to formulae, conditions have been provided for the automatic processing of prescription information. In this paper, the classification system and prescription data of the seventh edition of National Medical Insurance Catalogue [6], GBT 31773-2015 Coding Rules and Coding of Traditional Chinese Medicine Prescriptions [7] and Prescription Textbook [8] are manually integrated to form a prescription efficacy classification data set, and a variety of deep learning text classification models are adopted to realize text description based on prescription name, composition and main efficacy, and automatically judge the classification of the prescription. The classification efficiency of various models is compared, analyzed and discussed, which provides a useful reference for the further realization of the automatic information processing of ancient formulae.

2   Related Work

Text classification refers to the process of automatically determining text categories based on text content under a given classification system. The core issue is to extract the features of the classified data from the text, and then select the appropriate classification algorithms and models to model the features, thereby achieving classification. Figure 1 shows the processing procedure of text classification.

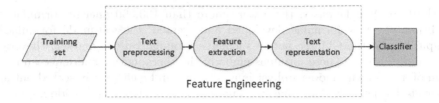

Fig. 1. Processing procedure of text classification.

Text classification algorithm model can be divided into traditional machine learning model and deep learning model. Among them, the commonly used traditional machine learning models include Naive Bayes [9], Support Vector Machine (SVM) [10], Decision Tree [11], etc. Deep learning models include TextCNN, TextRNN, TextRCNN, HAN [12], Bert [13], etc.

At present, text classification has been widely used in many fields, such as spam filtering [14], public opinion analysis [15] and news classification [16]. In the medical field, Yu [17] proposed a named entity recognition model to automatically identify the time and place information in the COVID-19 patient trajectory text. Li [18] proposed a three-stage hybrid method based on gated attention bidirectional long short-term memory (ABLSTM) and regular expression classifier for medical text classification tasks to improve the quality and transparency of medical text classification solutions. Prabhakar [19] proposed a medical text classification paradigm, using two novel deep learning architecture to alleviate human efforts. Cui [20] developed a new text classifier based on regular expressions. Machine-generated regular expressions can effectively combine machine learning technologies to perform medical text classification tasks, and have potential practical application value. Zheng [21] proposed a deep neural network model called ALBERT-TextCNN for multi-label medical text classification. The overall F1 value of the model classification reached 90.5 %, which can effectively improve the multi-label classification effect of medical texts. Li [22] proposed a two-level text classification model based on attention mechanism, which is used to classify biomedical texts effectively.

Generally, text classification has also been widely used in the medical field. In this paper, we use the deep learning model to classify the efficacy of prescription. The paper introduces the model used and analyzes the experimental results in detail.

3   Methodology

In this study, we aim to classify the efficacy of prescription into multiple categories using deep learning method which includes TextCNN, TextRCNN, RNN Attention, Bert and their combination models.

3.1 Convolutional Neural Network Model

CNN (Convolutional Neural Network) [23] is widely used in the field of image recognition. The structure of CNN model can be divided into three layers, i.e., convolutional layer, pooling layer and fully connected layer. The main function of convolution layer is to extract features. The pooling layer aims to down sampling but with no damage of the recognition results. The main role of the fully connected layer is classification.

TextCNN is a model proposed by Kim [24] in 2014, which pioneered the use of CNN to encode n-gram features for text classification. The convolutions in the image is two-dimensional, whereas TextCNN uses one-dimensional convolution (filter_size * embedding_dim), with one dimension equal to embedding. This can extract the information of filter_size grams. After inputting data, the Embedding layer converts words into word vectors and generates a two-dimensional matrix. Then, sentence features are extracted in one-dimensional convolution layer. After that, the sentences of different lengths are represented by fixed length in Max-Pooling layer. Finally, the probability distribution is obtained in Fully connected layer. The structure of TextCNN is shown in Fig. 2.

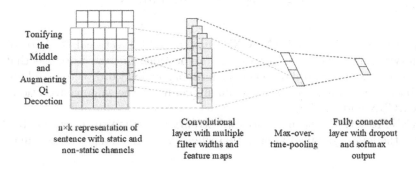

Tonifying the Middle and Augmenting Qi Decoction

| n×k representation of sentence with static and non-static channels | Convolutional layer with multiple filter widths and feature maps | Max-over-time-pooling | Fully connected layer with dropout and softmax output |

Fig. 2. The structure of TextCNN.

3.2 Recurrent Neural Network Model

RNN (Recurrent Neural Network) [25] is a kind of neural network with shortterm memory. In the RNN models, neurons cannot only accept the information of other neurons, but also accept their own information to form a network structure with loops. Compared with feedforward neural networks, RNNs are more in line with the structure of biological neural networks. RNN has been widely used in speech recognition, language models and natural language generation tasks.

TextRNN [26] takes advantages of RNN to solve text classification problems, trying to infer the label or label set of a given text (sentence, document, etc.). TextRNN has a variety of structures and its classical structure includes embedding layer, Bi-LSTM layer, concat output, FC layer and softmax. The structure of TextRNN is shown in Fig. 3.

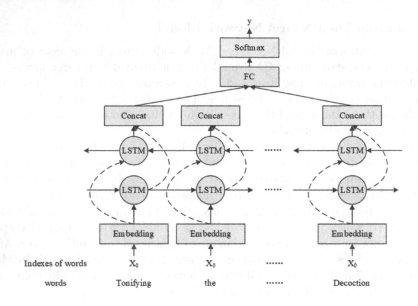

Fig. 3. The structure of TextRNN.

TextRCNN [27] (TextRNN+CNN) first uses bi-directional RNN to obtain the forward and backward context representations of each word, so that the representation of the word becomes the form of splicing word vectors and forward and backward context vectors, and finally connects the same convolution layer as TextCNN and pooling layer.

RNN_Attention [28] introduces an attention mechanism based on RNN to reduce the loss of detailed information when processing long text information.

3.3   Bidirectional Encoder Representation from Transformers Model

BERT (Bidirectional Encoder Representation from Transformers) [13] is a pretrained language representation model that uses MLM (Masked Language Model) for pre-training and deep bidirectional Transformer component to construct the entire model to generate deep bidirectional language representations that fuse contextual information.

Each token of the input information (the yellow block in Fig. 4) has a corresponding representation, including three parts, i.e., Token Embedding, Segment Embedding, and Position Embedding. The vectors of the final input model are obtained by adding their corresponding positions, and then classified. The structure of BERT-classify model is shown in Fig. 4.

Bert-CNN, Bert-RNN, Bert-RCNN and Bert-DPCNN are combination models using Bert pre-trained models. First, the data is put into Bert model for pretraining, and then the output of BERT (that is, the output of the last layer of the transformer) is used as the input of the convolution (embedding_inputs) to access other models.

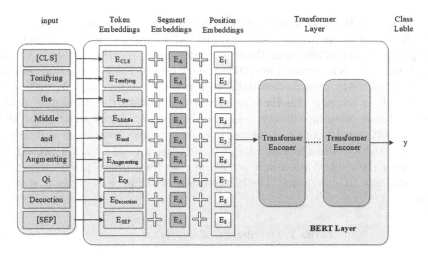

Fig. 4. The structure of BERT-classify model. (Color figure online)

4 Experiment

4.1 Data Sources

The experimental data were collected from Chinese patent medicines [6], national standard formulae [7] and the seventh editions of Chinese formula textbooks [8]. A total of 2,618 prescription data were manually integrated. Among them, there are 1,391 pieces of Chinese patent medicine data, 1,089 pieces of national standard formulae data and 138 pieces of Chinese formula textbook data. Each data contains the name, composition, efficacy and indications of the prescription. The experimental data format is as follow: {Prescription: Bufei Huoxue Capsule} {Composition: Membranous milkvetch root 720 g, radix paeonia rubra 720 g, malaytea scurfpea fruit 360 g} {Efficiency: benefiting qi for activating blood circulation, invigorating lung and nourishing kidney} {Indication: cor pulmonale (remission stage) diagnosed as qi deficiency and blood stasis syndrome. The symptoms include cough, shortness of breath, asthma, chest tightness, palpitation, cold and weak limbs, soreness and weakness of waist or knees, cyanosis of lips, pale tongue with white coating or dark purple tongue.} {Real label: 1} {Category: reinforcing agents}

4.2 Data Integration and Standardization

We exclude Tibetan medicine, Mongolian medicine, Uygur medicine and other ethnic medicines. We refer to the national classification standard and integrate the efficacy classification based on departmental division. For example, some traditional Chinese patent medicines are classified according to the names of modern diseases, such as drugs for nasal diseases, ear diseases, and anti-tumor drugs. Their efficacy indications are mainly described by modern diseases, which deviates greatly from the terminology of efficacy indications of conventional formula. Referring to the primary and secondary classifications in the national standard prescription classification, the fine curative efficacy classification of Chinese patent medicines is integrated into the primary classification of the same efficacy.

A total of 2,618 prescription data were manually integrated. We retain the data based on efficacy classification, remove duplicate data and the data of modern disease naming efficacy, resulting in total of 2,368 prescription data. Then we sift out 12 items of incomplete efficacy and treatment and 2 items of emetic (too few data in this category), leaving a total of 2,354 prescription data. After data integration and screening, the final prescription efficacy can be divided into 21 categories. The distribution of sample categories is shown in Table 1.

Table 1. Distribution of sample categories.

Label	Category	Sample size
0	tranquilizing formulae	21
1	reinforcing formulae	411
2	astringent preparations	55
3	settlement formulae	42
4	diaphoretic formulae	150
5	resuscitating formulae	25
6	regulating qi formulae	228
7	blood regulating formulae	253
8	antipyretic agents	402
9	anthelmintic	8
10	dampness-dispelling formulae	227
11	summer-heat clearing formulae	20
12	interior-warming formulae	76
13	digestive formulae	37
14	purgative formulae	53
15	carbuncle therapeutic formulae	25
16	wind-calming medicine	113
17	moistening formulae	14
18	turbidity lipid-lowering formulae	16
19	detumescence formulae	28
20	phlegm, cough, asthma formulae	150

4.3   Experimental Parameters

The 2,354 prescription data were randomly disrupted and divided into training set, validation set and test set at a ratio of 8: 1: 1.

The model parameters used in this paper are shown in Table 2. To reduce the risk of over-fitting of the model, the training follows the principle of early stopping, and sets the detection parameter detect_imp = 1,000, dropout = 0.1.

Table 2. Model parameters.

Model	epoch	batch size	pad size	learning rate	num filters	hidden size
TextCNN	20	32	128	1e−3	256	128
TextRCNN	20	32	128	1e−3	256	256
RNN_Attention	20	32	128	1e−3	256	128
Bert, Bert-CNN and Bert-RNN	7	32	128	5e−3	256	768
Bert-RNN	7	32	128	5e−3	250	768

4.4   Statistics for Model Evaluation Measures

To evaluate the performance, our experiment used the classic evaluation indexes in text classification, namely Accuracy, Precision, Recall and F1-Measure. Due to the limited number of multi category data in the research, and the imbalanced samples in each category, weighted recall and weighted F1 are used as the overall evaluation indicators. The formulas are as follows:

$$Accuracy = \frac{TP + TN}{TP + TN + FP + FN} \times 100\% \tag{1}$$

$$Precision = \frac{TP}{TP + FP} \times 100\% \tag{2}$$

$$Recall = \frac{TP}{TP + FN} \times 100\% \tag{3}$$

$$F1 - Measure = 2 \times \frac{Precision \times Recall}{Precision + Recall} \times 100\% \tag{4}$$

TP, TN, FP and FN are positive samples with correct prediction, negative samples with correct prediction, positive samples with wrong prediction and negative samples with wrong prediction, respectively. In multi-classification task, the currently tested classification is treated as a positive sample, and other classifications are treated as negative samples. Firstly, the accuracy rate, recall and F1 value of each category are measured, and then the weighted value is obtained (different weights are given according to the proportion of each category).

4.5   Results

Included the 3 non-pretrained models, 8 models were used to classify the efficacy of formula. The experimental results are shown in Table 3.

In the non-pre-trained model, the TextRCNN model outperforms the other two models. The accuracy is 3 to 8% points higher, reaching 72.77%, with a loss value of 0.87, slightly worse than TextCNN's 0.82. Weighted-Precision, Weighted-Recall and Weighted-F1 are the best of the three models. RNN_Attention model performed worst, with an accuracy of only 64.68%.

Table 3. Experimental results.

Type	Model	Accuracy	Loss	Weighted Precision	Weighted Recall	Weighted F1
Non-pre-trained model	TextCNN [24]	70.64%	**0.82**	71.03%	70.64%	69.63%
	TextRCNN [27]	72.77%	0.87	73.83%	72.77%	71.64%
	RNN_Attention [28]	64.68%	1.2	64.79%	64.68%	63.35%
Bert and Bert combination model	Bert [13]	77.87%	0.83	77.06%	77.87%	76.75%
	Bert-CNN	**77.87%**	0.84	**79.46%**	**77.87%**	**77.44%**
	Bert-RNN	67.66%	1.2	63.84%	67.66%	63.19%
	Bert-RCNN	69.79%	0.9	69.00%	69.79%	68.42%
	Bert-DPCNN	73.19%	1.1	68.44%	73.19%	69.50%

The combination models using the Bert pre-trained language model outperform non-pre-trained models. Among them, the Bert-CNN model has the best effect in the efficacy classification of experimental prescriptions in this paper. Accuracy reaches 77.87% and Loss value is slightly inferior to the TextCNN model. Based on the above results, Bert-CNN has the best effect and is most suitable for this experiment.

The Precision of each efficacy category in the Bert-CNN model is as follows. Tranquilizing formulae: 66.67%; reinforcing formulae: 73.17%; astringent preparations: 100.00%; settlement formulae: 20.00%; diaphoretic formulae: 63.16%; resuscitating formulae: 50.00%; regulating qi formulae: 77.27%; blood regulating formulae: 88.00%; antipyretic agents: 76.60%; anthelmintic: 100.00%; dampnessdispelling formulae: 90.91%; summer-heat clearing formulae: 100.00%; interiorwarming formulae: 100.00%; digestive formulae: 50.00%; purgative formulae: 100.00%; carbuncle therapeutic formulae: 100.00%; wind-calming medicine: 85.71%; moistening formulae: 0.00%; turbidity lipid-lowering formulae: 100.00%; detumescence formulae: 75.00% and phlegm, cough, asthma formulae: 90.00%.

5   Discussion

5.1   Analysis of the Overall Results of Experimental Classification

Model Performance. The classification prediction accuracy of most models is above 70%. In non-pre-trained models, Text_RCNN performs best and RNN_Attention performs worst. The attention mechanism of RNN_Attention model is usually combined with RNN, which relies on the historical information of t-1 to calculate the information at time t, so it cannot be implemented in parallel, resulting in low prediction efficiency. In Text_RCNN model, the convolutional layer + pooling layer in CNN is replaced by two-way RNN + pooling layer, so the prediction result is better. In general, Bert pre-trained language model and its combination models are superior to the non-pre-trained models, and Bert-CNN has the best classification effect. The main model structure of Bert is Transformer encoder, which has stronger text editing ability than RNN and LSTM.

Classification Data. The accuracy of the larger sample size is about 80%, while the smaller sample size is unstable. In addition to the sample size, it is also related to the classifications' clear feature words, which are better for machine learning. 7 classifications out of the 21 categories had accuracy of 100%. The primary reason is that these classifications' defining phrases are comparatively plain, and machines are generally simple to learn. Moistening formulae has the lowest accuracy, followed by digestive formulae and resuscitating formulae (50%), which is easily confused with other classifications.

5.2 Summary and Cause Analysis of Typical Misclassified Data

By analyzing the cases of machine classification errors, we can draw the following conclusions, as is shown in Table 4. It shows that although the model features, recognition features, and learning features are different, the results of some models are very similar. But each model usually has different characteristics, and the characteristics of recognition and learning also have their own emphasis. Taking Dingkun Dan as an example, the results of the model recognition are mainly divided into "reinforcing formulae" (label is 1) and "blood regulating formulae" (label is 7), and the number of both is also relatively close.

Table 4. Typical error cases.

Model	Prescription(Real Label)		
	Bolus for Woman Disease (1)	Decoction of Ophiopogonis (17)	Xinsusan Decoction (17)
TextCNN	1	17	20
TextRCNN	7	17	20
RNN_Attention	7	14	4
Bert	7	8	4
Bert-CNN	1	8	4
Bert-RNN	1	8	20
Bert-RCNN	7	8	4
Bert-DPCNN	1	8	4

Text classification algorithm based on text feature is difficult to understand text connotation without considering TCM theory knowledge. For example, Maimendong Decoction, as a "moistening formulae" (label is 17), has fewer characteristic words related to "dryness" than "heat" and "fire", which causes most models to recognize it as a "antipyretic agents" (label is 8). Similarly, the model cannot understand that "Qingbanxia" can dry and damp phlegm, nor "ophiopogon" can nourish yin and moisten dryness.

The classification of existing standards is also controversial. For example, the standard classification of Xingsusan Powder is "moistening formulae" (label is

17), while most models classify it as "phlegm, cough, asthma formulae" (label is 20) and "diaphoretic formulae" (label is 4). From the perspective of traditional Chinese medicine theory, the classification results are also acceptable.

6 Conclusion

In this paper, a standard data set of prescription efficacy classification was constructed. By comparing and analyzing the efficiency of several deep learning text classification models, it is found that the Bert-CNN model is the best, which improves the accuracy of automatic classification of prescription function.

Future work aims to address the following issues: Collect more data. Using methods such as K-fold cross validation to improve data utilization; Adding medical pre-training model to improve the accuracy of the model to identify medical terms, and increase prior knowledge in the field of traditional Chinese medicine, ultimately improve the accuracy of model classification; More experts in relevant fields will be invited to participate in the discussion on the efficacy classification of prescription. On the one hand, the efficacy classification criteria could be optimized. On the other hand, the standard classification results could be optimized and adjusted.

Acknowledgements. This work is supported by the National Natural Science Foundation of China (82174534, 61772249), and the Fundamental Research Funds for the Central Public Welfare Research Institutes (ZZ160311).

References

1. Yan, Z., Bo, G., Meng, C.: Design and implementation of the analysis system of TCM prescription. China J. Tradit. Chin. Med. Pharm. **29**(5), 1543–1546 (2014)
2. Cui, Y., Gao, B., Liu, L., Liu, J., Zhu, Y.: AMFormulaS: an intelligent retrieval system for traditional Chinese medicine formulas. BMC Med. Inform. Decis. Mak. **21-S**(2), 56 (2021). https://doi.org/10.1186/s12911-021-01419-8
3. Yan, Z., Bo, G., Meng, C.: Statistical analysis on phenomenon of homonym and synonym of Chinese materia medica. China J. Tradit. Chin. Med. Pharm. **30**(12), 4422–4425 (2015)
4. Deshan, G., Wenyu, L., Bingzhu, Z., Xingguang, M.: Application of named entity recognition in the recognition of words for Chinese traditional medicines and Chinese medicine formulae. Chin. Pharmac. Affairs (2019)
5. Yan, Z., Ling, Z., Jun-Hui, W., Meng, C.: An efficient approach of acquiring knowledge from ancient prescriptions and medicines based on information extraction. China J. Tradit. Chin. Med. Pharm. **30**(5), 316–321 (2015)
6. Administration N H S, China M O H R a S S O T P S R O. Medicine catalogue of national basic medical insurance. Work-Related Injury Insurance and Maternity Insurance (2020). 2020-12-28 [2022-03-02]
7. Medicine N a O T C. Coding rules for Chinese medicinal formulae and their codes: GB/T 31773-2015. China Quality and Standards Publishing, Beijing (2015)
8. Zuo, Y., Li, Z.: Prescriptions of Chinese materia medica. China Press of Traditional Chinese Medicine (2021)

9. Xiao, L., Wang, G., Liu, Y.: Patent text classification based on naive Bayesian method. In: 11th International Symposium on Computational Intelligence and Design, ISCID 2018, Hangzhou, China, 8–9 December 2018, vol. 1, pp. 57–60. IEEE (2018). https://doi.org/10.1109/ISCID.2018.00020
10. Xia, H., Min, T., Yi, W.: Sentiment text classification of customers reviews on the Web based on SVM. In: International Conference on Natural Computation. IEEE (2010). https://doi.org/10.1109/ICNC.2010.5584077
11. Cañete-Sifuentes, L., Monroy, R., Medina-Pérez, M.A.: A review and experimental comparison of multivariate decision trees. IEEE Access **9**, 110451–110479 (2021). https://doi.org/10.1109/ACCESS.2021.3102239
12. Yang, Z., Yang, D., Dyer, C., He, X., Smola, A.J., Hovy, E.H.: Hierarchical attention networks for document classification (2016). https://doi.org/10.18653/v1/n16-1174
13. Devlin, J., Chang, M., Lee, K., Toutanova, K.: BERT: pre-training of deep bidirectional transformers for language understanding. CoRR abs/1810.04805 (2018). http://arxiv.org/abs/1810.04805
14. Nagwani, N.K., Sharaff, A.: SMS spam filtering and thread identification using bi-level text classification and clustering techniques. J. Inf. Sci. **43**(1), 75–87 (2017). https://doi.org/10.1177/0165551515616310
15. Hai-Bing, M.A., Jiu-Yang, B.I., Xin-Shun, G.: Applications of text classification in network public opinion system. Inf. Sci. **33**(05), 97–101 (2015)
16. Kapusta, J., Drlík, M., Munk, M.: Using of N-grams from morphological tags for fake news classification. PeerJ Comput. Sci. **7**, e624 (2021). https://doi.org/10.7717/peerj-cs.624
17. Yu, H., Pan, X., Zhao, D., Wen, Y., Yuan, X.: A hybrid model for spatio-temporal information recognition in COVID-19 trajectory text. In: Zhao, X., Yang, S., Wang, X., Li, J. (eds.) WISA 2022. LNCS, vol. 13579, pp. 267–279. Springer, Cham (2022). https://doi.org/10.1007/978-3-031-20309-1_23
18. Li, X., Cui, M., Li, J., Bai, R., Lu, Z., Aickelin, U.: A hybrid medical text classification framework: Integrating attentive rule construction and neural network. Neurocomputing **443**, 345–355 (2021). https://doi.org/10.1016/j.neucom.2021.02.069
19. Prabhakar, S.K., Won, D.: Medical text classification using hybrid deep learning models with multihead attention. Comput. Intell. Neurosci. **2021**, 9425655:1–9425655:16 (2021). https://doi.org/10.1155/2021/9425655
20. Cui, M., Bai, R., Lu, Z., Li, X., Aickelin, U., Ge, P.: Regular expression based medical text classification using constructive heuristic approach. IEEE Access **7**, 147892–147904 (2019). https://doi.org/10.1109/ACCESS.2019.2946622
21. Zheng, C., Wang, X., Wang, T., et al.: Multi-label classification for medical text based on ALBERT-TextCNN model. J. Shandong Univ. (Sci. Edn.) **57**(04), 21–29 (2022)
22. Li, Q., Liao, W.: Biomedical text classification model based on attention mechanism. Chin. J. Med. Phys. **39**(04), 518–523 (2022)
23. Zhou, F., Jin, L., Dong, J.: Review of convolutional neural network. Chin. J. Comput. **40**(6), 23 (2017)
24. Kim, Y.: Convolutional neural networks for sentence classification. (2014). https://doi.org/10.3115/v1/d14-1181
25. Yang, L., Wu, Y., Wamg, J., Liu, Y.: Research on recurrent neural network. J. Comput. Appl. **38**(A02), 7 (2018)
26. Liu, P., Qiu, X., Huang, X.: Recurrent neural network for text classification with multi-task learning (2016)

27. Lai, S., Xu, L., Liu, K., Zhao, J.: Recurrent convolutional neural networks for text classification. In: National Conference on Artificial Intelligence (2015)
28. Zhou, P., Shi, W., Tian, J., Qi, Z., Xu, B.: Attention-based bidirectional long short-term memory networks for relation classification. In: Proceedings of the 54th Annual Meeting of the Association for Computational Linguistics (2016)

A Relation Extraction Model
for Enhancing Subject Features
and Relational Attention

Jiabao Wang[1,2], Weiqun Luo[1,2(✉)], Xiangwei Yan[1,2], and Zijian Zhang[1]

[1] College of Information Engineering, Xizang Minzu University,
Xianyang 712082, Shanxi, China
2826698337@qq.com
[2] XiZang Key Laboratory of Optical Information Processing and Visualization
Technology, Xianyang 712082, Shanxi, China

Abstract. In existing relation extraction methods, there are often issues such as Error propagation or insufficient attention, which limits the improvement of extraction performance. To this end, this paper proposes a relation extraction model that enhances subject features and relational attention (SFRARE). SFRARE uses the pre-trained model for word embedding, while using the LSTM neural network to enhance the semantic features of the subject to reduce the impact of error propagation on extraction performance, and uses multi-head attention mechanism to make the model pay more attention to the relation related to the subject. The experimental results of the model are better than the baseline model CasRel on two English public datasets NYT and WebNLG, and also better than the baseline model in complex contexts. Finally, all the experimental results show that the proposed model SFRARE outperforms the current mainstream models on both Chinese and English public data sets, proving that the model has certain generalization ability and alleviates the problems of error propagation and insufficient attention to relations.

Keywords: Relation Extraction · Subject Features · Attention mechanism · Pre-training model

1 Introduction

Knowledge graph is a technical method that uses graph models to describe knowledge and model the relations between everything in the world [1]. The key to structuring a knowledge graph is the relational facts between two entities and their relations, which are generally represented in the form of triplets of (subject, relation, object). The relation extraction task can help us extract entities and relations from unstructured text to form relational triplets.

Early relation extraction tasks generally used a pipeline model [2–4] which typically divided the extraction task into two sub tasks: named entity recognition and relation classification: first, identifying potential entities in the text

© The Author(s), under exclusive license to Springer Nature Singapore Pte Ltd. 2023
L. Yuan et al. (Eds.): WISA 2023, LNCS 14094, pp. 259–270, 2023.
https://doi.org/10.1007/978-981-99-6222-8_22

to form entity pairs; Then assign relational labels to each entity pair through classification. Such models often has the following three limitations:

1. Error propagation, which means that the errors generated during the entity extraction process are transmitted to the relation extraction, and gradually increase during the transmission process;
2. Lack of interaction between subtasks, treating the extraction of entities and relations as two independent parts and ignoring the association between the two subtasks;
3. There are redundant entity pairs, and during the entity extraction process, all entities in the text are extracted and paired. For some unrelated entities, it is also necessary to determine the relation between the two, which affects efficiency.

In recent years, the rise of joint extraction models for entity and relation extraction through unified modeling has to some extent broken these limitations. The model CasRel [5] regards relation as a function that maps subject to object, which breaks the traditional view that relation is separated from entity pair, and makes the model make a breakthrough in relation extraction. However, CasRel only uses four simple dense layers to extract subjects and objects, so there is still a certain error propagation problem inside the model. At the same time, because of this reason, the model does not pay more attention to the relation related to the subject and the extraction performance is affected. CasRel's method of extracting relational triples and its shortcomings inspired the model SFRARE(Subject Feature and Relationship Attention enhancement Relation Extraction) proposed in this paper. We believe that enhancing the feature representation of subjects can reduce error propagation caused by inaccurate recognition of subjects; and when extracting objects, calculating attention weights for different relations based on the subject features and text vector representations can enable the model to focus more attention on relations that are more closely related to the subject, thereby improving the extraction performance of the model. The main work of this paper is as follows:

1. This paper proposes a end-to-end relation extraction model that enhances subject features and relational attention (SFRARE);
2. The model SFRARE proposed in this paper achieved better results than the baseline model on the common dataset NYT and WebNLG, and was obtained on the NYT dataset in complex contexts;

2   Related Work

Extracting relational triples from unstructured text data is a key step in building large knowledge maps such as Wikidata [6] and DBpedia [7]. The current relation extraction models are mainly divided into pipeline model and joint model.

The early task of relation extraction usually adopts pipeline model. These models are divided into two steps to extract relational triples: 1) Named entity recognition of text to identify entities in the text and form entity pairs; 2) Classify the relation between entity pairs, that is, assign a relational label to entity pairs. This two-step method of extracting relation triples often leads to the problem of error transmission and ignores the relationship between entity extraction and relation extraction. In order to solve these problems, a joint extraction method is proposed in the subsequent research. Because neural network has certain advantages in feature learning, the current mainstream joint relation extraction models are generally based on neural network models.

The joint relation extraction model based on neural network generally implements the joint extraction of entity and relation in the form of parameter sharing or joint decoding. The end-to-end model integrating LSTM and tree structure proposed by Miwa in 2014 [8] realizes the joint extraction of entity and relation through parameter sharing, but the form of shared parameters often leads to information redundancy. The model proposed by Zheng et al. in 2017 [9] transforms relation extraction into tagging problem, fuses LSTM neural network to form an end-to-end model, and realizes joint extraction of entities and relations through joint decoding. Wei et al. viewed the relation extraction task from a new perspective, defined the relation as a function that maps the subject to the object, and proposed a cascading binary tagging framework, CasRel for the extraction of relational triplets. Zhang et al. designed an information booster [10] based on CasRel, which improved the extraction performance of the model by integrating location information and context semantics.

In general, the joint relation extraction model, which fuses the task information of the previous stage into the next task, can solve the error transmission problem of pipeline method, and can better connect entity extraction and relation extraction tasks through parameter sharing or joint decoding. Among them, the joint extraction model proposed by Wei et al. solves the problem of triple overlap while extracting relational triples, which inspired this paper to propose the SFRARE model for the extraction of relational triples.

3   The SFRARE Model

The structure of the SFRARE model is shown in Fig. 1. It is an end-to-end cascade extraction model, which includes a subject recognizer, a subject features enhancement module, a relational attention enhancement module and a relation-object recognizer.

3.1   Text Embedding Layer

The BERT pre-training model [11] can combine context information to learn the deep representation of words and has performed well in many natural language processing tasks since it was proposed, so the BERT pre training model was selected as the text embedding layer. In order to accelerate convergence and prevent gradient explosion or disappearance, normalization layer is used to process vector representation of text. As shown in Eq. (1):

$$H = LN(BERT(X)) \tag{1}$$

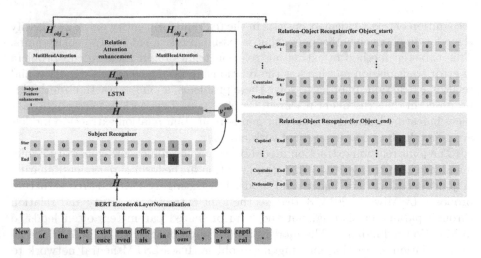

Fig. 1. Overall framework of SFRARE model. As shown in the figure, the BERT embedding layer and normalization layer obtain text vector. The subject recognizer extracts the subject by marking its start and end positions as 1, and the extracted subject is Sudan. The input of the subject feature enhancement module is the text vector fused with the subject features vector and the output is the enhanced text vector. The input of the relational attention enhancement module is the enhanced text vector and the output is two text vectors with relational attention. The relation-object recognizer takes two text vectors with relational attention as inputs to extract the start and end of the object, and extracts the object Khartoum corresponding to Sudan in the Capital and Contain relation.

In the equation, $X = x_1, x_2, ..., x_n$ represents the word sequence of the text, x represents the words that constitute the text, n represents the number of words in the text, LN represents the normalization layer, and H is the text vector representation obtained after the pre-trained model embedding and normalization layer processing. $H \in R^{b*l*d}$, where b is the amount of data in one training, l is the length of the text, and d is the dimension of the output of the last hidden layer of the pre-trained model.

3.2 Subject Recognizer

The subject recognizer is used to extract the potential subjects in the text. The input of subject recognizer is the vector representation of the text after BERT embedding. The recognizer treats subject's extraction as a binary classification task, and uses two binary classifiers to label each word in the text to determine whether it is the head or tail of the subject. If it is the head or tail of subject, it is marked as 1, otherwise it is marked as 0. The details are shown in Eq. (2) to (3):

$$p_i^{sub_s} = Sigmoid(W_{sub_s} h_i + b_{sub_s}) \tag{2}$$

$$p_i^{sub_e} = Sigmoid(W_{sub_e} h_i + b_{sub_e}) \tag{3}$$

In the equation, $p_i^{sub_s}$ and $p_i^{sub_e}$ respectively represents the probability of whether each word in the text is the start and end of the subject, W_{sub_s} represents the weight matrix used in the judgment of the start of the subject, and b_{sub_s} is its corresponding bias; W_{sub_e} represents the weight matrix used to judge the end of the subject, b_{sub_e} and b_{sub_e} is its corresponding biased term, h_i represents the vector representation of each word in the text.

3.3 Subject Features Enhancement Module

In CasRel, the inaccuracy of subject recognition may affect the accuracy of the subsequent task of object identification under a given relation, and then affect the extraction of the whole relational triplet. In order to alleviate this impact, this paper adds a subject features enhancement module to SFRARE. This module uses LSTM neural network to combine the semantic information of the context to calculate a richer representation of the subject features, and strengthen the memory of the semantic features of the subject, as shown in Eq. (4) to (5):

$$H_{sub} = H + v_i^{sub} \tag{4}$$

$$H_{sub} = LSTM(H_{sub} + b) \tag{5}$$

In the equation, H represents the text vector representation, H_{sub} represents the new text vector representation obtained after subject features fusion and enhancement, v_i^{sub} represents the feature vector of the subject and it is obtained by averaging the vectors of the start and end of the recognized subject, and b is the biased term of $LSTM$ neural network.

3.4 Relational Attention Enhancement Module

The relational attention enhancement module uses the multi-head attention mechanism to calculate the attention weights of different relations according to the text vector representation that enhances the features of the subject. By assigning weight values to different relations according to the subject features information and text vector representation, the model can pay more attention to the relations related to the above information, thus improving the extraction performance of relational triples, as shown in Eq. (6) to (9):

$$head_{s_i} = softmax(\frac{Q_{obj_s}K_{obj_s}^T}{\sqrt{d_k}})V_{obj_s} \tag{6}$$

$$H_{obj_s} = Concat(head_{s_1}, head_{s_2}, ..., head_{s_m})W_{obj_s} \tag{7}$$

$$head_{e_i} = softmax(\frac{Q_{obj_e}K_{obj_e}^T}{\sqrt{d_k}})V_{obj_e} \tag{8}$$

$$H_{obj_e} = Concat(head_{e_1}, head_{e_2}, ..., head_{e_m})W_{obj_e} \tag{9}$$

In the equation, Q_{obj_s}, K_{obj_s} and V_{obj_s} are obtained by linear transformation of the vector representation of the text enhanced by the semantic features of the subject, $head_{s_i}$ represent the attention weight assigned to each relation for recognizing the start of the object, and $head_{e_i}$ represent the attention weight assigned to each relation for recognizing the end of the object. W_{obj_s} and W_{obj_e} a and b are two learnable weight matrices respectively. H_{obj_s} and H_{obj_e} is the text vector representation obtained by merging relational attention and subsequently used to recognizer the start and end of the object, respectively.

3.5 Relation-Object Recognizer

Relation-object recognizer is used to extract the object corresponding to the subject under a given relation. As with the subject recognizer, for each relation, the object recognizer uses two binary classifiers to determine whether each word of the text is the start or end of the object corresponding to the subject under the current relation, as shown in Eq. (10) to (11) :

$$p_i^{obj_s} = Sigmoid(W_{obj_s}^r h_{i_s} + b_{obj_s}^r) \tag{10}$$

$$p_i^{obj_e} = Sigmoid(W_{obj_e}^r h_{i_e} + b_{obj_e}^r) \tag{11}$$

In the equation, $p_i^{obj_s}$ and $p_i^{obj_e}$ respectively represents the probability of whether each word in the text is the start and end of the object, similar to formula (2) (3), $W_{obj_s}^r$ represents the weight matrix used to judge the start of the object under the relation r, $W_{obj_e}^r$ represents the weight matrix used to judge the end of the object under the relation r, $b_{obj_s}^r$ and $b_{obj_e}^r$ is the biased term, $h_{i_s} \in H_{obj_s}$ and $h_{i_s} \in H_{obj_e}$ come from the relational attention enhancement module that used to recognize the start and end of the object respectively.

3.6 Loss Function

Since the SFRARE model regards the task of extracting the subject and the object as a binary classification labeling task, binary cross-entropy loss function is used. Moreover, since the model first extracts the subject and then extracts the corresponding object under the given relation, the losses during the extraction are calculated separately and summed. The details are shown in Eq. (12) to (14):

$$l_s ub = \sum_{j \in \{sub_s, sub_e\}} (-\frac{1}{L} \sum_{i=1}^{L} (y_i^j ln p_i^j + (1 - y_i^j) ln(1 - p_i^j)) \tag{12}$$

$$l_o bj = \sum_{j \in \{obj_s, obj_e\}} (-\frac{1}{L} \sum_{i=1}^{L} (y_i^j ln p_i^j + (1 - y_i^j) ln(1 - p_i^j)) \tag{13}$$

$$Loss = l_s ub + l_o bj \tag{14}$$

In the equation, $l_s ub$ represents the loss caused by the extraction of the subject, $l_o bj$ represents the loss caused by the extraction of the object, and $Loss$ represents the final loss of the entire model. L represents the length of the text, y_i^j represents the start and end of the real subject and object in Eq. (12) and (13), and p_i^j represents the start and end of the subject and object predicted by the model.

4 Experiments

4.1 Datasets

This paper uses English public dataset NYT and WebNLG for experiments. Specific composition is shown in Table 1:

Table 1. Data Statistics for NYT Dataset and WebNLG Dataset

Category	NYT			WebNLG		
	Train	Valid	Test	Train	Valid	Test
Normal	37013	3306	3266	1596	182	246
EPO	9782	849	978	227	16	26
SEO	14735	1350	1297	3406	318	457
All	56195	4999	5000	5019	500	703

The NYT dataset is an English public dataset built using distant supervision methods. The WebNLG dataset is built through natural language generation tasks and subsequently adapted for relation extraction tasks. In the table, Normal, SEO and EPO represent three different overlaps of the relational triplet. Normal indicates that it is normal, that is, the subject or object only participates in the construction of one triplet; EPO indicates the overlap of entity pairs, that is, a group of entity pairs participate in the construction of multiple triples at the same time; SEO represents the overlap of a single entity, that is, a subject or object participates in the construction of multiple triples at the same time. These three overlaps exist in both NYT and WebNLG datasets.

4.2 Implementation Details

The specific experimental settings of this paper are shown in Table 2:

Table 2. Experimental details

Items	Specific situation
Word embedding layer	bert-base
Recognition threshold	0.5
Hidden layer dimension of LSTM	768
Number of heads of attention mechanism	num-rels
Learning rate	1e−5
Batch-size	16
Dropout	0.2
Experimental device	NVIDIA GeForce RTX 3090(24G)

The accuracy rate, recall rate and F1-score of the extraction results were evaluated in the experiment, as shown in Eq. (15) to (17):

$$Pre = \frac{TT}{PT} \tag{15}$$

$$Rec = \frac{TT}{GT} \tag{16}$$

$$F1 = \frac{2 * (Pre * Rec)}{Pre + Rec} \tag{17}$$

In the equation, TT is the number of correct triples predicted, PT is the number of all predicted triples, and GT is the number of actually correct triples.

4.3   Experimental Result

In order to verify the performance of SFRARE model for relation extraction, this paper selects some representative relation extraction models in recent years, and takes CasRel as the baseline model for comparison:

- CopyR$_{OneDecoder}$ [12]: An end-to-end relation extraction model with replication mechanism proposed by Zeng et al., a single decoder is used at the decoding layer.
- CopyR$_{MultiDecoder}$: An end-to-end relational extraction model with replication mechanism proposed by Zeng et al., multiple decoders are used at the decoding layer.
- GraphRel$_{1p}$ [13]: The first phase of relational extraction model based on relational graph structure proposed by Tsu-Jui Fu et al.
- GraphRel$_{2p}$: The second phase of relational extraction model based on relational graph structure proposed by Tsu-Jui Fu et al.
- CopyR$_{RL}$ [14]: Zeng et al. proposed an end-to-end relation extraction model based on CopyR that applies reinforcement learning to the generation of relational triples.
- CasRel: A novel cascade binary tagging framework for relational triple extraction proposed by Wei et al.
- PATB: Inspired by CasRel, Zhang et al. proposed an information aggregator for joint entity and relationship extraction in 2022.

The experimental results on the public dataset are shown in Table 3:

Table 3 shows that SFRARE achieves better results than the baseline model CasRel on both public datasets. On the NYT dataset, the accuracy rate of SFRARE is 90.5%, the recall rate is 91.7%, and the F1-score is 91.1%, which is 1.5% higher than the baseline model CasRel. On the WebNLG dataset, SFRARE had an accuracy of 92.6%, a recall rate of 92.7%, and a F1-score of 92.6%, with an F1-score improvement of 0.8% compared to the baseline model CasRel. The experimental results show that the SFRARE model has a certain generalization ability and the extraction performance is better.

Table 3. Experimental results on NYT and WebNLG

Methods	NYT			WebNLG		
	Pre	Rec	F1	Pre	Rec	F1
CopyR$_{OneDecoder}$	59.4%	53.1%	56.0%	32.2%	28.9%	30.5%
CopyR$_{MultiDecoder}$	61.0%	56.6%	58.7%	37.7%	36.4%	37.1%
GraphRel$_{1p}$	62.9%	57.3%	60.0%	42.3%	39.2%	40.7%
GraphRel$_{2p}$	63.9%	60.0%	61.9%	44.7%	41.1%	42.9%
CopyR$_{RL}$	77.9%	67.2%	72.1%	63.3%	59.9%	61.6%
CasRel	89.7%	89.5%	89.6%	93.4%	90.1%	91.8%
PATB	90.3%	91.5%	90.9%	93.6%	91.4%	92.5%
SFRARE	90.5%	91.7%	91.1%	92.6%	92.7%	92.6%

Ablation Experiment. In order to verify whether enhancement of subject features and enhancement of relational attention can improve model extraction performance and to what extent, this paper conducted ablation experiments on NYT and WebNLG datasets. The experimental results are shown in Table 4, in which SFRARE$_{SubF}$ means only enhanced subject features and SFRARE$_{RelA}$ means only enhanced relational attention, SFRARE is the model that enhances both aspects.

Table 4. The results of the ablation experiments on NYT and WebNLG.

Methods	NYT			WebNLG		
	Pre	Rec	F1	Pre	Rec	F1
CasRel$_{Baseline}$	89.7%	89.5%	89.6%	93.4%	90.1%	91.8%
SFRARE$_{SubF}$	90.1%	89.8%	89.9%	92.2%	91.8%	92.0%
SFRARE$_{RelA}$	89.9%	91.0%	90.5%	93.7%	90.7%	92.2%
SFRARE	90.5%	91.7%	91.1%	92.6%	92.7%	92.6%

The experimental results show that when the subject features are enhanced, the F1 values obtained by SFRARE on the two datasets increase by 0.3% and 0.2%, respectively, compared with the baseline model. When the relationship attention was enhanced, SFRARE obtained F1 values on the two datasets increased by 0.9% and 0.4%, respectively, compared with the baseline model. When both are enhanced, the increase is 1.5% and 0.8% respectively. The results of the above ablation experiments show that enhancing the subject features and relational attention has a positive effect on improving the extraction performance of the model.

Experimental Results of Extracting from Texts Containing Triples of Different Numbers. In order to verify whether the extraction performance of SFRARE can be improved when there are multiple relational triples in the text compared with the baseline CasRel, we divided the dataset and conducted experiments. The experimental results are shown in Table 5:

Table 5. The results of the ablation experiments on NYT and WebNLG.

Methods	NYT					WebNLG				
	$N = 1$	$N = 2$	$N = 3$	$N = 4$	$N \geq 5$	$N = 1$	$N = 2$	$N = 3$	$N = 4$	$N \geq 5$
CopyR$_{OneDecoder}$	66.6%	52.6%	49.7%	48.7%	20.3%	65.2%	33.0%	22.2%	14.2%	13.2%
CopyR$_{MultiDecoder}$	67.1%	58.6%	52.0%	53.6%	30.0%	59.2%	42.5%	31.7%	24.2%	30.0%
GraphRel$_{1p}$	69.1%	59.5%	54.4%	53.9%	37.5%	63.8%	46.3%	34.7%	30.8%	29.4%
GraphRel$_{2p}$	71.0%	61.5%	57.4%	55.1%	41.1%	66.0%	48.3%	37.0%	32.1%	32.1%
CasRel	88.2%	90.3%	91.9%	94.2%	83.7%	89.3%	90.8%	94.2%	92.4%	90.9%
SFRARE	88.9%	91.9%	93.9%	95.5%	88.9%	87.4%	90.9%	94.8%	94.0%	92.3%

The experimental results show that the experimental results obtained by SFRARE model on NYT are comprehensively superior to the baseline model CasRel, while in WebNLG, when the number of triples in the text exceeds 3, SFRARE will greatly improve compared with CasRel. The analysis shows that when the number of triples in the text increases, the amount of subject features that the model needs to deal with will increase. At this time, the subject features enhancement module of SFRARE can help the model obtain more accurate subject features, and reduce the error propagation caused by the error of subject identification and the mutual interference caused by the continuous identification of multiple subjects.

Experimental Results of SFRARE on Overlapping Relation Triples. To verify whether the performance of SFRARE is improved when extracting relational triplets with different overlaps, comparative experiments are conducted on the NYT and WebNLG datasets. The experimental results are shown in Fig. 2:

The experimental results show that SFRARE model performs better than the baseline model CasRel on NYT and better than the baseline model under Normal and EPO on WebNLG for all three triples overlapping. Analysis shows that when entity pairs overlap, the model needs to consider the subject and the object, and the relation between them. The enhancement of subject feature enables the model to have more abundant information about the feature information when extracting the object according to the relation. According to the above features, all the object corresponding to the current subject under different relations can be extracted more accurately and comprehensively. The enhancement of relational attention can make the model pay more attention to

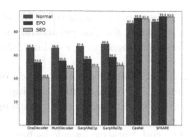
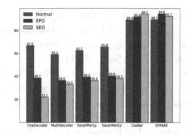

(a) The results of the experiment on NYT

(b) The results of the experiment on WebNLG

Fig. 2. Experimental Results of SFRARE on Overlapping Relation Triples

the relations related to the vector representation of the text and the features of the subject, thus improving the extraction performance.

5 Conclusions

This paper proposes a relation extraction model to enhance the subject features and relation attention (SFRARE). Inspired by CasRel, the model alleviates the impact of error propagation caused by inaccurate subject recognition by enhancing subject features and improves the performance of relation extraction tasks by enhancing relational attention. The experimental results show that SFRARE achieves better results than the baseline model CasRel on both NYT and WebNLG. In the complex context of multiple relational triples and overlapping relational triples in the text, the model also outperforms the baseline model on NYT datasets. It is proved that the enhancement of subject features and relation attention can comprehensively improve the extraction performance of the model. In the future, we plan to continue to improve the model by analyzing the semantic structure of single-entity overlapping text to improve the extraction performance of SFRARE when dealing with single-entity overlapping. The optimized model is planned to be used in the subsequent work related to Chinese knowledge graph construction.

References

1. Guo, Q., et al.: Constructing Chinese historical literature knowledge graph based on BERT. In: Xing, C., Fu, X., Zhang, Y., Zhang, G., Borjigin, C. (eds.) WISA 2021. LNCS, vol. 12999, pp. 323–334. Springer, Cham (2021). https://doi.org/10.1007/978-3-030-87571-8_28
2. Zelenko, D., Aone, C., Richardella, A.: Kernel methods for relation extraction. J. Mach. Learn. Res. **3**(3), 1083–1106 (2003)
3. Zhou, G., Su, J., Zhang, J., Zhang, M.: Exploring various knowledge in relation extraction. In: 43rd Annual Meeting on Association for Computational Linguistics on Proceedings, Ann Arbor, Michigan, USA, pp. 427–434 (2005)

4. Chan, Y.S., Su, J., Dan, R.: Exploiting syntactico-semantic structures for relation extraction. In: 49th Annual Meeting of the Association for Computational Linguistics: Human Language Technologies - Volume 1 on Proceedings, Portland, Oregon, USA, pp. 551–560 (2011)

5. Wei, Z., Su, J., Wang, Y., Tian, Y., Chang, Y.: Exploiting a novel cascade binary tagging framework for relational triple extraction. In: 58th Annual Meeting of the Association for Computational Linguistics on Proceedings, Seattle, Washington, USA, pp. 1476–1488 (2020)

6. Vrandecic, D., Krtoetzsch, M.: WikiData: a free collaborative knowledgebase. Commun. ACM **57**(10), 78–85 (2014)

7. Auer, S., Bizer, C., Kobilarov, G., Lehmann, J., Cyganiak, R., Ives, Z.: DBpedia: a nucleus for a web of open data. In: Aberer, K., et al. (eds.) ASWC/ISWC -2007. LNCS, vol. 4825, pp. 722–735. Springer, Heidelberg (2007). https://doi.org/10.1007/978-3-540-76298-0_52

8. Miwa, M., Sasaki, Y.: Modeling joint entity and relation extraction with table representation. In: 2014 Conference on Empirical Methods in Natural Language Processing on Proceedings, Doha, Qatar, pp. 1858–1869 (2014)

9. Zheng, S., Wang, F., Bao, H., Hao, Y., Zhou, P., Xu, B.: Joint extraction of entities and relations based on a novel tagging scheme. In: 55th Annual Meeting of the Association for Computational Linguistics on Proceedings, Vancouver, Canada, pp. 1083–1106 (2017)

10. Zhang, L., Lu, L., Wang, A., Yang, W.: PATB: an information booster for joint entity and relation extraction. J. Chin. Comput. Syst. **3**(3), 1–9 (2022)

11. Jacob, D., Chang, M.W., Kenton, L., Kristina, T.: BERT: pre-training of deep bidirectional transformers for language understanding. In: 2019 Conference of the North American Chapter of the Association for Computational Linguistics: Human Language Technologies Volume 1 (Long and Short Papers) on Proceedings, Minneapolis, USA, pp. 4171–4186 (2017)

12. Zeng, X., Zeng, D., He, S., Liu, K., Zhao, J.: Extracting relational facts by an end-to-end neural model with copy mechanism. In: 56th Annual Meeting of the Association for Computational Linguistics (Volume1: Long Papers) on Proceedings, Melbourne, Australia, pp. 1409–1418 (2018)

13. Fu, T.J., Li, P.H., Ma, W.Y.: Modeling text as relational graphs for joint entity and relation extraction. In: 57th Annual Meeting of the Association for Computational Linguistics on Proceedings, Florence, Italy, pp. 1858–1869 (2014)

14. Zeng, X., He, S., Zeng, D., Liu, K., Liu, S., Zhao, J.: Learning the extraction order of multiple relational facts in a sentence with reinforcement learning. In: 2019 Conference on Empirical Methods in Natural Language Processing and the 9th International Joint Conference on Natural Language Processing (EMNLP-IJCNLP) on Proceedings, Hong Kong, China, pp. 367–377 (2019)

An Approach of Code Summary Generation Using Multi-Feature Fusion Based on Transformer

Chao Yang and Junhua Wu[⊠]

College of Computer and Information Engineering,
Nanjing Tech University, Nanjing 211816, China
wujh@njtech.edu.cn

Abstract. The automatic generation of code comments is an important research in program understanding. Code summary describes the function and purpose of the code. It helps developers comprehend the program, and reduces the cost of software maintenance. Many previous approaches only use token sequence of source code as input or only rely on learning code structure information from Abstract Syntax Tree (AST), which will result in extracting source code information and obtaining richer semantic information ineffectually. We propose an approach of code summary generation using Multi-Feature Fusion based on Transformer (MFFT). The Multi-way Tree-LSTM is used to process the AST structure to obtain the semantic information of the code. To extract more code information, we also obtain code token features and combine them with a fusion method to generate a high-quality code summary. The experimental results show that our MFFT method has a good performance improvement.

Keywords: Code comments · Deep learning · Neural network · Transformer

1 Introduction

The key to automatic code comment generation lies in describing the source code using natural language. It will improve the readability and comprehensibility of code and help developers understand programs more effectively. However, in real project development, absence of or mismatching code caused by developers' lack of awareness, insufficient constraints, or time constraints are common problems. These will result in low quality code. Therefore, good code comments are critical to improving the efficiency of developers' program comprehension [1]. In order to alleviate the cognitive effort of developers in understanding programs, generation of source code comments has been proven useful [2,3]. Good code comments have the following points: a) correctness: code comments elucidate the functionality of the code, b) fluency: with comments written in fluent and natural language,

© The Author(s), under exclusive license to Springer Nature Singapore Pte Ltd. 2023
L. Yuan et al. (Eds.): WISA 2023, LNCS 14094, pp. 271–283, 2023.
https://doi.org/10.1007/978-981-99-6222-8_23

making the code easy to read and understand, c) consistency: code comments adhere to a standard style to facilitate code comprehension.

In the past, researchers have proposed a number of approaches to generate automatic comments for code. The methods range from template-based generation methods to generative methods for information retrieval. These methods define heuristic rules and synthesize the comments prototypes of classes and methods, and then apply the Information Retrieval (IR) [4] to generate summaries. The IR approach employs Vector Space Models (VSM) and Latent Semantic Indexing (LSI) to search for options from similar code fragments. Recently, data-driven deep learning methods have been used for generating comments for source code snippets. Based on deep neural networks, the algorithms for comment generation can be mainly divided into two categories: those based on Recurrent Neural Networks (RNN) and other neural network algorithms. These methods usually use the encoder-decoder framework, also known as the sequence to sequence model [3,5]. In this framework, the encoder takes the source code into a fixed-size vector, and the decoder responsible for decoding source code vectors and making predictions about source code comments. These works have achieved some good results.

However, in some studies, such as Loyola et al. [6] only used source code as input, ignoring the structural information of code. In contrast, usually there are some loops, conditional branching in source code, which are crucial in understanding the source code. Therefore, the researcher parsed source code into AST, which contains information about the structure of the code. Hu et al. [7] used the method to obtain the AST sequence with the semantic information of the code, which significantly improved the accuracy of code comments. Tai et al. [8] mentioned two types of Tree-LSTM that is a neural network architecture that deals with tree structures: Child-sum Tree-LSTM structure can handle trees with any number of child nodes, and N-ary Tree-LSTM structure can handle trees with a fixed number of ordered child nodes. But they are difficult to be apply into ASTs. Since ASTs are trees with any number of ordered sub-nodes, so, Multiway Tree-LSTM [9] is used to process the information of nodes in ASTs in order to obtain the semantic information of the context.

Code summarization based on the sequence to sequence is a kind of framework viewed as Neural Machine Translation (NMT). Iyer et al. [10] proposed an Long Short Term Memory (LSTM)-based comments generation model CODE-NN. Zheng et al. [11] used the encoder-decoder framework with Gated Recurrent Unit (GRU), and a global attention mechanism. Allamanis et al. [5] introduced CNN into the attention mechanism to generate short descriptive summaries for source code fragments. Although these neural network algorithms have achieved some success in code comments, they have limitations in handling complex information and long-term dependencies. Ahmad et al. [12] used the Transformer model to generate code summaries and addressed point of long-term dependencies. They achieved good results using only the code tokens as input, but they ignored the syntactic structure. Hu and Li [3] argued that AST sequences generated through Structural-Based Traversal (SBT) contain more structural

information about the source code. Wan and Zhao [13] improved the performance of automatic code summarization tasks by using reinforcement learning
to incorporate AST structural information. However, relying on AST structure
as input may not be enough in expressing semantic information.

In this paper, we propose a novel method MFFT for code comment generation. The encoder-decoder framework based on the RNN, presents long-term
dependence. So, we consider Transformer model to address the problem. To
improve the performance of the model, we use a mechanism similar to the self-
attention fuse the encoded semantic feature and token feature, and send the
fused vector along with the code comments to the decoder to generate code
summaries. Experiments show that our proposed method is more effective.

2   Approach

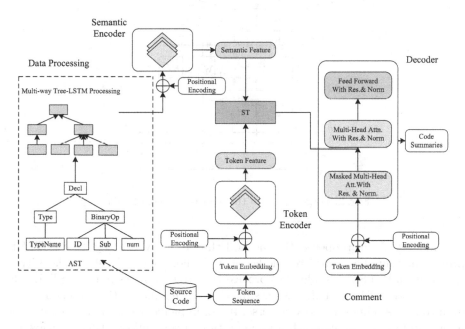

Fig. 1. The overall framework of MFFT

As shown in Fig. 1, the MFFT framework model consists of three main
components: data processing, fusion of Semantic and Token (ST) features, and
decoder. MFFT is based on the Transformer architecture, mainly consisting
of an encoder and a decoder. As in Transformer, all the words in the source
code are input to the network at once, which causes the loss of sequence order
information. Therefore, positional encoding is added to the model before encoding, which effectively preserves the sequence order information and enables the

model to learn the code characteristics better. To comprehensively capture the semantic information in the source code, we initially parse source code into an AST structure, which is then processed by a Multi-way Tree-LSTM model to obtain semantic vectors. The resulting vectors are subsequently combined with token embedding vectors and inputted into separate encoders to derive semantic and token features. ST fusion module aims to organically combine the semantic feature and token feature of the source code to produce a more accurate and comprehensive representation of the code information. The fused feature vectors are sent to the decoder for generating code summaries.

2.1   Data Processing

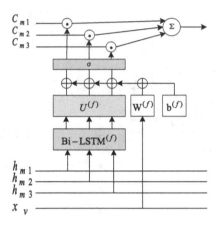

Fig. 2. Forget gates in the Multi-way Tree-LSTM

Shido et al. [9] proposed the method that using Multi-way Tree-LSTM to process any number of ordered nodes of AST. For each gate of standard Tree-LSTM [8], a Bi-LSTM is added to process the ordered nodes in the AST. By the characteristics of Bi-LSTM, it is possible to send the information of forward children to reverse children and vice versa. It can implement the interaction of child node information and complete representation of code information. As shown in Fig. 2, since the forgetting gate of each child node is calculated independently, all child nodes share the parameter $U^{(f)}$, without additional parameters. Therefore, Multi-way Tree-LSTM can handle any number of ordered nodes in AST.

For source code, we first use a standard AST parser to convert the source code into ASTs. Then we use Multi-way Tree-LSTM to process the nodes of the ASTs. For the input node x_v, $C_{(v)}$ is represented as the child node of v, and n_v is represented as the number of children. At each time step t, the memory cell c_t and the hidden state h_t are updated as follows:

$$\tilde{f}_v = \pounds^{(f)}(h_{v_1}..., h_{v_{n_v}}) \tag{1}$$

$$\tilde{u}_v = \pounds^{(u)}(h_{v_1}..., h_{v_{n_v}})v_{n_v} \tag{2}$$

$$\tilde{i}_v = \pounds^{(i)}(h_{v_1}..., h_{v_{n_v}})v_{n_v} \tag{3}$$

$$\tilde{o}_v = \pounds^{(o)}(h_{v_1}..., h_{v_{n_v}})v_{n_v} \tag{4}$$

$$f_{v_m} = \sigma(W^{(f)}x_v + U^{(f)}(\tilde{f}_{v_m}) + b^{(f)}) \tag{5}$$

$$i_v = \sigma(W^{(i)}x_v + U^{(i)}(\tilde{i}_v) + b^{(i)}) \tag{6}$$

$$o_v = \sigma(W^{(o)}x_v + U^{(o)}(\tilde{o}_v) + b^{(o)}) \tag{7}$$

$$u_v = \sigma(W^{(u)}x_v + U^{(u)}(\tilde{u}_v) + b^{(u)}) \tag{8}$$

$$c_v = \sum_{m \in C_{(v)}} c_m \odot f_{v_m} + i_v \odot u_v \tag{9}$$

$$h_v = o_v \odot tanh(c_v) \tag{10}$$

where $\pounds^{(*)}$ represents the standard chain-like LSTMs, $\tilde{f}_v$ is a sequence of n_v vectors, $\tilde{u}_v$, $\tilde{i}_v$, $\tilde{o}_v$ are the last vectors in the sequence of $\pounds^{(u)}$, $\pounds^{(i)}$ and $\pounds^{(o)}$, respectively. i, f and o are input, forget and output gate respectively, $W^{(*)}$ and $U^{(*)}$ represent are weight matrices.

For the input AST node x_v, the hidden state is calculated as follows:

$$h_v^{(E)} = f(x_v, h_{C_{(v)}}^{(E)}) \tag{11}$$

where, f stands for Multi-way Tree-LSTM encoding, $h_{C_{(v)}}^{(E)} = \{h_m | m \in C_{(v)}\}$ is the hidden state of the child node.

2.2   Encoder

The semantic encoder is responsible for encoding the semantic embedding vectors obtained from the data processing part. The token encoder does not compress the entire source code sentence $X = (x_1, x_2, ..., x_n)$ into a single holistic context vector. Instead, it produces a series of context vectors. The input embedding vectors pass through a stack of encoder layers, where each encoder layer consists of a multi-head attention layer and a feedforward layer.

As Transformer does not have recurrence, it is unaware of the order of tokens. Positional encoding is added to solve this problem. The formula for calculating positional encoding is as follows:

$$PE_{(pos,2i)} = sin(pos/10000^{2i/d_{model}}) \tag{12}$$

$$PE_{(pos,2i+1)} = cos(pos/10000^{2i/d_{model}}) \tag{13}$$

Here, PE denotes the position embedding, pos denotes the position of the word in the sentence, d_{model} is the dimension of the token embedding, $2i$ is the even dimension, and $2i + 1$ is the odd dimension.

2.3   Multi-head Attention

Multi-head attention is built on top of self-attention, which can extract and process different relationships between query Q, key K, and value V simultaneously, and finally concatenate them, which is beneficial for capturing more comprehensive information. Self-attention is a mechanism for calculating the interdependence between each element in a sequence. Its calculation formula is as follows:

$$Attention(Q, K, V) = softmax(\frac{QK^T}{\sqrt{d_k}})V \tag{14}$$

In the formula, the inner product of each row vector of matrix Q and K is calculated. In order to prevent the inner product from being too large, it is divided by the square root of d_k, where d_k is the vector dimension.

Each head in multi-head attention is an attention layer. In order to improve the difference, each head will pay attention to a specific feature:

$$head_i = Attention(QW_i^Q, KW_i^K, VW_i^V) \tag{15}$$

$$MultiHead(Q, K, V) = Concat(head_1, ..., head_i)W^O \tag{16}$$

where $W_i^{(*)}$ and W^O are learning parameter weight matrices.

2.4   Feature Fusion Module

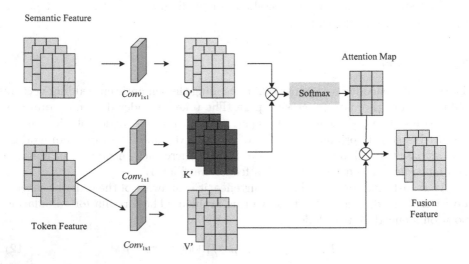

Fig. 3. ST fusion module

AST contains the structural information of the source code. After processing the AST information with Multi-way Tree-LSTM, the semantic information is obtained. In order to extract more information from the code, we need to fuse the semantic features and the token features of the source code. Before fusion, these two types of features need to be encoded, resulting in semantic feature vector Z and code tokens feature vector Z'. We fuse them into a unified representation using ST fusion module. As shown in Fig. 3, ST fusion module is similar to the self-attention mechanism in the Transformer. The purpose of the query vector in code summarization is to guide the allocation of attention, while semantic features capture the overall meaning and context of the code. As a query vector, they help the model focus on the overall semantic information relevant to the current position when generating comments, resulting in more accurate summarizations. On the other hand, token features only represent specific lexical and syntactic information, which are also crucial for generating comments but typically only involve local information within the code, lacking a comprehensive semantic overview. Using token features as query vectors may cause the model to overly focus on local features when generating comments, thereby disregarding overall semantic consistency. Therefore, the semantic feature Z is passed through $Conv_{1\times1}$, it is regarded as Q' vector, and the tokens feature Z' is treated as a K' and V' vector after being passed through $Conv_{1\times1}$. The calculation formula is as follows:

$$Q' = Conv_{1\times1}(Z),\ K' = Conv_{1\times1}(Z'),\ V' = Conv_{1\times1}(Z') \qquad (17)$$

After getting Q', K' and V', calculate their fusion feature as follows:

$$F_{st} = softmax(\frac{Q'(K')^T}{\sqrt{d_k}})V' \qquad (18)$$

where d_k is set to 64.

2.5   Decoder

The decoder is responsible for generating a summary based on the fused feature vector from the encoder's ST fusion module. The decoder is composed of multiple decoder layers, each of which has three sub-layers: a masked multi-head self-attention layer, a multi-head self-attention layer, and a fully connected layer. Unlike the encoder layers, there is an additional masked multi-head self-attention layer. Masking is used to hide certain values. Sequence masking is used to prevent the decoder from seeing future information. Specifically, for a sequence with a time step of t, the decoder output only depends on the output before t, not after t. The output of the decoder layer is processed by a softmax layer to calculate the probability of generating a summary. Our goal is to minimize the cross-entropy loss function and update the parameters of the model using optimization algorithms, such as gradient descendent. The formula for computing the loss function is as follows:

$$L_{(loss)} = -\frac{1}{N}\sum_{i=1}^{N}\sum_{j=1}^{n}logp(y_i^{(j)}) \qquad (19)$$

where N is the total number of training data, n is the length of each target summary, $y_i^{(j)}$ is the jth word in the ith sentence, and $p(y_i^{(j)})$ denotes the probability of generating the jth word.

3   Experiment

3.1   Dataset

In this experiment, we choose to adopt the Java dataset used by Hu et al. [7] as our research object. In this dataset, there are a total of 485,812 pairs of data, among which 88.63% of the samples with code length less than 200 and 96.96% of the samples with comment length less than 30.

3.2   Evaluation Metrics

In the evaluation of code summarization, we use performance metrics from NMT research to automatically assess the quality of code summarization. The metrics include BLEU, METEOR, ROUGE-L. BLEU [14] measures the n-gram overlap between the candidate document and the reference translation. The higher the overlap, the better the translation quality is considered to be. Follow Hu et al.'s study [7], we use sentence-level BLEU (S-BLEU) and corpus-level BLEU (C-BLEU) to evaluate the generated comments. S-BLEU calculates synthetic BLEU score from sentences, while C-BLEU calculates the composite BLEU score from the entire corpus. METEOR [15] is based on single-precision weighted harmonic means, unigram recall, aiming to address some inherent deficiencies in certain BLEU standards. ROUGE-L [16] calculates the length of the longest common subsequence between the candidate summary and the reference summary. The longer the length, the higher the score.

3.3   Experimental Settings

MFFT model is trained on GPU based on the PyTorch framework. We use the Adam optimizer to train the model, using cross-entropy as the loss function. We set the initial learning rate to 0.0001, the epoch value to 100, and dropout rate is 0.2. The embedding dimension of word vectors is set to 512, and the batch size is 64. We set the heads of the multi-head attention mechanism to 8, and the number of encoder-decoder layers to 6. When generating code summaries, use beam search and set the beam size to 5.

3.4   Result

We compared our proposed MFFT method with six baseline methods published by Lin et al. [17]. CODE-NN [10] uses an LSTM encoder on the attention framework, Hybrid-DRL [13], AST-attendgru [18], and ASTNN [19] methods all utilize the structural information of the source code, while Ahmad et al. [12] used a

Table 1. Our proposed method (MFFT) is compared with other baselines

Method	S-BLEU(%)	C-BLEU(%)	METEOR(%)	ROUGE-L(%)
CODE-NN	24.22	14.68	14.79	33.18
Hybrid-DRL	20.16	10.99	13.15	29.22
Hybrid-DeepCom	29.19	19.82	18.21	38.70
AST-attendgru	27.50	17.85	16.44	35.32
ASTNN	25.40	15.94	16.92	36.10
Transformer	30.98	23.63	20.30	42.20
MFFT	**33.24**	**24.89**	**21.68**	**44.57**

Transformer model, but only takes the source code as input. These methods either use the semantic information of the code or the token feature of the source code, but none of them consider a method for fusing the semantic and tokens feature of the code.

The comparison results between MFFT and baseline are in Table 1. From Table 1, we can see that our proposed MFFT method is superior to the other baselines. A S-BLEU score is 33.24, a C-BLEU score is 24.89, a METEOR score is 21.68, and a ROUGE-L score is 44.57. Specifically, MFFT outperforms Transformer by 2.26% in terms of S-BLEU, by 1.26% in terms of C-BLEU, by 1.38% in terms of METEOR, and by 2.37% in terms of ROUGE-L. Obviously, the fusion of semantic feature and tokens feature by MFFT will improve the performance of code summarization. Compared with Transformer, which only takes source code as a single input, the performance of MFFT model is better.

3.5 Ablation Experiments

Table 2. The results of ablation experiments

Method	S-BLEU(%)	C-BLEU(%)	METEOR(%)	ROUGE-L(%)
w/o MTL	29.85	22.84	19.57	41.39
w/o STF	32.37	24.35	21.22	43.64
MFFT	**33.24**	**24.89**	**21.68**	**44.57**

We conducted ablation experiments in order to demonstrate the Multi-way Tree-LSTM and feature fusion on the effects of MFFT. The results are shown in Table 2. Firstly, we only take the source code and AST as input, and encode the AST without using Multi-way Tree-LSTM (w/o MTL), and use only the dual encoder for encoding. The score in ROUGE-L is 41.39%, which is 3.18% lower than the MFFT performance, and other metrics are also significantly lower.

This shows that Multi-way Tree-LSTM can handle semantic information well and improve the performance of the model. Secondly, we do not perform feature fusion (w/o STF) between semantically encoded results and code token encoded results in the MFFT model. It can be seen that by fusing these features, the key information in terms of semantics and code tokens are enhanced, the accuracy of the model is improved, and a better summary is generated. Finally, examples from the test set are selected to compare the generated comments of MFFT and several other models. The comparison results are shown in Table 3. It is observed that MFFT comments results better.

Table 3. Examples of code comments generated by each method

```
public static void rename (int id, String path, String newPath, AsyncCallback
<VoidResult>cb) {
    Input in = Input.create ();
    in.oldPath (path);
    in.newPath (newPath);
    ChangeApi.edit (id).post (in, cb);
}
```

w/o MTL: rename the file on external storage.

w/o STF: rename the file on pending edit.

MFFT: rename the file in the pending edit.

Reference: rename a file in the pending edit.

```
private void notifyListeners (String key, Object newValue) {
    List <ModuleCommunicationListener>list;
    if (this.listeners == null) {
        return;
    }
    list = this.listeners.get (key);
    if (list == null) {
        return;
    }
    for (ModuleCommunicationListener mcl : list) {
        mcl.moduleValueChanged (key, newValue);
    }
}
```

w/o MTL: notifies all listeners of the update.

w/o STF: notifies all listeners that have subscribed a key.

MFFT: notifies all listeners that have subscribed a given key.

Reference: notifies all listeners that have subscribed to the given key.

3.6   Head-Size Analysis

MFFT model is designed based on Transformer architecture, where encoder and decoder are mainly composed of multi-head attention and a fully connected network. Compared with other attention mechanisms, the heads in multi-head attention are its unique part, which has an important impact on the performance of the model. To explore the role of heads in MFFT, we fixed other hyperparameters in the experiment and set the number of heads as {2, 4, 6, 8, 10}. Table 4 presents the results of automatic metric evaluation. From the experimental results, it is observed that the metrics show a general trend of increasing and then decreasing. When heads is set to 8, the model obtains the highest metric score. As the number of heads increases, the performance metric scores show a downward trend. This indicates that multi-head attention is not that the number of heads is as good as possible. The reason for this result may be that increasing the number of heads can extract features from different angles, but setting too many heads will cause the dispersion of attention and destroy the correlation between data.

Table 4. Comparison of different heads on the Java dataset

Approach	Head-size	S-BLEU(%)	C-BLEU(%)	METEOR(%)	ROUGE-L(%)
	2	30.12	23.75	20.63	43.25
	4	32.03	23.95	20.87	43.14
MFFT	6	32.53	24.17	21.04	43.68
	8	**33.24**	**24.89**	**21.68**	**44.57**
	10	32.15	23.86	20.69	43.20

4   Conclusion and Future Work

In this paper, we propose a neural network approach called MFFT based on the Transformer architecture, which aims to generate code summaries by effectively learning the semantic information of source code. In terms of code structure, this method uses Multi-way Tree-LSTM process the AST and extract the code structure information effectively. In addition, to integrate the token features of source code with its semantic features, we propose a feature fusion method to obtain more structural information and semantic correlations between source code tokens. The effectiveness of the method is validated through experiments. In future work, we will consider introducing more information such as API knowledge to learn better semantics of programs. Additionally, we can consider adopting advanced deep learning techniques to enhance the model and thus improve the performance of code summarization.

References

1. He, H.: Understanding source code comments at large-scale. In: Proceedings of the 2019 27th ACM Joint Meeting on European Software Engineering Conference and Symposium on the Foundations of Software Engineering, pp. 1217–1219 (2019)
2. Zhou, W., Wu, J.: Code comments generation with data flow-guided transformer. In: Zhao, X., Yang, S., Wang, X., Li, J. (eds.) Web Information Systems and Applications. WISA 2022. LNCS, vol. 13579, pp. 168–180. Springer, Cham (2022). https://doi.org/10.1007/978-3-031-20309-1_15
3. Hu, X., Li, G., Xia, X., Lo, D., Jin, Z.: Deep code comment generation. In: Proceedings of the 26th International Conference on Program Comprehension (ICPC), pp. 200–210. IEEE (2018)
4. Haiduc, S., Aponte, J., Moreno, L., Marcus, A.: On the use of automated text summarization techniques for summarizing source code. In: 2010 17th Working Conference on Reverse Engineering, pp. 35–44. IEEE (2010)
5. Allamanis, M., Peng, H., Sutton, C.: A convolutional attention network for extreme summarization of source code. In: International Conference on Machine Learning, pp. 2091–2100. PMLR (2016)
6. Loyola, P., Marrese-Taylor, E., Matsuo, Y.: A neural architecture for generating natural language descriptions from source code changes. arXiv preprint arXiv:1704.04856 (2017)
7. Hu, X., Li, G., Xia, X., Lo, D., Jin, Z.: Deep code comment generation with hybrid lexical and syntactical information. Empir. Softw. Eng. **25**, 2179–2217 (2020)
8. Tai, K.S., Socher, R., Manning, C.D.: Improved semantic representations from tree-structured long short-term memory networks. arXiv preprint arXiv:1503.00075 (2015)
9. Shido, Y., Kobayashi, Y., Yamamoto, A., Miyamoto, A., Matsumura, T.: Automatic source code summarization with extended Tree-LSTM. In: 2019 International Joint Conference on Neural Networks (IJCNN), pp. 1–8. IEEE (2019)
10. Iyer, S., Konstas, I., Cheung, A., Zettlemoyer, L.: Summarizing source code using a neural attention model. In: 54th Annual Meeting of the Association for Computational Linguistics 2016, pp. 2073–2083. ACL (2016)
11. Zheng, W., Zhou, H.Y., Li, M., Wu, J.: Code attention: translating code to comments by exploiting domain features. arXiv preprint arXiv:1709.07642 (2017)
12. Ahmad, W.U., Chakraborty, S., Ray, B., Chang, K.W.: A transformer-based approach for source code summarization. arXiv preprint arXiv:2005.00653 (2020)
13. Wan, Y., et al.: Improving automatic source code summarization via deep reinforcement learning. In: Proceedings of the 33rd ACM/IEEE International Conference on Automated Software Engineering, pp. 397–407 (2018)
14. Papineni, K., Roukos, S., Ward, T., Zhu, W.J.: BLEU: a method for automatic evaluation of machine translation. In: Proceedings of the 40th Annual Meeting of the Association for Computational Linguistics, pp. 311–318 (2002)
15. Banerjee, S., Lavie, A.: Meteor: an automatic metric for MT evaluation with improved correlation with human judgments. In: Proceedings of the ACL Workshop on Intrinsic and Extrinsic Evaluation Measures for Machine Translation and/or Summarization, pp. 65–72 (2005)
16. Lin, C.Y.: ROUGE: a package for automatic evaluation of summaries. In: Text Summarization Branches Out, pp. 74–81 (2004)

17. Lin, C., Ouyang, Z., Zhuang, J., Chen, J., Li, H., Wu, R.: Improving code summarization with block-wise abstract syntax tree splitting. In: 2021 IEEE/ACM 29th International Conference on Program Comprehension (ICPC), pp. 184–195. IEEE (2021)
18. LeClair, A., Jiang, S., McMillan, C.: A neural model for generating natural language summaries of program subroutines. In: 2019 IEEE/ACM 41st International Conference on Software Engineering (ICSE), pp. 795–806. IEEE (2019)
19. Zhang, J., Wang, X., Zhang, H., Sun, H., Wang, K., Liu, X.: A novel neural source code representation based on abstract syntax tree. In: 2019 IEEE/ACM 41st International Conference on Software Engineering (ICSE), pp. 783–794. IEEE (2019)

Combines Contrastive Learning and Primary Capsule Encoder for Target Sentiment Classification

Hang Deng[1], Yilin Li[1], Shenggen Ju[1], and Mengzhu Liu[2(✉)]

[1] College of Computer Science, Sichuan University, Chegndu 610005, China
[2] Computer and Information Engineering College, Guizhou University of Commerce, Guiyang 550014, China
136997915@qq.com

Abstract. Target Sentiment Classification (TSC) aims to judge the sentiment polarity of the specific target appearing in a sentence. Most of the existing TSC algorithms use Recurrent Neural Network (RNN) to encode and model sentences, which can mine the semantic features of sentences, but there are still some short-comings. RNN cannot fully capture long-distance semantic information, nor can it perform parallel processing calculations. At present, some research attempts to solve the problems of RNN as an encoder, but the generalization ability and prediction ability of these models are low, and there are certain limitations. In view of the above problems, this paper proposes a PC-SCL model that combines contrastive learning and primary capsule encoder. The primary capsule encoder network is designed to extract the hidden state of the word vector of the embedding layer, which can parallel calculate and fully extract and integrate the sentiment features between context and target. In addition, the model uses supervised contrastive learning, which enables the model to extract feature representations more accurately, and improves the generalization and prediction capabilities of the model. The model is tested on three general datasets, and the experimental results prove the effectiveness of the proposed model.

Keywords: target sentiment classification · primary capsule · supervised contrastive learning · RNN

1 Introduction

Text sentiment classification tasks are often differentiated based on the granularity, including document-level sentiment classification, sentence-level sentiment classification, and target sentiment classification. Document-level and sentence-level sentiment classification can only identify the overall sentiment tendency of a text and belong to coarse-grained sentiment classification. People often express their opinions from multiple perspectives, as a result, the text exhibits sentiment diversity. Target sentiment classification tasks can also be referred to as aspect-level sentiment classification tasks, where the objective is to determine the sentiment expressed by a given entity in a given

© The Author(s), under exclusive license to Springer Nature Singapore Pte Ltd. 2023
L. Yuan et al. (Eds.): WISA 2023, LNCS 14094, pp. 284–296, 2023.
https://doi.org/10.1007/978-981-99-6222-8_24

context. This type of classification belongs to fine-grained sentiment classification. For example, in the sentence "Comfortable bed but the ornament is really ugly.", the sentiment polarity of "bed" is positive, for the target word "ornament", the sentiment polarity is negative.

Methods for TSC can be categorized into three types: sentiment lexicon-based methods, machine learning-based methods, and deep learning-based methods. Sentiment Lexicon-based methods determine the sentiment category of a text based on lexicon and certain computational rules [1]. Cruz et al. [2] constructed domain-specific sentiment lexicons for specific classification tasks, which significantly improved the accuracy of sentiment classification. Zhu et al. [3] proposed a segmentation model that divides a sentence into multiple short phrases, with each phrase containing one target. Then, they analyzed the sentiment tendency of each short phrase using the sentiment lexicon. However, with the development of the internet, a vast number of novel internet words emerge, and the sentiment polarity of some words may change.

Machine learning-based methods have also achieved promising results in early target sentiment classification tasks. As early as 2002, Pang et al. [4] applied machine learning techniques to sentiment classification tasks. They used three machine learning methods, namely Naive Bayes, Maximum Entropy, and Support Vector Machine, to classify movie review data. Jiang et al. [5] used external syntactic parse trees to obtain multiple target-related dependency features and performed target sentiment classification based on these features. Kiritchenko et al. [6] utilized the Support Vector Machine method in their model, combined with N-gram models, lexicon, and other features for TSC, achieving good results. However, the identification of sentiment polarity in machine learning methods relies on the quality of the constructed feature engineering, and it fail to capture the complete sentiment information.

Deep learning-based methods have achieved better results in TSC. Dong et al. [7] first applied the adaptive RNN model to TSC, using syntactic parsing to extract semantic information between context and target words. Wang et al. [8] proposed the RNN-Capsule model, which utilizes an RNN model to obtain hidden vectors containing semantic information. The data of different sentiment are separated and trained separately in corresponding capsules, achieving good classification results. However, RNN models suffer from gradient vanishing and gradient exploding problems. Sundermeyer et al. [9] proposed the Long Short-Term Memory (LSTM) model, which uses gated mechanisms and memory units to enhance the network's memory capacity, alleviating the gradient issues in RNN. Tang et al. [10] introduced the TD-LSTM model and the TC-LSTM model, the TD-LSTM model uses two LSTM to model the left and right halves to obtain the semantic features of the entire sentence. The TC-LSTM model appends the target feature vector behind the embedding vector and performs average pooling on the target feature vector. Hu et al. [11] combined Convolutinal Neural Network (CNN) and LSTM in their model, allowing it to capture local and contextual features of the text. Hou et al. [12] proposed an self-adaptive context reasoning mechanism, which utilizes multi-channel CNN to extract text features of different lengths. Then, self-attention is used to learn context reasoning between semantic features of different lengths and adaptively adjust the weights of the fused convolution. Chen et al. [13] proposed the RAM model, which uses bidirectional LSTM models to construct context features and then applies multiple attention mechanisms to extract relevant information on the context features, combined with gated recurrent units [14] for sentiment classification. Wang et al. [15] proposed the ATAE-LSTM model, which connects the target word embedding vector with word representations and involves the target word in attention weight calculation. Song et al. [16] proposed an encoding network to replace the RNN as the encoder, but this method has lower generalization and prediction capabilities. The above-mentioned models rely on RNN and their variants as encoders to calculate the hidden states vector of the text. However, these sequential models have limited ability to capture long-distance semantic dependencies. Additionally, sequential models are difficult to parallelize, leading to high memory and time consumption during model training.

This paper proposes a model called the PC-SCL model. PC-SCL first utilizes BERT to obtain embedding of the target word and the context. In embedding layer, supervised contrastive learning (SCL) can bring samples with the same label closer together in the semantic space, while pushing samples with different labels farther apart, thereby enhancing the ability of the embedding layer to capture sentiment features in the context. The primary capsule (PC) encoding network is applied to perform semantic modeling on the embedding layer, addressing the limitation of RNN models as encoders, which cannot perform parallel computation. This module includes multi-head attention and primary capsule modules. The multi-head attention mechanisms enable to capture global information features in different subspaces and assigning effective weights. The PC modules can extract key sentiment features in the context that are more contributive and valuable. Finally, the semantic features are fused to achieve TSC.

The main contributions of this paper are as follows: (1) Designing the primary capsule encoding network as encoder, which enables parallelized computation, extracts and fuses sentiment features between the context and the target word. (2) Introducing supervised contrastive learning to enhance the accuracy of embedding feature extraction, thereby assisting in better performing classification tasks.

2 Methodology

2.1 Task Description

The objective of TSC can be mathematically defined as follows: Given a sequence of sentences $S = \{w_1, w_2, \cdots, w_{t-1}, w_t, \cdots, w_{t+m}, w_{t+m+1}, \cdots, w_n\}$, where the target sequence is $T = \{w_t, \cdots, w_{t+m}\}$, and the context sequence is $C = \{w_1, w_2, \cdots, w_{t-1}, w_{t+m+1}, \cdots, w_n\}$, the objective of sentiment classification task is to determine the sentiment polarity P of the target T based on the semantic information of its context C.

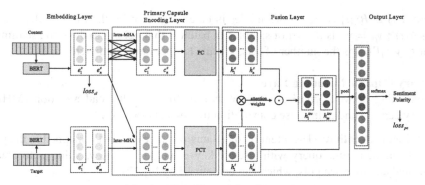

Fig. 1. PC-SCL model architecture.

2.2 PC-SCL Model Architecture

Figure 1 illustrates the overall architecture of the proposed model in this paper, which consists of an embedding layer, a primary capsule encoding layer, a fusion layer, and an output layer. In the following, each layer will be introduced separately.

Embedding Layer

In this paper, the pre-trained model BERT is used to obtain word embedding vectors. The input format of context and target are transformed into "[CLS] + context + [SEP]" and "[CLS] + target + [SEP]". This process produces context word vectors $e^c = \{e_1^c, e_2^c, \cdots, e_m^c\}$ and target word vectors $e^t = \{e_1^t, e_2^t, \cdots, e_k^t\}$.

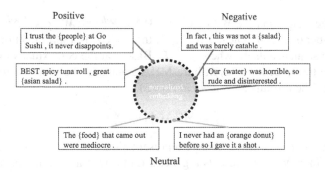

Fig. 2. Schematic diagram of supervised contrastive learning.

In this paper, SCL [17] is employed to capture sentiment-related features in the context. As shown in Fig. 2, samples belonging to the same label are distributed closely in the embedding space, while samples with different labels are farther apart. The supervised contrastive loss for batch B is computed based on the context embedding and the sentiment polarity, as shown in Eq. (1).

$$loss_{cl} = \sum_{i \in B} -\frac{1}{|P(i)|} \sum_{p \in P(i)} \log \frac{\exp(e_i^c \cdot e_p^c / \tau)}{\sum_{a \in A(i)} \exp(e_i^c \cdot e_a^c / \tau)} \quad (1)$$

where $A(i) \equiv B\backslash\{i\}$ is the set of samples in batch B except for the i-th sample, $P(i) \equiv \{p \in A(i) : y_p = y_i\}$ is the set of samples in batch B that belong to the same sentiment polarity y_i, $|P(i)|$ is the number of sets, τ is the temperature coefficient.

Primary Capsule Encoding Layer

PC encoding network consists of three sub-modules: Multi-head attention (MHA), primary capsule and point-wise convolution transformation (PCT).

Multi-head Attention: The attention mechanism can be abstractly mapped as the process of interacting the query with a set of key-value pairs < Key, Value > to obtain additional attention weights, which can be defined by Eq. (2).

$$\text{attention}(Q, K, V) = \text{softmax}(\frac{QK^T}{\sqrt{d_k}})V \tag{2}$$

where Q is the query matrix, K is the key matrix, V is the value matrix, and $\sqrt{d_k}$ acts as a scale factor to ensure that the dot product does not become too large.

MHA was proposed by Vaswani et al. [18]. In this mechanism, Q, K, V undergo different linear transformations with a distinct parameter matrix W. Then the similarity between Q and K is computed. Finally, the results of different heads are concatenated. The formulas are given in Eqs. (3) and (4).

$$head_i = \text{attention}(QW^q, KW^k, VW^v) \tag{3}$$

$$\text{MHA}(Q, K, V) = \text{concat}(head_1, \cdots head_h) \tag{4}$$

where W is the weight matrix, $W^q \in \mathbb{R}^{d_h \times d_q}$, $W^k \in \mathbb{R}^{d_h \times d_k}$, $W^v \in \mathbb{R}^{d_h \times d_v}$ and h represents the number of attention heads. In Fig. 1, Intra-MHA is a self-attention mechanism applied to the sequence itself, where $Q = K$. Based on the contextual word vectors e^c obtained from the embedding layer, Intra-MHA calculates the semantic representation of the context. The formula are shown in Eq. (5).

$$c_i^c = \text{MHA}(e_i^c, e_i^c, e_i^c) \tag{5}$$

where c^c is the context semantic feature representation, $c^c = \{c_1^c, c_2^c, \cdots, c_i^c, \cdots, c_n^c\}$.

In Fig. 1, Inter-MHA represents a general multi-head attention mechanism where Q is not equal to K. In this module, Q corresponds to the contextual word vectors e^c, while K corresponds to the target word vectors e^t. The fused semantic features c^t of the context and target can be calculated using Eq. (6).

$$c_i^t = \text{MHA}(e_i^c, e_i^t, e_i^t) \tag{6}$$

The attention mechanism allows for direct comparisons between text sequences and can capture global text information. To further capture local feature information, the primary capsule encoding layer introduces the point-wise convolution transformation and primary capsule modules to explore deeper local features in the text.

Point-wise Convolution Transformation: Since the target is typically a single word or a few words, it may not contain rich semantic information. Therefore, a simple PCT is employed to transform the collected target information. Given the input sequence h, the formula for PCT is shown as Eq. (7).

$$\text{PCT}(h) = \sigma(h * W_{pc}^1 + b_{pc}^1) * W_{pc}^2 + b_{pc}^2 \tag{7}$$

where σ is Relu activation function, $*$ refers to the convolution operation, $W_{pc}^1 \in \mathbb{R}^{d_{hid} \times d_{hid}}$, $W_{pc}^2 \in \mathbb{R}^{d_{hid} \times d_{hid}}$ are the weight matrix of the two convolution kernels. $b_{pc}^1 \in \mathbb{R}^{d_{hid}}$, $b_{pc}^2 \in \mathbb{R}^{d_{hid}}$ are the bias term of the two convolution kernels, d_{hid} is the dimension of the hidden state vector.

The target word vectors are transformed into context-aware target features c^t after passing through the Inter-MHA. Then, through the PCT module, the hidden state of the target word $h^t = \{h_1^t, h_2^t, \cdots, h_m^t\}$ can be obtained. The formula is given as Eq. (8).

$$h^t = \text{PCT}(c^t) \tag{8}$$

Primary Capsule: The context contains rich semantic and sentiment information. Inspired by Zhao et al. [19], this paper's model adopts the PC to extract semantic features from the context. Its structural diagram is illustrated in Fig. 3. PC differ from ordinary neurons in that they perform internal calculations on the input and encapsulate scalar outputs into tensors that contain rich information. Tensors have magnitude and direction, enabling them to capture more features of the text.

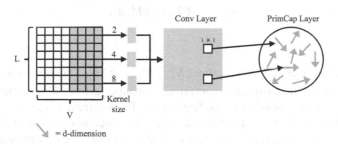

Fig. 3. Overall view of the primary capsule.

In the PC module, features are first extracted from the contextual embedding using convolutional operations, as shown in Eq. (9).

$$m_i = \sigma(W^a \circ X_{i:i+l-1}^T) \tag{9}$$

where $X \in \mathbb{R}^{l \times v}$ is a stacked v-dimension vector matrix with l marked input features, and $W^a \in \mathbb{R}^{l \times k}$ is a convolution filter of width k. Applying the filter to a local region $X_{i:i+k-1}^T \in \mathbb{R}^{k \times l}$ to generate a feature m_i, the symbol $\circ$ represents the multiplication of elements, σ is the nonlinear activation function. Then, through sliding filters, all the features are collected into a feature map $(m_1, \cdots, m_{(v-k+1)/2})$. To increase the diversity of feature extraction, multiple feature maps extracted by three filters with different window sizes (2, 4, 8) are concatenated and passed to the PC layer. Subsequently, group convolutional operations are applied to transform the feature maps into PC. Each scalar m_i in the feature map is transformed into a capsule p_i using d group 1×1 filters, denoted as $W^b = \{w_1, \cdots, w_d\} \in \mathbb{R}^d$, as shown in Eq. (10).

$$p_i = g(p_{i1} \oplus p_{i2} \oplus \cdots \oplus p_{id}) \in \mathbb{R}^d \tag{10}$$

where $p_i = m_i \cdot w_j \in \mathbb{R}$, $\oplus$ is the concatenation operator. g is a squashing function. The length of each capsule p_i represents the probability that it will be useful for the current task, and the length of the capsule is limited to between unit size [0,1] by the function g, and the squashing function formula is shown in Eq. (11).

$$g(x) = \frac{\|x\|^2}{1 + \|x\|^2} \frac{x}{\|x\|} \tag{11}$$

The contextual word vector e^c of the embedding layer can get the context semantic feature c^c after Intra-MHA, and then get the context hidden state $h^c = \{h_1^c, h_2^c, \cdots, h_n^c\}$ after the PC module.

Fusion Layer
The hidden state of the target word and the context are obtained in the encoding layer, then, another MHA function is used to obtain target-specific contextual semantic features $h^{tsc} = \{h_1^{tsc}, h_2^{tsc}, \cdots, h_i^{tsc}, \cdots, h_m^{tsc}\}$,the formula is shown in Eq. (12).

$$h_i^{tsc} = \text{MHA}(h_i^c, h_i^t, h_i^t) \tag{12}$$

Output Layer
First, the average feature representations h_{avg}^{tsc}, h_{avg}^c, and h_{avg}^t are obtained by applying average pooling to target-specific contextual semantic features h^{tsc}, context semantic features h^c, and target word semantic features h^t. These three average feature representations are then concatenated to obtain the final integrated representation h^o, as shown in Eq. (13). Next, a fully connected layer is used to project the concatenated vector into the target polarity space, as shown in Eq. (14). Finally, a softmax classifier is used to calculate the probabilities of target sentiment polarities, as shown in Eq. (15).

$$h^o = [h_{avg}^{tsc}; h_{avg}^c; h_{avg}^{ct}] \tag{13}$$

$$x = (W^o)^T \cdot h^o + b^o \tag{14}$$

$$y = \text{softmax}(x) = \frac{\exp(x)}{\sum_{k=1}^C \exp(x)} \tag{15}$$

where $W^o \in \mathbb{R}^{1 \times C}$ is the weight matrix, $b^o \in \mathbb{R}^C$ is the bias term, C is the number of sentiment polarity, $y \in \mathbb{R}^C$ is the distribution of predicted sentiment polarity.

The optimization objective function of the output layer is the regularized cross entropy loss function, defined as Eq. (16), and the back propagation algorithm is used to update the parameters and weights of the model.

$$loss_{pc} = L(\theta) = -\sum_{i=1}^{C} \hat{y} log(y_i) + \lambda \sum_{\theta \in \Theta} \theta^2 \tag{16}$$

where i represents the i-th sample, y_i is the predicted sentiment distribution, $\hat{y}$ is the one-hot vector of the target true sentiment, λ is the coefficient of the adjustment term, Θ is the parameter set.

The final loss function of the model is formula (17), where α, β are the learnable parameter, the initial value is 1.5 and 1, $loss_{cl}$ is the loss function of SCL.

$$loss_{sum} = \alpha \cdot loss_{pc} + \beta \cdot loss_{cl} \tag{17}$$

3 Experiment and Analysis

3.1 Dataset Information

PC-SCL was evaluated based on the restaurant review dataset (14Rest) and the laptop review dataset (14Lap) from SemEval 2014 Task 4 [20], and the restaurant review dataset (15Rest) from SemEval 2015 Task 12 [21]. The detailed are shown in Table 1.

Table 1. Experimental Dataset Statistics.

Category	14Rest		14Lap		15Rest	
	Train	Test	Train	Test	Train	Test
Positive	2164	728	994	341	808	340
Neutral	637	196	464	169	29	28
Negative	807	196	870	128	228	195
Total	3608	1120	2328	638	1065	563

3.2 Experimental Environment and Parameter Settings

This paper conducted experiments using pytorch deep learning framework. The programming language was python 3.6, the operating system was Ubuntu. The GPU was GeForce RTX3090. The temperature coefficient for SCL was set to 0.5. The hidden state dimension was set to 300. The learning rate was set to 2×10^{-5}. The batch size for training was set to 16. The dropout rate was set to 0.3. The regularization term λ was set to 1×10^{-5}. The Adam optimizer [22] was used to update parameters.

Table 2. Results from Comparative Experiment.

Models	14Rest(%)		14Lap(%)		15Rest(%)	
	Acc	Macro-F1	Acc	Macro-F1	Acc	Macro-F1
ATAE-LSTM [15]	77.20	-	68.70	-	-	-
IAN [23]	79.26	70.09	72.05	67.38	78.54	52.65
TNet-LF [24]	80.42	71.03	74.61	70.14	78.47	59.47
TNet-ATT [25]	81.53	72.90	77.62	73.84	-	-
ASCNN [26]	81.73	73.10	72.62	66.72	78.48	58.90
ASGCN [26]	80.86	72.19	74.14	69.24	79.34	60.78
MCRF-SA [27]	82.86	73.78	77.64	74.23	80.82	61.59
BERT-SPC [16]	84.46	76.98	78.99	75.03	-	-
AEN-BERT [16]	83.12	73.76	79.93	**76.31**	83.66#	58.58#
SK-GCN [28]	83.48	75.19	79.00	75.57	83.20	66.78
CF-CAN [29]	84.48	**77.48**	79.44	75.37	-	-
PC-SCL	**84.55**	75.94	**79.98**	75.40	**84.01**	**67.76**

3.3 Comparison of Models

To further illustrate the superiority of the model presented in this paper, the model has been compared with some typical Baseline models and mainstream models, Table 2 shows the results. The "-" indicates experiment results were not reported in the original papers. The "#" indicates results reproduced in our experimental environment.

3.4 Analysis of Comparative Experimental Results

As shown in Table 2, accuracy of our model obtained in the 14 Rest, 14Lap and 15Rest data sets are 84.55%, 79.98% and 84.01%, and F1 are 75.94%, 75.40% and 67.76%.

Compared to the AEN-BERT model, PC-SCL model achieves better performance in all dataset except for a slight decrease of 0.91% in F1 on the 14Lap dataset. Notably, the F1 on the 15Rest dataset is improved by 9.18%. This improvement can be attributed to the SCL employed in the proposed model, which enhances the model's predictive ability and generalization capability.

Compared to the SK-GCN model, the proposed model in this study achieves higher performance in terms of all metrics except for a slight decrease of 0.17% in F1 on the 14Lap dataset. This difference can be attributed to the fact that the SK-GCN model incorporates external dependency syntactic tree information, which enhances the model's performance. In contrast, the proposed model is trained only on the dataset without utilizing any additional external information.

Compared to the CF-CAN model, PC-SCL achieves a slight increase of 0.07% in accuracy but a decrease of 1.54% in F1 on the 14Rest dataset. Upon analyzing the dataset, it is observed that the average sentence length in the 14Res dataset is 20.05, which is shorter than the 22.06 of that in the 14Lap dataset. As a result, PC-SCL did not fully leverage the advantages of the PC encoding network in handling long-distance text.

In summary, the proposed PC-SCL model achieves the best results on most of the datasets, demonstrating the effectiveness of the model proposed in this study.

3.5 Analysis of Ablation Experiments

Table 3. Results from Ablation Experiment.

Models	14Rest(%)		14Lap(%)		15Rest(%)	
	Acc	Macro-F1	Acc	Macro-F1	Acc	Macro-F1
w/o both	80.63	69.84	77.12	73.43	83.66	58.58
w/o PC	81.61	70.53	77.59	73.14	82.42	62.28
w/o SCL	83.13	74.13	79.09	75.20	83.01	60.68
PC-SCL	**84.55**	**75.94**	**79.98**	**75.40**	**84.01**	**67.76**

Three ablation models are designed to reveal the effectiveness of component modules: w/o PC (primary capsule), w/o SCL (supervised contrastive learning) and w/o both.

The w/o PC model replaced the PC module with the PCT module to verify the effectiveness of the PC module. The w/o SCL model removed the SCL module to verify the effectiveness of SCL. In the w/o both model, SCL module is removed on the basis of the w/o PC model.

Table 3 reports the results of the ablation models. From the w/o both model, it can be observed that after replacing the PC module and removing the SCL module, the accuracy of three datasets are lower than those of the PC-SCL model by 3.92%, 2.86%, and 0.35%, the F1 are lower by 6.1%, 1.97%, and 9.18%. From the w/o PC model, the accuracy are lower by 2.94%, 2.39%, and 1.59%, and the F1 are lower by 5.41%, 2.26%, 5.48%. From the w/o SCL model, the accuracy on the three datasets are lower by 1.42%, 0.89%, and 1%, and the F1 are lower by 1.81%, 0.2%, and 7.08%. Therefore, it can be concluded that both the PC module and the SCL module adopted in this study contribute to the TSC task.

3.6 Analysis of the Number of Multi-Head Attention Heads

In order to investigate the impact of numbers of attention heads, we evaluating the model's performance with varying numbers of attention heads. The experimental results are shown in Figs. 4 and 5.

From Fig. 4, it can be observed that when the number of heads is 8, the model achieves the highest accuracy on the three datasets, with accuracy of 84.55%, 79.98%, and 84.01%. From Fig. 5, it can be seen that when the number of heads is 8, the model achieves the highest F1 on the 14Lap and 15Rest datasets, with scores of 75.40% and

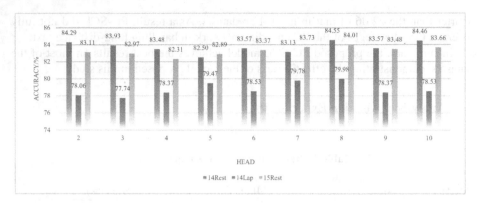

Fig. 4. The accuracy of the model under different heads.

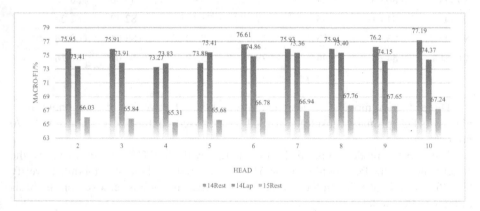

Fig. 5. The F1 of the model under different heads.

67.76%. When the number of heads is 10, the model achieves the highest F1 score of 77.19% on the 14Rest dataset.

4 Conclusion

This paper proposes the PC-SCL model, which combines SCL and PC encoder. PC encoder enable parallel computation, extracting and integrating semantic features between context and target. SCL enable to obtain contrastive learning loss for embedding layer. The final experimental results validate the effectiveness of PC-SCL. However, the model still has some areas for improvement, which will be addressed in future work.

First, to comprehensively capture implicit sentiment features in text, further research and experimentation can be conducted on how to mine implicit sentiment.

Second, this paper does not utilize external knowledge, such as syntactic dependency trees. In future work, research can be conducted on reducing the noise in the generated dependency relationships after introducing syntactic dependency trees.

References

1. Taboada, M., Brooke, J., Tofiloski, M., et al.: Lexicon-based methods for sentiment analysis. Comput. Linguist. **37**(2), 267–307 (2011)
2. Cruz, F.L., Troyano, J.A., Enríquez, F., et al.: 'Long autonomy or long delay?' The importance of domain in opinion mining. Expert Syst. Appl. **40**(8), 3174–3184 (2013)
3. Zhu, J., Wang, H., Tsou, B.K., et al.: Multi-aspect opinion polling from textual reviews. In: Proceedings of the 18th ACM Conference on Information and Knowledge Management, pp. 1799–1802 (2009)
4. Pang, B., Lee, L., Vaithyanathan, S.: Thumbs up? Sentiment classification using machine learning techniques. In: Proceedings of the ACL-02 Conference on Empirical Methods in Natural Language Processing, vol. 10, pp. 79–86 (2002)
5. Jiang, L., Yu, M., Zhou, M., et al.: Target-dependent twitter sentiment classification. In: Proceedings of the 49th annual meeting of the association for computational linguistics: human language technologies, pp. 151–160 (2011)
6. Kiritchenko, S., Zhu, X., Cherry, C., et al.: Nrc-canada-2014: Detecting aspects and sentiment in customer reviews. In: Proceedings of the 8th International Workshop on Semantic Evaluation (SemEval 2014), pp. 437–442 (2014)
7. Dong, L., Wei, F., Tan, C., et al.: Adaptive recursive neural network for target-dependent twitter sentiment classification. In: Proceedings of the 52nd annual meeting of the association for computational linguistics (volume 2: Short papers), pp. 49–54 (2014)
8. Wang, Y., Sun, A., Han, J., et al.: Sentiment analysis by capsules. In: Proceedings of the 2018 world wide web conference, pp. 1165–1174 (2018)
9. Sundermeyer, M., Schlüter, R., Ney, H.: LSTM neural networks for language modeling. In: Thirteenth annual conference of the international speech communication association (2012)
10. Tang, D., Qin, B., Feng, X., et al.: Effective LSTMs for target-dependent sentiment classification. In: Proceedings of COLING 2016, the 26th International Conference on Computational Linguistics: Technical Papers, pp. 3298–3307 (2016)
11. Hu, Z., Zhao, X.: Sentiment analysis based on word vector technology and hybrid neural network . Appl. Res. Comput. **35**(12), 42–45+60 (2018)
12. Hou, S., Zhao, X., Liu, N., Shi, X., Wang, Y., Zhang, G.: Self-adaptive context reasoning mechanism for text sentiment analysis. In: Zhao, X., Yang, S., Wang, X., Li, J. (eds.) Web Information Systems and Applications: 19th International Conference, WISA 2022, Dalian, China, September 16–18, 2022, Proceedings, pp. 194–205. Springer International Publishing, Cham (2022). https://doi.org/10.1007/978-3-031-20309-1_17
13. Chen, P., Sun, Z., Bing, L., et al.: Recurrent attention network on memory for aspect sentiment analysis. In: Proceedings of the 2017 conference on empirical methods in natural language processing, pp. 452–461 (2017)
14. Cho, K., van Merriënboer, B., Gülçehre, Ç,, et al.: Learning Phrase Representations using RNN Encoder–Decoder for Statistical Machine Translation. In: Proceedings of the 2014 Conference on Empirical Met-hods in Natural Language Processing, pp. 1724–1734 (2014)
15. Wang, Y., Huang, M., Zhu, X., et al.: Attention-based LSTM for aspect-level sentiment classification. In: Proceedings of the 2016 conference on empirical methods in natural language processing, pp. 606–615 (2016)
16. Song, Y., Wang, J., Jiang, T., Liu, Z., Rao, Y.: Targeted sentiment classification with attentional encoder network. In: Tetko, I.V., Kůrková, V., Karpov, P., Theis, F. (eds.) ICANN 2019. LNCS, vol. 11730, pp. 93–103. Springer, Cham (2019). https://doi.org/10.1007/978-3-030-30490-4_9
17. Khosla, P., Teterwak, P., Wang, C., et al.: Supervised contrastive learning. Adv. Neural. Inf. Process. Syst. **33**, 18661–18673 (2020)

18. Vaswani, A., Shazeer, N., Parmar, N., et al.: Attention is all you need. In: Advances in Neural Information Processing Systems, pp. 5998–6008 (2017)
19. Zhao, W., Peng, H., Eger, S., et al.: Towards scalable and reliable capsule networks for challenging NLP applications. arXiv preprint arXiv:1906.02829 (2019)
20. Pontiki, M., Galanis, D., Pavlopoulos, J., et al.: Semeval-2014 task 4: Aspect based sentiment analysis . In:Proceedings of the 8th International Workshop on Semantic Evaluation (SemEval 2014), pp. 27–35 (2014)
21. Pontiki, M., Galanis, D., Papageorgiou, H., et al.: Semeval-2015 task 12: Aspect based sentiment analysis. In: Proceedings of the 9th International Workshop on Semantic Evaluation (SemEval 2015), pp. 486–495 (2015)
22. Kingma, D.P., Ba, J.A.: A method for stochastic optimization. arXiv preprint arXiv:1412. 6980 (2014)
23. Yang, M., Tu, W., Wang, J., et al.: Attention based LSTM for target dependent sentiment classification. In: Proceedings of the AAAI Conference on Artificial Intelligence, vol. 31, issue 1 (2017)
24. Li, X., Bing, L., Lam, W., et al.: Transformation Networks for Target-Oriented Sentiment Classification. In: Proceedings of the 56th Annual Meeting of the Association for Computational Linguistics, vol. 1: Long Papers, pp. 946–956 (2018)
25. Tang, J., Lu, Z., Su, J., et al.: Progressive self-supervised attention learning for aspect-level sentiment analysis. In: Proceedings of the 57th Annual Meeting of the Association for Computational Linguistics, pp. 557–566 (2019)
26. Zhang, C., Li, Q., Song, D.: Aspect-based sentiment classification with aspect-specific graph convolutional networks. In: Proceedings of the 2019 Conference on Empirical Methods in Natural Language Processing and the 9th International Joint Conference on Natural Language Processing (EMNLP-IJCNLP), pp. 4568–4578 (2019)
27. Xu, L., Bing, L., Lu, W., et al.: Aspect based sentiment analysis with aspect-specific opinion spans. In: Proceedings of the 2020 Conference on Empirical Methods in Natural Language Processing (EMNLP), pp. 3561–3567 (2020)
28. Zhou, J., Huang, J.X., Hu, Q.V., et al.: SK-GCN: modeling syntax and knowledge via graph convolutional Network for aspect-level sentiment classification. Knowl.-Based Syst. **205**(3), 106292 (2020)
29. Cheng, L.C., Chen, Y.L., Liao, Y.Y.: Aspect-based sentiment analysis with component focusing multi-head co-attention networks. Neurocomputing **489**, 9–17 (2022)

An Entity Alignment Method Based on Graph Attention Network with Pre-classification

Wenqi Huang[1], Lingyu Liang[1], Yongjie Liang[2], Zhen Dai[1], Jiaxuan Hou[1], Xuanang Li[1], Xin Wang[2(✉)], and Xin Chen[2]

[1] China Southern Power Grid Digital Grid Research Institute Co., Ltd., Guangzhou, China
`{huangwq,lianglyl,daizhen,houjx}@csg.cn`
[2] Zhejiang University-China Southern Power Grid Joint Research Centre on AI, Zhejiang University, Hangzhou, China
`wangxin2009@zju.edu.cn`

Abstract. Entity alignment is the process of identifying entities that point to the same object in different knowledge graphs. Entity alignment is a key step in building knowledge graphs, and the result of entity alignment directly affects the quality of the knowledge graphs. Most of the current entity alignment methods learn the feature vectors of entities or entity attributes based on representation learning. While the study of the interaction learning between entity attributes and entity relationship features is not sufficient, and the data division is not specific enough to confuse the feature information of different categories of entities. That reduces the quality of the finally learned entity vectors. To solve that problem, we propose an entity alignment method that uses graph attention mechanism after entity classification. Firstly, we classify the entities according to their semantics of the source data, and then classify the entities with different entity attributes to complete the dual entity classification. Based on the dual entity classification, the graph attention mechanism is used to complete the aggregation of the feature information of different categories of entities, which is used to learn the final entity vector representation for the calculation of similarity in the entity alignment task. On the general dataset for entity alignment, DBP15K, our model achieves hits@1 scores 80.82, 80.17 and 93.13 on its three subdatasets. The results show our method are better than the compared methods.

Keywords: Knowledge Graph · Entity Alignment · Knowledge Representation Learning · Graph Attention Network

1 Introduction

In the context of the rapid development of Internet technology, the amount of data in the virtual and real world has shown an explosive growth. How to effectively integrate and apply the massive amount of data has become a research hot spot. Knowledge graph (KG) provide a structured representation of knowledge in the form of a triple. With its powerful semantic representation, the knowledge graph provides an efficient way to manage and utilize the massive [1], heterogeneous and dynamic data in health care, finance and transportation areas.

L. Yuan et al. (Eds.): WISA 2023, LNCS 14094, pp. 297–308, 2023.
https://doi.org/10.1007/978-981-99-6222-8_25

As the application domain becomes more extensive, a single KG may not meet practical requirements. For instance, consider a multilingual Q&A system that needs to handle tasks in different languages and cover multiple areas of expertise. Hence, it is essential to fuse knowledge from different sources or languages to construct a unified, large-scale KG. Entity alignment plays a crucial role in completing knowledge fusion and building a high-quality KG.

Early entity alignment methods were designed based on similarity calculation models, which relied heavily on manually designed features and required different features to be designed for different data and different tasks [2] with high development costs. In recent years, knowledge representation learning has been widely used in entity alignment tasks. The knowledge representation learning model embeds entities into a low-dimensional and dense vector space to learn entity feature vectors that have rich semantic information [3], and the model computes the similarity between these feature vectors to find equivalent entities. Knowledge representation learning reduces the impact of structural and domain differences between different knowledge graphs, and its main goal is to encode entities as vectors, thereby simplifying the process of knowledge inference and improving the efficiency of entity similarity calculation. The translation distance model was the first knowledge representation learning method applied to entity alignment tasks. Which treats the relationship between entities as a translation operation between entity vectors [4]. Specific constraints need to be satisfied between the relationship vector and the vectors of head entity and tail entity, from these constraints the model can learn some semantic relationship features between the entities. Graph neural network (GNN) based entity alignment methods focus on the topology of entities. Graph convolutional network (GCN), which is one kind of CNN, applies convolutional operations to the graph, aggregating features of entities as well as neighboring entities to learn a vector of entities with richer feature information. GNN-based approaches have now become the dominant approach in entity alignment tasks. However, it still have shortcomings, such as insufficient segmentation of data and the duplicate or poorly correlated information are easily mixed in the process of feature aggregation.

In order to address the shortcomings of previous methods, this paper proposes an entity alignment method that uses graph attention mechanism after entity classification. The main contributions of this paper are as follows:

- Classification of data in two steps. In the first step, classification of the source data is based on the different semantic referent objects of the entities, as result four subgraphs are obtained, namely, subgraph of entities of people, subgraph of entities of place, subgraph of entities of organizations, subgraph of entities in the category of others. In the second step, we perform attribute classification embedding based on the above subgraphs.
- For feature interaction learning of different classes of entity attributes, Graph attention network (GAT) is used to calculate the weights of neighboring entities and entities to be aligned to complete feature aggregation of entities.

2 Related Work

2.1 Knowledge Representation Learning

Knowledge graph representation learning is the basis of most current entity alignment methods, where low-dimensional vectors of entities are obtained through learning, and entity similarity calculations are performed through these vectors to obtain potentially aligned entity pairs. In TransE [5], the vectors of entity triples (h, r, t) need to satisfy specific constraints, that $h + r \approx t$, while TransE is less capable of handling complex relationships. To solve this problem, improved models such as TransH [6] and TransR [7] were proposed. TransH maps relational vectors into a hyperplane for modeling. TransR embeds entities of head and tail and relations into two different vector spaces for representation learning.

2.2 Translational Distance Model-Based Method

MTransE [8] uses TransE to learn the vector space of one single knowledge graph and then learns linear transformation to map them to the same vector space. AttrE [9] performs character-level embedding of entity attribute values, and optimizes the entity representation vector based on the scoring function of TransE. The above model uses a single attribute triplet or relationship triplet for representation learning, while the obtained entity vector contains insufficient feature information. Therefore, many models that integrate multiple knowledge are proposed. JAPE [10] uses Skip-grim [11] to learn entity attribute vectors and perform joint embedding of entity structure information and entity attribute information. IPTransE [12] uses the newly discovered equivalent entities to iteratively update the original entities to obtain a higher quality entity vector. TKGE [13] utilizes self-attention mechanisms to aggregate the context of relevant entities, integrating both static structural information and dynamic temporal information. By doing so, TKGE is able to generate high-quality embedding vectors for entities and relations, enabling tasks such as link prediction and entity alignment.

2.3 Graph Neural Network-Based Method

GCN-Align [14] uses GCN learning to obtain the feature vectors of entity structure and entity attributes, and then weights and sums the two to obtain the final entity vector representation. MHGCN [15] considers entity alignment in terms of entity semantics, relationship semantics and entity attributes. It also weights and fuses multiple entity views based on importance to obtain better entity embeddings. SG-CIM [16] can efficiently calculate the association degree between entities to achieve comparison and alignment between entities. The construction of entity relationship knowledge graphs by RDGCN [17] involves utilizing the original knowledge graphs to derive an entity vector and subsequently computing a relationship vector to be added to the feature aggregation of the entity.

In AVR-GCN [18], the graph convolution operation acts both on entities and relations and learns simultaneously to obtain a representation of entities and relations. AliNet [19] optimizes the aggregation method of domain features by considering the local structure information among multi-level neighbor entities.

The attention mechanism was first proposed in the field of natural language processing to focus the model on more important information and to improve the computational efficiency and robustness of the model. The GAT introduces the attention mechanism into the GNN, which makes up for the defect that GCN cannot exactly represent the importance of neighboring nodes. GAT uses the attention mechanism to replace the conventional graph convolution operation, and learns different attention weights for each neighbor node to achieve effective aggregation of neighbor node features. Jiang S et al. [20] proposed a method that combines graph attention with translation models. This method uses graph attention mechanisms to propagate information from neighboring nodes and integrates relationship information into entity representations. The method combines graph attention mechanisms with knowledge translation representation models to maintain consistency between different knowledge graphs and ensure accuracy of embedding vectors in low-dimensional space.

3   Model Structure Design

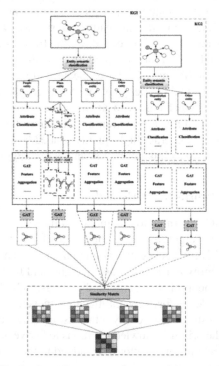

Fig. 1. Overall structure diagram of the model.

Our model that is specially designed to conduct meticulous classification on the primary knowledge graph. This is achieved by using a GAT to amalgamate entity features. The four main modules of this model comprise a subgraph partitioning module, graph attention feature aggregation module, vector space unification module, and similarity matrix integration module. The comprehensive structure of this model is depicted in Fig. 1.

3.1 Subgraph Division Module

Entities have different formats and features based on their semantics, and they possess various types of attributes and attribute values. These attributes exhibit differing similarity measures, such as numerical similarity determined by the difference in value size, and textual similarity based on semantic values. Firstly, we classify the source data into four subgraphs, namely: subgraphs of people, places, organizations, and other classes. Within the people, places, and organizations subgraphs, we further divide them into subgraphs of digital attribute value triples, literal attribute value triples, and entity name attribute value triples, respectively. It is important to note that entity classification and subgraph division are reliant on pre-annotated information within the dataset. Based on these label information, we can integrate entities of different categories together to form a subgraph. The process of subgraph division is illustrated in Fig. 2.

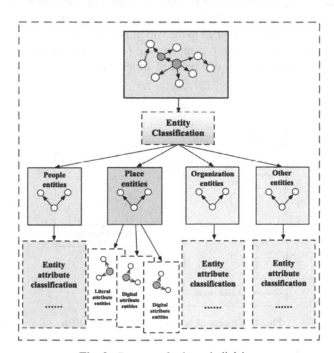

Fig. 2. Process of subgraph division.

3.2 Feature Aggregation Module

On the subgraphs of different attribute entities, we designed feature aggregation channels based on GAT for feature aggregation operations for entities with three different attribute values, respectively. Residual connections are added for the entity name, digital and literal channels, and the feature vectors of textual and numeric attribute values are obtained by initializing them using BERT [21]. The corresponding feature subgraphs are generated for each attribute subgraph aggregation as shown in Fig. 3.

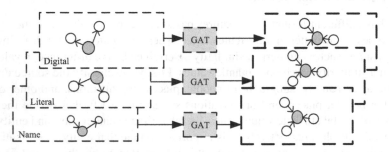

Fig. 3. Process of entity feature aggregation.

The feature aggregation process of the graph attention mechanism is shown in Fig. 4. The graph attention framework GAT1 is applied to the subgraph composed of entities with literal attributes, GAT2 is applied to the subgraph composed of entities with digital attributes, and then GAT3 is used to complete the feature aggregation operation of the entities derived from the previous two types of subgraphs.

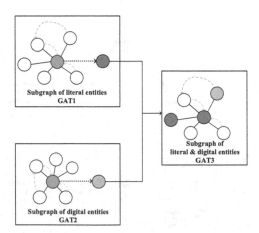

Fig. 4. Process of graph attention mechanism.

The calculation formulas are as follows:

$$h_e^l = \sigma\left(\sum_{j=1}^{n} \alpha_j W_1[a_j; v_j]\right) \tag{1}$$

$$a_j = \text{softmax}(o_j) = \frac{\exp(o_j)}{\sum_{k=1}^{n} \exp(o_k)} \tag{2}$$

$$o_j = \text{LeakyReLU}(u^T [h_e^0; a_j]) \tag{3}$$

In these formulas above, h_e^l denotes the output vector of the graph attention layer, h_e^0 is the input vector of the graph attention layer, α_j is the attention coefficient of the normalized entity attribute values, and W_1 and u are learnable parameter matrices, n is the number of attributes associated with the entity, s is the activation function, and the ELU($\cdot$)function is chosen here.

The entity vectors of different KGs are unified into the same vector space by reducing the distance between entity pairs in the seed set of pre-equivalent entities. In which training is performed and attention coefficients and other parameters are continuously updated. The lost function is defined as following:

$$Loss = \sum_{(e,e') \in A} (\sum_{e^* \in NS(e)} [d(e, e') - d(e^*, e') + \gamma]_+ + \sum_{e^{*'} \in NS(e')} [d(e, e') - d(e, e^{*'}) + \gamma]_+) \tag{4}$$

where S is the seed set of equivalent entities, $NS(e)$ denotes the negative samples of e, $d(\cdot,\cdot)$ is the cosine distance, $a[\cdot]_+ = \max\{\cdot, 0\}$ and γ is a margin hyperparameter.

3.3 Similarity Matrix Integration Module

The similarity matrices of the entity sub-graphs of people, places, organizations and other entities are $S^k (k = 1, 2, 3, 4)$. The final similarity matrix is obtained by the mean fusion calculation of the above four similarity matrices.

$$S^* = \frac{1}{4} \sum_{k=1}^{4} S^k \tag{5}$$

4 Experiment Analysis

4.1 Dataset

The experiments in this paper use DBP15K [14] as the experimental dataset. DBP15K is a cross-language dataset generated by extracting from DBpedia [22]. Three cross-language sub-datasets are included, namely, the Chinese-English, Japanese-English, and French-English datasets. The above three subdatasets are respectively represented as DBP$_{\text{ZH-EN}}$, DBP$_{\text{JA-EN}}$, and DBP$_{\text{FR-EN}}$. We will conduct multiple experiments on these three subdatasets to validate the effectiveness of our model.Each sub-dataset has about 15,000 pairs of pre-aligned seed entities.

4.2 Evaluation Indicators

Hits@k (k = 1,10) is utilized as evaluation metrics in this study. These metrics are commonly used in entity alignment tasks. Hits@k evaluates the ability of the model to identify the top k entities with the highest similarity to the reference entity among a set of candidates. A hit increment is recorded if the correct aligned entity is among the k entities. Finally, the total hit count is divided by the total number of entities in the test set to derive the Hits@k value. The effectiveness of the entity alignment process is proportional to higher values of Hits@k.

4.3 Parameter Setting

We utilized the Adagrad optimizer to train the proposed model with entity, relationship, and property vectors initialized with dimensions of 128. The learning rate was set to 0.004 while the hyperparameter γ was set to 1.0. For experimentation on the DBP15K dataset, a total of 20 negative entity samples were considered, and the training and testing sets were partitioned at a ratio of 7:3.

4.4 Overall Comparison Experiment

To verify the effectiveness of our proposed approach, we selected five different methods for comparison in this paper. Including two entity alignment models based on the translational distance model, MTransE and JAPE, and three entity alignment models based on GNN, GCN-Align, RDGCN, and NMN, respectively. The final experimental comparison results on the DBP15K dataset are shown in Table 1.

Table 1. Results of overall comparison experiment.

Model	DBP15K$_{ZH-EN}$		DBP15K$_{JA-EN}$		DBP15K$_{FR-EN}$	
	Hits@1	Hits@10	Hits@1	Hits@10	Hits@1	Hits@10
MTransE	30.83	61.41	27.86	57.45	24.41	55.54
JAPE	41.18	74.46	36.25	68.50	32.39	66.68
GCN-Align	41.25	74.37	39.91	74.46	37.29	74.49
RDGCN	70.75	84.55	76.74	89.54	88.64	95.72
NMN	73.30	86.90	78.50	91.20	90.20	96.70
Our-Model	80.82	90.51	80.17	91.79	93.13	95.74

Based on the experimental results in Table 1, the following conclusions can be drawn.

- Classifying entities according to semantics followed by attribute classification embedding. After that representation learning in the segmentation dimension helps to reduce the noise interference generated by irrelevant entities.

- Comparing with the alignment methods based on the translational distance model, the improvement of every index of our model is obvious, which illustrates the effectiveness of GNN. Comparing with the other three types of GNN-based methods, the indexes of our model are also improved, indicating that GNN can learn entity vector representations with richer feature information by better feature aggregation, and GAT can further improve the effect on the basis of GCN due to the fact that the graph attention mechanism is more adapted to the graph structure of knowledge graph.

4.5 Subgraph Division Module Ablation Experiment

To verify the enhancement effect of entity fine-dimensional classification for feature aggregation embedding of entities and attributes. Using GCN as the basic framework, we conducted entity classification ablation experiments. Model-nopeople means remove the classification of people entitiies subgraph and put all people entities into other entities subgraphs; Model-noplace means remove the classification of place entities subgraph and put all place entities into other entities subgraphs; Model-noorg means remove the classification of organization entities subgraph and put all organization entities into other entities subgraphs; Model-all means model without removing the classification of any entities subgraph. For the above four types of models, on the basis of the semantic classification of the respective source data, secondary classification is performed according to the different entity attributes to form different attributes using GCN for entity attribute vectors to complete feature aggregation and avoid the influence of attention mechanism. The calculated similarity matrix of each type of entity subgraph is used to learn the weights of each subgraph channel by LS-SVM, and the weighted sum is integrated to derive the final similarity matrix to calculate each index of entity alignment. The four models mentioned above were compared with Model-none without classification for experimental analysis. The experimental results obtained on DBP15K$_{ZH-EN}$ and DBP15K$_{JA-EN}$ cross-linguistic datasets in DBP15K using Hits@1 as the comparison index are shown in Table 2.

Table 2. Results of subgraph division module ablation experiment.

Model	DBP15K$_{ZH-EN}$	DBP15K$_{JA-EN}$
	Hits@1	Hits@1
Model-noname	66.75	65.21
Model-noplace	67.38	66.85
Model-noorg	67.26	67.17
Model-all	70.13	70.04
Model-none	65.84	64.25

From the experimental results in Table 2, we get the following conclusions:

- With the addition of the entity classification module, the alignment results of each model are better than those of Model-none without entity classification, indicating that classifying entities according to their semantic referents and performing attribute classification embedding on top of this is a significant improvement in the quality of entity feature vectors. The results are excellent when our model is applied to the alignment tasks.

- The model model-nopeople after removing the subgraphs of person-like entities has the most significant decrease compared with model-all, which may be related to the fact that the number of person-like entity triples in the ZH-EN dataset is the largest. The number of entities of each category is one of the key factors affecting the entity feature learning and entity alignment effect. The number of entities of corresponding categories should be considered when designing the classification method.

4.6   Graph Attention Mechanism Ablation Experiment

To demonstrate the role of graph attention mechanism for entity importance assignment during feature aggregation on different classes of entities and entity attributes, and the effect of noise removal. In this paper, an ablation analysis of the graph attention mechanism is performed. The graph attention mechanism in this model mainly acts in two key steps. The first step is the feature aggregation of entity subgraphs with the same attribute type, i.e., applying GAT on each of the three types of subgraphs, i.e., entity subgraphs with digital attributes, subgraphs with literal attributes and subgraphs with name attributes, here marked with the model name GAT1.The second step is the use of GAT aggregation among the three subgraphs, i.e., entity subgraphs with digital attributes, entity subgraphs with literal attributes and entity subgraphs with name attributes, here marked with the model name GAT2. Firstly the similarity matrices of each people entity subgraph, place entity subgraph, organization entity subgraph and other entity subgraph are calculated, after that the final entity similarity matrix is obtained based on LS-SVM integration based on the calculated similarity matrices. The model GAT_none with no GAT introduced (using GCN instead of GAT), the model GAT_1 with only GAT1 added, the model GAT_2 with only GAT2 added, and the model GAT_1&2 with GAT1 and GAT2 added were subjected to comparative experiments on the DBP15K$_{ZH-EN}$ dataset.

Table 3. Results of graph attention mechanism ablation experiment.

Model	DBP15K$_{ZH-EN}$	
	Hits@1	Hits@10
GAT_none	62.31	67.47
GAT_1	67.25	70.53
GAT_2	65.81	69.41
GAT1&2	68.63	73.27

From the experimental results in Table 3, we get the following conclusions. Firstly, all models with the graph attention mechanism introduced have improved metrics compared with the models without the graph attention mechanism, which proves that the graph attention mechanism can indeed improve the quality of feature aggregation and entity alignment as the GAT fully considers the weight distribution between each entity and its neighbor entities. Secondly, using two GAT modules can improve alignment even further than using a single GAT module.

5 Conclusion

In order to solve the entity alignment problem more effectively in knowledge graphs, build a high-quality and large-scale knowledge graph, optimize knowledge storage space, reduce entity ambiguity, and better apply knowledge graph technology to various fields, we propose an entity alignment method that incorporates the graph attention mechanism after entity classification. We demonstrate the effectiveness of our model through two ablation experiments and an overall comparison experiment.

In future work, we will focus on improving our model in the following directions. Incorporate entity relationship feature information into entity embeddings. Reduce the model's dependence on pre-aligned seed entities in the dataset. Extend the applicability of entity alignment methods from single textual knowledge graphs to multi-modal knowledge graphs that contain image, audio, and video information. Design language-specific entity classification methods for different languages.

Acknowledgment. The work was supported by the National Key R&D Program of China (2020YFB0906000, 2020YFB0906005, 2020YFB0906004).

References

1. Wang, Y., Luo, S., Yang, Y.: A survey on knowledge graph visualization. J. Comp.-Aided Des. Comp. Graphs **31**(10), 1666–1676 (2019)
2. Zhuang, Y., Li, G., Zhong, Z.: Hike: a hybrid human-machine method for entity alignment in large-scale knowledge bases. In: 26th ACM International Conference on Information and Knowledge Management, pp. 1917–1926 (2017)
3. Liu, Z., Sun, M., Lin, Y.: Research progress in knowledge representation learning. Comp. Res. Dev. **53**(2), 247–261 (2016)
4. Tian, J., Li, J., Liu, Q.: GCN-based entity alignment algorithm combined with attribute structure. Appl. Res. Comp. **38**(7), 1979–1992 (2021)
5. Bordes, A., Usunier, N., Garcia-Durán, A.: Translating embeddings for modeling multi-relational data. In: 26th International Conference on Neural Information Processing Systems, pp. 2787–2795 (2013)
6. Wang, Z., Zhang, J., Feng J.: Knowledge graph embedding by translating on hyperplanes. In: 28th AAAI Conference on Artificial Intelligence, pp. 1112–1119 (2014)
7. Lin, Y., Liu, Z., Sun, M.: Learning entity and relation embeddings for knowledge graph completion. In: 29th AAAI Conference on Artificial Intelligence, pp. 2181–2187 (2015)
8. Chen, M., Tian, Y., Yang, M.: Multilingual knowledge graph embeddings for cross-lingual knowledge alignment. In: 26th International Joint Conference on Artificial Intelligence, pp. 1511–1517 (2017)
9. Trisedya, B., Qi, J., Zhang, R.: Entity alignment between knowledge graphs using attribute embeddings. In: 33rd AAAI Conference on Artificial Intelligence, pp. 297–304 (2019)

10. Sun, Z., Hu, W., Li, C.: Cross-lingual entity alignment via joint attribute-preserving embedding. In: 16th International Semantic Web Conference, pp. 628–644 (2017)
11. Mikolov, T., Sutskever, I., Kai, C.: Distributed representations of words and phrases and their compositionality. In: 26th International Conference on Neural Information Processing Systems, pp. 3111–3119 (2013)
12. Zhu, H., Xie, R., Liu, Z.: Iterative entity alignment via joint knowledge embeddings. In: 26th International Joint Conference on Artificial Intelligence, pp. 4258–4264 (2017)
13. Zhang, Y., Deng, Z., Meng, D.: Temporal knowledge graph embedding for link prediction. In: 19th International Conference on Web Information Systems and Applications, pp. 3–14 (2022)
14. Wang, Z., Lv, Q., Lan, X.: Cross-lingual knowledge graph alignment via graph convolutional networks. In: Proc of Conference on Empirical Methods in Natural Language Processing, pp. 349–357 (2018)
15. Gao, J., Liu, X., Chen, Y.: MHGCN: multiview highway graph convolutional network for cross-lingual entity alignment. Tsinghua Sci. Technol. **27**(4), 719–728 (2022)
16. Li, F., Yan, R., Su, R.: An automatic model table entity alignment framework for SG-CIM model. In: IEEE 8th Intl Conference on Big Data Security on Cloud, pp. 81–86 (2022)
17. Wu, Y., Liu, X., Feng, Y.: Relation-Aware entity alignment for heterogeneous knowledge graphs. In: 28th International Joint Conference on Artificial Intelligence, pp. 6477–6487 (2019)
18. Ye, R., Li, X., Fang, Y.: A vectorized relational graph convolutional network for multi-relational network alignment. In: 28th International Joint Conference on Artificial Intelligence, pp. 4135–4141 (2019)
19. Sun, Z., Wang, C., Hu, W.: Knowledge graph alignment network with gated multi-hop neighborhood aggregation. In: 34th AAAI Conference on Artificial Intelligence, pp. 6849–6856 (2020)
20. Jiang, S., Nie, T., Shen, D.: Entity alignment of knowledge graph by joint graph attention and translation representation. In: 18th International Conference on Web Information Systems and Applications, pp. 347–358 (2021)
21. Devlin, J., Chang, M., Lee, K.: BERT: pre-training of deep bidirectional transformers for language understanding. In: 2019 Conference of the North American Chapter of the Association for Computational Linguistics: Human Language Technologies, pp. 4171–4186 (2019)
22. Auer, S., Bizer, C., Lehmann, J.: DBpedia: a nucleus for a web of open data. In: 6th International Semantic Web Conference, pp. 722–735 (2007)

Text-Independent Speaker Verification Based on Mutual Information Disentanglement

Chao Wang, Yulian Zhang, and Shanshan Yao[✉]

Institute of Big Data Science and Industry, Shanxi University, Taiyuan 030006, Shanxi, China
yaoshanshan@sxu.edu.cn

Abstract. In text-independent speaker verification, the speaker features extracted by the speaker model contain features of semantic content. Entanglement of speaker features and semantic features can lead to poor performance of speaker verification systems. To solve this problem, we propose the MI-TISV method to disentangle the semantic content features from the speaker features. The minimization of mutual information between speaker embedding and semantic content embedding can help us retain more useful information for speaker features. We conducted experiments in CN-Celeb and VoxCeleb, and found that MI-TISV method improves the generalization performance of speaker models while maintaining the accuracy of the existing models.

Keywords: Speaker verification · Text-independent · Mutual information · Disentanglement

1 Introduction

Speaker recognition (SR) [1] is one kind of biometric recognition, which has been widely used for tasks in telecommunication anti-fraud, criminal investigation and mobile payment. SR includes two types of tasks, speaker verification (SV) and speaker identification (SID). SV is used to determine whether two given utterances belong to the same person, while SID is used to determine which person in the database the given utterance belongs to. SR can also be classified into text-independent (TI) and text-dependent (TD) methods, based on the application scenario. Since TI methods are more challenge as there are no requirements for the content of the utterances, as well as SV methods can be easily extended to SI ones, text-independent speaker verification (TISV) [2–8] has become a hot and major research topic in recent years and is the focus of this paper. TISV is a complex SR problem that does not rely on specific semantic information, and can determine whether the enrolled utterance and the test utterance belong to the same speaker in a noisy environment.

Since utterances contain both the identity information of the speaker and the semantic content, existing TISV algorithms are often influenced by the semantic content in the process of training the model to obtain the speaker embedding. For example, the Vox-Celeb2 [9] dataset contains speeches of celebrities from different professional fields, and the words used in the speeches are often closely related to the industries they work

L. Yuan et al. (Eds.): WISA 2023, LNCS 14094, pp. 309–318, 2023.
https://doi.org/10.1007/978-981-99-6222-8_26

in. For example, cricketer Adam Gilchrist mentions the word "cricket" several times in his speeches, but never mentions the word "president". In contrast, politician Nancy Pelosi frequently uses politically relevant words such as "president" and "Democrat" in her speeches, but never mentions the word "cricket" [10]. Another example, the model trained on VoxCeleb2 (mainly English) would be poorly tested on the Cn-Celeb [11] dataset (Chinese). As can be seen, the trained TISV model may use the semantic content as a cue to identify the identity. Although there is inevitably some coupling between semantic content and identity information, over-reliance on semantic content may reduce the accuracy of the TISV model and affect its performance when generalizing to a new dataset.

Therefore, separating the identity information from the semantic content and eliminating the influence of semantic content on the TISV model can greatly help to improve both the accuracy and generalization of the TISV model. But most researchers currently do not consider the negative impact of speech content on speaker recognition modeling. Consequently, we proposed a text-independent speaker verification method based on mutual information disentanglement, shorted as MI-TISV. When training the speaker model, we use deep speech 2 [12] to obtain the semantic features of utterance, and use the mutual information mechanism to disentangle semantic features from speaker embeddings. Experiments show that the MI-TISV method improves the generalization performance of speaker models while maintaining the accuracy of the existing models.

2 Related Works

Traditional TISV methods are represented by i-vectors [13], which use probabilistic linear discriminant analysis as the back-end to calculate the similarity score between i-vectors. However, since these i-vector methods [14] rely on hand-crafted feature engineering, the performance has been gradually surpassed by deep learning approaches. With the widespread use of deep learning [15, 16], the focus of SR research has shifted towards deep learning. Deep neural network-based TISV methods usually consist of three components: frame-level feature, utterance-level feature, and loss function. Frame-level feature means short-span acoustic features, which can be extracted by time-delay neural networks or convolutional neural networks. Utterance-level feature is speaker representation which obtained based on frame-level features, by using pooling layers such as statistical pooling [17], maximum pooling [18] and so on, to aggregate frame-level features into utterance-level features. Commonly used loss functions are cross-entropy loss and triplet loss, with cross-entropy loss-based approaches focusing on reducing confusion among all speakers in the training data and triplet loss-based approaches [19] focusing on increasing the gap between similar speakers. Recent works [20, 21] have proposed some more complex and effective network architectures and loss functions.

Since utterances contain both speaker identity information and semantic information, as well as TISV methods aim to distinguish different speakers by their identity information rather than semantic information, some researchers have also considered the role played by semantic information in studying TISV problem. Initially, researchers only applied semantic information to speaker recognition for the purpose of improving speaker recognition performance. References [2, 3] uses the posterior values obtained through the phonetic-aware DNN to compute the Baum Welch statistics extracted by the i-vector, which improves the performance of the i-vector system. A multi-task network jointly trained by phoneme discriminative and speaker discriminative networks improves the performance of d-vector-based systems in TISV [4]. However, a recent part of researchers argue that semantic information should be enhanced at the frame level and suppressed at the utterance level. Reference [5] shows that adding semantic information by frame-level multitask training before the statistical pooling layer for x-vector systems [6] is still quite effective. Reference [7] applies multitask learning at the frame-level layer to enhance semantic information in frame-level features and used adversarial training at the utterance-level layer to learn speech-independent representations, and both operations lead to performance improvements. Reference [8] argues that in TISV, semantic information should be suppressed to work with frame-level or very short utterances.

We believe that disentanglement of semantic and identity information is a good idea to improve the performance of speaker recognition systems. At present, disentanglement methods have been widely used in fields such as images and time series. Independent component algorithm (ICA) method [22] aims to represent how the measured signal is linearly superimposed by multiple independent components. However, ICA is generally only applicable to the decoupled representation of linear system measurement data. For the decoupled representation of complex nonlinear system measurement data, reference [23, 24] proposes that auto-encoder model can gradually mine the relevant features that are more effective for reconstructed data and discard the irrelevant features. Most of the existing representation learning networks are based on the variational auto-encoders (VAE) model [25], which models the representation of real data from the perspective of maximum likelihood. In the field of speaker recognition, reference [26] proposes a domain adaptation framework to separate domain-specific features from domain-invariant speaker features. Reference [27] applies unsupervised adversarial invariant (UAI) architecture to decompose the discriminative information of speakers. Reference [26] proposes two different approaches to increase the uncertainty of the nuisance attributes inherent in speaker embedding vectors in order to separate non-speaker information from speaker embedding vectors. One approach is to train the embedding network to extract speaker embedding vectors to maximize the entropy of nuisance attribute identification, and the other approach is to reduce the correlation between speaker and nuisance embedding vectors by minimizing the mean absolute Pearson correlation (MAPC) [28].

Since current TISV methods do not consider the semantic information of utterance to be disentangled from speaker model, we propose a TISV method based on mutual information disentanglement, which introduces mutual information (MI) as a relevant metric to achieve proper decoupling of semantic content and speaker representations by reducing the interdependence in mutual information theory. Our experimental results reflect the superiority of the proposed approach in learning to effectively separate speech representations to remove semantic content while capturing the features of the target speaker.

3 Proposed Approach

3.1 Architecture of the MI-TISV System

As shown in Fig. 1, the proposed MI-TISV system consists of three modules: content encoder, speaker encoder, and MI module. The first two modules extract the content and speaker representation from the input speech, respectively; the third module, the MI module, disentangle the potential semantic content of the input speech from the speaker representation. Assuming that there are K utterances $X_k = \{X_{k,1}, X_{k,2}, \ldots, X_{k,T}\}$. We use FilterBank (Fbank) and mel-frequency Cepstral Coefficients (MFCC) as acoustic features and randomly select T frames from each utterance for training. The features of kth Fbank and MFCC are represented as $F_k = \{F_{k,1}, F_{k,2}, \ldots, F_{k,T}\}$ and $M_k = \{M_{k,1}, M_{k,2}, \ldots, M_{k,T}\}$, respectively.

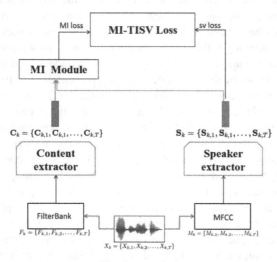

Fig. 1. MI-TISV

Content encoder θ_c. The content encoder strives to extract semantic content information from F_k by using deep speech 2 as shown in Fig. 2. Deep speech 2 is an RNN-CTC (Recurrent Neural Network-Connectionist Temporal Classification) model consisting of three parts: a fully connected layer, a recurrent layer and finally a CTC loss for training. The aim of the proposed method is to obtain robust speaker embeddings, so the speech recognition model is not trained, using a pre-trained model, and instead of obtaining a specific output, the last fully connected layer is extracted as an embedding of the semantic content. The semantic features extracted by the content encoder can be described as $C_k = \{C_{k,1}, C_{k,2}, \ldots, C_{k,T}\}$. The semantic content was encoded as $C_k = f(F_k; \theta_c)$, which is the content representation used to accurately remove the speaker's semantic content.

Speaker encoder θ_s. The model architecture of the speaker encoder is shown in Fig. 3. Speaker embeddings are obtained using the ECAPA-TDNN [21] model, which is improved on basis of x-vector to create more robust speaker embeddings. The pooling layer uses a channel- and context-dependent attention mechanism, which allows the network to attend different frames per channel. The one-dimensional Squeeze-and-Excitation Block (SE-Block) rescales the channels of the intermediate frame-level feature maps according to the global properties of the records in order to insert global contextual information in the convolutional block of local operations. The ECAPA-TDNN model achieves the best performance in the SdSV challenge 2020 [29] for the Chinese TISV task. In this paper, we mainly focus on separating speaker embeddings from semantic content to extract more robust speaker information. The speaker feature $S_k = \{S_{k,1}, S_{k,2}, \ldots, S_{k,T}\}$ is generated by $M_k = \{M_{k,1}, M_{k,2}, \ldots, M_{k,T}\}$ through the speaker encoder θ_s. $S_k = f(M_k; \theta_s)$ capture global speech characteristics to control the speaker identity of the generated speech.

MI module. we adopt vCLUB [30] to compute the upper bound of MI. In this module, there is a variational approximation networks $\theta_{u,v}$ The variational approximation networks can be trained by the semantic content embedding C_k and the speaker embedding S_k. After maximizing the log-likelihood, it is possible to train the mutual information network to obtain the mutual information upper bound and thus calculate the mutual information loss.

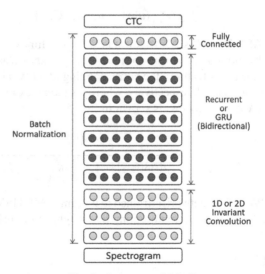

Fig. 2. Deep speech2 [12]

3.2 Optimization of MI-TISV

We assume that u and v are given random variables, then we can set the MI as the Kullback-Leibler(KL) [28] divergence between their joint and marginal distributions, $I(u, v) = D_{KL}(P(u, v); P(u)P(v))$. We use vCLUB to compute the upper bound of MI as:

$$I(u, v) = E_{P(u,v)}\big[\log Q_{\theta_{u,v}}(u|v)\big] - E_{P(u)}E_{P(v)}\big[\log Q_{\theta_{u,v}}(u|v)\big] \tag{1}$$

In Eq. (1), u, v $\in$ {C, S}, C and S are content and speaker representations respectively. Furthermore, $Q_{\theta_{u,v}}(u|v)$ is the variational approximation of ground-truth posterior of u given v, which can be parameterized by a network $\theta_{u,v}$. The unbiased estimation for vCLUB between semantic content representations and speaker representations is given by:

$$\hat{I}(C,S) = \frac{2}{K^2 T} \sum_{k=1}^{K} \sum_{l=1}^{K} \sum_{t=1}^{T/2} \left[logQ_{\theta_{C,S}}(C_{k,t}|S_k) - logQ_{\theta_{C,S}}(C_{l,t}|S_k) \right] \quad (2)$$

Equation (1) provides a reliable MI upper bound through a good variational approximation. Therefore, reducing the correlation between content representations and speaker representations can be achieved by minimizing Eq. (2), then the MI loss can be expressed as:

$$L_{M,I} = \hat{I}(C,S) \quad (3)$$

During the model training process, the optimization process of the variational approximation network and the speaker network is carried out alternately. Where the training objective of the variational approximation network is to maximize the log-likelihood:

$$L_{u,v} = \log Q_{\theta_{u,v}}(u|v), \quad u,v \in \{C,S\} \quad (4)$$

The loss of ECAPA-TDNN is AAM-softmax. AAM-softmax is originally used in face recognition tasks and is derived from softmax loss, also known as ArcFace loss. Let z_i, y_i, and W be the speaker's features, the corresponding identity labels and the weight matrix of the classification head, respectively, where i is the index in a batch data of size K, $0 < i < K$. The AAM-softmax loss is shown in Eq. (5):

$$L_{SV} = -\frac{1}{K} \sum_{i=1}^{K} \log \frac{e^{s\left(\cos\left(\theta_{y_i,i}\right)+m\right)}}{e^{s\left(\cos\left(\theta_{y_i,i}\right)+m\right)} + \sum_{j=1,j\neq y_i}^{K} e^{s\cos(\theta_{j,i})}} \quad (5)$$

While the ECAPA-TDNN network is trained to minimize MI-TISV loss: $L_{MI-TISV} = L_{SV} + \lambda L_{MI}$, where λ is a constant representing the weight of mutual information loss.

4   Experiments

4.1   Experimental Setup

Dataset. In order to validate the effectiveness of the proposed methodological framework, VoxCeleb2 is used for training and VoxCeleb1 and CN-Celeb datasets for testing. Most experiments are conducted on VoxCeleb dataset, which consist of interviews of celebrities extracted from YouTube. The VoxCeleb2 development set, containing approximately 1.1 million audio files from 6,000 speakers, is used for training. A validation set is created based on 2% of the development set data, which included all speakers, but no overlap in the recordings. In our evaluation, we use the cleaned original test set from VoxCeleb1 (Vox1-O, 40 speakers, approximately 37,000 trials), the extended test set

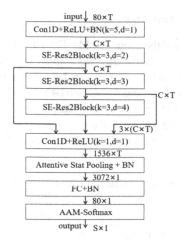

Fig. 3. ECAPA-TDNN [19]

(Vox1-E, 1251 speakers, approximately 580,000 trials) and the hard test set (Vox1-H, 1190 speakers, approximately 550,000 trials, same nationality and gender). CN-Celeb is a Chinese speech dataset designed by Tsinghua University, which includes about 130,000 speech samples from 1,000 Chinese celebrities. 11 speech scenes are covered in CN-Celeb, such as entertainment programs, interviews, singing, drama, movies, video blogs, live broadcasts, speeches, recitations, and advertisements. The CN-Celeb dataset is more challenging for current speaker recognition methods than VoxCeleb, which only has interview-based speech. The audio format of both datasets is 16 kHz sampling rate, 16-bit resolution, mono, PCM-WAV format.

Evaluation metrics. The model performance is measured by the Equal Error Rate (EER) and the minimum normalized detection cost (MinDCF) with $P_{\text{target}} = 10^{-2}$ and $C_{FA} = C_{Miss} = 1$. Smaller EER and MinDCF means better performance of the model.

System description. The proposed MI-TISV network consists of the content encoder, speaker encoder and MI module. Content Encoder is a mature speech recognition framework of deep speech 2. The input of Deep Speech 2 is Fbank and the output is a (101, 29)-dimensional embedding, where 101 represents the speech frames. During the experiments, we use a pre-trained model for speech recognition, so we do not describe too much about the architecture of the model. The speaker encoder follows reference [21], and the ECAPA-TDNN model has an 80-dimensional MFCC as input and a 256-dimensional embedding as output.

The variational approximation $Q_{\theta_{u,v}}(u|v)$ for all MI is parameterized in Gaussian distribution as $Q_{\theta_{u,v}}(u|v) = \mathcal{N}(u|\mu(v)), diag(\sigma^2(v))$, with mean $\mu(v)$ and variance $\sigma^2(v)$ inferred by a two-way fully-connected network u, v that is composed of four 256-dim hidden layers. The MI-TISV network is trained with Adam optimizer. The learning rate is 1e-3 and batch size is 256. 200 frames are randomly selected from each utterance for training per iteration. Variational approximation networks are also trained with the Adam optimizer with a learning rate of 3e-4.

4.2 Experimental Results and Analysis

We trained ECAPA-TDNN and the proposed MI-TISV method on the VoxCeleb2 development set respectively. Then we test the two trained models on several test sets, the results are shown in Table 1 and Table 2. It can be seen that the EER and minDCF of MI-TISV do not change much on the VoxCeleb1-O, VoxCeleb1-E and VoxCeleb1-H test sets, while have decreased on CN-Celeb compared to ECAPA-TDNN. This confirms that the MI-TISV method improves the generalization performance of the speaker model while maintaining the accuracy of the existing model.

Table 1. Evaluation results of VoxCeleb1-O, VoxCeleb1-E, VoxCeleb1-H datasets

Evaluation set	Method	EER/%	minDCF
VoxCeleb1-O	ECAPA-TDNN	1.08	0.1448
	MI-TISV	1.09	0.1435
VoxCeleb1-E	ECAPA-TDNN	1.28	0.1528
	MI-TISV	1.27	0.1519
VoxCeleb1-H	ECAPA-TDNN	2.39	0.2447
	MI-TISV	2.38	0.2450

Table 2. Evaluation results of the Cn-Celeb dataset

Method	EER/%	minDCF
ECAPA-TDNN	16.36	0.5888
MI-TISV	14.13	0.5126

To determine the effect of the weight hyperparameter λ on the performance of MI-TISV, experiments were conducted with λ as 0, 1, 10, and 100, and the optimal weight hyperparameter was selected based on the experimental results. Table 3 shows the evaluation results on CN-Celeb when the training set is the VoxCeleb2 development set. The experimental results show that the best model performance is obtained when λ is taken as 10.

Table 3. Evaluation results when the hyperparameter λ takes different values

λ	EER/%	minDCF
0	16.36	0.5888
1	14.81	0.5624
10	14.13	0.5126
100	15.1	0.5826

5 Conclusion

In this paper, we proposed a text-independent speaker verification method based on representation decoupling, which obtains embeddings representing semantic content and speaker identity by encoding the semantic content and speaker identity simultaneously, and then introduces a mutual information mechanism to decouple the two embeddings. The performance of the proposed method on the speaker verification task is evaluated using the VoxCeleb dataset and the CN-Celeb dataset, and the experiments show that the proposed method improves the generalization performance while maintaining accuracy.

References

1. Campbell, J.P.: Speaker recognition: a tutorial. Proc. IEEE **85**(9), 1437–1462 (1997)
2. Yun, L., Scheffer, N., Ferrer, L., Mclaren,M.: A novel scheme for speaker recognition using a phonetically-aware deep neural network. In: 2014 IEEE International Conference on Acoustics, Speech and Signal Processing (ICASSP), pp. 1695–1699. IEEE, Piscataway (2014)
3. Kenny, P., Gupta, V., Stafylakis, T., Ouellet, P., Alam, J.: Deep neural networks for extracting Baum-Welch statistics for speaker recognition. Odyssey (2014)
4. Qian, Y., Fu, T., Zhang, Y., Yu, K., Chen, N.: Deep feature for text-dependent speaker verification. Speech Commun. **73**, 1–15 (2015)
5. Liu, Y., He, L., Liu, J., Johnson, M.T.: Speaker embedding extraction with phonetic information (2018)
6. Snyder, D., Garcia-Romero, D., Sell, G., Povey, D., Khudanpur, S.: X-vectors: robust DNN embeddings for speaker recognition. In: 2018 IEEE International Conference on Acoustics, Speech and Signal Processing (ICASSP), pp. 5329–5333. IEEE, Piscataway (2018)
7. Wang, S., Huang, Z., Qian, Y., Yu, K.: Discriminative neural embedding learning for short-duration text-independent speaker verification. In: IEEE/ACM Transactions on Audio, Speech, and Language Processing, vol. 27, no. 11, pp. 1686–1696 (2019)
8. Tawara, N., Ogawa, A., Iwata, T., Delcroix, M., Ogawa, T.: Frame-level phoneme-invariant speaker embedding for text-independent speaker recognition on extremely short utterances. In: ICASSP 2020 – 2020 IEEE International Conference on Acoustics, Speech and Signal Processing (ICASSP), pp. 6799-6803. IEEE, Piscataway (2020)
9. Nagrani, A., Chung, J. S., Xie, W., Zisserman, A.: Voxceleb: large-scale speaker verification in the wild. Comp. Speech Lang. **60**(Mar.), 101027.1–101027.15 (2020)
10. Nagrani, A., Chung, J.S., Albanie, S., Zisserman, A.: Disentangled speech embeddings using cross-modal self-supervision. In: ICASSP 2020 – 2020 IEEE International Conference on Acoustics, Speech and Signal Processing (ICASSP), pp. 6829–6833. IEEE, Piscataway (2020)

11. Fan, Y., Kang, J.W., Li, L.T., Li, K.C., Chen, H.L., Cheng, S.T., et al: CN-Celeb: a challenging Chinese speaker recognition dataset. In: ICASSP 2020 – 2020 IEEE International Conference on Acoustics, Speech and Signal Processing (ICASSP), pp. 7604–7608. IEEE, Piscataway (2020)

12. Amodei, D., Ananthanarayanan, S., Anubhai, R., Bai, J., Zhu, Z.: Deep speech 2: end-to-end speech recognition in English and Mandarin. Comp. Sci. (2015)

13. Dehak, N., Kenny, P.J., Dehak, R., Dumouchel, P., Ouellet, P.: Front-end factor analysis for speaker verification. IEEE Trans. Audio Speech Lang. Process. **19**(4), 788–798 (2011)

14. Hansen, J., Hasan, T.: Speaker recognition by machines and humans: a tutorial review. IEEE Signal Process. Mag. **32**(6), 74–99 (2015)

15. Wang, R., Liu, J., Zhang, H.: Research on target detection method based on improved YOLOv5. In: Zhao, X., Yang, S., Wang, X., Li, J. (eds.) WISA 2022. LNCS, vol. 13579. Springer, Cham (2022). https://doi.org/10.1007/978-3-031-20309-1_40

16. Liu, J., Zuo, F., Wang, G.: An improved monte carlo denoising algorithm based on kernel-predicting convolutional network. In: Zhao, X., Yang, S., Wang, X., Li, J. (eds). WISA 2022. LNCS, vol. 3579. Springer, Cham (2022). https://doi.org/10.1007/978-3-031-20309-1_39

17. Snyder, D., Garcia-Romero, D., Povey, D., Khudanpur, S.: Deep neural network embeddings for text-independent speaker verification. In: Interspeech 2017 (2017)

18. Novoselov, S., Shulipa, A., Kremnev, I., Kozlov, A., Shchemelinin, V.: On deep speaker embeddings for text-independent speaker recognition (2018)

19. Li, C., Ma, X., Jiang, B., Li, X., Zhang, X., Liu, X., et al: Deep speaker: an end-to-end neural speaker embedding system. arXiv (2017)

20. Kwon, Y., Heo, H.S., Lee, B.J., Chung, J.S.: The ins and outs of speaker recognition: lessons from voxsrc 2020. In: 2021 IEEE International Conference on Acoustics, Speech and Signal Processing (ICASSP), pp. 5809–5813 (2020)

21. Desplanques, B., Thienpondt, J., Demuynck, K.: Ecapa-tdnn: emphasized channel attention, propagation and aggregation in TDNN based speaker verification (2020)

22. Locatello, F., Poole, B., Rtsch, G., Schlkopf, B., Tschannen, M.: Weakly-supervised disentanglement without compromises (2020)

23. Zhai, Z., Liang, Z., Zhou, W., Sun, X.: Research overview of variational auto-encoders models. Comput. Eng. Appl. **55**(3), 1–9 (2019)

24. Schmidhuber, J.: Learning factorial codes by predictability minimization. Neural Comput **4**(6), 863–879 (1992)

25. Kingma, D.P., Welling, M.: Auto-encoding variational bayes. arXiv.org (2014)

26. Kang, W.H., Mun, S.H., Han, M.H., Kim, N.S.: Disentangled Speaker and Nuisance Attribute Embedding for Robust Speaker Verification. IEEE Access, Piscataway (2020)

27. Peri, R., Pal, M., Jati, A., Somandepalli, K., Narayanan, S.: Robust speaker recognition using unsupervised adversarial invariance. In: 2020 IEEE International Conference on Acoustics, Speech and Signal Processing (ICASSP), pp. 6614–6618 (2019)

28. Morgen, O.: Representation learning for natural language. Ph.D. dissertation, Dept. Comput. Sci. Eng., Univ. Gothenburg, Gothenburg, Sweden (2018)

29. Zeinali, H., Lee, K., Alam, J., Burget, L.: SdSV challenge 2020: large-scale evaluation of short-duration speaker verification. Interspeech (2020)

30. Cheng, P., Hao, W., Dai, S., Liu, J., Gan, Z., Carin, L.: CLUB: a contrastive log-ratio upper bound of mutual information. In: International Conference on Machine Learning (2020)

FocusCap: Object-Focused Image Captioning with CLIP-Guided Language Model

Zihan Kong, Wei Li, Haiwei Zhang(✉), and Xiaojie Yuan

College of Computer Science, Nankai University, Tianjin 300350, China
kzh@mail.nankai.edu.cn, liwei@dbis.nankai.edu.cn,
{zhhaiwei,yuanxj}@nankai.edu.cn

Abstract. Image captioning refers to generating a corresponding natural language caption for a given image. This task is usually thought to be a large amount of supervised training to align the two modalities of image and text, which can be suffered from expensive occupation of time and resources. Recently, unsupervised methods based on multi-modal models for image captioning have attracted more attentions. However, the generated captions are often limited by the structure of the multi-modal model and cannot accurately capture essential information within the images. To address these limitations, we propose a method named FocusCap, which combines the CLIP multi-modal model with a pre-trained language model, for image captioning in an unsupervised way. FocusCap uses CLIP to calculate the similarity between image and text, as well as between objects and text. The calculated similarity score is used as visual information to control the language model generation process. FocusCap leverages the multi-modal feature information extracted in CLIP to guide the language model for text generation, thereby simplifying the costly process of large-scale training for image-text feature extraction and modality alignment. Experimental results show that our proposed FocusCap outperforms existing zero-shot methods on the Microsoft COCO and Flickr30K datasets.

Keywords: Image captioning · Multi-modal features · Zero-shot

1 Introduction

Image Captioning is an important task in Deep Learning, which is a multi-modal problem that combines natural language processing and computer vision. Its task is to generate a natural language caption of a given image. Image captioning can be applied to image indexing, social media platforms, and helping visually impaired individuals. Figure 1 shows two examples of image captioning.

Recently, Contrastive Language-Image Pre-training (CLIP) [7] based on multi-modal representation learning of text and images has received extensive research attention. CLIP uses large-scale image-text pairs with weak alignments

L. Yuan et al. (Eds.): WISA 2023, LNCS 14094, pp. 319–330, 2023.
https://doi.org/10.1007/978-981-99-6222-8_27

(a) A herd of horse riders riding down a trail.

(b) A couple standing in front of a statue of a man at nighttime.

Fig. 1. Examples of image captioning by FocusCap and Magic

for contrastive learning, and the learned joint model achieves strong zero-shot performance on image classification and image-text retrieval tasks. Since the introduction of pre-trained language model GPT-2 [8], generative language models trained on large amounts of unstructured text have shown excellent performance on various natural language processing tasks. Given text prompts, a language model can generate text by using a decoding scheme that predicts the next token. It has become possible to control the output of a language model by inserting text prompts.

Encouraged by CLIP, which guides the decoding process for text generation and simplifies the process of visual feature extraction, we propose FocusCap model. Our contributions are as follows:

(1) We propose the FocusCap model, which is a zero-shot model that utilizes multi-modal features from pre-trained models for generating image caption and introduces an image-text matching module without requiring extensive supervised training to align image and text features.
(2) We propose a more efficient and object-focused decoding method. This method has a higher computational efficiency for text decoding by avoiding supervised training and gradient updates. Additionally, the inclusion of the object-text matching module makes the generated text more focused on the primary information in images, which conforms to the human's habits of describing images.
(3) Experiments on two datasets, Microsoft COCO and Flick30K, show that FocusCap achieves the best performance on 11 out of 12 metrics, demonstrating the clear advantages of our proposed method.

2 Related Work

Recently, the specific models designed for image captioning are constantly improving: from the first deep learning method using Recurrent Neural Networks (RNN) to the breakthroughs using Transformer models with self-attention mechanisms and non-autoregressive BERT models, and the creation of automated evaluation metrics. Despite many improvements and innovations in recent years, image captioning is still far from being considered a solved task. This section introduces supervised and unsupervised image captioning methods, with a focus on three unsupervised methods.

CLIP. CLIP is a large-scale pre-trained multi-modal model proposed by the Open AI [7], which learns from a set of large-scale weakly aligned image-text pairs using contrastive learning. It has achieved strong zero-shot performance in image classification and image-text retrieval tasks. CLIP consists of an image encoder and a text encoder. Given an input image, CLIP's image encoder produces image embeddings. CLIP's text encoder encodes a set of candidate sentences to obtain the text embeddings. Then calculate the similarity between text embeddings and image embeddings, and select the sentence with the highest scores as image caption.

Supervised Image Captioning Methods. To generate textual captions from a given input image, traditional image captioning tasks usually rely on carefully curated sets of image-text pairs to train encoder-decoder models. Some early attempts [19,21] use CNN-based encoders to extract image features and RNN/LSTM-based decoders to generate output sentences. To better understand the visual information in images, some approaches [17,18] use object detectors to extract attentive image regions. Recently, large-scale image-language pre-training models [22,23] have been established and have exhibited significant performance in various downstream tasks, including image captioning task. However, these methods [20,22] still require supervised fine-tuning on human-annotated datasets.

Unsupervised Image Captioning Methods. In the research fields of computer vision and natural language processing, large-scale pre-trained models have demonstrated great potential in transferring knowledge to tasks from unsupervised training data, and their zero-shot ability has received widespread attention. However, the zero-shot ability of large models has not been fully utilized in the image captioning task. Tewelet et al. [2] proposed ZeroCap, which uses pre-trained GPT-2 and updates the context cache to generate each word by minimizing the image-text matching loss computed by CLIP during inference. Su et al. [5] proposed an unsupervised framework named MAGIC, which introduces a CLIP-induced score to influence the GPT-2 generation. Zeng and Zhang et al. [3] proposed ConZIC, which, due to its bidirectional attention, can generate and continuously refine each word with flexible generation ordering.

3   Method Framework

3.1   Pre-process

In order to adapt the language model to the text in the dataset and improve the effectiveness of its generation, we conduct unsupervised learning for the language model using the text corpus in the dataset. The learning objective $\mathcal{L}$ of the language model is:

$$\mathcal{L} = \mathcal{L}_{\text{MLE}} + \mathcal{L}_{\text{CL}}. \tag{1}$$

One of the learning objectives is to maximize the likelihood function $\mathcal{L}_{\text{MLE}}$, where x is the given textual sequence, and θ is the language model. The maximum likelihood objective is defined as:

$$\mathcal{L}_{\text{MLE}} = -\frac{1}{|\boldsymbol{x}|} \sum_{i=1}^{|\boldsymbol{x}|} \log p_\theta\left(x_i \mid \boldsymbol{x}_{<i}\right). \tag{2}$$

The second learning objective is the contrastive objective $\mathcal{L}_{\text{CL}}$. Su et al. proposed in [1] to include the contrastive objective in the training of the language model to calibrate the representation space of the model and reduce its perplexity. Given the textual sequence x, ρ is a pre-defined boundary used to regulate the distribution of the model's representation space, h_{x_i} is the token x_i in the textual sequence x, and s is a function used to calculate the cosine similarity between tokens. The contrastive objective $\mathcal{L}_{\text{CL}}$ is defined as:

$$\mathcal{L}_{\text{CL}} = \frac{1}{|\boldsymbol{x}| \times (|\boldsymbol{x}| - 1)} \sum_{i=1}^{|\boldsymbol{x}|} \sum_{j=1, j \neq i}^{|\boldsymbol{x}|} \max\{0, \rho - s(h_{x_i}, h_{x_i}) + s(h_{x_i}, h_{x_j})\}. \tag{3}$$

3.2   Framework

As a zero-shot image captioning model, FocusCap does not require training on supervised datasets. Given an image I, it can generate a piece of text containing n words $x_{<1,n>}$. The region of interest in the image is I_o, and at time t, the model searches for x_t through maximum likelihood $p(\boldsymbol{x}_{<1,t>}|I, I_o)$. By applying Bayes' rule, this can be derived as:

$$\begin{aligned} &\log p(\boldsymbol{x}_{<1,t>}|I, I_o) \\ &\propto \log p(\boldsymbol{x}_{<1,t>}, I, I_o) \\ &= \log p(I|\boldsymbol{x}_{<1,t>}) + \log p(I_o|\boldsymbol{x}_{<1,t>}) + \log p(\boldsymbol{x}_{<1,t>}). \end{aligned} \tag{4}$$

This means that the three basic rules guiding the decoding process are implemented by three modules. Specifically, the language model evaluates $p(\boldsymbol{x}_{<1,t>})$, which helps to generate text with high fluency. The image-text matching network calculates the similarity between the input image and the generated text, i.e., $p(I|\boldsymbol{x}_{<1,t>})$, which helps to generate a piece of text highly relevant to the input image. The object matching network calculates the similarity between the object in the image and the generated text, i.e., $p(I_o|\boldsymbol{x}_{<1,t>})$, which helps generate text that focuses on the key object in the image. These three modules constitute the multi-modal feature-based decoding strategy proposed in this paper. Figure 2 shows the overview of FocusCap.

Language Model. To model $p(\boldsymbol{x}_{<1,t>})$, this paper adopts the autoregressive GPT-2 model as the language generation model for FocusCap, generating text step by step from left to right using the following method:

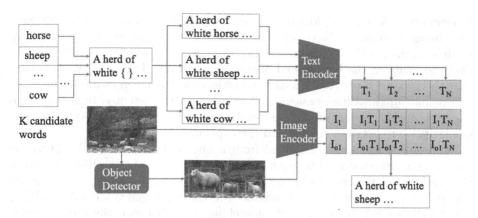

Fig. 2. An overview of our approach. FocusCap consists of three modules: language model, image-text matching module, and object-text matching module.

$$p(\boldsymbol{x}_{<1,t>}) = p(x_t|\boldsymbol{x}_{<t}) \cdots p(x_2|x_1)p(x_1). \qquad (5)$$

Recently, Su et al. proposed a decoding method called contrastive search in their paper [6] to improve the generation performance of GPT-2 language model. It was pointed out in [6] that the token distinctiveness generated by GPT-2 is low, which leads to text repetition during decoding. The FocusCap model proposed in this paper uses GPT-2 as the language generation model and proposes a decoding strategy based on multi-modal features based on contrastive search.

Image-Text Matching Module. To ensure that the generated text is highly relevant to the image, we introduce an image-text matching module to calculate the similarity between the image and text. Recently, CLIP has been pre-trained on a large number of image-text pairs, making it well-suited for modelling this task.

When selecting the next word, GPT-2 first provides the top K candidate words based on their predicted word distribution in the vocabulary. Then, these K candidate words are concatenated with the sentence already generated to create K candidate sentences $\{s_k = (x_1, ..., x_{n-1}, x_{n_k})\}_{k=1}^{K}$. CLIP is used to calculate the image-text matching score $CLIP(s_k, I)$, which represents the degree of matching between the image and text. The higher the matching score, the higher the image-text relevance. Normalize the similarity scores to obtain the predicted distribution of K candidate words:

$$p(I|\{s_k\}_{k=1}^{K}) \propto softmax[CLIP(s_k, I)]. \qquad (6)$$

The word with the highest probability is selected as x_t, which is combined with other previously generated words $\boldsymbol{x}_{<1,t-1>}$ to form a complete sentence.

Object-Text Matching Module. Currently, the framework can perform zero-shot image captioning. Next, we will explain how to add an object matching module to the existing framework. Many studies have found that CLIP is more likely to focus on the background of the image rather than objects. To ensure that the generated text focuses more on objects in the image, we introduces an object matching module to calculate the similarity between objects in the image and the caption. In this module, pre-trained CLIP is also used to calculate the similarity score between objects and the caption.

The object regions are obtained from a object detector, with YOLOv5 [4] used as the detector in this paper. Only the position of the object region is obtained, the detection results are not used. The text is similar to the previous $p(I|\boldsymbol{x}_{<1,t>})$ modelling. K candidate sentences are generated using GPT-2 and CLIP is used to calculate the matching score between the object region and the text, represented as $CLIP(sk, I_o)$, with higher scores indicating better alignment between the object region in the image and the text. Normalize the similarity scores to obtain the predicted distribution of K candidate words:

$$p(I_o|\{s_k\}_{k=1}^K) \propto softmax[CLIP(s_k, I_o)]. \tag{7}$$

Combining the predicted distribution of candidate words from the two modules of image-text matching and object-text matching, we select the word with the highest probability as x_t and form a sentence with other previously generated words, $\boldsymbol{x}_{<1,t-1>}$.

3.3 Decoding Method

We propose a decoding strategy based on CLIP multi-modal features to guide the language model's decoding process. Given an image I and previously generated text prefix $x_{<1,t-1>}$, we select the next word x_t and represent K candidate words as h_v.

$$x_t = argmax[(1-\alpha) \times p(\boldsymbol{x}_{<1,t>}) - \alpha \times (max\{s(h_v, h_{\boldsymbol{x}_{<1,t-1>}})\}+ \\ \beta_1 \times softmax[CLIP(s_k, I)] + \beta_2 \times softmax[CLIP(s_k, I_o)]] \tag{8}$$

Inspired by Su et al. [6], we incorporate model confidence and degradation penalty in candidate selection to encourage the model to generate smooth text and avoid degradation. Moreover, we propose a novel scoring metric, Focus score, to describe the similarity distribution between image and text over K candidate words. By applying Focus score, we insert multi-modal visual control into the language model's decoding process, encouraging it to generate text semantically relevant to image content. The hyperparameters β_1 and β_2 in the formula are used to adjust the strength of visual control. When $\beta_1 = \beta_2 = 0$, multi-modal visual control is disabled, and the decoding strategy based on multi-modal features degenerates to regular contrastive search [6].

Since FocusCap directly inserts visual control into the language model's decoding process without additional supervised training [16] or gradient updates on additional features [2,16], it has high computational efficiency.

4 Experimental Results and Analysis

4.1 Experiment Settings

Datasets. This section experiments based on two datasets, Microsoft COCO [14] and Flickr30K [15]. The experiments use the version of Microsoft COCO released in 2014. In the experiments, 5,000 images from Microsoft COCO and 1,000 images from Flickr30K are selected to evaluate the results. Each image has five corresponding captions to describe the content of the image.

Evaluation Metrics. The metrics used in the experiments are divided into two categories: text evaluation metrics and textual caption evaluation metrics. In the field of Natural language processing, some approaches [24] use text evaluation metrics to evaluate the text generated by language models. Textual caption evaluation metrics assess the textual captions generated by the model and are specifically designed for image captioning tasks. We select BLEU [9], METEOR [10], ROUGE [11], CIDEr [12] and SPICE [13] as evaluation metrics.

Implementation Details. We choose CLIP ViT - B/32 as the image encoder for CLIP, masked self attention Transformer as the text encoder for CLIP, the smallest version of GPT-2 with 124M parameters as the language model, and YOLOv5 as the object detector. In the baseline model comparison experiment, hyperparameters $\alpha, \beta_1, and \beta_2$ are set to 0.1, 1.36 and 1.28 respectively. All experiments are conducted on a single RTX2080 Ti GPU.

4.2 Baselines

To demonstrate the effectiveness of the experimental method, several zero-shot models are selected as baseline models. The results generated by the proposed FocusCap model are compared with those generated by three baseline models, namely, ZeroCap, MAGIC, and ConZIC. Details of the three baseline methods are as follows:

1. **ZeroCap** [2] implements zero-shot image captioning through gradient-oriented search without the need for training, but the time cost of its iterative gradient update is quite high.

Table 1. Image Captioning Results on Microsoft COCO and Flickr30K. The ZeroCap+ data utilizes the results from paper [5]. The experimental results for the MAGIC* model were reproduced based on the original paper code.

Model	Microsoft COCO						Flickr30K					
	B@1	B@4	M	R-L	C	S	B@1	B@4	M	R-L	C	S
ConZIC(sequential)	–	1.31	11.54	–	12.84	5.17	–	–	–	–	–	–
ConZIC(shuffle)	–	1.29	11.23	–	13.26	5.01	–	–	–	–	–	–
ZeroCap+	49.8	7.0	15.4	31.8	34.5	9.2	44.7	5.4	11.8	27.3	16.8	6.2
MAGIC*	56.6	12.4	17.4	39.6	48.8	11.1	42.6	5.6	12.8	30.6	19.3	6.8
FocusCap	**57.0**	**13.0**	**17.5**	**40.0**	**49.5**	**11.1**	**46.2**	**6.8**	**13.4**	**32.7**	**22.0**	**7.2**

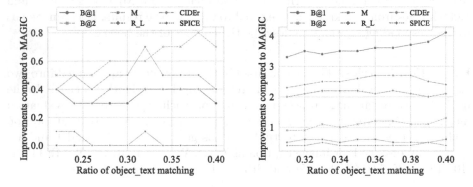

Fig. 3. Hyperparameter analysis on Microsoft COCO and Flickr30K datasets.

2. **MAGIC** [5] directly combines the language model GPT-2 with the contrastive image-text pre-training model CLIP, inserting CLIP's visual control into the language model's text generation process, thereby achieving zero-shot image captioning.

3. **ConZIC** [3] is a sampling-based non-autoregressive language model Gibbs BERT, which can generate and continuously update each word.

4.3   Results

This section compares the accuracy of image caption generated by the FocusCap model with three other models, including ConZIC, ZeroCap, and MAGIC, on two datasets, Microsoft COCO and Flickr30K. The evaluation metrics used are BLEU@1, BLEU@4, METEOR, ROUGE-L, CIDEr, and SPICE. The comparison results of the FocusCap model with the ConZIC, ZeroCap, and MAGIC models on the Microsoft COCO and Flickr30K dataset are shown in the Table 1. From the experimental results, it can be seen that the proposed FocusCap achieves the best performance on 11 out of 12 metrics, demonstrating the clear advantages of our proposed approach.

Table 2. Manual evaluation results on Microsoft COCO and Flickr30K. People are asked to score the degree of conformity between the caption and the image, with a score range of 1–5, with higher scores indicating greater conformity between the text and image.

Model	Microsoft COCO		Flickr30k	
	Human score	Var	Human score	Var
MAGIC	2.83	1.21	2.82	1.28
FocusCap	4.11	1.00	4.07	1.22

4.4 Hyperparameter Optimization

The proposed FocusCap model in this paper employs three hyperparameters, namely α, β_1, and β_2. Following the configuration in the paper [5], this paper sets α to 0.1, and the sum of β_1 and β_2 to 2.0. This section mainly conducts experiments and adjustments on hyperparameters β_1 and β_2 on two datasets, Microsoft COCO and Flickr30K. The two hyperparameters are used to adjust the control strengths of the image-text matching and object-text matching modules during the caption generation process.

Figure 3 shows the hyperparameter experiment results of the FocusCap model on the Microsoft COCO and Flickr30K. It can be seen that when the control strength ratio of the image-text matching and object-text matching modules is 68:32, the BLEU_1, METEOR, ROUGE_L, and CIDEr metrics reach the best results on Microsoft COCO. Therefore, the ratio of β_1 and β_2 on Microsoft COCO is set to 68:32. When the control strength ratio of the image-text matching and object-text matching modules is 64:36, the METEOR and CIDEr metrics reach the best results, and the BLEU_4, ROUGE_L, and SPICE metrics reach the second-best results on Flickr30K. Therefore, the ratio of β_1 and β_2 is set to 64:36 on Flickr30K.

4.5 Manual Evaluation

In order to verify that the image captions generated by FocusCap are more closely related to the content of the image and better reflect human's habits of describing images, this section designs human evaluation metrics.

Table 2 shows the human evaluation results on the Microsoft COCO and Flickr30K. On both datasets, the FocusCap performs better on the human evaluation metrics than the MAGIC model. And the variances of the results of the FocusCap on the two datasets is lower than MAGIC. The human evaluation metrics are not meant to objectively and accurately evaluate the generation performance of the models, but rather to reflect human preferences for different model generation results. The experimental results show that the proposed FocusCap model performs better than the MAGIC model on the human evaluation metrics.

(a) **MAGIC:** A large, white tent is about to be demolished.

FocusCap: A group of people sitting outdoors eating.

(b) **MAGIC:** A bathroom with a toilet, shelf, and sink.

FocusCap: A bathroom with a brown cabinet and a white toilet.

(c) **MAGIC:** A skateboard hitting a white and orange kite.

FocusCap: A group of boys playing soccer on a field.

(d) **MAGIC:** Group playing an event outdoors.

FocusCap: Group of musicians performing on stage.

Fig. 4. Examples of image captioning by FocusCap and MAGIC. The four examples in the table demonstrate the advantages of FocusCap from four aspects: grasping key image information, object feature description, object determination, object behavior determination. Red indicates incorrect, blue indicates more accurate. (Color figure online)

4.6 Case Study

This section presents some examples of captions generated by the MAGIC and FocusCap models. Figure 4 shows some image caption examples generated by the two models on images from the Microsoft COCO and Flickr30K datasets. The first column shows the input image, the second column shows the object regions determined by the object detector, and the third column shows the captions generated by the two models respectively. It can be seen that the FocusCap model pays more attention to objects in the image, and provides more detailed captions of them.

In the Fig. 4(a), FocusCap better grasps the key information in the image as a person, while MAGIC focuses too much on the background. In the Fig. 4(b), FocusCap provides a more specific description of the objects in the image, indicating the colors of the toilet and the cabinet. In the Fig. 4(c), FocusCap accurately identifies the object in the image as a boy playing soccer, rather than a skateboard or kite. In the Fig. 4(d), FocusCap provides a more specific description of the object's behavior in the image, not simply describing it as "playing an event outdoors", but as "performing on stage".

5   Conclusion

In this paper, we propose an efficient and flexible framework for zero-shot image captioning, named FocusCap. FocusCap utilizes visual information from multimodal models to control the generation process of the language model. To evaluate the generative effect of FocusCap, we conduct evaluations on the generated image captions using the Microsoft COCO and Flickr30k datasets. Experimental results demonstrate that our approach outperforms baseline methods in both automated evaluation metrics and human evaluation. The case study shows that the object-text matching module we designed allows FocusCap to pay more attention to the objects in the image, accurately determine the objects, and provide more specific descriptions of the features of the objects and the relationships between them.

References

1. Su, Y., Lan, T., Wang, Y., Yogatama, D., Kong, L., Collier, N.: A contrastive framework for neural text generation. arXiv preprint arXiv:2202.06417 (2022)
2. Tewel, Y., Shalev, Y., Schwartz, I., Wolf, L.: ZeroCap: zero-shot image-to-text generation for visual-semantic arithmetic. In: Proceedings of the IEEE/CVF Conference on Computer Vision and Pattern Recognition, pp. 17918–17928 (2022)
3. Zeng, Z., Zhang, H., Wang, Z., Lu, R., Wang, D., Chen, B.: Conzic: controllable zero-shot image captioning by sampling-based polishing. arXiv preprint arXiv:2303.02437 (2023)
4. Zhu, X., Lyu, S., Wang, X., Zhao, Q.: TPH-YOLOv5: improved YOLOv5 based on transformer prediction head for object detection on drone-captured scenarios. In: Proceedings of the IEEE/CVF International Conference on Computer Vision, pp. 2778–2788 (2021)
5. Su, Y., et al.: Language models can see: plugging visual controls in text generation (2022)
6. Su, Y., Collier, N.: Contrastive search is what you need for neural text generation. Trans. Mach. Learn. Res. (2023). https://openreview.net/forum?id=GbkWw3jwL9
7. Radford, A., et al.: Learning transferable visual models from natural language supervision. In: International Conference on Machine Learning, pp. 8748–8763. PMLR (2021)
8. Radford, A., Wu, J., Child, R., Luan, D., Amodei, D., Sutskever, I., et al.: Language models are unsupervised multitask learners. OpenAI blog **1**(8), 9 (2019)
9. Papineni, K., Roukos, S., Ward, T., Zhu, W.J.: BLEU: a method for automatic evaluation of machine translation. In: Proceedings of the 40th Annual Meeting of the Association for Computational Linguistics, pp. 311–318 (2002)
10. Banerjee, S., Lavie, A.: Meteor: an automatic metric for MT evaluation with improved correlation with human judgments. In: Proceedings of the ACL Workshop on Intrinsic and Extrinsic Evaluation Measures for Machine Translation and/or Summarization, pp. 65–72 (2005)
11. Lin, C.Y.: ROUGE: a package for automatic evaluation of summaries. In: Text Summarization Branches Out, pp. 74–81 (2004)

12. Vedantam, R., Lawrence Zitnick, C., Parikh, D.: Cider: consensus-based image description evaluation. In: Proceedings of the IEEE Conference on Computer Vision and Pattern Recognition, pp. 4566–4575 (2015)
13. Anderson, P., Fernando, B., Johnson, M., Gould, S.: SPICE: semantic propositional image caption evaluation. In: Leibe, B., Matas, J., Sebe, N., Welling, M. (eds.) ECCV 2016. LNCS, vol. 9909, pp. 382–398. Springer, Cham (2016). https://doi.org/10.1007/978-3-319-46454-1_24
14. Lin, T.-Y., et al.: Microsoft COCO: common objects in context. In: Fleet, D., Pajdla, T., Schiele, B., Tuytelaars, T. (eds.) ECCV 2014. LNCS, vol. 8693, pp. 740–755. Springer, Cham (2014). https://doi.org/10.1007/978-3-319-10602-1_48
15. Young, P., Lai, A., Hodosh, M., Hockenmaier, J.: From image descriptions to visual denotations: new similarity metrics for semantic inference over event descriptions. Trans. Assoc. Comput. Linguist. **2**, 67–78 (2014)
16. Dathathri, S., et al.: Plug and play language models: a simple approach to controlled text generation. arXiv preprint arXiv:1912.02164 (2019)
17. Anderson, P., et al.: Bottom-up and top-down attention for image captioning and visual question answering. In: Proceedings of the IEEE Conference on Computer Vision and Pattern Recognition, pp. 6077–6086 (2018)
18. Cornia, M., Stefanini, M., Baraldi, L., Cucchiara, R.: Meshed-memory transformer for image captioning. In: Proceedings of the IEEE/CVF Conference on Computer Vision and Pattern Recognition, pp. 10578–10587 (2020)
19. Donahue, J., et al.: Long-term recurrent convolutional networks for visual recognition and description. In: Proceedings of the IEEE Conference on Computer Vision and Pattern Recognition, pp. 2625–2634 (2015)
20. Fang, Z., et al.: Injecting semantic concepts into end-to-end image captioning. In: Proceedings of the IEEE/CVF Conference on Computer Vision and Pattern Recognition, pp. 18009–18019 (2022)
21. Gu, J., Wang, G., Cai, J., Chen, T.: An empirical study of language CNN for image captioning. In: Proceedings of the IEEE International Conference on Computer Vision, pp. 1222–1231 (2017)
22. Hu, X., et al.: Scaling up vision-language pre-training for image captioning. In: Proceedings of the IEEE/CVF Conference on Computer Vision and Pattern Recognition, pp. 17980–17989 (2022)
23. Jia, C., et al.: Scaling up visual and vision-language representation learning with noisy text supervision. In: International Conference on Machine Learning, pp. 4904–4916. PMLR (2021)
24. Zhou, W., Wu, J.: Code comments generation with data flow-guided transformer. In: Zhao, X., Yang, S., Wang, X., Li, J. (eds.) Web Information Systems and Applications. WISA 2022. LNCS, vol. 13579, pp. 168–180. Springer, Cham (2022). https://doi.org/10.1007/978-3-031-20309-1_15

Chinese Nested Named Entity Recognition Based on Boundary Prompt

Zhun Li, Mei Song[✉], Yi Zhu, and Lingyue Zhang

School of Computer Science and Technology, Jiangsu Normal University, Xuzhou 221116, China
msong@jsnu.edu.cn

Abstract. The paper proposes a novel method called BPCNNER that enhances boundary information for Chinese nested named entity recognition using prompt learning. The method first involves populating the nested entities into a boundary prompt template based on predefined rules and combining it with the original text. The pre-trained BERT model is then used to obtain semantic features of the text, while the BiLSTM-CRF framework captures contextual information and calculates label dependencies. The experimental results demonstrate that the proposed method achieves an F1 score of 93.02% on the People's Daily dataset, outperforming other models. Therefore, the method is effective in improving Chinese nested named entity recognition, which is a challenging task due to the lack of effective boundary supervision.

Keywords: Chinese Nested Named Entity Recognition · Prompt Learning · Boundary Information · Prompt Template · The Pre-trained Model

1 Introduction

Named Entity Recognition (NER) is a fundamental task in Natural Language Processing (NLP) that aims to identify and extract entities such as people (PER), locations (LOC), and organizations (ORG) from text data [1]. Nested named entities are a specific type of entities that appear in overlapping structures with complex and variable structures, often contain two or more entities of the same or different types [2]. The recognition of nested named entities poses a significant challenge for traditional NER models, as they struggle to accurately identify entity boundary information when faced with complex structured nested named entities. As shown in Fig. 1, such as the organization "[[德国] [大众汽车公司]]" (LOC-Germany) (ORG-Volkswagen Group) can be challenging due to entities being nested within one another. This difficulty makes the task of Chinese Nested Named Entity Recognition (CNNER) even more challenging and requires specialized techniques to address this problem. While traditional NER models perform well in identing non-nested named entities, they may not provide accurate and complete results when faced nested named entities. As a result, thus CNNER remain an active area of research in NLP.

L. Yuan et al. (Eds.): WISA 2023, LNCS 14094, pp. 331–343, 2023.
https://doi.org/10.1007/978-981-99-6222-8_28

Fig. 1. Examples of Chinese nested entity

Due to the growing popularity of Pre-trained Language Model (PLM), many researchers have utilized them to tackle NER tasks. Yu et al. [3] implemented a BERT-based end-to-end approach for NER in the Chinese Twenty-Four Histories. Ma et al. [4] used two BERTs for text and label encoding in NER tasks. The Prompt Learning (PL) based approach has emerged as a solution to bridge the gap between the pre-trained and the objectives of various downstream tasks caused by the increasing scale of PLM and the amount of data. This strategy creates boundary prompt template (BPT) based on task characteristics, which effectively enhances the model's performance without adding more parameters. By utilizing prompt learning, the effectiveness of the PLM is improved, making it easier to adapt to different downstream tasks such as NER. The use of templates allows the model to capture task-specific information, which enhances its ability to perform well on Chinese nested NER.

Motivated by the above observations, we propose a boundary prompt-based Chinese nested named entity recognition (BPCNNER) method to resolve the existing challenges. To our knowledge, this is the first study that use prompt learning to enhance boundary for CNNER. The main contributions of our work can be summarized as follows:

1. This paper propose a boundary prompt learning method for Chinese nested named entity recognition.
2. Two Boundary Prompt Templates (BPTs) construction methods are designed, the BPTs are added to fuse the original data with the final training samples. This method provides more effective identification of nested named entity boundary.
3. We compare with other different models on the same dataset. Experimental results indicate that our method achieve better results, outperforms other methods.

2 Related Work

2.1 A Study of CNNER

In the early stages of nested NER research, researchers employed rule-based, dictionary-based, and machine-learning-based techniques to detect nested named entities. Zhang et al. [5] and Zhou et al. [6] utilized hidden Markov model and SVM in combination with rule-based methods to identify nested named entities. Zhou et al. [7] proposed a cascading conditional random field (CRF) model for automatic Chinese organization name recognition that hierarchically recognized simple entities and complex entities. Liu et al. [8] used CRF and SVM to fuse contextual semantic features and hierarchical information coding methods. Fu et al. [9] improved the detection of Chinese nested named entities by combining hybrid annotation rules with a two-layer CRF. Although these approaches achieve not too bad acceptable results in identifying nested named entities, they rely heavily on rule generation and feature engineering, have highly time complexity, and may lack generality.

As deep learning techniques have shown great performance in NLP tasks, several researchers have employed deep learning approaches to recognize nested named entities. Ju et al. [10] introduced a Layered-BiLSTM-CRF model that dynamically overlays flat NER layers to detect nested named entities. Zhang et al. [11] proposed a boundary-aware cascaded neural network model and achieved improved performance in experimental results. Jin et al. [12] suggested a hierarchical annotation-based technique. With the growing popularity of nested NER research, new models and approaches for identifying nested named entities, such as hypergraph-based techniques [13, 14] and span-based methods [15, 16]. While these methods have demonstrated good results in the CNNER task, however, the capacity to recognize entity boundaries needs to be improved further. Therefore, in current CNNER's research, there is still a need to improve the extraction of nested named entity boundary information.

2.2 Prompt Learning

In recent years, the GPT-3 model [17] has introduced a new training paradigm known as "prompt learning", which breaks the traditional "pre-training + fine-tuning" approach. This novel approach involves executing multiple downstream tasks using "prompt" without adding new parameters. PLM trained on large corpora provide a wealth of information, but when utilized for diverse NLP downstream tasks, the model must be fine-tuned in terms of parameters and run alongside task-specific datasets. The PLM loses part of its capabilities in order to "accommodate" the downstream activities. The PL-based approach avoids for fine-tuning of parameters according to tasks. Instead, it reconfigures different tasks to adapt PLM, exploiting the potential knowledge of PLM and improving the effectiveness of the model. The PL-based approach is also referred to the "Fourth Paradigm" in the development of NLP technology [18], and has achieved better results in tasks such as event extraction [19] and text classification [20].

Prompt-based learning methods have attracted more attention. Cui et al. [21] first applied the idea of prompt learning to the NER task, scoring the prompt templates with the BART model to predict the entity categories. Lee et al. [22] proposed an entity-oriented and sentence-oriented example construction method, but this method has a single type of template. Chen et al. [23] transformed sequence labeling into sequence-to-sequence generation by integrating prompt information into the attention mechanism. Ma et al. [24] converted the sequence annotation task into a basic pre-trained language model task, bridging the gap between PLM and downstream tasks to enhance model recognition. The results of previous experiments demonstrate that incorporating prompt learning into the NER tasks can effectively utilize the information given by PLM. Furthermore, it introduces a new research potential to the NER problem. However, prompt-based learning strategy have not been used in the Chinese nested named entity identification challenges. Therefore, this paper introduces prompt learning into the Chinese nested NER challenge, to enhance the boundary information of nested named entities by constructing BPTs. This approach further uncovers the potential knowledge of PLM, improving the accuracy of Chinese nested NER.

3 CNNER Model Based on Prompt Learning

The figure shows the architecture of BPCNNER as Fig. 2. The input sentences are first subdivided according to entities and checked against a entity library. The matching entities are then added to templates and combined with the sentence before being passed to the sequence modeling layer. This layer generates a contextual semantic vector that incorporates template information using the pre-trained model BERT. The BiLSTM network captures the contextual semantic information and obtains prompt information from the BPTs. Finally, the CRF outputs the results based on the label dependencies.

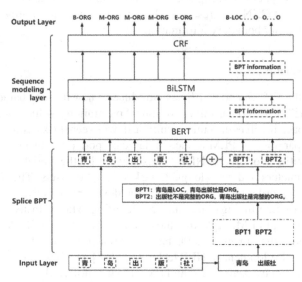

Fig. 2. Architecture of BPCNNER.

3.1 CNNER Framework Incorporating BPTs

This paper notes that while PLM has a wealth of information, there are limitations when applied to NLP downstream tasks. In particular, PLM may not perform well when dealing with CNNER tasks involving complex structures and unclear entity boundaries. To address this challenge, the research proposes a CNNER framework that incorporates BPTs. The BPT designed in this paper is a natural language text containing slots to be filled. Two types of BPTs are given here, BPT1 and BPT2.

BPT1: When the nested named entity split produces the same entity as the entity in the category entity library, declare the template type as "[Entity-Split] 是 [Entityspl-Type] and [Entity-Ori] 是 [Entity-Type]。". In this notation, [Entity-Split] refers to the outcome of the nested named entity split, [Entity-Ori] denotes the type of entity matched to, and [Entity-Type] refers to the nested named entity type. As shown in Fig. 3, in the sentence "有意收购它的除宝马外, 还有德国大众汽车公司" (In addition to BMW, Volkswagen of Germany is interested in buying it), "德国大众汽车公司" is an ORG entity with the word "德国" in the split result. The word "德国" also exists in the LOC entity library, so the BPT reads: "德国是LOC, 德国大众汽车公司是ORG." (Germany is LOC, Volkswagen of Germany is ORG). The prompt template is then spliced with the original text, resulting in the final sample: "有意收购它的除宝马外, 还有德国大众汽车公司。德国是LOC, 德国大众汽车公司是ORG。"(In addition to BMW, Volkswagen of Germany is interested in buying it. Germany is LOC, Volkswagen of Germany is ORG).

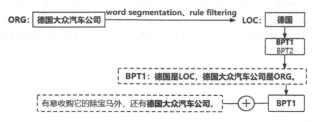

Fig. 3. BPT1 filling diagram.

BPT2: If the split result of a nested named entity is identical to an entity in another entity library, delete it as well as the special symbols; the remaining split result is prone to recognition mistakes; specify the prompt template as "[Entity-Split]不是完整的[Entity-Type], [Entity-Ori]是完整的[Entity-Type]。". As shown in Fig. 4, "李政道先生主编的《绿色战略》青岛出版社出版一书。"(Green Strategy edited by Li Zhengdao was published by Qingdao Publishing House). In this sentence, "青岛出版社" (Qingdao Publishing House) belongs to ORG entity, and the result of the participle is "青岛"(Qingdao), "出版社"(Publishing house), while "青岛" belongs to the LOC entity and only the word "出版社" is retained. The BPT is "出版社不是完整的ORG, 青岛出版社是完整的ORG。"(Publishing house is not a complete ORG, Qingdao Publishing House is a complete ORG). Splicing the prompt template with the original text, the final sample reads: "李政道先生主编的《绿色战略》青岛出版社出版一书。出版社不是完整的ORG, 青岛出版社是完整的ORG。"(Green Strategy edited by Li Zhengdao was published by Qingdao Publishing House. Publishing house is not a complete ORG, Qingdao Publishing House is a complete ORG).

Fig. 4. BPT2 filling diagram.

3.2 BERT-BiLSTM-CRF Model

The pre-trained model BERT (Bidirectional Encoder Representation from Transformers) [25] consists of a bidirectional Transformer encoder, trained using a large-scale unlabeled corpus. BERT has strong linguistic representation and feature extraction capabilities. The BPTs constructed enhance the boundary information of nested named entities, while can help BERT to better extract the semantic information of the text, and its output contains semantically complete template information.

Long Short Term Memory (LSTM) networks use a "gate" control mechanism that include a forgetting gate, an input gate, an output gate, and a memory unit. The formulae for calculating the individual gates of the LSTM network are shown below.

$$f_t = \sigma(W_f \cdot [h_{t-1}, x_{t-1}] + b_f) \tag{1}$$

$$i_t = \sigma(W_i \cdot [h_{t-1}, x_t] + b_i) \tag{2}$$

$$\tilde{C}_t = \tanh(W_C \cdot [h_{t-1}, x_t] + b_C) \tag{3}$$

$$C_t = f_t * C_{t-1} + i_t * \tilde{C}_t \tag{4}$$

$$o_t = \sigma(W_o \cdot [h_{t-1}, x_t] + b_o) \tag{5}$$

$$h_t = o_t * \tanh(C_t) \tag{6}$$

where f_t, i_t, o_t, C_t represent the forgetting gates, input gates, output gates and memory cells, $\tilde{C}_t$ represents the candidate vector for storing the current cell state information, x_t, h_t represent the network input and output at the time, $\tanh()$ and σ are the activation functions, W and b represent the weight and bias term of each gate.

Because LSTM networks can only extract semantic characteristics in one way, and Chinese nested named entities have a complicated structure, a Bi-directional Long Short Term Memory (BiLSTM) network is utilized. The information acquired in each channel is stitched together to enhance the BPTs information, and increases the model's sensitivity to the boundary information of nested named entities.

Although the BiLSTM model can obtain more the semantic information from contexts, it cannot capture the dependencies between labels, especially when the number of entities will increase after using the BPTs in this paper. It may lead to confusing label output, whereas the CRF can learn the dependencies between labels during the training process and output the labels of entities in the correct order.

For a given sequence of text $x = (x_1, x_2, ..., x_n)$ and its corresponding sequence of labels $y = (y_1, y_2, ..., y_n)$, the CRF model predicts the transfer fraction of the labels of the input sequence by the following formula:

$$Score(x, y) = \sum_{i=1}^{n} (W_{yi-1,yi} + P_{i,yi}) \tag{7}$$

where $W_{yi-1,yi}$ is the label transfer probability and the $P_{i,yi}$ probability value that the output at the ith position is the label y_i. After obtaining all the label sequence transfer scores, they are normalised using a softmax function with the following equation:

$$P(y|x) = \frac{e^{Score(x,y)}}{\sum_{i=0}^{n} e^{Score(x,y)}} \tag{8}$$

The maximum probability of the corresponding label is found during the training process using the maximum likelihood estimation with the following equation:

$$\log(P(y|x)) = Score(x, y) - \log(\sum_{i=0}^{n} e^{Score(x,y)}) \tag{9}$$

4 Experiments

4.1 Dataset

In the study, the corpus of People's Daily, which is commonly used in the CNNER task, containing three types of entities, Person (PER), Location (LOC) and Organization (ORG). The structure is shown in Table 1.

Table 1. The structure of People's Daily dataset.

Type	Train	Dev	Test	Total
Statements	22219	2710	2889	27818
Characters	1038574	129499	137577	1305650
Entities	35948	4876	4694	45618

The boundary prompt-based methods proposed in this study were all used for nested named entities in the training set. These entities were constructed according to the framework in Sect. 3.1, and information on the various types of nested named entities that satisfied the construction rules is shown in Table 2.

Table 2. Candidate nested named entity.

Type	BPT1	BPT2	Total
PER	0	0	0
LOC	0	720	720
ORG	756	3909	4665

4.2 Evaluation Indicators

The precision, recall and summed mean values F_1 were chosen as the evaluation metrics. TP is the number of true positive instances. FP is the number of false positive instances. FN is the number of false negative instances. The formula is shown below.

$$P = \frac{TP}{TP + FP} \times 100\% \tag{10}$$

$$R = \frac{TP}{TP + FN} \times 100\% \tag{11}$$

$$F_1 = \frac{2 \times P \times R}{P + R} \times 100\% \tag{12}$$

4.3 Experimental Environment and Parameter Settings

To choose the best hyperparameters of the proposed model, we experiment with the selection of hyperparameters, including the maximum sequence length, the hidden size and the learning rate. The experimental results are shown in Table 3. The best performance of hyperparameters is obtained using 256 hidden sizes, 512 maximum sequence length and 1e−5 learning rate. We keep the best hyperparameters configuration for the next experiments.

Table 3. Experimental results of hyperparameters setting.

Hidden Size	Max Seq-length	Learning Rate	P	R	F_1
512	256	5e−5	93.75	90.26	91.97
512	512	1e−5	94.38	90.51	92.40
256	256	5e−5	93.82	89.76	91.75
256	**512**	**1e−5**	**95.11**	**91.02**	**93.02**

The experimental environment was based on Python 3.8 and the deep learning framework PyTorch version 1.7.1. The pre-training model BERT uses the common BERT-base version with a maximum sequence length of 512, the optimizer uses Adam, the number of iterations is 100, the hidden size is 256, the training sample batch is 16, the learning rate is set to 1e−5, and the Dropout layer is set to 0.5 to prevent overfitting.

4.4 Experimental Results and Analysis

4.4.1 Boundary Prompt Template Tuning

It has been shown that the construction of prompt templates can have a large impact on the final recognition effect of the model [22]. It's due to the characteristics of Chinese text, where templates with the same meaning can be expressed in different ways. To explore the type of template the best recognition performance for named entities, the study constructed various BPTs for comparison experiments. Due to the limited number of entities that satisfy the BPT1 condition, only BPT2 is tuned here. The different template types and the experimental results are shown in Table 4.

Table 4. Prompt template type and experimental results.

NO	Template type	F_1
Type I	[Entity-Split]不是完整的[Entity-Type], [Entity-Ori]是完整的[Entity-Type]。	**93.02**
Type II	[Entity-Split]不能被识别为完全的[Entity-Type], [Entity-Ori]能被识别为完全的[Entity-Type]。	92.05
Type III	[Entity-Split]不能视为完备的[Entity-Type], [Entity-Ori]能视为完备的[Entity-Type]。	91.65
Type IV	[Entity-Split]是部分的[Entity-Type], [Entity-Ori]不是部分的[Entity-Type]。	91.83

The trial findings revealed that template Type I was the most successful, with an F1 value of 93.02%, a 1.37% increase over the least effective template Type III. The results show that different model types have a certain influence on the recognition effect of the model. In the following experiments, BPT2 used template Type I.

4.4.2 Model Comparison Experiments and Analysis

To verify the effectiveness of the method BPCNNER, the model in this paper is compared with the remaining four models in the experiments. Three of them are the methods proposed by related research scholars [7, 11, 26] as well as the pre-trained model BERT, and an additional set of comparison experiments are conducted here using the original data. The experimental results are shown in Table 5.

Analysis of experimental results:

(1) Zhou et al. [7] transfer the results of simple entities recognized in the low-level CRF to the high-level CRF model to provide corresponding knowledge for recognizing complex institutional NER, and the F1 value of the experimental results of this method is 89.07%, but the ability to accurately recognize entities is weak, and there is some risk of over-fitting.
(2) Zhang et al. [11] use a location-aware self-attention mechanism to obtain the features and entity boundary information in this paper. The method achieves an F1 value of

Table 5. Experimental results of each model.

Model	P	R	F_1
Zhou [7]	88.12	90.05	89.07
Zhang [11]	92.26	90.58	91.41
Li [26]	94.22	84.58	89.13
Ori-BERT	92.71	89.75	91.20
BP-BERT	93.67	90.86	92.24
BPCNNER	**95.11**	**91.02**	**93.02**

(Note: "Ori-BERT" uses the raw data, "BP-BERT" uses the double BPTs data.)

91.41% for the experimental results, but the method is only used for two levels of nested NER and does not consider multi-level nested named entities.

(3) Li et al. [26] use automatic extraction and feature construction to re-label nested named entities on the original corpus, and nested NER was performed using the method of joint labeling and cascading model, but the recall rate was only 84.58%, indicating that the method is weak for the recognition.

(4) The F1 value of Ori-BERT on the original data was 91.20%, which was better than the literature [7, 26], and the accuracy was higher than the literature [11], indicating that BERT has a strong semantic characterization capability. When trials were carried out utilizing the data contained in the double BPTs developed in this paper, the BP-BERT experimental outcome, the F1 value was 92.24%, 1.04% higher than the Ori-BERT value. Demonstrating that BPTs may successfully harness the potential knowledge of the BERT and increase the model recognition effectiveness.

(5) The BPCNNER approach in this paper considered the structural properties of nested named entities and is applicable to $N(N >= 2)$ levels of nested named entities, with finer granularity and greater assurance that the model can accurately identify nested named entity boundary information. Compared with the experimental results in the literature [7, 11, 26], the method improves by 1.61% over the highest F1 value, indicating stronger feature extraction ability to obtain more nested named entity boundary information. When compared to BERT, the F1 value of the method improves 0.78%, indicating that the BiLSTM-CRF can enhance the model's ability to capture semantic features and label dependencies [27]. The final experimental results show that the F1 value of this method is 93.02%, which is the best among all models. In short, the method proposed can effectively improve the recognition of Chinese nested named entities without the assistance of external knowledge.

4.5 Ablation Experiments

In order to verify the effectiveness of model BPCNNER in combining two BPTs, the following ablation experiments were set up in this study:

(1) Original text + BPT1: incorporates BPT1 on top of the original text, considering only the case where different entity categories merge.

(2) Original text + BPT2: incorporates BPT2 on top of the original text, considering only the entity overlap of the nested named entities themselves.

(3) Original text BPT1 + BPT2: incorporates two BPTs on top of the original text.

The results of the ablation experiments are shown in Table 6.

Table 6. Ablation results.

Model	P	R	F_1
Original text	93.46	89.84	91.61
Original text + BPT1	93.98	90.12	92.01
Original text + BPT2	94.78	90.82	92.76
Original text + BPT1 + BPT2	**95.11**	**91.02**	**93.02**

Analysis of the experimental results:

(1) The experimental results of using BPT1 or BPT2 increased by 0.4% and 1.15% respectively compared to the F1 values of the original text. It shows that both BPTs have improved the recognition effect, with the improvement of BPT2 being more apparent. There are excessive number of BPT2, and BPT2 offers more boundary information, which can effectively aid the model in precisely identifying entities.

(2) The experimental results of using BPT1 and BPT2 improved the F1 value by 1.41%. It indicates that the two BPTs constructed can provide the model with more information on nested named entities and improve model recognition results.

5   Conclusions

The research proposes a Chinese nested named entity recognition based on boundary prompt method, which can improve boundary information extraction in CNNER by using prompt learning. The method involves creating two BPTs based on the boundary information within the nested named entities and compatibility with each other. This template is then combined with the original text to enhance the recognition of entity boundaries. This method performs well across multiple comparison trials, proving its effectiveness in recognizing Chinese nested named entities.

This method is only applicable to the task of CNNER, and may incur significant costs in practical applications. In the future, we will explore more universal methods and reduce costs by using automatic template generation method. Next, as conditions allow, we will conduct experiments on extra datasets to comprehensively evaluate the effectiveness of BPCNNER. Additionally, we will innovate the model structure to improve its combination effect.

Acknowledgement. This work was supported by the National Natural Science Foundation of China (71503108, 62077029), the Research and Practice Innovation Project of Jiangsu Normal University (2022XKT1533).

References

1. Grishman, R., Sundheim, B.: Message Understanding C-onference-6: a Brief History. In: Proceedings of the 16th Conference on Computational Linguistics, pp. 466–471(1996)
2. Yu, S.Y., Guo, S.M., Huang, R.Y., et al.: Overview of Nested Entity Recognition. Comput. Sci. **48**(S2),1–10+29 (2021)
3. Yu, P., Wang, X.: BERT-based named entity recognition in Chinese twenty-four histories. In: Wang, G., Lin, X., Hendler, J., Song, W., Xu, Z., Liu, G. (eds.) WISA 2020. LNCS, vol. 12432, pp. 289–301. Springer, Cham (2020). https://doi.org/10.1007/978-3-030-60029-7_27
4. Ma, J., Ballesteros, M., Doss, S., et al.: Label semantics for few shot named entity recognition. In: Annual Meeting of the Association for Computational Linguistics, pp. 1956–1971 (2022)
5. Zhang, J., Shen, D., Zhou, G., et al.: Enhancing HMM-based biomedical named entity recognition by studying special phenomena. J. Biomed. Inform. **37**(6), 411–422 (2001)
6. Zhou, G.D.: Recognizing names in biomedical texts using mutual information independence model and SVM plus sigmoid. Int. J. Med. Informatics **75**(6), 456–467 (2006)
7. Zhou, J.S., Dai, X.Y., Yin, C.Y., et al.: Automatic recognition of Chinese organization name based on cascaded conditional random fields. Acta Electron. Sin. **34**(05), 804–809 (2006)
8. Liu, F.F., Zhao, J., Xu, B.: Study on multi-scale nested entity mention recognition. J. Chinese Inform. Process. **21**(2), 14–21 (2007)
9. Fu, C., Fu, G.: Morpheme-based Chinese nested named entity recognition. In: Proceedings of the 8th International Conference on Fuzzy Systems and Knowledge Discovery, pp. 1221–1225 (2011)
10. Ju, M., Miwa, M., Ananiadou S.: A Neural Layered Model for Nested Named Entity Recognition. In: North American Chapter of the Association for Computational Linguistics, pp. 1446–1459 (2018)
11. Zhang, R.J., Dai, L., Guo, P., et al.: Chinese nested named entity recognition algorithm based on segmentation attention and boundary-aware. Comput. Sci. **50**(01), 213–220 (2023)
12. Jin, Y.L., Xie, J.F., Wu, D.J.: Chinese nested named entity recognition based on hierarchical tagging. J. Shanghai Univ. **28**(02), 270–280 (2022)
13. Muis, A.O., Lu, W.: Labeling gaps between words: recognizing overlapping mentions with mention separators. In: Proceedings of the 2017 Conference on Empirical Methods in Natural Language Processing, pp. 2608–2618 (2017)
14. Wang, B., Wei, L.: Neural segmental hypergraphs for overlapping mention recognition. In: Proceedings of the 2018 Conference on Empirical Methods in Natural Language Processing, pp. 204–212 (2018)
15. Fisher, J., Vlachos, A.: Merge and label: a novel neural network architecture for nested NER. In: Annual Meeting of the Association for Computational Linguistics, pp. 5840–5850 (2019)
16. Luan, Y., Wadden, D., He, L., et al.: A general framework for information extraction using dynamic span graphs. In: North American Chapter of the Association for Computational Linguistics, pp. 3036–3046 (2019)
17. Brown, T., Mann, B., Ryder, N., et al.: Language models are few-shot learners. Adv. Neural. Inf. Process. Syst. **33**, 1877–1901 (2020)
18. Liu, P., Yuan, W., Fu, J., et al.: Pre-train, prompt, and predict: A systematic survey of prompting methods in natural language processing. arXiv preprint arXiv. 2107.13586 (2021)

19. Chen, N., Li, X.H.: An Event Extraction Method Based on Template Prompt Learning. Data Analysis and Knowledge Discovery, pp. 1–17, 14 Jan2023
20. Zhang, B.X., Pu, Z., Cheng, X.: Research on Uyghur Text Classification Based on Prompt Learning. Comput. Eng. **49**(06), 292–299+313 (2023)
21. Cui, L., Wu, Y., Liu, J., et al.: Template-based named entity recognition using BART. In: Findings of the Association for Computational Linguistics, pp. 1835–1845 (2021)
22. Lee, D.H., Agarwal, M., Kadakia, A., et al.: Good Examples Make A Faster Learner: Simple Demonstration-based Learning for Low-resource NER. arXiv preprint arXiv: 2110.08454 (2021)
23. Chen, X., Zhang, N., Li, L., et al.: LightNER: A Lightweight Generative Framework with Prompt-guided Attention for Low-resource NER. arXiv preprint arXiv: 2109.00720 (2021)
24. Ma, R., Zhou, X., Gui, T., et al.: Template-free Prompt Tuning for Few-shot NER. arXiv preprint arXiv: 2109.13532 (2021)
25. Devlin, J., Chang, M.W., Lee, K., et al.: BERT: Pre-training of Deep Bidirectional Transformers for Language Understanding. arXiv preprint arXiv: 1810.04805 (2018)
26. Li, Y.Q., He, Y.Q., Qian, L.H., et al.: Chinese nested named entity recognition corpus construction. J. Chinese Inform. Process. **32**(08), 19–26 (2018)
27. Xu, L., Li, S., Wang, Y., et al.: Named entity recognition of BERT-BiLSTM-CRF combined with self-attention. In: Xing, C., Fu, X., Zhang, Y., Zhang, G., Borjigin, C. (eds.) Web Information Systems and Applications: 18th International Conference, WISA 2021, Kaifeng, China, September 24–26, 2021, Proceedings, pp. 556–564. Springer International Publishing, Cham (2021). https://doi.org/10.1007/978-3-030-87571-8_48

Security, Privacy and Trust

An Analysis Method for Time-Based Features of Malicious Domains Based on Time Series Clustering

Gezhi Yan[1], Kunmei Wen[2(✉)], Jianke Hong[1], Lian Liu[1], and Lijuan Zhou[1]

[1] Network and Computation Center, Huazhong University of Science and Technology, Wuhan, China
yangezhi@hust.edu.cn
[2] Network and Informatization Office, Huazhong University of Science and Technology, Wuhan, China
kmwen@hust.edu.cn

Abstract. Malicious domains are widely used in network attacks. As DNS traffic features related to domain access time, time-based features of domains can describe the regularity of malicious activities but cannot be circumvented by attackers. These features are commonly used in malicious domain detection. Traditional detection methods generally use static statistical values in analyzing time-based features, but they ignore the temporal regularity of the features, thus resulting in inaccurate feature extraction. In view of this, in this paper, we proposed an analysis method for time-based features of malicious domains based on time series clustering. Firstly, density clustering is used to divide time intervals of time series to preserve the integrity of consecutive requests of malicious domains. Secondly, multiple time-based features are selected to depict malicious activity patterns. Thirdly, dynamic time warping and hierarchical clustering are applied as series similarity measure and clustering method respectively. The proposed method explores malicious domains by analyzing the similarity of time-based features series of different domains. Experimental results show that compared with the detection method using static statistical values, the accuracy and precision in this method improves from 88.49% to 96.30% and from 59.21% to 92.43% respectively, which proves that it can help detect malicious domains effectively.

Keywords: Malicious domain · Time-based feature · Time series · Clustering

1 Introduction

With the development of Internet, the means of malicious attacks continue to evolve. As one of the most serious network security threats, Botnet is often used to launch Distributed Denial of Service (DDoS) attacks. Botnets spread through massive malicious web pages [1] and communicate with a large number of compromised hosts through Command and Control (C&C) servers. In the early days, botnets wrote the domain/IP of the C&C server into the malicious code planted on victim hosts, but fixed addresses

L. Yuan et al. (Eds.): WISA 2023, LNCS 14094, pp. 347–358, 2023.
https://doi.org/10.1007/978-981-99-6222-8_29

could be easily discovered. Afterwards, Domain Generation Algorithm (DGA) [2] and Fast-Flux Service Network (FFSN) [3] emerged. Attackers use DGA to generate a large number of domains in conjunction with FFSN and Domain Name System (DNS) to constantly change the matching relation between domains and IP addresses. Therefore, the domains are resolved into different IP addresses in a short period of time to avoid being blocked by a blacklist way. Consequently, detecting malicious domains is of great significance for discovering and preventing from network attacks performed by DGA domains.

At present, detection methods for malicious domains are mainly based on literal features of domain name [4–6] or DNS traffic [7–9]. Researchers usually combine features of these two types to accomplish the detection of multi-feature fusion. Lots of studies have applied DNS traffic features that are changing with time, including the number of requests [10–12], timestamp of requests [13] and repeating patterns [12, 14]. Researchers pointed out that the time-based features describe regularity of malicious activities, which cannot be circumvented by attackers. Previous work usually adopts a fixed interval to split the timeline into time series, then extracts feature values of each interval and calculates the total, mean or maximum value to generate statistical values.

In this paper, we propose a method to apply time series clustering to analyze time-based features of malicious domains. Although it is common to use clustering algorithms in the intermediate step of a detection method, the method based on time series clustering has not been revealed in previous work. Our contributions are summarized as follows.

(1) We propose the method to analyze time-based features of domains based on time series clustering. Density clustering is used to split time intervals. Therefore, each period of the time series does not have to be fixed, which helps preserve the integrity of consecutive requests of malicious domains.
(2) We extract multiple temporal features of domains to describe the regularity of malicious activities. The features are from the DNS log system of our campus network, and there is no need to rely on third-party data such as passive DNS or Whois records, thus making it more applicable.
(3) We conduct three mainstream clustering algorithms to cluster the time series, including partition clustering, density clustering and hierarchical clustering. The results reveal that hierarchical clustering is more suitable for analyzing time-based features.

2 Related Work

Requests of DGA domains are initiated by malware in batches, which is likely to produce a mutation in time-based feature values. Exposure [10] was the first to apply time-based features to detect malicious domains. It extracted the number of DNS requests of different domains in hourly units to form time series for each day, and explored traffic mutation or daily similar requests patterns from the series. In order to identify daily patterns, it used Euclidean distance to measure the similarity of time series on different dates. Similarly, based on the thesis that Fast-flux domains are queried consecutively during short time intervals, Fast-flucos [8] divided DNS traffic into 5 min intervals to obtain the DNS requests series of a given domain in one day, and generated 288 dimensional vectors as features for each domain.

Li et al. [11] analyzed the DNS sequences requested by each host and their time-related features to identify compromised hosts. They extracted the total number of requests, domains requested, frequent requests respectively in a fixed interval of different hosts, and the variation coefficient (the maximum value of each day minus the average value of the last five days) of each feature to constitute feature vectors. After clustering by K-means, they sorted all clusters based on feature averages and considered the cluster with each average value higher than the others as the list of compromised hosts.

IMDoC [12] used the number of aggregated DNS requests per day to constitute time series. It calculated a Spearman's rank correlation coefficient between the series of a candidate domain and a set of malicious domains from the same malware campaign. If the sum of the correlation coefficient value was above a predetermined threshold, the candidate domain was considered to be related to the malware campaign. The quality of detection is directly related to threshold selection.

Niu et al. [13] observed that the C&C request traffic of malwares exhibited time periodicity. They extracted the sequence of intervals among DNS requests of one domain and converted each time interval into a letter to form a letter vector for each domain. They used 5 different periodicity detection algorithms based on string matching to calculate periodic confidence as features of the random forest classifier to obtain the final recognition result. However, we found in practice that lots of malicious domains were only requested once in their lifetime, which didn't show obvious time periodicity.

The previous methods have the following shortcomings:

(1) In exploring time-based features, the time series are divided by fixed intervals [8, 10, 11], which cannot preserve the integrity of typical consecutive requests of malicious domains that cross the intervals.
(2) The selected time-based features are generally the number of requests [8, 10, 12] or time intervals [13]. It is insufficient to describe the regularity of malicious domains, requiring the comprehensive application of other non-time-based features, or cannot detect malicious domains that do not exhibit regularity in the single feature.
(3) After obtaining the time series, the statistical values across the time series such as total, mean or maximum value are calculated to be the feature values [11], ignoring the value of each element in the series, resulting in inaccurate feature extracting.

3   Methodology

To address the above issues, we propose an analysis method based on time series clustering. A time series is essentially classified as dynamic data because its feature values change as a function of time, which means that the value(s) of each point of a time-series is/are one or more observations that are made chronologically [15]. Time series clustering explores the regular patterns of series. The typical components are feature extraction, similarity measure and clustering. In the feature extraction stage, we use density clustering to cluster DNS request timestamps, which divides the timeline into non-fixed intervals to preserve the integrity of requests. Multiple time-based features are extracted to describe the regularity of malicious activities. In the similarity measure stage, all elements of time series are incorporated into analysis. In the clustering stage, we evaluate the quality of three mainstream classification algorithms. This method distinguishes malicious domains and benign domains according to the similarity of the series of time-based features for domains. Moreover, it is able to explore unknown malicious domains.

3.1 Feature Extraction

Feature extraction consists of two steps: dimensionality reduction and feature representation. The raw requests recorded in DNS log is in seconds, so there are 86,400 dimensions of raw time series in one day. Clustering among raw data is computationally expensive. Dimensionality reduction transforms time-series to a lower dimensional space. Representation extracts the corresponding elements in dimensionality reduced series to represent the key information of the raw series.

Dimensionality Reduction. As discussed in Section 2, many researchers have found that DNS requests of malicious domains exhibit clustering behavior [8, 10–12]. We use density clustering algorithm DBSCAN [16] to cluster timestamps of DNS requests. DBSCAN uses two parameters, neighborhood radius ε and neighborhood density threshold *MinPts*. Parameter ε describes the minimum distance between different clusters. Parameter *MinPts* describes the minimum number of samples per cluster. Let $T = \{t_1, t_2, ..., t_m\}$ represent the distribution of request timestamps on the timeline, DBSCAN clusters a set of timestamps that are within a distance no greater than ε. After clustering, suppose that T is divided into N segments, where each $T_i = \{t_{is}, ..., t_{ie}\}$, $1 \leq i \leq N$ and (t_{is}, t_{ie}) represent the start and end timestamps of segment i, then each pair of timestamps perform the division of the raw series. Compared with previous methods that divided time series by fixed intervals, density clustering preserves the integrity of the batch requests of domains, which helps avoid the same batch of malicious requests being separated into different intervals. Even if some different batches of malicious requests are clustered together due to constantly requests of benign domains among the batches, it will only result in the overlay of statistical data of a part of batches. For the time series, it doesn't affect the consistency of series of malicious domains or the difference between series of malicious and benign domains.

Feature Representation. We use three fields of raw requests in DNS log to extract features, including timestamp, source IP and domain name. For each source IP, we collect all domains it queries on a daily basis and calculate five feature values for each domain in each time interval of the dimensionality-reduced series. It forms a set of time series with N elements, each of which is a five-dimensional feature vector. Meanings of the features are described in Table 1. Assuming there are M unique domains queried by the source IP, then we get M time series containing a total of $M \cdot N$ feature vectors.

To eliminate the influence of dimensions of different features, the $M \cdot N$ feature vectors $f_1, ..., f_{M \cdot N}$ are first normalized. If we consider each vector as a row in a matrix, then the elements of each column of the matrix are $f_{1,j}, ..., f_{M \cdot N, j}, 1 \leq j \leq 5$. The L2 norm of the columns is calculated as $\text{norm}(f_j) = \sqrt{f_{1,j}^2 + f_{2,j}^2 + ... + f_{M \cdot N, j}^2}$, and the normalized value of each dimension $f_{i,j}$ of f_i is calculated as $h_{i,j} = \frac{f_{i,j}}{\text{norm}(f_j)}$, $1 \leq i \leq M \cdot N, 1 \leq j \leq 5$.

Table 1. Time-based features and their meanings

Feature	Meaning
Access_IPs	Total number of unique IP addresses that accessed the domain
Other_access_times	Total number of times the domain accessed except for the source IP
Access_times	Total number of times the domain accessed by the source IP
Timestamp_center	The median value of multiple timestamps when the domain accessed by the source IP
Timestamp_num	Total number of unique timestamps the domain accessed by the source IP

Then we use entropy weight method to assign weights to the dimensions of h_i. We first calculate the proportion of h_{ij} as $g_{i,j} = \frac{h_{i,j}}{\sum_{i=1}^{M \cdot N} h_{i,j}}$ and the entropy of information:

$$e_j = -K \cdot \sum_{i=1}^{M \cdot N} g_{i,j} \cdot \ln g_{i,j}, K = \frac{1}{\ln M \cdot N}$$

So the weight of $h_{i,j}$ is $w_j = \frac{1-e_j}{\sum_{j=1}^{5}(1-e_j)}$. Afterwards, we form a weight vector w with w_j, and calculate the weighted vector for each h_i as $f_i' = h_i \cdot w$. Finally, each f_i is replaced by f_i' to form the series of statistical features.

3.2 Similarity Measure

We use Dynamic Time Warping (DTW) [17] to measure the similarity of time series. DTW is a common method for similarity measure in time series clustering, with which clusters of time series with similar patterns of change are constructed regardless of time points. Given the two series $S_p = (f_{p1}', f_{p2}', ..., f_{pN}')$ and $S_q = (f_{q1}', f_{q2}', ... f_{qN}')$, DTW establishes the original distance matrix $A_{(N+1) \times (N+1)}$. Except for the first row and the first column, each $A_{m,n}$ is calculated as the Euclidean distance of f_{pm}' and f_{qn}'. For the multi-dimension feature vector described in Sect. 3.1, it is

$$\text{dist}(f_{pm}', f_{qn}') = \sqrt{\sum_{j=1}^{5}(f_{pm,j}' - f_{qn,j}')^2}.$$ The element of A is calculated as follows:

$$A_{m,n} = \begin{cases} 0, & m=0, n=0 \\ inf, & m=0, n=1,2,...,N \\ inf, & m=1,2,...,N, n=0 \\ \text{dist}(f_{pm}', f_{qn}'), & else \end{cases}$$

DTW then establishes the cumulative distance matrix $D_{N \times N}$. Starting from $A_{0,0}$, each element is calculated as $D_{m,n} = A_{m+1,n+1} + \min(A_{m,n}, A_{m,n+1}, A_{m+1,n})$. This step matches the next two elements in S_p and S_q by minimizing the sum of distances among the current matching relationship. After each computation, $A_{m+1,n+1}$ is updated to $D_{m,n}$

synchronously. After all the calculations, matrix D records an optimal matching path between S_p and S_q. The value of the last element $D_{M-1, N-1}$ is called DTW distance. The smaller the value is, the higher the similarity between the series is.

Generally, similarity between equally-sized series can be measured with Euclidean distance. However, DTW achieves shape matching by warping and shifting the series, which is able to handle the case where the source IP requests malicious domains from the same malware campaign in batches but domains in each batch are totally different. We found this situation is very common in practice. While the requests are distributed across different bathes, there is not a one-to-one matching relationship between the series. It is hardly for Euclidean distance to explore the similarity properly. After the DTW distances for each pair of the M series are calculated, M^2 similarity values are generated to form the similarity matrix $Z_{M \times M}$, which is the basis of clustering.

3.3 Clustering

We use hierarchical clustering algorithm to perform the clustering work. We also apply the partitioning clustering algorithm K-medoids and density clustering algorithm DBSCAN in practice. The quality of each algorithm will be discussed in Sect. 4. We've found that hierarchical clustering is more suitable for our method.

Hierarchical clustering [18] can be thought of as a set of flat clustering methods organized in a tree structure. Initially, each object represents a cluster of its own, then similar sub-clusters are iteratively merged until the distances between each pair of the sub-clusters exceed the predefined threshold δ. According to the similarity matrix $Z_{M \times M}$, we use the ward variance minimization algorithm in Python library "science.cluster.hierarchy" to measure the distance between sub-clusters. It is calculated as $d(C_u, C_v) = \sqrt{\frac{|C_v|+|C_s|}{T}d(C_v, C_s)^2 + \frac{|C_v|+|C_t|}{T}d(C_v, C_t)^2 - \frac{|C_v|}{T}d(C_s, C_t)^2}$, where C_u is a new cluster composed of C_s and C_t, C_v is an unused cluster, $|C|$ is the number of elements in cluster C and $T = |C_v|+|C_s|+|C_t|$. Each time, the two sub-clusters that minimize the increase in variance within all the sub-clusters are selected for merging.

We set the value of δ as the maximum distance among the time series of statistical features of known malicious domains in each dataset. In other words, for the dataset P consisting of the statistical time series of all domains, if the series of the known malicious domains constitute dataset $Q = \{q_1, ..., q_R\}$, $Q \subseteq P$, then $\delta = \max(\{\text{DTW}(q_i, q_j) \mid q_i \in Q, q_j \in Q\})$, where DTW (q_i, q_j) is the DTW distance of q_i and q_j. After clustering, we assume that the domains with time series belonging to the same cluster tend to be homomorphic. If there are known malicious domains in the cluster, all domains of the cluster are considered malicious.

4 Experiments

4.1 Experiment Setup

Datasets. Our data is collected from the DNS logs of the campus network of Huazhong University of Science and Technology. In order to prevent botnet attacks, a monitoring module has been deployed on the campus network for a long time. According to the malicious domain dataset provided daily by the Netlab DGA Project [19], the monitoring module detects malicious domains that are active in the campus network. The monitoring results from May 2021 to December 2022 were collected for this study. It was found that there were malicious domains in the logs of 145 days, involving 131 source IP addresses. For each source IP, all DNS request records on the day of accessing malicious domains were extracted to form 145 sets of experimental data, including 13,648 matches to known malicious domains, 11,569 matches to known benign domains, and 6,088 matches to domains with unknown properties. For the matching of benign domains, we selected the top 20,000 domains from the list of 1 million domains with the highest global traffic provided daily by the Alexa [20] as the whitelist. Domains with suffixes listed in the whitelist were considered as benign domains.

Experiment Environment. The experiments were conducted on a server (CentOS/CPU @2.0GHZ × 32/128 GB RAM) from Public Service Platform of High Performance Computing of Huazhong University of Science and Technology. The code was written in Python 3.7.3. The clustering algorithms were implemented using the hierarchical clustering library scipy.cluster.hierarchy, DBSCAN library sklearn.cluster.DBSCAN, and K-medoids library sklearn_extra.cluster.kmedoids.

Clustering Algorithms. We evaluated the quality of three clustering algorithms: K-medoids, DBSCAN and hierarchical clustering. K-medoids is a common algorithm in time series clustering. For a given sample set $D = \{x_1, x_2, ..., x_m\}$, K-medoids minimizes the squared error $E = \sum_{i=1}^{K} \sum_{x \in C_i} \|x - \mu_i\|_2^2$ of the partitions $C = \{C_1, C_2, ..., C_K\}$ of D, where the centroid μ_i is an object in C_i and $\|\cdot\|_2$ is the L_2 norm. K-medoids improves the selection of cluster centroid compared to K-means. For a set of values of $K = 1, 2, ..., M - 1$, we calculated the square error E_K to form a two-dimensional line chart of $K - E_K$. According to the "elbow rule" which is commonly used to determine the value of K, we examined the $K - E_K$ chart and manually selected the K value when the decrease rate of E_K significantly slowed down as K increased for each set of experimental data.

DBSCAN and hierarchical clustering do not need to predefine the number of clusters, but corresponding thresholds need to be set. In the feature extraction stage, when clustering the timestamps, DBSCAN neighborhood radius ε is set to 300 and the neighborhood density threshold *MinPts* is set to 1. In the clustering stage, the value of *MinPts* in DBSCAN is set to 1. For ε in DBSCAN and the inter-cluster distance δ in hierarchical clustering, the values are both set to the maximum distance among the time series of known malicious domains in each set of experimental data as described in Sect. 3.3. For DBSCAN, the series with distances not greater than ε are clustered together. For hierarchical clustering, δ determines the minimum distance between clusters.

Evaluation Metrics. To evaluate the classification quality, we labeled known malicious domains as 1, which are positive samples, and known benign domains as 0, which are negative samples. After clustering, we assume that the domains with time series belonging to the same cluster tend to be homomorphic. If the number of domains in the cluster is greater than 1 and there are known malicious domains in the cluster, all domains of the cluster are considered malicious with a predicted value of 1, otherwise the predicted value is 0. Unknown property domains are not included in the evaluation metrics. After the classification of all experimental data is completed, accuracy, recall, false positive rate, and F1-score values are evaluated based on label values and predicted values. The definitions of each indicator are shown in Table 2.

Table 2. Evaluating indicators

Indicator	Meaning
Accuracy	The Number of correctly classified samples / The Number of total samples
Precision	The Number of correctly classified positive samples / The Number of classified positive samples
Recall	The Number of correctly classified positive samples / The Number of positive samples
False positive rate	The Number of misclassified negative samples / The Number of negative samples
F1-score	Formed by precision and recall: $2 \times \frac{precision \times recall}{precision + recall}$

4.2 Results and Analysis

Comparison of Different Clustering Algorithms. The evaluation results of three clustering algorithms are listed in Column 2 to 4 of Table 3. For K-medoids, we find that in practice not all the datasets show a clear "elbow" in the K-E_K chart, or the selected K value leads to a large number of clusters containing only one malicious domain, resulting in inaccurate classification results. In DBSCAN, since the neighborhood radius ε is the maximum distance among the time series of known malicious domains, some series of benign domains with distances smaller than ε from series of malicious domains are inevitably assigned to the cluster of malicious domains, resulting in low accuracy and precision and a high false positive rate. In hierarchical clustering, the distance between clusters is calculated using statistical value. Even though the inter-cluster distance δ is the same as ε, for the series of benign domains with distances smaller than δ from series of malicious domains, if they are more closely related to other benign domains, i.e., the distances are smaller, they may not be assigned to the cluster of malicious domains. Therefore, the quality of hierarchical clustering is better. Its accuracy, precision, recall and F1-score values are all above 90%, which is significantly better than other algorithms. The false positive rate is also lower, making it more suitable for the application scenario of our method.

We further analyze the domains in the detection results that are in the malicious clusters but labeled with unknown properties. The time-based features of these domains are homomorphic with known malicious domains, so they are classified into the same cluster. But they have not been included in the malicious domain list. Since the lifetime of malicious domains are typically short, they are difficult to be fully monitored. Table 4 shows 5 sets of comparison examples in which known malicious domains and

Table 3. Evaluation results of different clustering algorithms

Algorithms	Hierarchical clustering	DBSCAN	K-medoids	Hierarchical clustering (previous method)
Accuracy	96.30%	75.13%	87.39%	88.49%
Precision	92.43%	62.12%	80.74%	59.21%
Recall	99.06%	100%	90.73%	99.61%
False positive rate	5.59%	42.02%	14.92%	13.74%
F1-score	95.63%	76.64%	85.44%	74.27%

unknown domains are clustered together. In data1 and data2, there are altogether 284 such unknown domains, which have not only highly consistent time-based features with known malicious domains but also very similar literal features. They are highly likely to be malicious domains that have not been included yet. In data3- data5, we list some domains with similar time-based features but different literal features to malicious domains. By querying the malware search engine VirusTotal [21], we confirm that these domains are all involved in malicious activities. It shows that in addition to the DGA domains, there are some inherent malicious domains with long lifetime. They can be detected through time-based features.

Table 4. Comparison examples in which known malicious domains and unknown domains are clustered together

	Data1	Data2	Data3	Data4	Data5
Malicious domains	yyuocy.com ulecha.com oyjory.com	hqauzt.com odwkyi.com euaaek.com	ukhizp.com ueahwa.com hdnght.com	ayatgh.com yyqqgj.com xndqqh.com	dkyhtgbb.biz rqofclji.org yhorpvdw.org
Unknown domains	weyjps.com vsjlyj.com forion.com	siaeoo.com qlyigb.com ocnywx.com	ilo.brenz.pl a-gwas-01.slyip.net	a.gwas.perl.sh ant.trenz.pl	euserv9p.ezddns.tk bos.pgzs.com

Comparison with Static Statistical Feature Analysis Method. Literature [11] adopts the previous common practice of analyzing time-based features of domains, which divides the time series with fixed interval, and calculates the total and mean value of DNS requests as static features. It aims to detect compromised hosts, thus, feature such as total number of domains requested by a single host are applied. In order to analyze the change over time of the features, it also quantifies changes for each feature by calculating variation coefficient. However, our purpose is to detect malicious domains, thus, we can only collect features from the perspective of domains. Moreover, we have already applied time series clustering to analyze the change over time of features. As a result, in order to keep the experimental comparison conditions consistent, we generate time series of the number of DNS requests per hour for each domain in each set of the experimental data, and calculate static statistical features such as total, mean, variance, maximum and minimum values to form a 5-dimension feature vector for each domain. We measure the distance between each pair of vectors by Euclidean distance. Since hierarchical clustering works best in our approach, we also apply hierarchical clustering to cluster the static feature vectors. Moreover, in the clustering results, when the number of domains in each cluster is greater than 1 and there are known malicious domains in the cluster, all domains of the cluster are considered malicious, otherwise they are considered benign. The 5th column of Table 3 lists the evaluation result. Compared with the result in the 2nd column, the accuracy and precision are significantly lower. The accuracy is only 59.21%. The false positive rate is 13.74%, indicating that the static feature analysis method misclassifies more benign domains as malicious. This is because the method cannot correctly split the timeline to preserve the integrity of DNS requests in batch, and the static features cover up the details of time series, which cannot extract the key information of time series well. Static features need to be combined with other domain features to achieve the detection of multi feature fusion.

Time Cost Analysis. Figure 1 shows the time cost for each set of experimental data. The main time-consuming stage of the static feature analysis method is to calculate the Euclidean distance between each pair of domains in hierarchical clustering, and the time cost increases synchronously with the number of domains. However, the time cost of our method is affected by multiple factors. In the feature extraction stage, the time cost of clustering the timestamps in dimensionality reduction is affected by the amount of data to be clustered, that is, the number of DNS requests of the source IP. In feature representation, the time-based features are extracted by querying various data from the DNS log database, and the time cost is affected by the complexity of query statements, that is, the number of domains in the query clause. In similarity measure stage, the main time-consuming step is to calculate the DTW distances. The time cost increases with the number of domains requested by the source IP and the length of time series. Since the distances between the time series to be clustered have been calculated in the distance measure stage, the time cost in clustering stage can be ignored. Due to the introduction of density clustering and DTW measurement, the total time cost of our method is higher than that of the static feature analysis method. The time costs of ours vary from 1 to 230 s in all experimental data, which is still acceptable. Through Fig. 1, we can see that when the number of domains further increases, the overall time cost will also increase. When the time cost is unacceptable, there are two ways to reduce it: collecting whitelists to pre filter known benign domains to reduce the number of domains to be analyzed; increasing the threshold of neighborhood radius of density clustering in dimensionality reduction stage to shorten the length of time series. As a result, the overall computational complexity is reduced. Our method does not apply the third-party data such as passive DNS or Whois records, and does not require the collection of a large number of labeled samples to continuously train and update machine learning models. Therefore, the time cost is relatively controllable.

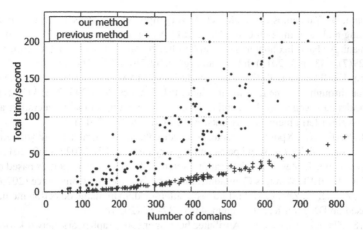

Fig. 1. Time cost analysis

5   Conclusion

In this paper, we propose an analysis method based on time series clustering to extract time-based features and explore malicious domains. In order to obtain the time series, we use density clustering to divide the timeline to preserve the integrity of consecutive requests of malicious domains. Each element of the series is a feature vector composed of five time-based features, detailing the regularity of malicious activities. By using time series clustering, all elements of the series are included in the analysis, thus preserving the temporal characteristics of the features. We use three clustering algorithms to conduct the experiments, and compare the results with that of the static statistical feature analysis method for time-based features applied in previous work. The experimental results show that hierarchical clustering works best and our method effectively detects malicious domains. In the future, we will examine the benign domains that have been misclassified by the existing method and introduce more time-based features to further improve the detection performance.

Acknowledgement. This work was supported by National Key R&D Program of China (2020YFB1805601) and the Research and Planning Project for Higher Education Science of China Association of Higher Education (22XX0403). The computing work in this paper was supported by the Public Service Platform of High Performance Computing by Network & Computation Center of HUST.

References

1. Jiang, Y., Di, W.: An integrated Chinese malicious webpages detection method based on pre-trained language models and feature fusion. In: Zhao, X., Yang, S., Wang, X., Li, J. (eds.) Web Information Systems and Applications: 19th International Conference, WISA 2022, Dalian, China, September 16–18, 2022, Proceedings, pp. 155–167. Springer International Publishing, Cham (2022). https://doi.org/10.1007/978-3-031-20309-1_14

2. Plohmann, D., Yakdan, K., Klatt, M., et al.: A comprehensive measurement study of domain generating malware. In: 25th USENIX Security Symposium, pp. 263–278 (2016)
3. Almomani, A.: Fast-flux hunter: a system for filtering online fast-flux botnet. Neural Comput. Appl. **29**(7), 483–493 (2016). https://doi.org/10.1007/s00521-016-2531-1
4. Iwahana, K., Takemura, T., Cheng, J., et al.: MADMAX: browser-based malicious domain detection through extreme learning machine. IEEE Access **9**, 78293–78314 (2021)
5. Woodbridge, J., Anderson, H., Ahuja, A., et al.: Predicting domain generation algorithms with long short-term memory networks. ArXiv 1611.00791 (2016)
6. Saxe, J., Berlin, K.: eXpose: a character-level convolutional neural network with embeddings for detecting malicious urls, file paths and registry keys. ArXiv 1702.08568 (2017)
7. Liang, Z., Zang, T., Zeng, Y.: Malportrait: sketch malicious domain portals based on passive DNS data. In: IEEE Wireless Communications and Networking Conference (2020)
8. Han, C., Zhang, Y., Zhang, Y.: Fast flucos: malicious domain name detection method for fast flux based on DNS traffic. J. Commun. **41**(5), 37–47 (2020)
9. Zhang, S., Zhou, Z., Li, D., et al.: Attributed heterogeneous graph neural network for malicious domain. In: 24th International Conference on Computer Supported Cooperative Work in Design, pp. 397–403 (2021)
10. Bilge, L., Sen, S., Balzarotti, D., et al.: Exposure: a passive DNS analysis service to detect and report malicious domains. ACM Trans. Inf. Syst. Secur. **16**(4), 1–28 (2014). https://doi.org/10.1145/2584679
11. Li, M., Li, Q., Xuan, G., et al.: Identifying compromised hosts under apt using DNS request sequences. J. Parallel Distrib. Comput. **152**, 67–78 (2021)
12. Lazar, D., Cohen, K., Freund, A., et al.: IMDoC: identification of malicious domain campaigns via DNS and communicating files. IEEE Access **9**, 45242–45258 (2021)
13. Niu, W., Xiao, J., Zhang, X., et al.: Malware on internet of UAVs detection combining string matching and fourier transformation. IEEE Internet Things J. **8**(12), 9905–9919 (2021)
14. Tomatsuri, T., Chiba, D., Akiyama, M., et al.: Time-series measurement of parked domain names and their malicious uses. IEICE Trans. Commun. **E104B**(7), 770–780 (2021)
15. Aghabozorgi, S., Shirkhorshidi, A.S., Wah, T.Y.: Time-series clustering – A decade review. Inf. Syst. **53**(16), 16–38 (2015)
16. Zhu, D., Li, Z., Hu, P., et al.: Improved DBSCAN algorithm based on relative mass of the data field. In: Proceedings of SPIE - The International Society for Optical Engineering, p. 12168 (2022)
17. Alaee, S., Mercer, R., Kamgar, K., et al.: Time series motifs discovery under DTW allows more robust discovery of conserved structure. Data Min. Knowl. Disc. **35**(3), 863–910 (2021)
18. Ran, X., Xi, Y., Lu, Y., et al.: Comprehensive survey on hierarchical clustering algorithms and the recent developments. Artif. Intell. Rev. **56**(8), 8219–8264 (2023)
19. NetLab DGA project: http://data.netlab.360.com/dga/. Last accessed 2 May 2023
20. Alexa's top ranked web sites: http://s3.amazonaws.com/alexa-static/top-1m.csv.zip. Last accessed 2 May 2023
21. Virustotal: https://www.virustotal.com/. Last accessed 2 May 2023

Vulnerability Detection Based on Unified Code Property Graph

Wei Li[1] , Xiang Li[1], Wanzheng Feng[1(✉)], Guanglu Jin[2], Zhihan Liu[1],
and Jing Jia[3]

[1] College of Computer Science and Technology, Harbin Engineering University,
Harbin 150001, China
{wei.li,lzhlzh}@hrbeu.edu.cn, 13170025216@163.com
[2] College of Communication Engineering, Jilin University, Jilin 130012, China
[3] Faculty of Arts, Design and Architecture, School of Built Environment, The University of
New South Wales, Sydney, NSW 2052, Australia
jingl2021@mails.jlu.edu.cn, jing.jia.1@unsw.edu.au

Abstract. As the number of source codes grows rapidly, detecting source code
vulnerabilities in current software has become an important study. Most current
deep learning-based vulnerability detection technologies treat source code as a
sequence, which loses the source code's structural information, leading to many
false positives in the detection results. We propose a novel source code vulnerability detection model, named UCPGVul, based on the Unified Code Property Graph
(UCPG). A new graph representation, UCPG, is proposed to extract semantic
features from the source code. By extracting features from UCPG, our proposed
UCPGVul model can capture more vulnerability features. Experimental results on
a publicly available dataset show that UCPGVul can achieve more accurate and
stable detection results compared to five state-of-the-art methods.

Keywords: Vulnerability detection · Source code analysis · Graph neural
network

1 Introduction

Before computer software is put into service, a code review process is typically performed. Traditional code vulnerability detection methods rely on manual work and
cannot meet large-scale detection requirements. As deep learning technology develops, automated vulnerability detection methods are gradually gaining attention. In deep
learning-based source code vulnerability detection, the program source code does not
directly serve as the input to the neural network model, so the code needs to be converted
into a processable vector representation.

The main problem of current research arises in the feature extraction of semantics,
large-scale pre-training models are starting to show some results, and experts in the vulnerability detection field are beginning to explore building program pre-training models
on Transformer [1–4]. If the source code is treated as natural language, the unique structural information of the source code may be ignored. Therefore, designing a reasonable

L. Yuan et al. (Eds.): WISA 2023, LNCS 14094, pp. 359–370, 2023.
https://doi.org/10.1007/978-981-99-6222-8_30

intermediate representation is an important means to improve the effectiveness of code vulnerability detection models.

For the selection of intermediate representations, Fabian Yamaguchi et al. [5] first proposed the concept of CPG, which has gradually been widely used. This approach represents the source code as a graph structure, with nodes and edges representing code fragments and the dependencies between them. Ghaffarian et al. [6] synthesized abstract syntax trees, control flow graphs, and program dependency graphs and proposed a special intermediate graph representation technique and based on this representation for vulnerability detection. Ghaffarian et al. continued to propose a new representation intermediate graph representation, using the PROGEX [11] tool to extract the intermediate graphs, choosing TF-IDF [12] with the SVD method to do the embedding of the graphs, and achieved good results on public datasets. Şahin [13] generated code attribute graphs from source code and proposed based on the code attribute graphs a new network structure GGS-NNs, which is experimentally proven to be better than traditional methods in cross-project detection, with the shortcoming that it cannot locate specific vulnerability locations during cross-project detection. Wi et al. [14] performed graph matching task by target sample CPG with CPG generated from known vulnerabilities in using CPG for graph isomorphism matching.

In general, the mainstream intermediate representation methods currently in use are graph representation methods, which are the most promising ways to characterize code. However, these methods still have drawbacks, for instance, the relationship between different function calls cannot be reflected in the underlying code property graph, and it is difficult to detect cross-functional vulnerabilities. Therefore, there is still room for improvement in the accuracy of intermediate representation and detection in the current stage of the methods.

Based on the above motivation, We uses the UCPG structure to retain the structural and feature information of the source code, and de-duplicates and vectorizes the generated UCPG, then performs feature extraction of the UCPG by graph neural network to complete the vulnerability detection.

Specifically, our main contributions are:

1. We propose a new graph representation UCPG. UCPG is generated based on the extension of code property graph, we add function call graph (FCG) and natural code sequence (NCS) to it in the way of edges, which can contain more information of vulnerability features.
2. We propose a new vulnerability detection model UCPGVul based on UCPG, whose data preprocessing module transforms the source code into the corresponding joint code attribute graph, thus incorporating the feature information of the source code into the nodes of the code attribute graph.
3. We conducted extensive experiments. The experimental results show that the detection accuracy of UCPGVul is above 90% and exceeds the current mainstream methods.

2   Methodology

As aforementioned, to improve the accuracy of vulnerability detection, We propose a UCPG-based model UCPGVul, whose overall architecture is shown in Fig. 1. It consists of three aspects: (1) UCPG generation. (2) Node feature vectorization, domain aggregation and vector readout. (3) Training of classifier and vulnerability classification. The overall structure of the model is shown in Fig. 1.

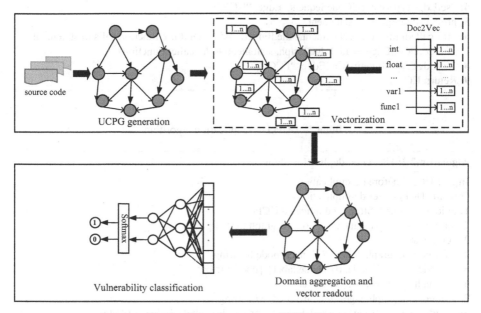

Fig. 1. General structure of UCPGVul.

2.1   Generation and De-duplication of UCGP

The code property graph is a directed graph, where a node represents a structure in the source code and its directed edges can represent different relationships, including AST, CFG, CDG, DDG. by analyzing the source code using the joern tool, we can generate the corresponding CPG. Based on this, the CPG is expanded by adding function call graph (FCG) and natural code sequence (NCS). Finally, the UCPG is stored with Neo4j. The united code attribute graph generation algorithm is shown in Table 1.

In practical applications, the source code is often highly complex and diverse. Therefore, duplicate samples are prone to occur when constructing UCPG datasets, leading to overfitting or underfitting of the model. For this reason, we propose a de-duplication algorithm: each UCPG is transformed into a unique string and hashed and stored in a hash table. If the same UCPG already exists in the hash table, it is directly discarded; otherwise, it is added to the training set. This method is more efficient and accurate compared to the method of de-duplication by neural network (Table 2).

Table 1. UCPG generation algorithm.

Algorithm2.1: UCPG generation algorithm
Input: CPG
Output: UCPG
1: For Node in CPG
2: if previousStatement is not None: // Add natural code sequence edge
3:   naturalCodeSequenceEdge = addEdge(graph, previousStatement, statement)
4:   setEdgeType(naturalCodeSequenceEdge, "NCS")
5: previousStatement = statement
6:   for calledFunction in statement.calledFunctions: // Handling function calls in statements
7:   functionCallEdge = addEdge(graph, statementNode, calledFunction)
8:   setEdgeType(functionCallEdge, "FCG")
9: Return UCPG

Table 2. UCPG de-duplication algorithm.

Algorithm2.2: UCPG de-duplication algorithm
Input: UCPG before de-duplication
Output: UCPG after de-duplication
1: unique_graphs // Store de-duplicated UCPG
2: for graph in graph_dataset: // Convert graph to string
3:   graph_str = " "
4:   for node in graph.nodes: // Convert node to string
5:     node_str = f"{node.id} {node.label} {node.attributes}"
6:     graph_str + = node_str + "\n"
7:   for edge in graph.edges: // Convert edge to string
8:     edge_str = f"{edge.source} {edge.target} {edge.label} {edge.attributes}"
9:     graph_str + = edge_str + "\n"
10:   if graph_str not in unique_graphs:
11:     unique_graphs.add(graph)
12: Return unique_graphs

2.2 Feature Vectorization

The node contents in the joint code property graph are usually of string type, which are not convenient to be processed directly using machine learning models. Therefore, we use the Doc2Vec tool to transform their node contents into initial vectors.

In the graph convolutional neural network, the information transfer messages between nodes are obtained by stitching the feature vectors of the nodes with the feature vectors of their neighbors. Specifically, the information transfer message for each node i is:

$$m_i = \sum_{j \in N(i)} \frac{1}{\sqrt{\deg(i)\deg(j)}} \left(h_i \oplus h_j \right) \tag{1}$$

where the set $N(i)$ contains all neighboring nodes of node i, $\deg(i)$ and $\deg(j)$ denote the degree of node i and node j, respectively, h_i and h_j denote the feature vectors of node i and node j, and $\oplus$ denotes the splicing operation of the vectors.

For each node i, the information transfer messages of its neighbor nodes are aggregated as:

$$a_i = \sigma\left(\sum_{j\in N(i)} m_{i,j}W\right)m_{i,j} \qquad (2)$$

where $m_{i,j}$ denotes the message transfer message between node i and node j, W denotes the weight matrix to be learned, and $\sigma(\cdot)$ denotes the nonlinear activation function.

Finally, the new node feature vector h'_i is defined as a concatenation of the node's own feature vector h_i and the aggregated neighbor node information a_i:

$$h'_i = [h_i \oplus a_i] \qquad (3)$$

Through information transfer and neighborhood aggregation operations, the feature vector of each node can be continuously obtained and updated, and finally h'_i is input to the model as a result vector of function blocks.

The UCPG proposed in this paper has seven types of edges, including AST, CFG, CDG, DDG, NCS, and FCG. Considering the interaction and influence between different edges, all types of edges cannot be simply summed up when generating the adjacency matrix. Therefore, when generating the adjacency matrix, the unidirectional edges are converted into bidirectional edges and multiple type edges are fused to obtain a symmetric adjacency matrix. This can preserve the important information in the code property graph and avoid information redundancy and noise.

2.3 Architecture of Graph Neural Network

The neural network architecture in the UCPGVul model is shown in Fig. 2. The convolutional layer is used to process node features, and the pooling layer is used to reduce the size of the graph by merging multiple nodes into one, thus reducing computational complexity and reducing the possibility of overfitting.

In our paper, we adopt a pooling approach with an attention mechanism, where a set of weights αj is obtained by calculating the inner product and normalization of node features and pooling vectors. Then, the feature vector of each node is multiplied by the corresponding weight, and the weighted feature vectors of all nodes are summed to obtain the pooled feature vector. This attention mechanism-based pooling can adaptively select the nodes that have the greatest impact on the global features, and therefore can better capture the structure and features of the graph.

The graph readout layer performs aggregation operations on node feature representations at different levels, over spreading the feature vectors of all nodes into one vector to obtain the feature representation of the whole graph, in addition, each convolutional pooling module of the UCPGVul model performs a readout operation and combines all readout results. This structure is similar to that of JK-net [17], which allows the retention of graph features at different levels for subsequent classification tasks.

$$r^{(l)} = Radout\left(H^{(l)}, u^{(l-1)}\right) = \left[Avg\left(H^{(l)}\right); Max\left(H^{(l)}\right)\right]u^{(l-1)} \qquad (4)$$

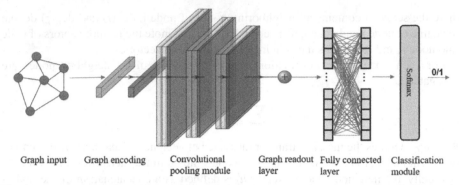

| Graph input | Graph encoding | Convolutional pooling module | Graph readout layer | Fully connected layer | Classification module |

Fig. 2. Neural Network Architecture. From left to right: graph input, graph encoding, convolutional pooling module, graph readout layer, fully connected layer, classification module.

where the node feature matrix of layer l is $H^{(l)}$, $Avg(H^{(l)})$ and $Max(H^{(l)})$ table the design of the average pooling and maximum pooling operations, respectively, $u^{(l-1)}$ is a weight vector for weighting the combination of readout results from different layers, and $r^{(l)}$ is the graph vector of layer l, which represents the features of the whole graph.

Eventually, the full graph feature vector output from the graph readout layer is output to the classifier. The role of the fully connected layer is to multiply the input vector r with the weight matrix W, and then add the bias vector b. Finally, the output vector h is obtained by the activation function. The Softmax layer normalizes the output vector into a binary probability distribution.

3 Experiments

3.1 Experimental Setup

Datasets. We use the dataset from the paper [15] as the base dataset, which contains data for five common vulnerability types, namely (1) incorrect input validation dataset CWE-020 (2) divide by zero exception dataset CWE-369 (3) resource depletion dataset CWE-400 (4) resource allocation exception dataset CWE-770 (5) resource used and not recovered dataset CWE-772.

After being processed by the UCPG de-weighting algorithm, we solve the problem of duplicate samples and avoids overfitting while maintaining the representativeness and diversity of the samples. Meanwhile, we remove some samples randomly to make the number of positive and negative samples the same, eliminating the impact of data imbalance on the experiment. The dataset is divided into training set, validation set and test set according to 8:1:1. The distribution of normal samples and loophole samples of the data set after the processing is shown in Table 3.

Hyper-Parameter Tuning. The initial feature vector length of the node is 64, the batch size is set to 64, the learning rate is set to 0.001, the optimizer is Adam, the epoch is set to 128, the number of node feature iterations is set to 5, and the graph convolution pooling layer in the feature update function is set to 3 layers. The experimental environment is

Table 3. Distribution of samples in the dataset.

Dataset	Vulnerabilities	Normal samples	Total
CWE-020	3684	3684	7368
CWE-369	916	916	1832
CWE-400	2041	2041	4082
CWE-770	412	412	824
CWE-772	1261	1261	2522

a computer running Ubuntu OS with Intel(R) Gold 5218R CPU and NVIDIA GeForce RTX 3080Ti 12GB graphics card. Using the TensorFlow framework in a Python 3.7.1 based environment.

Metrics. The evaluation metrics used in this experiment are as follows: (1) Accuracy (Acc) $= \frac{TP+TN}{TP+TN+FP+FN}$. (2) Precision (Pre) $= \frac{TP}{TP+FP}$. (3) Recall (Rec) $= \frac{TP}{TP+FN}$. (4) False positive rate (FPR) $= \frac{FP}{TN+FP}$. (5) False negative rate (FNR) $= \frac{FN}{TP+FN}$. (6) F1 $= \frac{2*Pre*Rec}{Pre+Rec}$.

Among them, the F1 value is the main performance indicator we consider, as it reflects the overall quality of the model.

3.2 Model Training

The training set is used to train the UCPGVul model, the validation set is used to sieve the model with the best effect, and the test set is used to test the model effect. In this paper, a cross-validation method is used, aiming to find the experimental parameters with the best model effects.

When using graph neural networks for node updating, the number of iterations will determine how much each node is affected by the collar nodes and also affect the computational effort. Therefore, we need to make a trade-off between computational efficiency and model performance. The results obtained by increasing the number of iterations one by one are shown in Fig. 3. The F1 score of the model tends to be stable after five iterations, so the number of iterations is finally set to five in the UCPGVul model.

In addition, the number of convolutional pooling layers also has an impact on the results, and when the number of layers is too large it may lead to overfitting of the model and also increase the computational effort. In order to determine the appropriate number of layers, we conduct experiments using samples from the dataset so that the convolutional pooling layers are added one after another. As shown in Fig. 4, the F1 value reaches the maximum when the number of convolutional pooling layers is 3, so the number of convolutional pooling layers is set to 3.

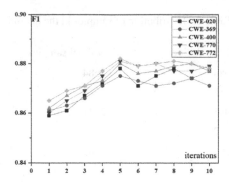

Fig. 3. Effect of the iteration number on the experimental results.

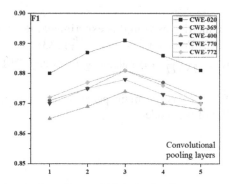

Fig. 4. Effect of the number of convolution pooling layers on experimental results.

3.3　Comparison Experiments

Eval-I: Comparison of UCPG and CPG. We use UCPG for vulnerability feature extraction of source code, while the control group uses the classical CPG. The experimental results are shown in Fig. 5 and Table 4.

According to Table 4, when using our proposed UCPGVul model, using UCPG as an intermediate representation of code can improve the accuracy of source code vulnerability detection better than using CPG. Although the highest accuracy of 88.2% can already be achieved using CPG, the accuracy of over 90% is achieved using UCPG, showing some degree of improvement. On the dataset CWE-369, our model achieved the highest accuracy of 93.4%. In addition, on dataset CWE-020, the UCPG intermediate representation not only achieves the best results in terms of accuracy, but also improves the F1 score by 7.5%. When the model is subjected to feature extraction, our UCPG can extract more effective feature vectors, which increases the accuracy of the model prediction. This result shows that the UCPG intermediate representation is a more suitable code representation for source code vulnerability detection tasks.

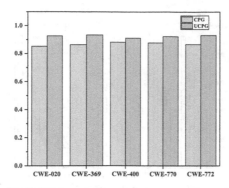

Fig. 5. F1 score of UCPGVul model in graph representation using UCPG or CPG.

Table 4. Performance of the UCPGVul model when using UCPG vs. CPG.

Dataset	Graph type	Acc	Pre	Rec	FPR	FNR	F1
CWE-020	CPG	0.853	0.858	0.845	0.140	0.155	0.852
	UCPG	0.928	**0.951**	0.903	**0.047**	0.097	0.926
CWE-369	CPG	0.863	0.859	0.868	0.142	0.132	0.864
	UCPG	**0.934**	0.939	**0.928**	0.060	**0.072**	**0.934**
CWE-400	CPG	0.882	0.883	0.880	0.117	0.120	0.881
	UCPG	0.913	0.917	0.907	0.082	0.093	0.912
CWE-770	CPG	0.882	0.920	0.837	0.073	0.163	0.877
	UCPG	0.924	0.923	0.925	0.078	0.075	0.924
CWE-772	CPG	0.875	0.928	0.813	0.063	0.187	0.867
	UCPG	0.933	0.941	0.924	0.058	0.076	0.932

Eval-II: Comparison of UCPGVul and Mainstream Models. We compared the UCPGVul model with five other different approaches as follows: (1) RAT [12]; (2) Flawfinder [13]; (3) VDDY [14]; (4) VulDeePecker [15]; (5) Devign [16]. The result is shown in the following Fig. 6.

As can be seen in Table 5, the UCPGVul model outperformed several other models in all metrics, with its detection accuracy above 90% and reached the highest value of 93.4% on the dataset CWE-369. This indicates that the UCPGVul model has excellent detection and generalization capabilities. In summary, the UCPGVul model has high potential and application value, and can provide strong support for vulnerability detection in software development and security.

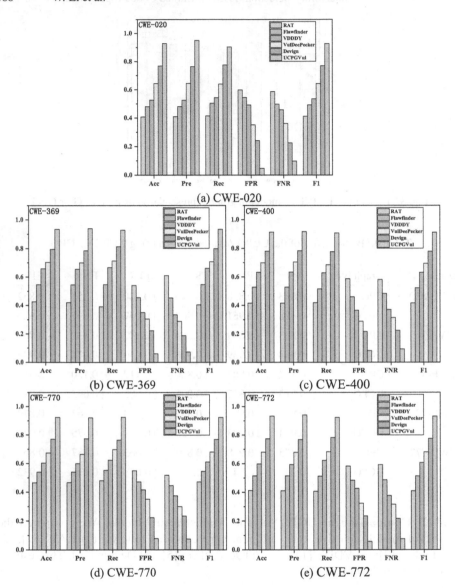

(a) CWE-020

(b) CWE-369

(c) CWE-400

(d) CWE-770

(e) CWE-772

Fig. 6. Comparison of experimental results of six models on different data sets.

Table 5. UCPGVul and other five models vulnerability detection performance comparison.

Dataset	Model	Acc	Pre	Rec	FPR	FNR	F1
CWE-020	RAT	0.408	0.409	0.414	0.598	0.586	0.412
	Flawfinder	0.479	0.480	0.503	0.545	0.497	0.492
	VDDDY	0.526	0.525	0.543	0.491	0.457	0.534
	VulDeePecker	0.644	0.645	0.639	0.352	0.361	0.642
	Devign	0.767	0.763	0.775	0.240	0.225	0.769
	UCPGVul	**0.928**	**0.951**	**0.903**	**0.047**	**0.097**	**0.926**
CWE-369	RAT	0.425	0.420	0.391	0.540	0.609	0.405
	Flawfinder	0.546	0.546	0.547	0.455	0.453	0.546
	VDDDY	0.658	0.655	0.666	0.350	0.334	0.661
	VulDeePecker	0.704	0.700	0.712	0.305	0.288	0.706
	Devign	0.795	0.785	0.812	0.223	0.188	0.798
	UCPGVul	**0.934**	**0.939**	**0.928**	**0.060**	**0.072**	**0.934**
CWE-400	RAT	0.415	0.416	0.418	0.587	0.582	0.417
	Flawfinder	0.528	0.529	0.516	0.460	0.484	0.523
	VDDDY	0.632	0.633	0.629	0.365	0.371	0.631
	VulDeePecker	0.699	0.704	0.685	0.288	0.315	0.695
	Devign	0.780	0.783	0.776	0.216	0.224	0.779
	UCPGVul	**0.913**	**0.917**	**0.907**	**0.082**	**0.093**	**0.912**
CWE-770	RAT	0.465	0.466	0.481	0.551	0.519	0.473
	Flawfinder	0.540	0.539	0.553	0.473	0.447	0.546
	VDDDY	0.603	0.599	0.624	0.417	0.376	0.611
	VulDeePecker	0.674	0.665	0.699	0.352	0.301	0.682
	Devign	0.771	0.774	0.765	0.223	0.235	0.769
	UCPGVul	**0.924**	**0.923**	**0.925**	**0.078**	**0.075**	**0.924**
CWE-772	RAT	0.412	0.411	0.408	0.584	0.592	0.409
	Flawfinder	0.514	0.514	0.513	0.485	0.487	0.514
	VDDDY	0.598	0.593	0.623	0.427	0.377	0.608
	VulDeePecker	0.681	0.680	0.684	0.323	0.316	0.682
	Devign	0.774	0.769	0.783	0.236	0.217	0.776
	UCPGVul	**0.933**	**0.941**	**0.924**	**0.058**	**0.076**	**0.932**

4 Conclusion

We proposed a new source code graphical representation UCPG and the corresponding sample de-duplication algorithm, which can reflect the structure and characteristics of source code more comprehensively and improve the accuracy and reliability of experimental results. We further designed the source code vulnerability detection model UCPGVul. This method converts the source code into UCPG, and then transforms the source code vulnerability detection into a graph classification problem. Our Experimental results on public datasets show that UCPGVul detection is more accurate and stable.

References

1. Feng, Z., Guo, D., Tang, D., et al.: Codebert: A pre-trained model for programming and natural languages. arXiv preprint arXiv:2002.08155 (2020)
2. Zhou, X., Han, D.G., Lo, D.: Assessing generalizability of codebert. In: 2021 IEEE International Conference on Software Maintenance and Evolution (ICSME), pp. 425–436. IEEE (2021)
3. Ahmad, W.U., Chakraborty, S., Ray, B., et al.: Unified pre-training for program understanding and generation. arXiv preprint arXiv:2103.06333 (2021)
4. Zhou, W., Junhua, W.: Code comments generation with data flow-guided transformer. In: Zhao, X., Yang, S., Wang, X., Li, J. (eds.) Web Information Systems and Applications: 19th International Conference, WISA 2022, Dalian, China, September 16–18, 2022, Proceedings, pp. 168–180. Springer International Publishing, Cham (2022). https://doi.org/10.1007/978-3-031-20309-1_15
5. Yamaguchi, F., Golde, N., Arp, D., et al.: Modeling and discovering vulnerabilities with code property graphs. In: 2014 IEEE Symposium on Security and Privacy, pp. 590–604. IEEE (2014)
6. Ghaffarian, S.M., Shahriari, H.R.: Neural software vulnerability analysis using rich intermediate graph representations of programs. Inf. Sci. **553**, 189–207 (2021)
7. https://github.com/ghaffarian/progex/
8. Manning, C.D.: An Introduction to Information Retrieval. Cambridge University Press (2009)
9. Şahin, C.B.: Semantic-based vulnerability detection by functional connectivity of gated graph sequence neural networks. Soft Comput. 1–17 (2023)
10. Wi, S., Woo, S., Whang, J.J., et al.: HiddenCPG: large-scale vulnerable clone detection using subgraph isomorphism of code property graphs. Proc. ACM Web Conf. **2022**, 755–766 (2022)
11. Wang, H., Ye, G., Tang, Z., et al.: Combining graph-based learning with automated data collection for code vulnerability detection. IEEE Trans. Inf. Forensics Secur. **16**, 1943–1958 (2020)
12. https://code.google.com/archive/p/rough-auditing-tool-for-security/
13. https://dwheeler.com/flawfinder/
14. Kim, S., Woo, S., Lee, H., et al.: Vuddy: A scalable approach for vulnerable code clone discovery. In: 2017 IEEE Symposium on Security and Privacy (SP), pp. 595–614 (2017)
15. Li, Z., Zou, D., Xu, S., et al. : Vuldeepecker: A deep learning-based system for vulnerability detection. arXiv preprint arXiv:1801.01681 (2018)
16. Zhou, Y., Liu, S., Siow, J., et al.: Devign: Effective vulnerability identification by learning comprehensive program semantics via graph neural networks. In: Advances in neural information processing systems, vol. 32 (2019)
17. Xu K, Li C, Tian Y, et al. Representation learning on graphs with jumping knowledge networks. In: International Conference on Machine Learning. PMLR, pp. 5453–5462 (2018)

Short Repudiable Ring Signature: Constant Size and Less Overhead

Ziqiang Wen, Xiaomei Dong$^{(\boxtimes)}$, Chongyang Yang, and Ning Zhang

School of Computer Science and Engineering, Northeastern University,
Shenyang 110169, China
xmdong@mail.neu.edu.cn

Abstract. Ring signature, introduced by Rivest, allows the signer to make valid signatures without the knowledge of the signer. But unconditional anonymity also protects malicious users, so that the reputation of the group can be affected. Lin et al. proposed a rejectable signature scheme, which solved this problem well, but it has a drawback of increasing the signature's size with the number of ring members. To address the problem, a new scheme is proposed in combination with the existing short associated ring signature scheme. This proposed scheme utilizes a dynamic accumulator to accumulate public keys and knowledge signatures for rapid identity verification without revealing one's private key or identity. By doing so, in our scheme we can maintain a constant signature length and minimize the signature overhead. Additionally, in this paper we prove that the resulting signature is both unforgeable and anonymous while also being able to repudiate. Furthermore, related experiments are conducted to demonstrate the effectiveness of this approach. Experimental results show reduced time overhead and shorter signal length compared to theoretical calculations, reinforcing the feasibility of our scheme.

Keywords: dynamic accumulator · short linkable ring signature · signatures based on proofs of knowledge

1 Introduction

Ring signature, as a type of digital signature scheme, allows a member of a group to sign a message indicating that the signer is a member of the group without revealing individual identity. Proposed by Rivest et al. [1] in 2001, it has gained significant attention in the research community due to its unconditional anonymity property and the lack of trusted centers. In cryptocurrencies like Monero [2], such a currency system uses ring signatures to ensure the anonymity of transactions. Recently, due to the active research on blockchain privacy protection [3], ring signatures, as an extremely important privacy protection mechanism for blockchains, have been recognized by scholars.

However, unconditional anonymity presents a significant challenge to the regulatory system [4–6]. Malicious signers can easily use anonymous signatures to authenticate false information. Park et al. [6] proposed a repudiable ring signature that allows honest signers to prove that they did not generate malicious

L. Yuan et al. (Eds.): WISA 2023, LNCS 14094, pp. 371–382, 2023.
https://doi.org/10.1007/978-981-99-6222-8_31

signatures. There is a detailed diagram in Fig. 1. Lin et al. [7] further developed this idea, introducing repudiation-unforgeability to prevent adversaries from pretending to be honest signers. However, both these methods suffer from expensive signature size and time, especially when the number of users in the ring is large, causing a significant performance issue. To solve these problems, we propose using dynamic accumulator and double discrete logarithm knowledge for fast verification of signatures to reduce time overhead, while keeping the signature constant overhead.

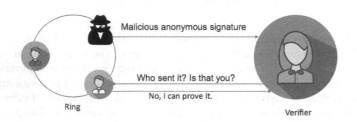

Fig. 1. Repudiable ring signature

The short repudiable ring signature scheme (SRRS) is then introduced, based on this approach, and its security is proven in the paper. Comparative experiments are conducted to evaluate the performance of the SRRS scheme. Overall, we suggest a viable solution in this paper for repudiable ring signatures that are both more efficient and secure, demonstrating the importance of continually improving such systems.

Although Wang et al. [8] used dynamic accumulators to shorten the signature length before, our scheme is completely different from Wang's scheme:

- The knowledge signature we constructed is completely different from the knowledge signature constructed by Wang. Our knowledge signature is constructed according to the more efficient double discrete logarithm relationship signature in the scheme [9], while Wang used single discrete logarithm relationship construction.
- The application background is different. We use signature of the proof of knowledge (SPK) in a repudiable signature, which has never been done before.
- We believe that adding the *link* tag suffix to the signature alone could make it easy for an adversary to forge a signed link by using the previous person's *link* tag. To prevent this, we propose the use of accumulator evidence to create association labels. If a malicious node attempts to generate a forged *link* label, the accumulator proof would fail to generate a signature, thereby deterring any fraudulent activity.

2   Related Work

Komano et al. [5] introduced an interactive deniable ring signature, but interactive deniable ring signatures are not popular. Park et al. [6] proposed a non-interactive repudiable ring signature scheme to generate signature denials through a non-interactive algorithm, but they did not consider repudiation-unforgeability and the time complexity of the signature algorithm is very high. Lin et al. [7] propose a logarithmically-sized rejectable signature scheme that improves the above problems. But spending still has room for improvement.

In 2004, Liu et al. [10] proposed the first linkable ring signature scheme. Tsang et al. [11] proposed a linkable ring signature with constant signature size using accumulator and knowledge signature technology. Au et al. [12] made a security optimization of the short signature technique. Wang et al. [8] applied short association ring signature technology to editable blockchain. However, the above schemes do not meet the latest requirement of ring signatures, repudiability.

In this paper, we combine the above problems and propose a deniable ring signature scheme SRRS based on dynamic accumulator and SPK. The paper is structured into three main parts. The first part describes the algorithm of SRRS, which outlines the steps involved in generating the signature. The second part is dedicated to the security proof, where it is demonstrated that our scheme can effectively generate signatures correctly. Lastly, the third part presents the experimental evaluation, which involves comparing our scheme to existing ones to show that it is more practical and cost-effective.

3   SRRS Scheme

Our scheme combines the construction of short associated ring signatures and repudiable ring signatures. We first accumulate the public keys of ring members, then construct new knowledge proof signatures based on the knowledge proof signatures of Camenisch and Stadler [9], and finally propose the SRRS scheme. It contains seven algorithms (*Setup, Gen, Sign, Verify, Repudiate, VerRepu, Link*), the specific description is as follows:

1. $SRRS.Setup(1^\lambda) \rightarrow (G, q, g, H)$. We input the security parameters, and according to the security parameters, select a GDH group G whose generator is g and order is q. The hash function H is a one-way map of $(0,1)^* \rightarrow G$.
2. $SRRS.Gen(G, q, g, H) \rightarrow (pk_i, sk_i)$. We input the system parameters, select a random number sk_i according to the security parameters, generate its public key $pk_i = g^{sk_i}$, and output the user's public-private key pair (pk_i, sk_i).
3. $SRRS.Sign(R, m, sk_\pi) \rightarrow (\sigma_R(m))$. Input the ring $R = \{pk_1, \cdots, pk_n\}$, message m, and the private key of the signer sk_π, and run the following algorithm to generate the SRRS signature of the message:

Compute the cumulative value of all elements in the collection using a dynamic accumulator generation algorithm.

$$v = f(u, R) = u^{\prod pk_i} \bmod N \tag{1}$$

Use the dynamic accumulator member affiliation generation algorithm to generate evidence pk_π belongs to the public key ring R, and the evidence is expressed as follows.

$$w_\pi = f(u, R/\{pk_\pi\}) = u^{\prod\limits_{i \neq \pi} pk_i} \bmod N \qquad (2)$$

Computing Knowledge Signatures

$$SPK\{(pk_\pi, sk_\pi) : v = w_\pi{}^{pk_i} \bmod N \wedge pk_i = g^{sk_i}\}(m) \qquad (3)$$

We think that we only need to prove that we have the private key of the member of the accumulator. We construct the knowledge signature according to the knowledge signature construction algorithm of Cam [9] for double discrete logarithms. Proving the double discrete logarithm relationship,

$$v = w_\pi{}^{g^{sk_\pi}} \qquad (4)$$

First, we have to randomly select k numbers: $r_1, r_2, \cdots\cdots, r_k \in Z_p$, and then compute the variable c and s

$$c = H(m||u||N||g||t_1||\cdots||t_n) \quad with \quad t_i = w_\pi{}^{g^{r_i}} \bmod N \qquad (5)$$

$$s_i = r_i - c[i] * sk_\pi \qquad (6)$$

When doing *link* tags, I found that we can easily obtain other users' *link* tags, so that we can use other users' *link* tags to make illegal signatures. Or when someone is linking, others cannot verify whether his *link* label is correct, then we have improved the *link* label.
Calculate $h = H(R)$, and the *link* label is

$$\tilde{y}_\pi = h^{w_\pi} \qquad (7)$$

Finally, we get the SRRS signature

$$\sigma_\pi(m) = (c, s_1, s_2, \cdots\cdots, s_k, v, \tilde{y}_\pi) \qquad (8)$$

4. SRRS.Verify$(R, m, \sigma(m)) \rightarrow (0 \text{ or } 1)$. Enter the signature message m, enter the public key ring R, sign $\sigma_\pi(m)$, and perform the following calculations

$$v = u^{\prod pk_i} \bmod N \qquad (9)$$

If the above formula is not established, deny the correctness of the signature and reject the signature, otherwise continue to compute c'

$$c' = H(m||u||N||g||t_1||\cdots||t_n||\tilde{y}_\pi) \quad with \quad t_i = \begin{cases} w_\pi{}^{g^{s_i}} & if \ c[i] = 0 \\ v_\pi{}^{g^{s_i}} & otherwise \end{cases} \qquad (10)$$

If $c = c'$ the signature is accepted, the *Verify* step is complete. Otherwise, the signature is refused.

5. $SRRS.Repudiate(sk_D, m, R, \sigma_\pi(m)) \rightarrow \xi(\sigma_D(m), \sigma_\pi(m))$. Having our own private key, we can generate our own different signatures for this message to deny that the previous message is signed by ourselves. Input the private key of the denier sk_D, the information represented by the signature, $m \in \{0,1\}^*$ the public key ring R, and output the negative evidence of the signature ξ, the calculation is as follows: Use the dynamic accumulator membership generation algorithm to generate evidence of membership in the public key ring.

$$w_D = f(u, R/\{pk_D\}) = u^{\prod\limits_{i \neq D} pk_i} \bmod N \tag{11}$$

Calculate the knowledge signature

$$SPK\{(pk_D, sk_D) : v = w_\pi^{pk_D} \bmod N \wedge pk_D = g^{sk_D}\}(m) \tag{12}$$

Prove the double discrete logarithm relationship like the Sign function

$$v = w_D^{g^{sk_D}} \tag{13}$$

First, randomly select k numbers: $r_1', r_2', \cdots\cdots, r_k' \in Z_p^*$, Calculate according to the formula

$$c_D = H(m||u||N||g||t_1|| \cdots ||t_n) \quad with \quad t_j - w_D^{g^{r_j}} \bmod N \tag{14}$$

$$s_{Dj} = r_j - c_D[i] * sk_D \tag{15}$$

The SPK of sk_D on $m \in \{0,1\}^*$ is

$$SKLOGLOG(sk_D) = (c_D, s_{D1}, s_{D2}, \cdots\cdots, s_{Dk}) \tag{16}$$

Then we get the SRRS deny signature

$$\sigma_D(m) = (c_D, s_{D1}, s_{D2}, \cdots\cdots, s_{Dk}, v, \tilde{y}_D) \tag{17}$$

We only need to prove that the proof C_D is valid, and if it is inconsistent with the proof C_π, we can prove that the signature is not generated by me. From this we can get the denial signature $\xi = (\sigma_D(m), \sigma_\pi(m))$.

6. $SRRS.VerRepu(\xi(\sigma_\pi(m)), v) \rightarrow (0 \text{ or } 1)$. From the above analysis we can get our negative confirmation algorithm.
Verify if

$$v = u^{\prod pk_i} \bmod N \tag{18}$$

If the above formula is not established, deny the correctness of the rejection, deny the denial, otherwise continue.
Calculate

$$c_D' = H(m||u||N||g||t_1|| \cdots ||t_n) \quad with \quad t_i = \begin{cases} w_D^{g^{s_i}} & if \ c_D[i] = 0 \\ v^{g^{s_i}} & otherwise \end{cases} \tag{19}$$

If $c_D' = c_D$ and $w_D \neq w_\pi$, we consider the denial to be correct, that is, the ring signature was not generated by user D.

7. $SRRS.Link(\tilde{y}_1, \tilde{y}_2) \rightarrow (0 \text{ or } 1)$, Input the *link* identifiers of two signatures, if the two *link* identifiers are equal, output 1, otherwise output 0.

4 Security of SRRS

4.1 Correctness

We have the correctness of our accumulated value according to the correctness of the dynamic accumulator

$$v = w_\pi{}^{pk_\pi} \bmod N = u^{\prod y_i} \bmod N = u^{\prod g^{x_i}} \bmod N \tag{20}$$

Then we need to prove the correctness of the signed value, when $c[i] = 0$, we have that

$$t_i(proof) = w_\pi{}^{g^{s_i}} = w_\pi{}^{g^{r_i - 0}} = w_\pi{}^{g^{r_i}} = t_i(verify) \tag{21}$$

when $c[i] = 1$, we have that

$$
\begin{aligned}
t_i(proof) &= v^{g^{s_i}} \\
&= v^{g^{r_i - sk_\pi}} \\
&= (w_\pi{}^{g^{sk_\pi}})^{g^{r_i - sk_\pi}} \\
&= w_\pi{}^{g^{sk_\pi + r_i - sk_\pi}} \\
&= w_\pi{}^{g^{r_i}} \\
&= t_i(verify)
\end{aligned}
\tag{22}
$$

Since we have $t_i(proof) = t_i(verify)$ a whether $c[i]$ equal to 0 or 1, we have

$$c(sign) = H(m||u||N||g||t_1||\cdots||t_n) = c(verify) \tag{23}$$

4.2 Unforgeability

Theorem 1. *Assuming strong RSA assumption and discrete logarithm problem is hard, then SRRS signature is unforgeable.*

Proof. Under the random number oracle model, let the adversary $\mathcal{A}$ be an attacker who chooses the ciphertext, and make at most q_H queries within the specified time. If the adversary wins the unforgeable game with an advantage of ε, then either the attacker breaks the security of the dynamic accumulator, or solves the discrete logarithm problem with a non-negligible advantage, which is impossible in the cryptography.

The unforgeability game of Challenger $\mathcal{C}$ and Adversary $\mathcal{A}$ is as follows:

1) *Initialization phase.* Given the security parameters, the challenger $\mathcal{C}$ runs $SRRS.Setup(1^\lambda)$ to generate the corresponding hash function H, the corresponding GDH group G, and sends all the parameters to the adversary $\mathcal{A}$.
2) *Query phase.* In addition to q_H hash queries, $\mathcal{A}$ can also make the following queries:
   (a) Key extraction query. $\mathcal{A}$ submits the user identity ID_{i*} to $\mathcal{C}$, and $\mathcal{C}$ returns the the corresponding private key sk_i.

(b) Signature query. $\mathcal{A}$ submits user public key ring $R^* = \{pk_1, pk_2, \cdots pk_{n*}\}$ to $\mathcal{C}$. It should be pointed out that here n^* can be any number of $(1, n)$ and a message m^*. $\mathcal{C}$ returns a ring signature σ^* of R^* on the message m^*.

3) Forgery phase. $\mathcal{A}$ needs to forge a SRRS signature σ^* of the user group $R^* = \{pk_1, pk_2, \cdots pk_{n*}\}(1 \leq n^* \leq n)$ on the message m^*, and the challenger $\mathcal{C}$ will not disclose any private key information to the opponent $\mathcal{A}$, or directly ask for any user's signature. If $\mathcal{A}$'s forged signature is legal, then Adversary $\mathcal{A}$ wins the game, SRRS is not unforgeable.

According to the Forking lemma [13], if the adversary $\mathcal{A}$ can forge a legal SRRS signature without knowing the private key, then the challenger $\mathcal{C}$ submits the same user to $\mathcal{A}$, and $\mathcal{A}$ can conduct two challenges with same output. It means that challenger $\mathcal{C}$ can solve the difficult problem based on SRRS. Suppose two valid signatures $\sigma_1^*(m^*) = (c^*, s_1{}^*, s_2{}^*, \cdots\cdots, s_k{}^*, v^*, \tilde{y}^*)$ and $\sigma_2^* = (c_0^*, s_{01}{}^*, s_{02}{}^*, \cdots\cdots, s_{0k}{}^*, v^*, \tilde{y}^*)$ output by $\mathcal{A}$ satisfy that $c^* \neq c_0^*$.

According to the signature mechanism, the two signatures output must meet the following conditions

$$s_i^* = r_i^* - c^*[i] * \delta_\pi^* \tag{24}$$

$$s_{0i}^* = r_i^* - c_0^*[i] * \delta_\pi^* \tag{25}$$

Subtracting the two equations

$$\delta_\pi^* = \frac{s_i^* - s_{0i}^*}{c^*[i] - c_0^*[i]} \tag{26}$$

we set that

$$y_\pi^* = y^{\delta_\pi^*} \tag{27}$$

According to the nature of the accumulator, we need to forge the public key must meet the condition of the accumulator, namely

$$w_\pi^* = f(u, R^*/\{pk_\pi^*\}) = u^{\prod_{i \neq \pi} pk_i} \mod N \tag{28}$$

$$v^* = w_\pi^{*pk_\pi^*} \tag{29}$$

At this time, we discuss in two cases.

1) If $pk_\pi^* \notin R^* = \{pk_1, \cdots, pk_{n*}\}$, then we believe that it can satisfy the judgment condition of the dynamic accumulator, but it is not a member of the accumulator, which violates the security of the dynamic accumulator.

2) If $pk_\pi^* \in R^* = \{pk_1, \cdots, pk_{n*}\}$, according to $s_i = r_i - c[i] * sk_\pi$ we have

$$pk^{\delta_\pi^* c^*[i]} = g^{r_i^* - s_i^*} \tag{30}$$

$$pk^{\delta_\pi^* c_0^*[i]} = g^{r_i^* - s_{0i}^*} \tag{31}$$

Divide the two expressions to calculate:

$$pk^{\delta_\pi^* c_0^*[i] - \delta_\pi^* c^*[i]} = g^{s_i^* - s_{0i}^*} \tag{32}$$

Equation (32) is reduced to:

$$pk = g^{\frac{s_i^* - s_{0i}^*}{\delta_\pi^* c_0^*[i] - \delta_\pi^* c^*[i]}} \tag{33}$$

Because we have $pk = g^{sk}$, we can directly get the value of the private key sk by comparison, in other words, we can get the private key sk by signing, which further solves the discrete logarithm problem.

To sum up, if the adversary $\mathcal{A}$ forges the signature successfully, then we can prove by reasoning that either $\mathcal{A}$ destroys the security of the cryptographic accumulator, or destroys the difficulty of the discrete logarithm problem. This is clearly impossible, thus proving the unforgeability of SRRS.

4.3 Anonymity

Theorem 2. *If the source of random numbers used by SRRS is uniformly distributed, the SRRS scheme has anonymity.*

Proof. Suppose there are two uniformly distributed random number sources Ω_1 and Ω_2, for any given $sk_1 \leftarrow_\$ \Omega_1$, $sk_2 \leftarrow_\$ \Omega_2$, sk_1 and sk_2 are statistically indistinguishable. If the adversary can find the anonymous with a non-negligible advantage, then the challenger $\mathcal{C}$ can distinguish sk_1 from sk_2 with a non-negligible advantage.

The anonymity game of Challenger $\mathcal{C}$ and Adversary $\mathcal{A}$ is as follows:

1) *Initialization phase.* Given the security parameters, the challenger $\mathcal{C}$ runs $SRRS.Setup(1^\lambda)$ to generate the corresponding hash function H, the corresponding GDH group G, and sends all the parameters to the adversary $\mathcal{A}$.
2) *Query phase.* In addition to q_H hash queries, $\mathcal{A}$ can also make the following queries:
   (a) Key extraction query. $\mathcal{A}$ submits the user identity ID_{i*} to $\mathcal{C}$, and $\mathcal{C}$ returns the corresponding private key sk_i.
   (b) Signature query. $\mathcal{A}$ submits a user ID group $R_{ID} = \{ID_1, ID_2, \cdots ID_{n*}\}$ to $\mathcal{C}$. It should be pointed out that here n^* can be any number of $(1, n)$ and a message m^* . $\mathcal{C}$ returns a ring signature σ^* of R^* on the message m^*.
3) *challenge phase.* $\mathcal{A}$ submits a message to $\mathcal{C}$, a user ID group R_{ID}, and two user group identity ID_{π_1} and ID_{π_2}, $\mathcal{C}$ randomly chooses to use ID_{π_1} or ID_{π_2} to perform SRRS ring signature on the message m^*, and returns the obtained signature to $\mathcal{A}$.
4) *guessing phase.* $\mathcal{A}$ returns ID_{π_1} or ID_{π_2}, and if the returned content is the content selected by $\mathcal{C}$, then $\mathcal{A}$ wins the ANON game.

Analyze the signature $\sigma^* = (c^*, s_1{}^*, s_2{}^*, \cdots\cdots, s_k{}^*, v^*, \tilde{y}^*)$ generated by $\mathcal{C}$, because $\mathcal{C}$ randomly chooses $\pi \in \{1, 2, \cdots, n\}$ to calculate the accumulation algorithm, so the generated w_π^* is random, and no matter which $\pi \in \{1, 2, \cdots, n\}$ $\mathcal{C}$ chooses, the final cumulative value v^* obtained is the same.

Then for c^* and s_i^*, and c^* is the content after the hash operation, which means it possesses randomness. Since r_i is randomly selected in Z_q^*, knowing the values of c^* and s_i^* does not provide any advantage for guessing the value of π.

To sum up, none of the parameters in the signature can provide any advantage for guessing the value of π, SRRS has anonymity.

4.4 Repudiability, Repudiation Unforgeability and Linkability

We briefly introduce Repudiability, Repudiation unforgeability and Linkability of SRRS.

Repudiability. If user D wants to deny the signature $\sigma_\pi(m)$, then he can definitely generate a corresponding signature of his own, and the evidence in his signature w_D is completely different from w_π. This signature can pass the Ver-Repu function, and the Repudiability is established.

Repudiation Unforgeability. User π wants to deny the signature $\sigma_\pi(m)$, the signature generated by him with his own evidence cannot pass the denial verification, and the signature cannot be forged with other people's evidence (the unforgeability has been proved before).

Linkability. We use proofs w_π when linking to calculate *link* parameters $\tilde{y}_\pi$. Adversary $\mathcal{A}$ will not pass the verification if he directly steals the *link* parameters of other users π. Users can't forge *link* parameters at will, so that they don't match the evidence they use to prove that they can't pass the verification.

5 Performance Evaluation

5.1 Performance Analysis

Since our scheme is an improvement upon the RRS scheme, we only compare the performance improvement of our solution with that of RRS [7]. Because the author of RRS did not instantiate it, so we instantiate it roughly by the following method: We use the VRF based on construction for strong RSA assumptions [14]. The SPB function was constructed using the SSB function of Tatsuaki et al. [15] NIWI is constructed using Pedersen Commitment and Schnorr signatures. Use n to represent the number of members in R, and use T_e, T_H and T_m to represent the time overhead of performing exponential operations, group multiplication operations, and hash operations, respectively. The less time-consuming operations are omitted, and the signature generation, verification, and repudiation phases of the two schemes are analyzed separately.

It can be seen from Table 1 that with the increase of ring members, the time overhead of both the SRRS scheme and the RRS scheme increases. Lin [7] have pointed out that the signature size of the RRS scheme is logarithmic in size, which is one of the three advantages of RRS. However, the signature size of the SRRS scheme constructed by our dynamic accumulator combined with the knowledge signature remains unchanged as same as wang [8]. From the table, we can roughly see that the overhead of the RRS scheme is less than that of the SRRS scheme in terms of signature time.

Table 1. Performance comparison of signature schemes

Cost	SRRS	RRS
Sign time	$(n+k)T_m + 2T_H + T_e$	$(2\lceil \log n \rceil + 6)T_m + (3n+7)T_H + 18T_e$
Verify time	$(n+k)T_m + 2T_H + (2k+1)T_e$	$2T_m + (2n+1)T_H + 6T_e$
Repudiate time	$(n+k)T_m + (2k+1)T_e$	$7T_m + (2n+8)T_H + 12T_e$
VerRepu time	$nT_m + T_H + (2k+1)T_e$	$2T_m + (2n+1)T_H + 6T_e$
Signature length	$poly(k) \cdot poly(\lambda)$	$\log(n) \cdot poly(\lambda)$

5.2   Experimental Evaluation and Limitations

The experiment in this paper is carried out on a laptop, which is configured as an Inter Core i5-11260H@2.60 GHz CPU, a 512G solid state drive, 16G RAM, and the operating system is Win10. In Figs. 2, 3, 4 and 5, we present the results of comparing the SRRS scheme and the RRS scheme in terms of signature time, verification time, repudiation time, and repudiate-verification time with varying numbers of ring members in this paper. The results show that both schemes experience an increase in time required for these four phases as the number of ring members increases. This trend is consistent with the theoretical analysis presented in the paper. However, the SRRS scheme has a slower growth rate and less overhead compared to the RRS scheme. It needs to be pointed out that our scheme also has limitations as we sacrifice a certain level of security in exchange for the performance of the signature algorithm. However, our solution is still able to meet the basic security requirements. Overall, these findings support the effectiveness of the SRRS scheme proposed in this paper.

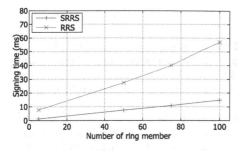

Fig. 2. Time comparison of each scheme in signature stage.

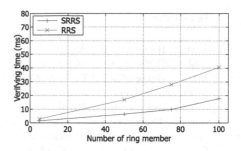

Fig. 3. Time comparison of each scheme in verification stage.

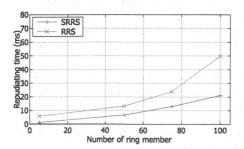

Fig. 4. Time comparison of each scheme in repudiation stage.

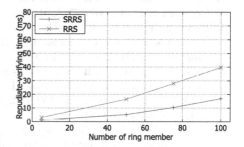

Fig. 5. Time comparison of each scheme in repudiate-verification stage.

6   Conclusion

In this paper we propose a new deniable ring signature scheme named SRRS, which is relatively short and effective. It demonstrates that the proposed scheme has unforgeability, anonymity, repudiability, and repudiation-unforgeability. The size of the signature in SRRS remains unchanged despite an increase in the number of ring members, which is not the case with other schemes. We conducted experiments to validate our scheme. In our experiments, we found that in the phases of signature, verification and repudiation, as the number of ring members increases, the speed of the SRRS signature scheme increases but remain lower overhead. Our findings align with theoretical calculations. In the future, it will be an exciting research direction to use the SRRS scheme in the privacy protection of blockchain.

Acknowledgement. This work is supported by the Major projects of National Social Science Foundation of China (No. 21&ZD124): research on community home care model and quality safety System based on blockchain. We also thank anonymous reviewers for the helpful reports.

References

1. Rivest, R.L., Shamir, A., Tauman, Y.: How to leak a secret. In: Boyd, C. (ed.) ASIACRYPT 2001. LNCS, vol. 2248, pp. 552–565. Springer, Heidelberg (2001). https://doi.org/10.1007/3-540-45682-1_32
2. Noether, S.: Ring signature confidential transactions for monero. IACR Cryptol. ePrint Arch., p. 1098 (2015)
3. Xu, X., Dong, X., Li, X., He, G., Xu, S.: Patient-friendly medical data security sharing scheme based on blockchain and proxy re-encryption. In: Zhao, X., Yang, S., Wang, X., Li, J. (eds.) Web Information Systems and Applications. WISA 2022. LNCS, vol. 13579, pp. 615–626. Springer, Cham (2022). https://doi.org/10.1007/978-3-031-20309-1_54
4. Zeng, S., Jiang, S.: A new framework for conditionally anonymous ring signature. Comput. J. **57**(4), 567–578 (2014)
5. Komano, Y., Ohta, K., Shimbo, A., Kawamura, S.: Toward the fair anonymous signatures: deniable ring signatures. IEICE Trans. Fundam. Electron. Commun. Comput. Sci. **90-A**(1), 54–64 (2007)
6. Park, S., Sealfon, A.: It wasn't me! - repudiability and claimability of ring signatures. In: Boldyreva, A., Micciancio, D. (eds.) Advances in Cryptology – CRYPTO 2019. LNCS, vol. 11694, pp. 159–190. Springer, Cham (2019). https://doi.org/10.1007/978-3-030-26954-8_6
7. Lin, H., Wang, M.: Repudiable ring signature: stronger security and logarithmic-size. Comput. Stand. Interfaces **80**, 103562 (2022)
8. Jiechang, W., Yuling, L., Ping, Z., Muhua, L., Jie, L.: Short linkable-and-redactable ring signature and its blockchain correcting application. J. Beijing Univ. Aeron. Astron **48** (2022)
9. Camenisch, J., Stadler, M.: Efficient group signature schemes for large groups. In: Kaliski, B.S. (ed.) CRYPTO 1997. LNCS, vol. 1294, pp. 410–424. Springer, Heidelberg (1997). https://doi.org/10.1007/BFb0052252
10. Liu, J.K., Wei, V.K., Wong, D.S.: Linkable spontaneous anonymous group signature for Ad Hoc groups. In: Wang, H., Pieprzyk, J., Varadharajan, V. (eds.) ACISP 2004. LNCS, vol. 3108, pp. 325–335. Springer, Heidelberg (2004). https://doi.org/10.1007/978-3-540-27800-9_28
11. Tsang, P.P., Wei, V.K.: Short linkable ring signatures for e-voting, e-cash and attestation. In: Deng, R.H., Bao, F., Pang, H.H., Zhou, J. (eds.) ISPEC 2005. LNCS, vol. 3439, pp. 48–60. Springer, Heidelberg (2005). https://doi.org/10.1007/978-3-540-31979-5_5
12. Au, M.H., Chow, S.S.M., Susilo, W., Tsang, P.P.: Short linkable ring signatures revisited. In: Atzeni, A.S., Lioy, A. (eds.) EuroPKI 2006. LNCS, vol. 4043, pp. 101–115. Springer, Heidelberg (2006). https://doi.org/10.1007/11774716_9
13. Pointcheval, D., Stern, J.: Security arguments for digital signatures and blind signatures. J. Cryptol. **13**(3), 361–396 (2000)
14. Micali, S., Rabin, M.O., Vadhan, S.P.: Verifiable random functions. In: FOCS'99, pp. 120–130. IEEE Computer Society (1999). https://doi.org/10.1109/SFFCS.1999.814584
15. Okamoto, T., Pietrzak, K., Waters, B., Wichs, D.: New realizations of somewhere statistically binding hashing and positional accumulators. In: Iwata, T., Cheon, J.H. (eds.) ASIACRYPT 2015. LNCS, vol. 9452, pp. 121–145. Springer, Heidelberg (2015). https://doi.org/10.1007/978-3-662-48797-6_6

Lazy Machine Unlearning Strategy
for Random Forests

Nan Sun⑩, Ning Wang$^{(\boxtimes)}$, Zhigang Wang, Jie Nie, Zhiqiang Wei,
Peishun Liu, Xiaodong Wang, and Haipeng Qu

Ocean University of China, Qingdao, China
sunnan@stu.ouc.edu.cn,
{wangning8687,wangzhigang,niejie,weizhiqiang,liups,wangxiaodong,
quhaipeng}@ouc.edu.cn

Abstract. Removing the impact of some revoked training data from the machine learning models, i.e., machine unlearning, is a non-trivial task, which plays a pivotal role in fortifying the privacy and security of ML-based applications. This paper focuses on the problem of machine unlearning for random forests efficiently, with the streaming setting of revocation requests. The existing works are all devoted to speeding up the unlearning process of a single revocation request, none of works target at the streaming scenario. A straightforward solution is to carry out the unlearning technique tailored for a single request immediately when a new revocation request arrives. Undoubtedly, that is time-inefficient, since the time cost is proportional to the number of requests involved in the streaming. To solve this problem, this paper proposes a lazy unlearning strategy to carry out the unlearning operations involved in different revocation requests in a single batch, so as to avoid redundant computations and implement computation sharing. In particular, we adopt a node level unlearning policy, which carries out the unlearning operations on demand of the testing request by checking whether the revocation requests on this node can affect the inference of the testing request. Experiments on several real datasets show that compared to the baseline, lazy unlearning strategy can improve the unlearning efficiency by 1.1X-4X on different datasets and the number of retraining times is reduced to 1/4 on average.

Keywords: machine unlearning · random forests · lazy unlearning

1 Introduction

With people's growing demand for privacy, relevant laws have been promulgated one after another to require model providers to satisfy consumers' "right to be forgotten" [12]. In order to meet this demand, machine unlearning came into being. Many algorithms have been proposed to improve the efficiency of data revocation, such as works [6] and [1].

The problem is that the testing and revocation requests can be submitted to the service or model provider at any time, and the latter needs to process

© The Author(s), under exclusive license to Springer Nature Singapore Pte Ltd. 2023
L. Yuan et al. (Eds.): WISA 2023, LNCS 14094, pp. 383–390, 2023.
https://doi.org/10.1007/978-981-99-6222-8_32

the two kinds of requests alternately. Actually, if the provider can answer one testing request based on an unlearned model, which has removed the impacts brought by the revocation instances submitted before this testing request, we can say that s/he fulfills the obligation stipulated by GDPR [12]. With such a fact, it is time-inefficient to carry out the unlearning frequently for each revocation request. This is because the same structures of the model are retrained multiple times, but some of them are not necessary to answer testing.

That inspires us to design a *lazy unlearning strategy*, in which the unlearning operation is not carried out immediately when a revocation request arrives. Instead, it is triggered by a testing request. In this way, the unlearning operations for some revocation requests submitted before the testing request can be accumulated and then processed in a batch. We thereby can save the time cost of recomputing the parameters and retraining sub-structures.

In a summary, we make the following contributions.

- At the streaming scenario, we propose the idea of lazy unlearning for the first time. The model does not process revocation requests immediately but caches them until a testing request arrives, to achieve computation sharing.
- For the particularity of the tree structure, we design a node level *lazy unlearning strategy*, to improve the unlearning efficiency to the greatest extent.
- Extensive experiments on 11 datasets are conducted to validate the superiority of our proposal, which illustrate that our proposal can improve the unlearning efficiency by 1.1X-4X on different datasets.

Organization. The rest of this paper is organized as follows. Section 2 reviews the existing studies about machine unlearning. Section 3 introduces the preliminary knowledge of the baseline. Section 4 gives the detailed design of *lazy unlearning strategy*. Section 5 evaluates the performance of our proposal, and Sect. 6 concludes this paper.

2 Related Work

The training of the model may involve the user's private information [11]. The general data protection regulation(GDPR) [12] in the European Union provides legal protection for users' "right to be forgotten". And the work [3] defines the concept of "machine unlearning" for the first time. In order to satisfy the machine unlearning, it is necessary to retrain the model from scratch, which is very inefficient. In order to solve this problem, a lot of works [1,6] focus on how to improve the efficiency of unlearning operations.

Aiming at the tree structure that this paper focuses on, there are few related works. In the work [15], the robust segmentation decisions are greedily selected to avoid retraining, and the subtree structures generated by candidate decisions are stored. The necessary structure replacement can be made quickly when deleting an instance. The disadvantage is that it is only suitable for a limited number of data deletions. Therefore it cannot work under the streaming scenario.

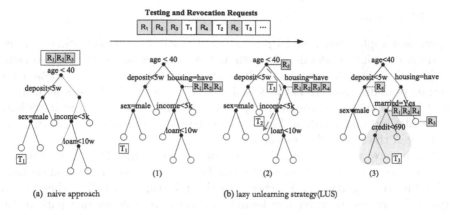

Fig. 1. An example to illustrate *lazy unlearning strategy*

The work [2] stores the statistical information of all candidate decisions. When deleting an instance, update the information of corresponding nodes, and only retrain a subtree if the optimal decision changes. At the same time, the concept of random node is put forward, in which the segmentation decision is randomly selected, independent of the database data. This work focuses on improving the efficiency of single revocation operation with not considering streaming scenarios.

3 Preliminary Knowledge of Baseline: DaRE

DaRE [2] backtracks the training process by caching intermediate information, and only retrains some sub-structures. The workflow of a decision tree in DaRE is as follows. Training starts from the root node. Each decision node randomly selects $\sqrt{p} * k$ candidate segmentation decisions (a, t). Statistical information of all candidate decisions is saved in this node. The (a^*, t^*) with the lowest Gini score is chosen as the segmentation decision. Then all instances are split to train corresponding child nodes.

When a testing instance arrives, it first visits the root. If it's value of the attribute a^* is less than t^*, it will continue to visit the left child node. Repeat the above operations, until the leaf node is reached. When deleting an instance, iterate over the nodes it affects and update the statistical information of all candidate decisions. If the optimal decision changes, retrain this subtree. Otherwise, continue to check other nodes until a leaf is encountered.

4 Lazy Unlearning Strategy(LUS)

Section 4.1 describes the motivation of our proposal. Section 4.2 introduces the algorithm in detail.

4.1   Motivation

In practical applications, model providers usually only provide testing services, but do not publish model details, such as Google forecasting API and Amazon ML [16]. This means that the result of the testing task does not contain the imprints of nodes that are not accessed by the test task. And Users with sensitive information make revocation requests more often, most of these users have similar attribute information. For the tree structure, this may lead to frequent updating of the same sub-structures.

This inspires us not process the revocation request immediately. As shown in Fig. 1(a) naive *lazy unlearning strategy* caches the obtained revocation requests without processing, triggers the whole model's unlearning operation when testing request T_1 arrives. However one testing request only accesses one path of the tree. So when answering a testing request, we try to strengthen the idea of lazy unlearning, make a node level update.

4.2   Algorithm Design

Figure 1 shows a concrete example, which can help to better understand our proposal. At the top of this figure is a data streaming, in which R_i stands for the instance requesting revocation and T_i stands for the instance requesting testing, arranged in time stamp order. In Fig. 1(b1), when the revocation requests arrive, they are stored in *root* without processing. For testing request T_1, its access path is the blue route in Fig. 1(b1). When T_1 visits the *root*, because the *root* still contains the sensitive imprints of the revocation instances. So it is necessary to update *root* and there is no need to retrain. Since the ages of R_1; R_2; R_3 are all greater than 40, append them to the associated right child node. Then the dotted blue line in Fig. 1(b2) is the path that T_3 will visit soon. In Fig. 1(b3), $R5$ is pushed to the left child node due to T_3's access to the *root*. And then T_3 visits the second-level related node, which accumulates four revocation instances R_1; R_2; R_3; R_4. By batch processing them, this node unlearns four instances through one update to realize computation sharing. Then for the third-level related node, the revocations of $R1$; $R2$ and $R3$ cause the optimal segmentation decision to change from *income* $< 5k$ to *married* $= Yes$, which triggers the retraining of the subtree. This retraining operation involves in three revocation requests, which reduces the number of retraining times and the cost of unlearning.

Given a node v and a testing sample x, Algorithm 1 is designed to eliminate the influence of all revocations submitted before x on node v. Specifically, if v is a leaf node, update the statistical information and return the new value of v: N_1/N (Line 1–4). If v is a decision node, firstly, update statistical information of all candidate decisions by $v.DQ$. Then compute corresponding Gini scores, deriving a new optimal segmentation decision (Line 7). If the optimal segmentation decision changes, all instance pointers contained in the subtree with v as the root will be recycled, except the instances in DQ. Note that DQ instances in lower nodes of this subtree should be appended to $v.DQ$ before recycle. And then the subtrees with v as root will be retrained. (Line 9–11). If the optimal

---

Algorithm 1: Testing a sample with *lazy unlearning strategy*

    Input   : sample to test: x, a node associated with x: v
    Output: the probability of sample x with label "+1"

1 **if** v *is a* LeafNODE **then**
2   **if** $v.DQ \neq \emptyset$ **then**
3   update v and let $v.\text{DQ} = \emptyset$;
4   **return** N_1/N;
5 **else**
6   **if** $DQ \neq \emptyset$ **then**
7   update information of each candidate decision based on DQ; **if** *split criterion changes* **then**
8            $D \leftarrow$ get instances without the ones in $v.DQ$ from v's leaf;
9   retrain sub-tree;
10   **else**
11   append $\forall (x, y, id) \in v.DQ$ to $v.r.DQ$ or $v.l.DQ$;
12           $v.DQ \leftarrow \emptyset$;
13   **if** $x_{v.a^*} \leq v.t^*$ **then**
14        $\text{Test}(x', v.l)$;
15   **else**
16        $\text{Test}(x', v.r)$;

---

decision doesn't change, the instances in DQ is split according to the chosen attribute-threshold pair(a^*, t^*) of v and appended to the DQ list of corresponding child nodes (Line 12–13). Finally, according to the segmentation decision of v, x continues to access the related child node (Line 15–18).

5 Experiments

In this section, we assess the performance of our method on different real-world and synthetic datasets.

5.1 Setup

Datasets. We use 11 datasets to evaluate our proposal and the competitor. They are real datasets *Surgical* [10], *Adult* [5], *Bank* [13], *Flight* [14], *Diabetes* [17], *No Show* [7], *Census* [5], *Credit* [9], *CTR* [4], *Olympics* [8] and a synthetic dataset *Synthetic*. For the dataset without tesing set, we randomly select 20% as the testing set. All data processing is referenced to DaRE [2].

Competitor and Parameters. The baseline is to immediately take the unlearning operation for each revocation on the DaRE model. For *lazy unlearning strategy*, we apply it on the DaRE to analyze its effectiveness. All parameters and are set according to the optimal selection of baseline to ensure fairness. The exact values are given in DaRE [2].

Metrics. We use unlearning efficiency and retraining cost to evaluate the effectiveness of our proposal. In particular, the unlearning efficiency is measured by the average time taken to process a revocation request. As for the retraining cost, we not only show the number of instances that need to be revisited for retraining, but also show the number of retraining times to better demonstrate the effectiveness of our proposal.

5.2 Evaluation of LUS(lazy Unlearning Strategy)

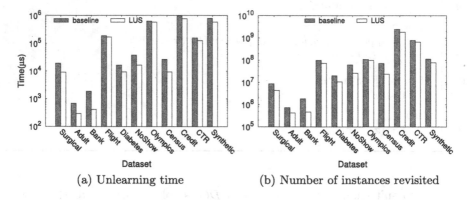

(a) Unlearning time   (b) Number of instances revisited

Fig. 2. The comparison between *lazy unlearning strategy* and baseline.

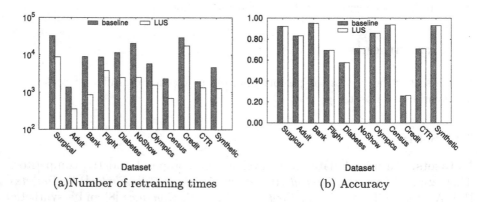

(a)Number of retraining times   (b) Accuracy

Fig. 3. The comparison between *lazy unlearning strategy* and baseline.

We compare LUS (*lazy unlearning strategy*) against the baseline, the evaluation case is constructed as follows. We construct a revocation set with 1000 samples from training set and use the pre-existing or randomly generated testing sets to access the model. As mentioned above, revocation instances are more likely have the similar attribute values. So we randomly select 10 representative samples as sensitive samples, and compute 100 samples that most similar to them respectively to form the revocation set with 1000 samples. Assume that the testing requests and revocation requests are submitted to the model alternately, to simulate the real scenario. The above process is repeated ten times, and the final experimental results are averaged

Figure 2(a) shows the average unlearning time when answering 1000 revocation requests. Obviously, LUS is always superior to baseline, and the speedup ratio varies from 1.1x to 4.5X on different datasets. This is because LUS can share the updating and retraining through batch processing, so improve the efficiency. Figure 2(b) shows the number of instances revisited when retraining. Similar to the trend of unlearning time, LUS is still superior to the baseline. The number of instances revisited is reduced by an average of 1/2. Figure 3(a) shows the number of retraining times triggered by 1000 revocation requests, and it can be found that the number of retraining times of LUS can be reduced to 1/4 of that of baseline on average. This is because LUS can merge multiple retraining through batch and delayed processing. Figure 3(b) describes the testing accuracy of the 1000 testing requests. As excepted, LUS provides the same accuracy with baseline. The main reason is that when using a node to answer a testing request, the LUS algorithm can generate the node's segmentation decision with the same probability as baseline.

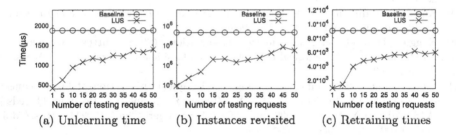

| (a) Unlearning time | (b) Instances revisited | (c) Retraining times |

Fig. 4. The performance by varying the density of testing requests on bank dataset.

Sometimes the testing density may be higher. Therefore, we evaluated the impact of testing requests density on LUS performance. We gradually increased the number of testing requests between two revocation requests in units of 5 requests. As shown in the Fig. 4, with the increase of testing requests density, all the metrics tend to increase. But in general, although the advantage of LUS is weakened, it can still ensure the improvement of unlearning efficiency.

6   Conclusion

This paper focuses on how to improve the unlearning efficiency at the streaming scenario. A lazy unlearning strategy triggered by testing requests is proposed to realize computation sharing. According to the structural characteristics of random forests, the algorithm is further improved. By node-level updating, LUS can improve the unlearning efficiency and publish the testing results without leaking privacy imprints. In the future work, we plan to extend LUS to more machine learning models to make it universal.

References

1. Bourtoule, L., et al.: Machine unlearning. In: IEEE Symposium on Security and Privacy, pp. 141–159 (2021)
2. Brophy, J., Lowd, D.: Machine unlearning for random forests. In: ICML. Proceedings of Machine Learning Research, vol. 139, pp. 1092–1104 (2021)
3. Cao, Y., Yang, J.: Towards making systems forget with machine unlearning. In: IEEE Symposium on Security and Privacy, pp. 463–480 (2015)
4. Criteo.: Criteo click-through rate prediction (2015). https://ailab.criteo.com/downloadcriteo-1tb-click-logs-dataset/. Accessed 25 Jan 2021
5. Dua, D., Graff, C.: UCI machine learning repository (2019). https://archive.ics.uci.edu/ml
6. Ginart, A., Guan, M.Y., Valiant, G., Zou, J.: Making AI forget you: data deletion in machine learning. In: NeurIPS, pp. 3513–3526 (2019)
7. Kaggle.: Medical appointment no shows (2016). https://www.kaggle.com/joniarroba/noshowappointments. Accessed 25 Jan 2021
8. Kaggle.: 120 years of olympic history: Athletes and events (2018). https://www.kaggle.com/heesoo37/120-years-of-olympic-history-athletes-and-results. Accessed 28 July 2020
9. Kaggle.: Credit card fraud detection (2018). https://www.kaggle.com/mlg-ulb/creditcardfraud/. Accessed 27 July 2020
10. Kaggle: Dataset surgical binary classification (2018). https://www.kaggle.com/omnamahshivai/surgical-dataset-binary-classification/version/1#. Accessed 29 July 2020
11. Liu, X., Zhao, R., Zhang, Y., Zhang, F.: Prognosis prediction of breast cancer based on CGAN. In: Xing, C., Fu, X., Zhang, Y., Zhang, G., Borjigin, C. (eds.) WISA 2021. LNCS, vol. 12999, pp. 190–197. Springer, Cham (2021). https://doi.org/10.1007/978-3-030-87571-8_16
12. Mantelero, A.: The EU proposal for a general data protection regulation and the roots of the 'right to be forgotten'. Comput. Law Secur. Rev. **29**(3), 229–235 (2013)
13. Moro, S., Cortez, P., Rita, P.: A data-driven approach to predict the success of bank telemarketing. Decis. Support Syst. **62**, 22–31 (2014)
14. Research, Administration, I.T.: Airline on-time performance and causes of flight delays (2019). https://catalog.data.gov/dataset/airline-on-time-performance-and-causes-of-flight-delays-on-time-data. Accessed 16 April 2020
15. Schelter, S., Grafberger, S., Dunning, T.: Hedgecut: maintaining randomised trees for low-latency machine unlearning. In: SIGMOD Conference, pp. 1545–1557 (2021)
16. Shokri, R., Stronati, M., Song, C., Shmatikov, V.: Membership inference attacks against machine learning models. In: 2017 IEEE Symposium on Security and Privacy (SP), pp. 3–18 (2017). https://doi.org/10.1109/SP.2017.41
17. Strack, B., et al.: Impact of HbA1c measurement on hospital readmission rates: analysis of 70,000 clinical database patient records. BioMed. Res. Int. **2014** (2014)

Will Data Sharing Scheme Based on Blockchain and Weighted Attribute-Based Encryption

Chongyang Yang, Xiaomei Dong[✉], Ning Zhang, and Ziqiang Wen

School of Computer Science and Engineering, Northeastern University, Shenyang 110169, China
xmdong@mail.neu.edu.cn

Abstract. Traditional testator-written wills can easily be lost, easily be tampered with, and forged when kept by the testator himself. The appearance of electronic wills provides convenience for will retention. However, centralized storage has risks such as malicious attacks and single point of failure. Due to the advantages of decentralization, immutable and traceable, applications based on blockchain can realize the sharing of will data but faces the problem of privacy leakage in the sharing. In this paper, we propose a weighted attribute-based encryption scheme to achieve access control of will data in blockchain while reducing the complexity of access policy. The Traceable Radix Tree index structure is designed to improve the query efficiency of users while supporting the quick query of wills' historical versions in reverse chronological order to ensure the legal validity of wills. The supervision chain, which is regulated through supervision institutions, is introduced to store the query records of users on the blockchain and prevent the privacy disclosure of users during the query process. Finally, the effectiveness of the proposed scheme is proved by relevant experiments.

Keywords: Electronic will record · Weighted attribute-based encryption · Blockchain · Will data sharing

1 Introduction

With the rapid development of economy and the growing trend of an aging population, more and more people are choosing to make a will to distribute their property according to their wishes after death [1]. Testator-written will has become the preferred method for most people due to its advantages such as convenient establishment and no cost. However, it usually faces problems such as difficulty in preservation and easy tampering [2]. Electronic will is a new form of will retention. However, the traditional centralized storage is faced with the risk of single point of failure and data leakage due to malicious attacks [3].

Blockchain is essentially a distributed shared database, which has been widely used in many scenarios due to its characteristics of decentralization, tamper-proof and traceability, and is well suited for will retention [4]. However, when using blockchain to share will data, there is a risk of privacy leakage due to its open and transparent data, as well as a lack of effective supervision [5]. Testator-written will data often need to contain both the will document and a real-time video that can confirm the civil capacity of the testator is not impaired. Since the storage capacity of each block is limited and storing all of them in the block increases the storage overhead, hybrid storage has been used in related fields [6]. Overall, there are three major challenges that still need to be overcome:

- Large amounts of wills and auxiliary validation data are stored on the blockchain, resulting in significant storage overhead. At the same time, there is a risk of privacy leakage when using blockchain to share wills.
- Using smart contracts for data queries is less efficient. At the same time, according to the Civil Code, when the contents of several wills made by a testator contradict each other, the last one shall prevail. Therefore, this poses the challenge of ensuring the validity of wills.
- There is a lack of supervisory measures for the blockchain that stores electronic will records. Along with supervision, protecting users' privacy is also a challenge.

Based on above problems, we propose a will data sharing scheme WSBA based on blockchain and weighted attribute-based encryption. The main contributions of this paper can be summarized as follows:

- We use hybrid storage to reduce the storage overhead of the blockchain. Access control of electronic will records is implemented by using ciphertext-policy attribute-based encryption (CP-ABE) [7] and weights are attached to attributes to reduce the complexity of access policy.
- We design a Traceable Radix Tree index structure. While improving the query efficiency of users, it supports fast query of historical versions of will records in reverse chronological order.
- We obfuscate the identifiers of users who query block data to protect privacy under supervision.

The rest of this paper is organized as follows. In Sect. 2, we discuss the work related to will and judicial data sharing based on blockchain. In Sect. 3, we describe the design of WSBA in detail. In Sect. 4, we perform safety analysis and experimental evaluation of the proposed scheme, and we summarize the full paper in Sect. 5.

2   Related Work

There has been some research on using blockchain for will data sharing. Sreehari et al. [8] proposed the concept of saving wills in the blockchain through smart contracts, increasing the speed of probating, but the scheme is only in the drafting stage. Chen et al. [5] built a testator-written will system using the Ethereum platform, which focuses on privacy protection during the transmission of the will, without implementing effective on-chain data access control.

Blockchain is also widely used in related areas like judicial data sharing. Jing et al. [6] used a hybrid storage of judicial data to reduce the storage burden of blockchain, but data access needs to be authorized by multiple departmental nodes, which is less flexible. Tian et al. [9] proposed Block-DEF, a digital evidence framework based on blockchain, which supports the storage and retrieval of evidence and guarantees the traceability of evidence but does not consider the attributes of users in the access stage of evidence. Yan et al. [10] applied CP-ABE to blockchain to achieve access control of electronic evidence, but it is less efficient and has a more complex access policy. Nyaletey et al. [11] proposed a blockchain-enabled Interplanetary File System BlockIPFS for forensics and trusted data tracking. The scheme can do file sharing with related activity traceability, but not access control in sharing. Chen et al. [12] proposed a blockchain-based CP-ABE framework called GovChain to address the issues of fine-grained access control of government data, but this shared model has a high computational overhead. Li et al. [13] implemented blockchain-based lawful evidence management in digital forensics, which covering the entire life cycle of evidence. The scheme uses CP-ABE to achieve fine-grained access control but does not consider the issue of attribute weights and blockchain storage overhead.

In this paper, we design a hybrid storage method based on blockchain and implement fine-grained access control of will records by using the weighted attribute-based encryption. We also design the Traceable Radix Tree index structure. To enhance the security supervision of data queries, we adopt a dual-chain structure and obfuscate the identities of data users to protect their privacy.

3   Design of WSBA

In this section, the proposed scheme WSBA is described in detail. First, we show the composition of the system (see Sect. 3.1). Then, we describe the design of weighted attribute-based encryption scheme (see Sect. 3.2), Traceable Radix Tree index structure (see Sect. 3.3), and supervision chain which supporting obfuscated identity (see Sect. 3.4).

3.1   System Model

The system comprises seven entities: data owner, data uploader, key generation center, data user, blockchain, supervision chain, and supervision institution. The system model is shown in Fig. 1.

The characteristics and functions of each entity are described as follows:

1. Key Generation Center (KGC): Responsible for generating the master key MK, the public key PK, and the weighted attribute private key for each data user.
2. Data Owner (DO): Mainly composed of testators. Write wills (year, month and day required) and take live videos. Send digitally signed electronic wills to the community for preservation along with live videos.
3. Data Uploader (DUP): Mainly composed of community workers. After verifying the integrity of the data, the community worker encrypts the electronic will, corresponding digital signature as well as the live video using a symmetric key and uploads the resulting ciphertext to the database. The symmetric key and the returned ciphertext storage address are encrypted using the access policy set by the corresponding testator and stored in the blockchain.
4. Data User (DU): Mainly composed of testators' relatives and judicial personnel. All have weighted attribute private keys corresponding to their identities.

5. Blockchain (BC): Used to store information related to will records, return query results based on query requests, and send query information to the supervision chain.
6. Supervision Chain (SC): Used to store data users' query information on the BC. It is mainly monitored and maintained by the SI.
7. Supervision Institution (SI): Responsible for maintaining the supervision chain. The supervision institution adds noise to the identity ID of the data user to generate an obfuscated identity ID and stores locally the correspondence between the real identity ID of the data user and the obfuscated identity ID.

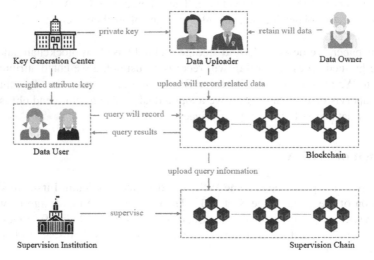

Fig. 1. System model

3.2 Weighted Attribute-Based Encryption

Inspired by [14], we implement access control of will records by using weighted attribute-based encryption. Suppose the age values of a testator's relatives include 20, 28 and 42, then we distribute weights of 1, 2 and 3 to each age value in turn. Therefore, these attribute values can be expressed as "Age: 1", "Age: 2" and "Age: 3". In this case, they can be denoted by one attribute which has just different weights. The binary states supported only on description attributes in CP-ABE, such as "1 – satisfying" and "0 – not-satisfying", are converted to handle arbitrary state attributes.

We here assume that a general access policy set by a testator can be represented as: T {("Daughter" OR "Son" OR "Wife") AND ("28 years" OR "42 years")}. According to our proposed scheme, T can be simplified as: T' {("Relative: 1") AND ("Age: 2")}, where the attribute "Age: 2" denotes the minimum level in the access policy and includes {"Age: 2", "Age: 3"} by default, and the access policy is simplified. These two structures are shown in Fig. 2.

Our weighted attribute-based encryption scheme consists of the following four algorithms, and the symbols involved in the algorithms are shown in Table 1.

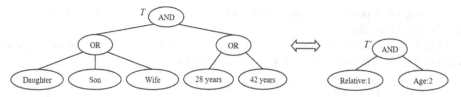

Fig. 2. Two equivalent access structures of a ciphertext

Let G_0 be a bilinear group of the prime order p, and let g be a generator of G_0. Let $e : G_0 \times G_0 \to G_T$ denote the bilinear map. A security parameter λ will determine the size of the groups. We also define the Lagrange coefficient: $\Delta_{i,S}(x) = \prod_{j \in S, j \neq i} \frac{x-j}{i-j}$. $i \in Z_p$ and S is a set of elements in Z_p. We employ a hash function $H : \{0, 1\}^* \to G_0$. In addition, a universe of attribute set $A = \{a_1, \ldots, a_n\}$ and a set of weights $W = \{\omega_1, \ldots, \omega_n\}(\omega_1 \leq \ldots \leq \omega_n)$ are defined. Thus, weighted attributes that the system contains are $\tilde{A} = \{a_1 : \omega_1, \ldots, a_1 : \omega_n, \ldots, a_n : \omega_1, \ldots a_n : \omega_n\}$, where the higher hierarchy of attributes is used, the bigger weighted value is distributed. The specific construction steps are as follows:

1. $Setup(1^\lambda) \to (PK, MK)$. KGC runs the algorithm which inputs a security parameter λ, then outputs the master key MK and the public key PK.

    KGC chooses two random exponents $\alpha, \beta \in Z_p$, then calculates and outputs:

$$PK = \big(G_0, g, h = g^\beta, e(g, g)^\alpha\big), \quad MK = \big(\beta, g^\alpha\big) \tag{1}$$

2. $KeyGen(MK, PK, S) \to SK$. KGC inputs the master key MK, the public key PK, and a set of weighted attributes S. For each weighted attribute $j \in S$, it possesses weighted value $\omega_j(\omega_j \in W)$. KGC outputs the private key SK corresponding to the weighted attributes S and sends it to the user.

    The algorithm first chooses a random $r \in Z_p$, and then random $r_j \in Z_p$ for each weighted attribute $j \in S$. Then it computes the private key as:

$$SK = \left(D = g^{\frac{\alpha+r}{\beta}}, \forall j \in S : D_j = g^r \cdot H(j)^{\omega_j r_j}, D'_j = g^{r_j}\right) \tag{2}$$

3. $Encrypt(PK, P, CK, T) \to CT$. The community worker inputs the public key PK, the plaintext P, the symmetric key CK and the access policy T, then outputs the encrypted will data CT.

    First, the algorithm uses AES algorithm to generate the ciphertext C. After the ciphertext C is stored to the database and the storage location information L is obtained, the algorithm generates M by concatenating L and CK, that is $M = L \| CK$.

    Second, the algorithm automatically selects a polynomial q_x for each node x in T. From the root node R, the polynomials are selected in a top-down manner. For each node x in T, degree of the polynomial d_x is set to $K_x - 1$, where K_x is the threshold value. Starting with the root node R, the algorithm chooses a random number $s \in Z_p$ and sets $q_R(0) = s$. Then, it chooses d_R other points of the polynomial q_R randomly to define it completely. For each non-root node x, it sets $q_x(0) = q_{parent(x)}(index(x))$ and randomly chooses d_x other points to completely define q_x. Meanwhile, each leaf node donates an attribute with weight.

Let Y be the set of leaf nodes in the access tree T, and ω_i be the minimum weight of each leaf node. The ciphertext CT is then computed as:

$$CT = \left(\begin{matrix} T, \tilde{C} = M \cdot e(g,g)^{\alpha \cdot s}, C = h^s \\ \forall y \in Y, i \in [1, n]: C_y = g^{q_y(0)}, C'_y = H(att(y))^{\omega_i q_y(0)}, \\ \forall j \in (i, n]: C_{y,j'} = H(att(y))^{(\omega_j - \omega_i)q_y(0)} \end{matrix} \right) \tag{3}$$

4. $Decrypt(PK, CT, SK) \rightarrow P/\bot$. The user inputs PK, CT, and SK. If SK satisfies the access policy of ciphertext CT, it can be decrypted successfully and output the plaintext P. Otherwise, decryption fails and the algorithm outputs $\bot$. The operation is a recursive algorithm which is defined as $DecryptNode(CT, SK, x)$, where x is a node from T.

If x is a leaf node, let $k = att(x)$, then ω_k be the weighted value of the user's node x, and ω_i be the weighted value of the access policy T's node. If $k \notin S$ or $k \in S$, $\omega_i > \omega_k$, $DecryptNode(CT, SK, x) = \bot$. If $k \in S$ and $\omega_i = \omega_k$., $DecryptNode\ (CT, SK, x)$ can be calculated as the formula (4). If $k \in S$, $\omega_i < \omega_k$ and $\omega_k = \omega_j$, $DecryptNode(CT, SK, x)$ can be calculated as the formula (5):

$$DecryptNode(CT, SK, x) = \frac{e(D_k, C_x)}{e(D'_k, C'_x)} = \frac{e(g^r \cdot H(k)^{\omega_k r_k}, g^{q_x(0)})}{e(g^{r_k}, H(k)^{\omega_i q_x(0)})} = e(g,g)^{r \cdot q_x(0)} \tag{4}$$

$$DecryptNode(CT, SK, x) = \frac{e(D_k, C_x)}{e(D'_k, C'_x \cdot C'_{x,j})} = \frac{e(g^r \cdot H(k)^{\omega_k r_k}, g^{q_x(0)})}{e(g^{r_k}, H(k)^{\omega_i q_x(0)} \cdot H(k)^{-(\omega_j - \omega_i)q_x(0)})}$$

$$= \frac{e(g^r \cdot H(k)^{\omega_k r_k}, g^{q_x(0)})}{e(g^{r_k}, H(k)^{\omega_j q_x(0)})} = e(g,g)^{r \cdot q_x(0)} \tag{5}$$

If x is not a leaf node, $DecryptNode(CT, SK, x)$ is defined: for all nodes z that are children of x, it runs $DecryptNode(CT, SK, z)$ and stores the output as F_z. Let S_x be an arbitrary k_x-sized set of child nodes z such that $F_z \neq \bot$, and let $k = index(z)$, $S'_x = \{index(z) : z \in S_x\}$.

$$F_z = \prod_{z \in S_x} F_z^{\Delta_{k, S'_x}(0)} = \prod_{z \in S_x} \left(e(g,g)^{r \cdot q_{parent(z)}(index(z))} \right)^{\Delta_{k, S'_x}(0)}$$

$$= \prod_{z \in S_x} \left(e(g,g)^{r \cdot q_x(k)} \right)^{\Delta_{k, S'_x}(0)} = e(g,g)^{r \cdot q_x(0)} \tag{6}$$

Then, we define the decryption algorithm by calling $DecryptNode(CT, SK, x)$ on the root node R of the access tree T. If T is satisfied by S, set $A = DecryptNode(CT, SK, r) = e(g,g)^{r q_R(0)} = e(g,g)^{rs}$. The algorithm decrypts by computing:

$$\frac{\tilde{C}}{\left(\frac{e(C,D)}{A} \right)} = M \cdot \frac{e(g,g)^{\alpha s}}{e\left(h^s, g^{\frac{\alpha + r}{\beta}} \right)} \cdot e(g,g)^{rs} = M \tag{7}$$

Now we have completed the decryption and got $M = L\|CK$. Then the algorithm uses the location information L to get C, and use the AES symmetric key CK to decrypt it to get the plaintext P.

Table 1. Symbols used in this scheme.

Symbols	Meaning of the symbols
PK	The public key
MK	The system master key
SK	The private key
CK	The AES symmetric key
P	The plaintext data
C	The AES encrypted ciphertext
M	The data to be weighted attribute encrypted
S	The set of weighted attributes of a user
Y	The set of weighted attributes in the access policy
T	The access policy
CT	The weighted attribute encrypted data

3.3 Traceable Radix Tree Index Structure

To improve the efficiency of will records query and support the query of wills' historical versions, we design the Traceable Radix Tree (TRT) index structure. Each will record has a unique WID. We use WID to build the index for all the data in blockchain. We stipulate WID = Hash (NID‖times), where NID represents the ID number of the testator and times represents the number of wills currently made by the testator. As shown in Fig. 3, except for the root node, TRT contains two types of nodes: index nodes and data nodes. An index node contains the common prefix of WID and a pointer to the next node in a different path. A data node contains the full value of WID, the storage block number of the will record represented by WID and the location of the data node in the tree corresponding to the WID of will record's previous version. The TRT is stored on the blockchain, and the TRT root node's address is stored in the block header.

Users get the TRT from the latest block and use the WID to get the storage block number of the corresponding will record. Users can also quickly search the historical versions of will records represented by WID in reverse chronological order to ensure that they have the most current version of the will record. It's more efficient than traversing the block one by one and the query efficiency is greatly improved.

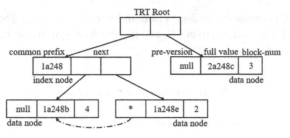

Fig. 3. Traceable Radix Tree

3.4 Design of the Supervision Chain Supporting Obfuscated Identity

Supervision chain is designed to store information about data users' queries in the blockchain and is supervised by the SI. When a user makes a data query on the blockchain, the smart contract is triggered to respond to his operation and the query information is transmitted to supervision chain for storage. The design of the data structure on the supervision chain is shown in Fig. 4.

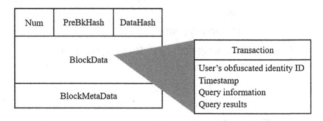

Fig. 4. The structure of the supervision chain

A user's obfuscated identity ID is formed by adding noise to his original identity ID. That is, the original identity ID has "ASCII plus n" noise added to each bit, where n is a randomly generated positive integer less than ten. SI locally stores the correspondence between the user's real identity ID and the obfuscated identity ID. When a query information leak occurs, SI can find out which user leaked the query information according to the correspondence. With the help of supervision chain, the query information leakage tracking is achieved.

4 Security Analysis and Performance Evaluation

4.1 Security Analysis

The proposed scheme has the following security properties.

1. Security of Will Data

In our solution, the secret sharing is embedded in the ciphertext instead of the user's private key. In order to decrypt, an attacker clearly must recover $e(g, g)^{\alpha s}$, and in order to do this the attacker must pair C from the ciphertext with the D component from the user's private key, i.e., $e(C, D) = e\left(g^{\beta s}, g^{\frac{\alpha+r}{\beta}}\right) = e(g, g)^{s(\alpha+r)} = e(g, g)^{\alpha s} \cdot e(g, g)^{rs}$. This will obtain the desired value $e(g, g)^{\alpha s}$, but blinded by value $e(g, g)^{rs}$. Only if the weighted attributes in the user's key component satisfy the secret sharing scheme embedded in the ciphertext, this value can be blinded out. Because the blinding value is randomized to the randomness from a particular user's private key, collusion attacks won't help. Thus, the scheme guarantees the security of plaintext.

2. Identity Security of Data Users

The identity IDs of data users are stored on the supervision chain after being obfuscated. After an illegal visitor obtains the query information of the data user on the supervision chain, he cannot infer the original identity ID from the obfuscated identity ID because he does not know the construction rule.

4.2 Evaluation of Performance

We evaluate the efficiency of the proposed scheme in this paper. The experimental environment is built using a PC with Intel(R) Core(TM) i5-10210U CPU @ 1.60GHz, and an operating system of Ubuntu 20.04.4, 64-bit, 4GB of RAM. Mainly using Hyperledger Fabric framework and writing SDK with Golang.

Table 2. Comparison with other schemes

Scheme	Paper [7]	Paper [13]	Paper [15]	Our scheme
Attribute-based encryption	Y	Y	Y	Y
Attribute weight	N	N	N	Y
Query index	N	N	Y	Y
Historical version query	N	N	Y	Y
Obfuscated identity	N	N	N	Y

There exist few research works that proposes blockchain-based will data sharing solutions to improve one or more aspects of will retention domain, so the performance of data access control in our scheme is compared with other related literatures. In this paper, we combine AES and CP-ABE for hybrid encryption. We introduce weights for attributes to reduce the complexity of access policy. The results of scheme comparison are shown in Table 2. In experiments, we assume the maximum value of each weighted attribute is set as 5, and the lowest value of each weighted attribute is chosen to be encrypted. We simulate will data using publicly available medical datasets. We implement our scheme and comparison schemes under the equivalent access policy encrypted in ciphertext.

Since AES execution time differs for files of different sizes and according to the actual size of will data, we use files with size from 10^4 KB to 10^5 KB to evaluate the computational overhead. Here we compare our scheme with the schemes in paper [7] and paper [13], and the number of weighted attributes is set to 10.

As shown in Fig. 5, the total computational overhead of encryption and decryption of all three schemes increase linearly with the increase of data size, but the computational overhead of our scheme is better than the scheme in [7] and [13] in general. The hybrid encryption method of AES and CP-ABE is used both in our scheme and the scheme in [13], so the performance is better than that in [7]. And compared with the scheme in paper [13], our scheme reduces the complexity of access strategy by considering attribute weights, which further reduces the computational overhead.

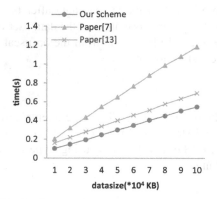

Fig. 5. Computational overhead for different file size

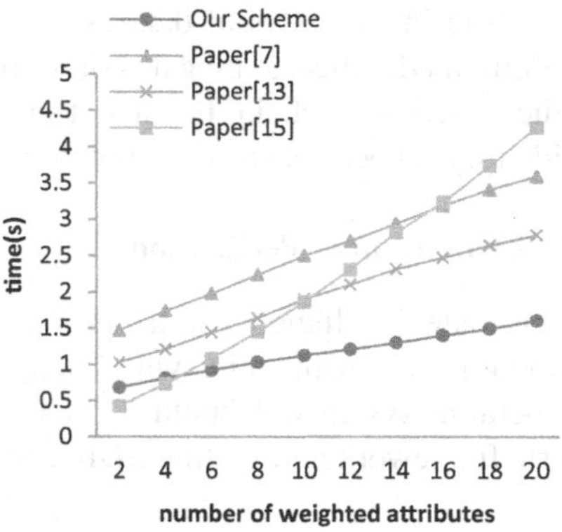

Fig. 6. Computational overhead for different number of weighted attributes

We then fix the file size at 10^5 KB and measure the total computational overhead of encryption and decryption by changing the number of weighted attributes in the access policy. The experimental results are shown in Fig. 6. We compare our scheme with the schemes in paper [7, 13] and [15]. In our scheme, CP-ABE is used to encrypt an AES key and the location information rather than a file of 10^4 KB size, which is much smaller. Attribute weights reduce the number of leaf nodes in the access tree, leading to a further reduction in the overhead of encryption and decryption. So, the total computational overhead is less than that in [7]. Compared with our scheme, the scheme in paper [13] has a relatively high computational overhead because the attribute weights are not considered, and the third-party node is required to verify whether the data users' attributes match the access policy. The computational overhead of the scheme in paper [15] is initially minimal because the access policy is used to encrypt an indexed keyword, which is smaller than an AES key. However, the computational overhead grows faster as it requires additionally generate search trapdoors and match them with keywords. Overall, our scheme has a better performance.

We conduct another set of experiments on different sizes of data to test the percentage of computational overhead to the overall system overhead when combined with the Fabric blockchain system. The test environment simulates three organizations consisting of three peer nodes and one orderer node, and the orderer node uses the raft consensus algorithm. As shown in Fig. 7, the computational overhead of encryption and decryption is a relatively small percentage to the overall system overhead for less than 10%. We can estimate from the experimental results that the computational overhead is less than 5% for the data size smaller than 10^4 KB. Combined with real life, the size of the will data may be much larger than 10^5 KB. However, the percentage of computational overhead grows very slowly, which is not obvious to the whole system.

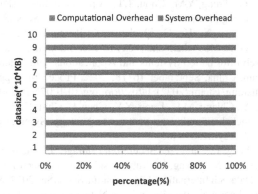

Fig. 7. Percentage of computational overhead

5 Conclusion

In this paper, we propose a will data sharing scheme WSBA based on blockchain and weighted attribute-based encryption. We combine AES with CP-ABE to implement fine-grained access control of electronic will records to protect the testators' data privacy. The weighted attribute is introduced to improve the expression of attribute, which can not only describe arbitrary state attributes, but also reduce the complexity of access policy. We design the Traceable Radix Tree index structure to improve the efficiency of will data sharing while supporting quick search of will records' historical versions to ensure the legal validity of wills. We also design a supervision chain that stores each user's query information on the blockchain and protects the privacy of user identities under supervision through obfuscated identity. Finally, we presented the security analysis and performance evaluation for the proposed scheme, in which the results demonstrate high efficiency of our scheme. However, there are also some limitations. For example, threshold gates in the form of "m of n" are not supported. In future research, we will optimize and improve the limitations of our scheme. Overall, our proposed scheme is suitable for the sharing of will data based on blockchain.

Acknowledgement. This work is supported by the Major projects of National Social Science Foundation of China (No. 21&ZD124): research on the community home-based elderly care model and quality safety system based on blockchain. We also thank anonymous reviewers for the helpful reports.

References

1. Zhang, M.M.: How to make a legally valid will. Financ. View (Wealth) **15**(9), 42–45 (2021). (In Chinese)
2. Sheng, Y.F.: Exploring the development path of our will field from the perspective of blockchain. Shanghai Bus. **29**(5), 181–184 (2022). (In Chinese)
3. Chen, C.L., Lee, C.C., Tseng, Y.M., Chou, T.T.: A private online system for executing wills based on a secret sharing mechanism. Secur. Commun. Netw. (SCN) **5**(7), 725–737 (2012)
4. Xinhao, X., Dong, X., Li, X., He, G., Shicheng, X.: Patient-friendly medical data security sharing scheme based on blockchain and proxy re-encryption. In: Zhao, X., Yang, S., Wang, X., Li, J. (eds.) Web Information Systems and Applications: 19th International Conference, WISA 2022, Dalian, China, September 16–18, 2022, Proceedings, pp. 615–626. Springer International Publishing, Cham (2022). https://doi.org/10.1007/978-3-031-20309-1_54
5. Chen, C.L., Lin, C.Y., Chiang, M.L., Deng, Y.Y., Chen, P., Chiu, Y.J.: A traceable online will system based on blockchain and smart contract technology. Symmetry. **13**(3), 466 (2021). https://doi.org/10.3390/sym13030466
6. Jing, Z., Cao, C., Wang, L., Sang, Y.: Secure data storage scheme of judicial system based on blockchain. In: Chen, X., Huang, X., Kutyłowski, M. (eds.) Security and Privacy in Social Networks and Big Data: 8th International Symposium, SocialSec 2022, Xi'an, China, October 16–18, 2022, Proceedings, pp. 339–350. Springer Nature Singapore, Singapore (2022). https://doi.org/10.1007/978-981-19-7242-3_22
7. Bethencourt, J., Sahai, A., Waters, B.: Ciphertext-policy attribute-based encryption. In: IEEE Symposium on Security and Privacy (S&P), pp. 321–334 (2007)
8. Sreehari, P., Nandakishore, M., Krishna, G., Jacob, J., Shibu, V.: Smart will converting the legal testament into a smart contract. In: International Conference on Networks & Advances in Computational Technologies (NetACT), pp. 203–207 (2017)
9. Tian, Z., Li, M., Qiu, M., Sun, Y., Su, S.: Block-DEF: a secure digital evidence framework using blockchain. Inf. Sci. **491**, 151–165 (2019)
10. Yan, W., Shen, J., Cao, Z., Dong, X.: Blockchain based digital evidence chain of custody. In: Proceedings of the 2020 The 2nd International Conference on Blockchain Technology (ICBCT 2020), pp. 19–23 (2020)
11. Nyaletey, E., Parizi, R.M., Zhang, Q., Choo, K.K.R.: Blockipfs-blockchain-enabled interplanetary file system for forensic and trusted data traceability. In: IEEE International Conference on Blockchain (Blockchain), pp. 18–25 (2019)
12. Chen, T., Yu, Y., Duan, Z., Gao, J., Lan, K.: Blockchain/abe-based fusion solution for e-government data sharing and privacy protection. In: Proceedings of the 2020 4th International Conference on Electronic Information Technology and Computer Engineering (EITCE 2020), pp. 258–264 (2020)
13. Li, M., Lal, C., Conti, M., Hu, D.: LEChain: a blockchain-based lawful evidence management scheme for digital forensics. Future Gener. Comput. Syst. (FGCS) **115**, 406–420 (2021)
14. Wang, S., Liang, K., Liu, J.K., Chen, J., Yu, J., Xie, W.: Attribute-based data sharing scheme revisited in cloud computing. IEEE Trans. Inf. Forensic. Secur. (TIFS) **11**(8), 1661–1673 (2016)
15. Yin, H., et al.: CP-ABSE: A ciphertext-policy attribute-based searchable encryption scheme. IEEE Access **7**, 5682–5694 (2019)

Secure Mutual Aid Service Scheme Based on Blockchain and Attribute-Based Encryption in Time Bank

Ning Zhang, Xiaomei Dong$^{(\boxtimes)}$, Ziqiang Wen, and Chongyang Yang

School of Computer Science and Engineering, Northeastern University, Shenyang 110169, China
xmdong@mail.neu.edu.cn

Abstract. Time bank is a new model intended to supplement traditional pension systems and alleviate social pension pressure. However, the existing time bank system only considers service demands from the elderly as public, ignoring the need for privacy protection. There is low public participation due to the lack of trust among users. To address these issues, we propose a scheme based on blockchain and attribute-based encryption in time bank, enabling fine-grained access control for shared service demands and promoting secure mutual aid service. We design a multi-authority authorization model to avoid the security implications of a central authority. In order to quickly respond to the secure mutual aid service, we design an outsourced decryption mechanism to reduce the computation overhead on the user side, making it more suitable for resource-constrained devices (e.g., cell phones, iPads). In addition, we introduce an anonymization mechanism to protect the identity privacy of users while regulating the circulation of time coins. Finally, the effectiveness of our proposed scheme is proved by relevant experiments.

Keywords: Time bank · Blockchain · Attribute-based encryption · Multi-authority · Outsourced decryption

1 Introduction

Time bank was originally developed from "Time Dollar", a complementary monetary system proposed by American economist Cahn in 1992 [1]. Specifically, it builds a platform for community residents who need senior care services. The elderly who needs help publish their service demands on the platform, and volunteers take orders and provide services for them, thus gaining a certain reward of time coins. In the future, volunteers can pay time coins for themselves or their loved ones in exchange for senior care services when they need them. In this way, time bank encourages the public to participate in volunteering activities and increase people's enthusiasm for volunteering.

Nowadays, China has tried out the voluntary service of time bank in some areas. However, there are certain shortcomings and deficiencies in the process of implementation. The transactions of time coins in most time banks are still mainly recorded manually resulting in inefficient information management. And they are stored in a centralized way, which may lead to data loss or even unrecoverable in case of an attack. Blockchain is

© The Author(s), under exclusive license to Springer Nature Singapore Pte Ltd. 2023
L. Yuan et al. (Eds.): WISA 2023, LNCS 14094, pp. 403–414, 2023.
https://doi.org/10.1007/978-981-99-6222-8_34

an emerging technology with decentralized, transparent and immutable features, which has a better application prospect in the future development of time bank pension model [2, 3].

Even so, there are still three major challenges need to be overcome:

- The existing time bank only considers service demands from the elderly to be public, ignoring the need for privacy protection. The service demands published by the elderly involves a large amount of private data (e.g., their name, cell phone number, address, health status, etc.). If it is leaked out, the elderly may be deceived.
- The users' secret attribute keys are centralized by a central authority, which poses a risk of single point of failure. Moreover, the central authority holds excessive power and may directly access the users' private data.
- Volunteers need to decrypt the elderly's private data on their own devices in order to provide services to the elderly, which can lead to wasted computing resources and delayed response time.

To address the above issues, we propose a secure mutual aid service scheme based on blockchain and attribute-based encryption in time bank. The main contributions are as follows:

- We adopt ciphertext-policy attribute-based encryption (CP-ABE) [4] based on linear secret sharing scheme (LSSS) to achieve fine-grained access control for shared service demands. And we introduce an identity anonymization mechanism to ensure users' identity privacy when regulating the circulation of time coins.
- We propose a multi-authority authorization model, which can guarantee the security of users' secret attribute keys and solve the problem of single point of failure.
- In order to quickly respond to the secure mutual aid service among people, we design an outsourced decryption mechanism to reduce the computation overhead on the user side, making it more suitable for resource-constrained devices.

The rest of the paper is organized as follows. In Sect. 2, we discuss related works on time bank and attribute-based encryption. In Sect. 3, we detail the design of our proposed scheme. In Sect. 4, we analyze the scheme and perform an experimental evaluation. In Sect. 5, we summarize our work.

2   Related Work

With the widespread use of blockchain technology, more and more scholars use its characteristics of decentralization, immutability, and traceability to build a time bank system to solve the trust problem among people. Lee et al. [5] proposed a blockchain-based time bank that the matching between service supply and demand can be done autonomously through smart contracts. However, it stores all data on the blockchain, which will undoubtedly increase the storage overhead of the blockchain network. Based on [5], Lin et al. [6] proposed a blockchain-based dynamic service matching mechanism in time bank to efficiently achieve service matching between the elderly and volunteers, but the scheme cannot guarantee the authenticity of both parties' identities. Cheng et al. [7] proposed a blockchain-based volunteer time bank (VOLTimebank) that uses Delegated Proof of Stake (DPoS) [8] consensus mechanism to select auditors to ensure the correctness of time coin transactions. However, the scheme lacked privacy protection mechanism.

In terms of privacy protection technology, attribute-based encryption (ABE) proposed by Sahai and Waters is widely used, which is gradually applied to different scenarios [9–11]. Li et al. [12] proposed an attribute-based encryption scheme for fog-enable IoT which supports outsourced decryption and attribute revocation. But the overall overhead of this scheme is too high due to the access tree used to build the access policy. He et al. [13] proposed a multi-authority attribute-based encryption scheme that supports collaborative decryption. But it can't solve the problem of single point of failure. Sun et al. [14] proposed an attribute-based encryption scheme for medical data sharing that enables patients to manage their personal medical data directly and avoid the risk of privacy leakage. However, the decryption overhead of this scheme is too large for resource-constrained devices.

In this paper, we propose a secure mutual aid service scheme based on blockchain and attribute-based encryption in time bank to achieve fine-grained access control for shared service demands. We create a multi-authority model with (t, n)-distributed key generation (t, n-DKG) [15] that guarantees the security of the users' secret attribute keys while solving the problem of single point of failure. To quickly respond to the service provisioning needs of both parties, we design an outsourced decryption mechanism. At the same time, anonymity mechanism is introduced to protect the identity privacy of users while regulating the circulation of time coins.

3   Scheme Design

In this section, we design the corresponding system model based on our proposed scheme (see Sect. 3.1), and then describe in detail the five phases of the scheme: system initialization, multi-authority authorization, publishing service demands, providing services and time coin tracking (see Sect. 3.2).

3.1   System Model

In our scheme, a time bank secure mutual aid service system includes seven entities: trusted authority, attribute authorities, cloud server, service demander, service provider, service demand blockchain and time coin transaction blockchain. The system model is shown in Fig. 1. The characteristics and functions of each entity are as follows:

1. Trusted Authority (TA): responsible for initializing the system, maintaining the time coin transaction blockchain.
2. Attribute Authorities (AAs): responsible for generating key pair for itself and users. All AAs manage the entire attribute set together, but no AA can assign a user's key pair individually.
3. Cloud Server (CS): stores the encrypted service demands, matches service provider's attributes with the access policy, and then performs outsourced decryption.
4. Service Demand Blockchain (SDBC): stores encrypted service demand index, storage address, encrypted symmetric key.

5. Time Coin Transaction Blockchain (TCTBC): stores the time coin transaction records between service demanders and service providers. When users have doubts about their time coin balance, they can track the historical information in TCTBC.
6. Service Demander (SD): mainly consists of the elderly. They can publish encrypted service demands with symmetric key to CS. After getting the storage location, they use CP-ABE to encrypt symmetric key and embed it in ciphertext with LSSS access policy, and then upload to SDBC.
7. Service Provider (SP): mainly consists of the volunteers. When the service provider's attributes satisfy the demander's access policy, he can use his secret attribute key to decrypt and provide service to the demander.

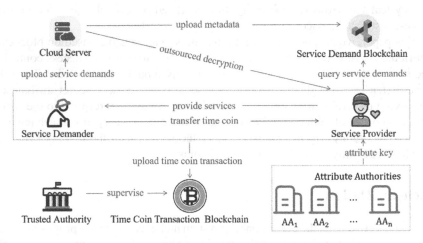

Fig. 1. System model

3.2 Description of Proposed Scheme

Our proposed scheme includes five phases with nine algorithms: system initialization, multi-authority authorization, publishing service demands, providing services and time coin tracking. The symbols involved in the algorithms are shown in Table 1.

System Initialization. This phase is divided into two parts: the initialization of attribute authorities and users who newly join to the system as well as identity anonymization.

1. $Setup(1^{\lambda}) \rightarrow (MSK, PK)$. TA inputs a security parameter λ and defines the bilinear map $e : G \times G \rightarrow G_T$ where g is the generator of G, and then selects two collision-resistant hash functions $H_1 : \{0, 1\}^* \rightarrow G, H_2 : G_T \rightarrow Z_p$. TA generates a certificate $AID.Cert$ and assigns an identifier $AID = H_1(AID.Cert) \in Z_p$ to each new legal attribute authority that joins the system. The n AAs with a threshold value of t jointly manage attribute set $U = \{Att_1, Att_2, \cdots, Att_U\}$. TA also generates $h_1, h_2, \cdots, h_U \in G$ for each attribute Att_i and selects a random number $a \in Z_p$. Each AA_i selects a sub-secret $\alpha_i \in Z_p$, and collectively form the master key $\alpha = \sum_{i=1}^{n} \alpha_i$. It returns the public key $PK = (g, g^{\alpha}, e(g, g)^{\alpha}, n, t, h_1, \cdots, h_U)$ and the master key $MSK = (a, \alpha)$.

2. $IdAnonymity(MSK, UID, PK_u) \rightarrow UID'$. Each user (SD or SP) sends a registration request to TA before joining the system, and upon successful certification, is assigned a certificate $UID.Cert$ and a unique identifier $UID = H_1(UID.Cert) \in Z_p$. The user randomly selects $k \in Z_p$ to generate a pair of secret and public key $\{SK_u, PK_u\}$, where $SK_u = k, PK_u = g^k$. In order to protect the user's identity privacy while regulating the circulation of time coins, the user sends the real UID and public key PK_u to TA, and both parties jointly calculate the negotiation key $R = (PK_u)^\alpha$ to generate the pseudonym $UID' = UID \oplus H_1(R)$. Create a form T to store the user's public key and the pseudonym in one-to-one correspondence.

Table 1. Symbols used in this scheme.

Symbols	Meaning of the symbols
UID	Global identifier for the user
AID	Global identifier for each AA_i
MSK	The system master key
PK	The system public key
$\{sk_i, pk_i\}$	The secret and public key pair for each AA_i
S	The user's attribute set
$\{SK_u, PK_u\}$	The secret and public key pair for the user
SK_{UID}	The secret attribute key for the user
TK_u	The transformation key
K_s	The AES symmetric key
(A, ρ)	The access policy
M	The service demand plaintext
$CT = \{CT_M, CT_K\}$	The ciphertext packet
CT'	The outsourced partially decryption ciphertext
R	The negotiation key
T	The form

Multi-authority Authorization. In this phase, the user's secret attribute key is jointly generated by multiple attribute authorities using (t, n-DKG), which ensures the security of the user's secret attribute key and solves the problem of single point of failure.

$MultiKeyGen(PK, \{sk_i, pk_i\}, S, UID) \rightarrow SK_{UID}$. Firstly, we use the DPoS consensus mechanism in the federated blockchain to select multiple consensus nodes AA_i to play the role of attribute authorities to credit-endorse the user's attribute set S, which increases the authenticity of the user's identity. Then each AA_i separately generates a random t-1th polynomial $f_i(x)$ that satisfies the formula $\alpha_i = f_i(0)$. Each AA_i calculates and sends the sub-share $s_{ij} = f_i(AID_j) mod q$ for each of the other $AA_j (j = 1, \cdots, n$ and $j \neq i)$. After receiving n-1 sub-shares s_{ij}, each AA_i computes its master key share $sk_i = \sum_{j=1}^{n} s_{ji}$ and relevant public key share $pk_i = e(g, g)^{sk_i}$. Then AA_i selects a random number $\beta_i \in Z_p$ based on the user's attribute set S to generate the secret key share as follows.

$$K_i = g^{sk_i} \cdot g^{a\beta_i}, L_i = g^{\beta_i}, \forall Att_i \in S : K_{Att_i} = h_{Att_i}^{\beta_i} \tag{1}$$

Finally, the user only needs to select t different AAs according to his preference to obtain t secret key shares, which can be used with (t, n)-DKG to generate complete secret attribute key $SK_{UID} = (TK_u, SK_u)$. $TK_u = (K, L, K_{Att})$ is the transformation key that can be used to outsourced decryption and $b = \sum_{i=1}^{t} \left(\beta_i \cdot \prod_{j=1, j\neq i}^{t} \frac{AID_j}{AID_j - AID_i} \right)$. By this means, we solve the problem of single point of failure.

$$
\begin{cases}
K = \prod_{i=1}^{t} K_i^{\prod_{j=1,j\neq i}^{t} \frac{AID_j}{(AID_j - AID_i) \cdot k}} = g^{\alpha/k} \cdot g^{a \cdot \sum_{i=1}^{t} \left(\beta_i \cdot \prod_{j=1,j\neq i}^{t} \frac{AID_j}{(AID_j - AID_i) \cdot k} \right)} = g^{\alpha/k} \cdot g^{ab/k} \\[2ex]
L = \prod_{i=1}^{t} L_i^{\prod_{j=1,j\neq i}^{t} \frac{AID_j}{(AID_j - AID_i) \cdot k}} = g^{\sum_{i=1}^{t} \left(\beta_i \cdot \prod_{j=1,j\neq i}^{t} \frac{AID_j}{(AID_j - AID_i) \cdot k} \right)} = g^{b/k} \\[2ex]
\forall Att \in S : K_{Att} = \prod_{i=1}^{t} K_{Att_i}^{\prod_{j=1,j\neq i}^{t} \frac{AID_j}{(AID_j - AID_i) \cdot k}} = h_{Att}^{\sum_{i=1}^{t} \left(\beta_i \cdot \prod_{j=1,j\neq i}^{t} \frac{AID_j}{(AID_j - AID_i) \cdot k} \right)} = h_{Att}^{b/k}
\end{cases}
\tag{2}
$$

Publishing Service Demands. In order to provide fine-grained access control for shared service demands, this phase is divided into two parts: symmetric encryption and attribute-based encryption.

1. $SymEncrypt(K_s, M) \rightarrow CT_M$. Service demander SD selects symmetric encryption algorithm AES to encrypt service demands M, and gets the ciphertext CT_M.
2. $ABEEncrypt(PK, K_s, (A, \rho)) \rightarrow CT$. Firstly, SD sets the LSSS access policy (A, ρ) according to his own demands to improve the flexibility of the access. A is a $l \times m$ matrix, where l is the scale of a specific attribute set and the function ρ maps each row of A to a specific attribute, marked as $\rho(i)(\in Att_1, Att_2, \cdots, Att_U)$. To encrypt the symmetric key K_s, a random secret parameter s is selected. A random vector $\overrightarrow{v} = (s, y_2, y_3, \ldots, y_m)^T \in Z_p^n$ is chosen to hide the parameter s. For $i = 1$ to l, SD calculates $\lambda_i = A_i \cdot \overrightarrow{v}$ where A_i denotes the ith row of the matrix A and randomly selects $r_i \in Z_p$ to calculate CT_K.

$$
CT_K = \big(C_0 = K_s \cdot e(g, g)^{\alpha s}, C_1 = g^s, C_2 = g^{as}, C_v = H_2(e(g, g)^{\alpha s}), \\
C_i = g^{a \cdot \lambda_i} \cdot h_{\rho(i)}^{-r_i}, D_i = g^{r_i}, \forall i = 1 \text{ to } l \big)
\tag{3}
$$

C_0 is the symmetric key ciphertext, C_1 is the attribute ciphertext, C_v is used to check the correctness of the outsourced decryption, and C_2, C_i, D_i are used for attribute matching. Finally, SD sends the ciphertext packet $CT = \{CT_M, CT_K\}$ and the verification key K_{ver} to CS for storage and sharing through the secure channel. The ciphertext packet index, storage address, and verification key are uploaded to SDBC.

Providing Services. In order to respond quickly to the service that provided by SP to SD, we designed the outsourced decryption mechanism. Matching first, then service. Then we outsource a large amount of ciphertext to CS.

1. $AttMatch(PK, CT_K, S) \rightarrow T/F$. SP sends a request to CS when he wants to provide services to SD. Then CS uses SP's attribute set S to match with the attributes in the ciphertext and calculates whether (4) holds. If it does not hold, CS considers that SP does not have access rights and denies his access request. Conversely, SP's attribute set S satisfy the ciphertext policy and he can continue the outsourced decryption.

$$e(C_2, L) = \prod_{i \in I} \left(e(C_i, L) \cdot e(D_i, K_{\rho(i)})\right)^{w_i} \tag{4}$$

2. $OutDecrypt(CT_K, TK_u) \rightarrow CT'$. After a successful match, CS uses the transformation key TK_u to calculate the partially decrypted ciphertext CT' and send it to SP.

$$CT' = \frac{e(C_1, K)}{\prod_{i \in I} \left(e(C_i, L) \cdot e(D_i, K_{\rho(i)})\right)^{w_i}} = e(g, g)^{s\alpha/k} \tag{5}$$

3. $SPDecrypt(CT', SK_u) \rightarrow CT$. Based on the partially decrypted ciphertext CT', SP perform only one power operation without any bilinear map to recover the message. Firstly, SP calculates $CT_v = (CT')^{SK_u}$ with his own secret key SK_u. The calculation result is compared with the hash value stored in SDBC. If $C_v \neq H_2(CT_v)$, it indicates that the outsourced calculation is incorrect. Otherwise, the decryption calculation $K_s = \frac{C_0}{CT_v}$ is performed. Finally, SP uses K_s to compute $Dec(CT_M, K_s)$ to obtain the service demand plaintext M. So far, SP can provide services to SD. Both parties record their pseudonyms and time coins transactions in TCTBC. The block structure of TCTBC is shown in Fig. 2.

Time Coin Tracking

$Track(MSK, UID', T) \rightarrow UID$. If a user has doubts about the time coin balance in his account, he can send his public pseudonym UID' to TA to apply for tracking. TA queries the user's public key in the form T by UID' and uses the master key MSK to calculate the negotiation key $R = (PK_u)^\alpha$ which can be used to recover the user's real identity $UID = UID' \oplus H_1(R)$. This enables tracing of a user's time coin transactions in TCTBC and determining his time coin balance.

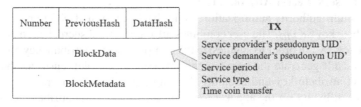

Fig. 2. The block structure of TCTBC

4 Performance Analysis and Evaluation

In this section, we analyze and evaluate this scheme in terms of correctness, security and performance.

4.1 Correctness Analysis

There are three places to prove of Sect. 3.2.

1. Correctness verification of attribute matching.

$$e(C_2, L) = e(g^{as}, g^{b/k}) = e(g, g)^{asb/k} \tag{6}$$

$$
\begin{aligned}
\prod_{i \in l} \left(e(C_i, L) \cdot e(D_i, K_{\rho(i)}) \right)^{w_i} &= \prod_{i \in l} \left(e\left(g^{a \cdot \lambda_i} \cdot h_{\rho(i)}^{-r_i}, g^{b/k}\right) \cdot e\left(g^{r_i}, h_{\rho(i)}^{b/k}\right) \right)^{w_i} \\
&= \prod_{i \in l} \left(e\left(g^{a \cdot \lambda_i}, g^{b/k}\right) \cdot e\left(h_{\rho(i)}^{-r_i}, g^{b/k}\right) \cdot e\left(g^{r_i}, h_{\rho(i)}^{b/k}\right) \right)^{w_i} \\
&= \prod_{i \in l} e(g, g)^{ab \cdot \lambda_i w_i / k} = e(g, g)^{abs/k}
\end{aligned} \tag{7}
$$

Obviously, the Eq. (4) holds when SP's attribute set S satisfies the access policy of the ciphertext.

2. Correctness verification of outsourced decryption.

$$CT' = \frac{e(C_1, K)}{\prod_{i \in l}\left(e(C_i, L) \cdot e(D_i, K_{\rho(i)})\right)^{w_i}} = \frac{e\left(g^c, g^{\alpha/k} g^{ab/k}\right)}{e(g, g)^{abs/k}} = e(g, g)^{s\alpha/k} \tag{8}$$

3. Correctness verification of SP decryption.

$$K_s = \frac{C_0}{CT_v} = \frac{K_s \cdot e(g, g)^{s\alpha}}{(e(g, g)^{s\alpha/k})^k} = K_s \tag{9}$$

4.2 Security Analysis

The scheme in this paper has the following security features.

Security of Users' Secret Attribute Keys
We build a multi-authority authorization model based on (t, n)-DKG to generate user's secret attribute key SK_{UID}. The user combines different t AAs' secret key shares with his own secret key SK_u in order to synthesize the complete secret attribute key SK_{UID}. Less than t AAs can't generate the user's secret attribute key, which ensuring the security of the secret attribute key and avoiding the single point of failure problem. With t/n constant, the difficulty for the attacker to obtain the user's secret attribute key increases linearly as n increases. Increasing the cost to the attacker is effective.

Security Against Collusion Attack

Some malicious users maybe collude with each other by sharing their secret keys to gain more privileges. Since random element $b = \sum_{i=1}^{t} \left(\beta_i \cdot \prod_{j=1, j \neq i}^{t} \frac{AID_j}{AID_j - AID_i} \right)$ is different in different users' secret attribute keys, they can't gain more privilege by combining their secret keys.

Confidentiality Guarantee

CS is honest and curious, and it does not participate in AAs' master key sharing or user's full secret attribute key generation. When outsourced decryption, it only gets partially decrypted ciphertext CT' and cannot decrypt the private data without the user's secret key SK_u.

4.3 Evaluation of Performance

In this subsection, we perform experiments to validate our scheme and compare this scheme with similar schemes or systems. In our scheme, AES and CP-ABE are combined to achieve fine-grained access control for shared data. Constructing LSSS access matrix enhances the flexibility of policy expression. Our multi-authority authorization model ensures the security of user's secret attribute key and solves the problem of single point of failure. The computational overhead at the user side is reduced by outsourced decryption mechanism. At the same time, we support identity anonymity and time coin tracking. The results of scheme comparison are shown in Table 2.

Table 2. Function comparison of different schemes.

	BSW [4]	Li [12]	He [13]	Our scheme
Access structure	Access Tree	Access Tree	LSSS	LSSS
Identity anonymity	✗	✗	✗	✓
Multi-authority	✗	✓	✓	✓
Single point of failure	✗	✗	✗	✓
Outsourced decryption	✗	✓	✗	✓
Verifiable mechanism	✗	✗	✓	✓
Tracking	✗	✗	✗	✓
Blockchain	✗	✗	✓	✓

Our experimental environment is under Ubuntu 20.04.2 LTS system with Intel (R) Core (TM) i7-10700 CPU @ 2.90 GHz × 4 and 16 GB RAM. The service demand data size is set to 10^3 KB, and the number of attribute authorities is fixed at 5. We measure the total time for key generation, client encryption and decryption by varying the number of attributes in the access policy and averaging the results of 10 runs. The comparison of our scheme with the schemes in [4, 12, 13] is shown in Figs. 3, 4 and 5.

As shown in Fig. 3, we compare our scheme with Li [12] and He [13] which are multi-authority in terms of key generation. The key generation cost of all schemes increases linearly with the number of attributes; however, our scheme has the slowest growth rate. This is because the multi-authority model of He [13] is built with BLS signatures, which adds additional key generation overhead. As the secret attribute key of user is allocated through the key encryption tree (KEK) in Li [12], an increase in the recursive depth of the tree leads to a rapid increase in computational overhead. In contrast, our multi-authority model only requires t sub-secret shares among n attribute authorities to generate user's secret attribute key. Our multi-authority model demonstrated significant advantages with increasing attributes and solved the issue of single point of failure.

As shown in Fig. 4, we compare the encryption time of different schemes. As the number of attributes increases, the encryption time for all schemes increases linearly. An access tree as access policy is used in BSW [4] and Li [12], which leads to recursive operations during encryption and consumes a large amount of computation when the recursion is reached a certain depth. LSSS is used to construct the access policy in both our scheme and He [13], but mapping operations are performed for each user attribute in He [13] during direct data encryption with CP-ABE, leading to multiple rounds of exponential operations. In contrast, our scheme combines AES with CP-ABE, avoiding this computational overhead and achieving more efficient encryption.

As shown in Fig. 5, we compare the client decryption time of different schemes. The client decryption time of our scheme and Li [12] are much lower than BSW [4] and He [13] because of the outsourced decryption introduced. But a larger number of bilinear map operations are still required by Li [12], resulting in a larger decryption time, while our scheme only requires multiplication and exponential operations, leading to a smaller decryption time. Therefore, the client decryption time of our scheme is slightly smaller than Li [12], and much smaller than BSW [4] and He [13]. Additionally, our scheme also ensures the correctness of the outsourced data while reducing the client decryption overhead.

As shown in Fig. 6, we evaluate the percentage of key generation, client encryption and decryption time in the overall operation when running with the Hyperledger Fabric federated blockchain system. The test environment consists of ten peer nodes and five orderer nodes using DPoS consensus mechanism. The key generation, client encryption and decryption operations account for less than 16% of the total system runtime as the size of the service demand data increases. Although specific service demand data may exceed 10^4 KB with pictures and videos, the experimental results in Fig. 6 show a slower trend of overall overhead growth as the size of the service demand data increases in the system, ensuring the feasibility of the scheme.

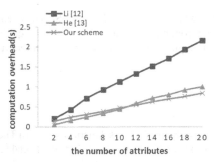

Fig. 3. Computation overhead of key generation

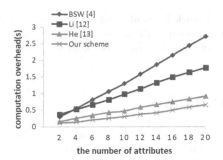

Fig. 4. Computation overhead of encryption

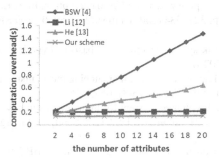

Fig. 5. Computation overhead of client decryption

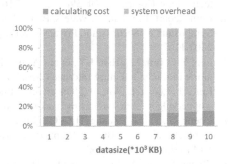

Fig. 6. Percentage of calculating cost

5 Conclusion

In this paper, we propose a scheme based on blockchain and attribute-based encryption to achieve secure mutual aid service in time bank. We combine AES with CP-ABE to encrypt privacy data to achieve fine-grained access control for shared service demands, and an LSSS access matrix enhances policy expression flexibility. To prevent excessive authority of a central authority, we design a multi-authority model to generate users' secret attribute keys, which guarantees the security of users' secret attribute keys and solves the problem of single point of failure. To quickly respond to service supply and demand, we design an outsourced decryption mechanism to reduce the client computational overhead. In addition, we introduce anonymity mechanism to protect the identity privacy of users while regulating the circulation of time coins. Finally, the experimental results proved the effectiveness and applicability of our scheme within the blockchain-based time bank system, while protecting the privacy of the elderly. However, our scheme also has certain shortcomings, which can be enhanced by designing a hidden time coin transaction mechanism in future research.

Acknowledgement. This work is supported by the Major projects of National Social Science Foundation of China (No. 21&ZD124): research on community home care model and quality safety System based on blockchain. We also thank anonymous reviewers for the helpful reports.

References

1. Cahn, E.S., Rowe, J.: Time Dollars: The New Currency that Enables Americans to Turn Their Hidden Resource-Time-into Personal Security and Community Renewal, 2nd edn. Rodale Pr, Pennsylvania, PA, USA (1992)
2. Cao, B., Li, Y., Zhang, L., et al.: When Internet of things meets blockchain: challenges in distributed consensus. IEEE Netw. **33**(6), 133–139 (2019)
3. Xiao, K., Wang, M., Tang, Y., et al.: A public welfare time bank system based on blockchain technology. Comput. Appl. **39**(7), 2156–2161 (2019). (in Chinese)
4. Bethencourt, J., Sahai, A., Waters, B.: Ciphertext-policy attribute-based encryption. In: IEEE Symposium on Security and Privacy (S&P), pp. 321–334 (2007). https://doi.org/10.1109/SP. 2007.11
5. Lee, Y.T., Lin, J.J., Hsu, J.Y., et al.: A time bank system design on the basis of hyperledger fabric blockchain. Future Internet **12**(5), 84 (2020)
6. Lin, J.J., Lee, Y.T., Wu, J.L.: The effect of thickness-based dynamic matching mechanism on a hyperledger fabric-based timebank system. Future Internet **13**(3), 65 (2021)
7. Cheng, S., Shi, W., Zhang, H.: VOLTimebank: a volunteer system for mutual pension based on blockchain. In: ICBCT 2019, pp 75–79. ACM, Honolulu, HI, USA (2019). https://doi.org/ 10.1145/3320154.3320160
8. Zhang, W., Ge, Y.: Improvement of DPoS consensus based on block chain. In: ICIIP 2019, pp. 352–355. ACM, New York, NY, USA, (2020). https://doi.org/10.1145/3378065.3378132
9. Li, W., Xue, K., Xue, Y., et al.: TMACS: a robust and verifiable threshold multi-authority access control system in public cloud storage. Trans. Parallel Distrib. Syst. **27**(5), 1484–1496 (2016)
10. Li, X., Dong, X., Xu, X., et al.: A blockchain-based scheme for efficient medical data sharing with attribute-based hierarchical encryption. In: Zhao, X., Yang, S., Wang, X., Li, J. (eds.) WISA 2022. LNCS, vol. 13579, pp. 661–673. Springer, Cham (2022). https://doi.org/10. 1007/978-3-031-20309-1_58
11. Xiong, H., Huang, X., Yang, M., et al.: Unbounded and efficient revocable attribute-based encryption with adaptive security for cloud-assisted internet of things. IEEE Internet Things J. **9**(4), 3097–3111 (2022)
12. Li, L., Wang, Z., Li, N.: Efficient attribute-based encryption outsourcing scheme with user and attribute revocation for fog-enabled IoT. IEEE Access **8**, 176738–176749 (2020)
13. He, Y., Wang, H., Li, Y., et al.: An efficient ciphertext-policy attribute-based encryption scheme supporting collaborative decryption with blockchain. IEEE Internet Things J. **9**(4), 2722–2733 (2022)
14. Sun, Y., Song, W., Shen, Y.: Efficient patient-friendly medical blockchain system based on attribute-based encryption. In: Wang, G., Lin, X., Hendler, J., Song, W., Xu, Z., Liu, G. (eds.) WISA 2020. LNCS, vol. 12432, pp. 642–653. Springer, Cham (2020). https://doi.org/10. 1007/978-3-030-60029-7_57
15. Gennaro, R., Jarecki, S., Krawczyk, H.: Secure distributed key generation for discrete-log based cryptosystems. In: Stern, J. (ed.) EUROCRYPT 1999. LNCS, vol. 1592, pp. 295–310. Springer, Heidelberg (1999). https://doi.org/10.1007/3-540-48910-X_21

Exact Query in Multi-version Key Encrypted Database via Bloom Filters

Guohao Duan, Siyuan Ma, and Yanlong Wen[(✉)]

College of Computer Science, Nankai University, Tianjin, China
{1911407,1911452}@mail.nankai.edu.cn, wenyl@nankai.edu.cn

Abstract. As people become more privacy conscious, the security requirements of databases are increasing, and thus the technology of encrypted databases is developing rapidly. Although many theoretical results have been produced in the academic community, only a few examples have been operated in the enterprise community, mainly due to the imbalance in the trade-off between the efficiency and security of encrypted databases.

In this paper, we focus on an example of the encrypted database, the multi-version key encrypted database, which loses the unique indexing function and has excessive exact query time overhead in achieving security. In order to solve this problem and fully optimize the database, we systematically analyze the implementation principle of the exact query algorithm of this database to reduce the redundant operations in the original query process, and draw on Blind Seer's idea of index design, use Bloom filters to construct unique indexes for this encrypted database, optimize the exact query process, and finally achieve high efficiency and security at the cost of tolerable space overhead and false alarm rate. Theoretical analysis shows that the new scheme is not less secure than the original scheme, the time complexity is reduced, and it is compatible with other operations; experimental results show that the efficiency of the exact query has been effectively improved.

Keywords: Encrypted database · Exact query · Bloom filter · Unique index

1 Introduction

In the information age, data has inevitably penetrated into people's lives. As the primary carrier of data, database management systems (DBMS), which store and process data, and ensure data security and integrity, play a vital role and are an integral part of the functioning of today's society.

However, most of today's DBMSs store data in plaintext, which has high-security risks and is susceptible to inference attacks [1], differential privacy attacks [2], frequency attacks [3], etc., and sensitive data is vulnerable to threats. If the data is corrupted or stolen, it may temporarily prevent users from accessing the service, make them victims of counterfeiting or surveillance, and bring them unpredictable losses. Therefore, the security of DBMS is widely concerned.

© The Author(s), under exclusive license to Springer Nature Singapore Pte Ltd. 2023
L. Yuan et al. (Eds.). WISA 2023, LNCS 14094, pp. 415–426, 2023.
https://doi.org/10.1007/978-981-99-6222-8_35

The encrypted database stores data in the form of ciphertext and establishes security mechanisms for the database from the inside and outside, which is highly secure; with the help of searchable encryption schemes, such as PPE [4], FHE [5], ORAM [6], SSE [7], etc., it realizes highly efficient access policies and rich functions. It provides a more efficient, secure, and reliable solution for storing and processing huge amounts of data in the cloud, which is of good practicability and huge development potential. It has good practicability and great development potential.

At the same time, it is designed based on cryptographic primitives, and since there are many types of cryptographic primitives, there are also more types of encrypted databases. However, the commercial encrypted database has a high standard of security and performance, the design process will exclude some algorithms that are too extreme in terms of indicators, such as OPE [8] - excellent performance and inferior security - leaking plaintext features, ORAM - very high security but significant performance overhead, and choose a more balanced algorithm - SSE - in terms of metrics, so as to achieve the two goals of efficiency and security.

In this paper, we take the mainstream multi-version key encrypted database as the research object and design an exact query strategy to achieve a balance between security and efficiency, hoping to provide some reference value for its development.

Related Work. With the goal of rich query function, secure query process, and efficient query performance, the encrypted database keeps exploring new feasible technical routes. In recent years, due to the rise of trusted hardware technologies (such as ARM TrustZone [9] and Intel SGX [10]), the construction strategy has also transitioned from the early pure cryptography scheme to a scheme [11] combining cryptography and trusted hardware technologies, and a large number of outstanding examples have emerged.

CryptDB [12] uses pure cryptography technology to build an onion model to achieve various query functions, but it also faces the problem of excessive space and time overhead; Blind Seer [13] uses Bloom filters to build indexes, which excel in performance and enable efficient sub-linear search, but there are some flaws in security, which are properly addressed in [14]; StealthDB [15] introduces trusted hardware technology that can be integrated with PostgreSQL, providing strong end-to-end security assurance at the cost of 30% throughput, supporting full SQL query functionality and large transactional workloads with high utility; Always Encrypt [16] is the first industrial-grade database system to protect against DMA attacks, integrating with SQL Server, which has already gained large-scale use, realizing ideal IND-CPA security and supporting rich query functionality.

It can be seen that the encrypted database has achieved fruitful theoretical results in the past ten years, but the practical results in the industry are few. At present, only AE has the advantage, and OpenGauss is emerging. The performance of the encrypted database in the industry still has huge room for development. The demand determines the type and optimization strategy of the encrypted database and promotes the development and application of the encrypted database.

Our Contributions. We discover and solve the problems existing in the existing exact query scheme of the encrypted database, and at the same time provide appropriate interfaces for subsequent development.

1. Using Blind Seer's index design idea for reference, the Bloom filter is introduced into the existing database to build a unique index. The Bloom filter uses the trapdoor idea, which is only affected by the plaintext, reflects the characteristics of the plaintext, and will not leak any plaintext information, so it can improve efficiency without reducing security.
2. Appropriately simplify the ciphertext space of IN operation and improve the efficiency of the accurate query. In the multi-version key mechanism, the diversity of keys is beneficial to improve system security, and the IN operation must exist to ensure the rationality of keys. The ciphertext space of the original IN operation is all the fields of the relevant attribute columns. Now it is processed by the Bloom filter, which greatly simplifies the ciphertext space and improves the efficiency of the IN operation.
3. Provide interfaces for other operations to improve algorithm compatibility and practicability. As a core function, database query needs to be compatible with other operations, such as adding operations to avoid primary key and foreign key conflicts, and need to implement query operations, etc. The exact query algorithm in this paper can return the ID slice of the query record to assist other operations to proceed smoothly.

2 Architecture

2.1 Original Model

The multi-version key-based encrypted database consists of three main components, namely, server, client, and key manager (KMS), where the server and KMS are operated and maintained by the DBA, the ciphertext information is stored in the server, the key information is stored in the KMS, and the client is controlled by the user. The interaction process of each component to execute the exact query algorithm is shown in Fig. 1.

In the server, only the plaintext information is encrypted, and no additional attributes are added. The plaintext table and the ciphertext table belong to a one-to-one mapping relationship, resulting in the following exact query scheme.

KMS Mechanism. The plaintext information is encrypted by the key group DEK, in order to get the plaintext query result, the server needs to get DEK beforehand; KMS stores the ciphertext form of DEK, EDEK, and its key group KEK, the server constructs DEK request according to the information of SQL statement and sends it to KMS; KMS receives and processes the request, queries

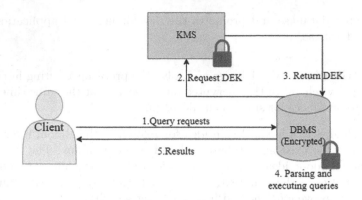

Fig. 1. The original interaction process of the encrypted database

KEK and decrypts EDEK, returns DEK to the server; the server loads the DEK into the memory. A key table example is shown in Table 1, which can query the relevant keys in DEK based on the table name and attribute information of the SQL statement, and use each key encryption field to get the cipher space for IN operation.

Table 1. A DEK instance

ID	TableName	AttrName	KeyVer
1	stu	name	4
1	stu	mail	5
2	stu	mail	6
3	tea	age	7

Query Mechanism. First, the client initiates a plaintext query request; second, the server parses, rewrites and executes the query request, obtains the DEK from the KMS and loads it in the cache, and calculates the ciphertext space based on the query value; then, the server performs the IN operation with all the fields of the specific attribute in the ciphertext space, and the decryption can be accomplished if the match is completed; finally, the query result is returned to the client.

Shortcomings. The original exact query process had the following problems:

1. Due to the complicated key allocation mechanism, the same plaintext may be assigned different keys and mapped to different ciphertexts due to different library names, table names, or attributes, and the unique index fails.

2. Unique index failure can have a negative impact on a series of operations. For example, for correlation queries, the plaintext has lost its characteristics and only exact queries have to be executed on multiple tables to complete the join operation, which brings a huge overhead that far exceeds its security cost.
3. Redundant operations are more frequent. the ciphertext space for IN operations is the entire field of the attribute column, and most of the fields will only lose computational performance with no real effect in terms of security.

2.2 Improved Model

To solve the above problem, a new component is introduced for this architecture - a proxy server, which is trusted to both rewrite query requests and filter query results to achieve transparency in client-server interaction, thus obtaining the new interaction process shown in Fig. 2.

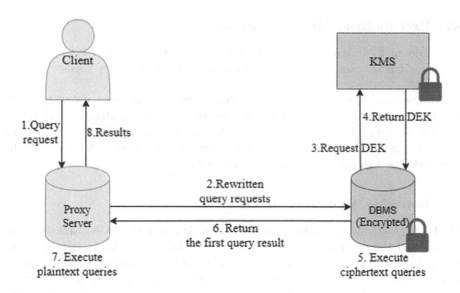

Fig. 2. The new interaction process of the encrypted database

3 Implementation

3.1 Add Auxiliary Attributes

In order to construct a unique index, this paper draws on the design idea of Blind Seer to add auxiliary attributes to the original ciphertext table structure, and field values are generated by Bloom filters. Before executing the exact query, the plaintext data table is pre-scanned, the ciphertext table structure is decided according to the query requirements, and the batch retrieval and encryption of records are migrated to the ciphertext table. Figure 3 gives a pre-processing example (assuming that there is only a query/index building requirement for ID and ATTR2).

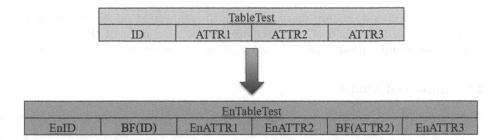

Fig. 3. The preprocessing example for TableTest

3.2   Execute Exact Query Algorithm

Since Bloom filter has a false alarm rate and each record of the database is sensitive, it is necessary to execute a secondary query to ensure the accuracy of the query.

After establishing the auxiliary attributes, during the query process, firstly, match the Blum filter query result to narrow down the matching field space, secondly, execute IN operation to obtain the record key, then decrypt the record to get the first query result, and finally, execute the secondary query in the plaintext space to return the final query result.

3.3   Extended Application of Algorithm

The DML (Data Manipulation Language) functions in the database are interconnected, and the implementation logic of other DMLs needs to be appropriately adapted after changing the exact query algorithm to maximize the optimization effect of exact query and avoid some errors. In industrial practice, foreign key constraint relations are often not used, so this section only focuses on primary key conflicts, and adapts the new exact query algorithm based on the add, modify and delete operations to improve the efficiency of related operations.

Insert Operation. In the insert operation, if the record does not contain a primary key field, then you can directly encrypt the record and insert it into the database; if the record involves a primary key field, you need to perform a query operation on the primary key field first and return the empty set before you can perform the insert operation; otherwise, reject this insert request to avoid primary key conflicts, as implemented below.

Algorithm 1. insert operation

Input: must enter non-incremental attributes and their field values, select to enter incremental attributes and their field values
Output: if the insertion is successful
1: **function** ISEXIST(*KeyAttr*)
2:     if QueRecord(KeyAttr) is able to return the query result:
3:         return true
4:     else return false
5: **end function**
6: **function** ADDRECORD(*Attr*)
7:     if the primary key *KeyAttr* is in *Attr*:
8:         if IsExist(KeyAttr).
9:             return false
10:     else
11:             encrypt the record and insert
12:             return true
13:     else
14:             encrypt the record and insert
15:             return true
16: **end function**

Delete Operation. In the deletion operation, due to the existence of the false alarm rate, we can first delete the Blum filter matching results in the ciphertext table, decrypt these records in the proxy server, perform the specific deletion operation in the proxy server, and finally re-insert the false alarm records into the ciphertext table with encryption processing, which is implemented as follows.

Algorithm 2. delete operation

Input: Related properties and their fields
Output: if the deletion is successful
1: **function** DELRECORD(*Attr*)
2:     Query BF(Attr) in the ciphertext table, delete the match result, and store it decrypted in the proxy server
3:     Perform the delete operation in the proxy server
4:     encrypt the remaining records and insert them into the cipher table
5:     return true
6: **end function**

Modify Operation. The modification operation needs to solve both the false alarm rate and the primary key conflict problem, which is implemented as follows.

Algorithm 3. modify operation

Input: retrieves the related attribute and its field $Attr1$, the modified related attribute and its field $Attr2$

Output: if the modification is successful

1: **function** REVRECORD($Attr1,Attr2$)
2:    Query BF(Attr1) in the ciphertext table, delete the match result, and store it in the proxy server by decrypting it
3:    if primary key $KeyAttr2$ in $Attr2$:
4:        if IsExist(KeyAttr2).
5:            encrypt the plaintext record and insert it into the ciphertext table
6:            return false
7:        else
8:            Perform plaintext modification operation
9:            Encrypt the plaintext record and insert it into the ciphertext table
10:            return true
11:    else
12:        Perform plaintext modification operation
13:        Encrypt the plaintext record and insert it into the ciphertext table
14:        return true
15: **end function**

---

4   Performance Evaluation

After designing the exact query algorithm and adapting some DML operations, this section will analyze the performance of the algorithm from both theoretical and practical aspects.

4.1   Theoretical Analysis

Time Perspective Analysis. The main time overhead of the algorithm exists in the IN operation, so we analyze this and assume that the total number of fields in the attribute column is N and the size of the ciphertext space is M.

Before establishing the unique index, each field needs to match the cipher space, and the query time complexity is O(M*N); after establishing the unique index, only the fields filtered by the Bloom filter need to perform IN operation, and the query time complexity is O(M*logN), so the new algorithm achieves greater optimization in performance.

Spatial Perspective Analysis. Spatially, the space size of the auxiliary attributes depends on the bit length of the Bloom filter. The length of the bit vector depends on the plaintext space size. In the application process, a shorter bit vector can satisfy the demand, and the space overhead is negligible compared to the ciphertext size.

Usefulness Analysis. Commercial encrypted databases generally have two characteristics, one is the huge data volume and the other is the sensitive data content. In this case, the Bloom filter has significant advantages, on the one hand, its optimization effect and data volume are positively correlated, on the other hand, it adopts the trapdoor idea, which can effectively protect sensitive information and retain the original IN operation without reducing the security of the original encryption scheme. Thus, it is a very feasible and practical choice to adopt it for index optimization and security.

4.2 Experimental Environment

The experimental environment for the implementation and operation of this algorithm is shown in Table 2.

Table 2. Environment

Programming Language	Go1.20.1
Development Environment	GoLand 2023.3.2
DBMS	MariaDB 10.4.27
Operating System	Windows 11 64bit
Memory	8G
CPU	Intel(R) Core(TM) i7-9750H CPU @ 2.60GHz

The test database is the employees database of MySQL instance, which contains 3 million rows of data. The main data table structure is shown in Table 3

Table 3. Test Database Table

employees	
Column Name	Data Type
emp_no	INT
first_name	VARCHAR(14)
last_name	VARCHAR(14)
gender	ENUM('M','F')
hire_date	DATE

4.3 Performance Test

When performing the exact query, it is necessary to simultaneously compare the query results of the plaintext data table and the ciphertext data table to test the accuracy of the ciphertext exact query algorithm. On this basis, explore the specific impact of various factors on the algorithm.

Dataset Size. There is a non-negligible impact of dataset size on query performance, we set up datasets with sizes of 100, 500, 1000, 5000, 10000, 20000, 30000, 40000, and 50000, respectively, to compare and analyze the average response time of executing exact queries in these datasets, and after obtaining the results in Fig. 4 (a), it is found that the curve is too smooth when the size of the dataset is lower than 5000, in order to facilitate the analysis, some of the results are selected and enlarged, and the results are shown in Fig. 4.

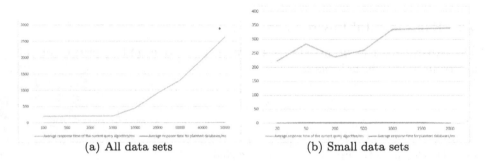

(a) All data sets   (b) Small data sets

Fig. 4. Impact of dataset size on query performance

The experimental results show that when the data set size is less than 10000, the increase in data set size will have a certain negative impact on the precise query efficiency, but not significant, the main performance overhead stems from the machine configuration being too low, even if the size of the data set is as small as 20, there still exists a 200ms time overhead; when the size of the data set is greater than 10,000, the impact of the data set size on the precise query efficiency increases, and the average response time of query statements increases significantly, even so, the overall curve roughly fits the theoretical analysis of the function curve, which indicates that the main framework of the algorithm is roughly realized.

Bloom Filter. Bloom filters make an important contribution to existing schemes, here, we choose the same dataset as the above experiments, compare and analyze the average response time of executing exact query statements in the original and existing schemes to explore the impact of Bloom filters on the performance of the exact query, and the results are shown in Fig. 5.

The experimental results show that the average response time of the new scheme is lower than the average response time of the original scheme under any size of dataset, and the introduction of Bloom filters indeed improves the efficiency of the exact query, in addition to less optimization effect under small datasets, and the optimization effect is significantly improved under larger datasets, which is in line with the theoretical expectation.

This scheme is also affected by other factors, such as the choice of encryption algorithms, the number, and type of hash algorithms, the length of the Bloom

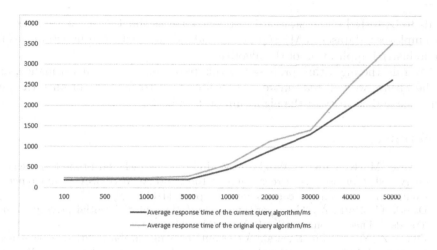

Fig. 5. Impact of Bloom filters on query performance

filter, etc., but these factors are closely related to the security, and can not choose the optimal efficiency of the solution at will, but need to be based on the practice of the demand for the decision by the DBA, the design of the comparative experiments is not very significant, so it is not considered here.

5   Conclusions and Future Directions

In this paper, we propose an exact query algorithm based on a multi-version key encrypted database and introduce Bloom filters in order to solve the problems of unique index failure and excessive redundant operations in the existing encrypted databases, and finally, it is confirmed by theories and experiments that the new scheme optimizes the querying efficiency and achieves the security which is not lower than the original scheme.

However, there is still some room for improvement in the algorithm.

1. The number of bits of the Bloom filter and the type of hash algorithm are highly flexible, and the appropriate parameters need to be determined by the DBA in practice to reduce the false alarm rate and space overhead of the Bloom filter.
2. Further optimizing the implementation details of the algorithm, such as enhancing the scheduling of memory to reduce the I/O overhead, we were able to reduce the query overhead to 20ms with a dataset of size 1000 after our initial attempts.
3. In terms of security, the rewriting mechanism can be improved appropriately, such as mapping information such as table names and column names to specific enumeration values to further hide sensitive content.

In addition, the practicality of the algorithm also needs to be explored, and the experimental data in this paper is only for reference. In the future, it still needs to be tested under standard test sets, such as the TPC-C benchmark and the employees' dataset of MYSQL, to provide a sufficient experimental basis for the industrial application of the algorithm.

Overall, the algorithm proposed in this paper has a good performance, and in the future, we will read more related literature to further optimize the performance and security of the algorithm on this basis.

References

1. Naveed, M., Kamara, S., Wright, C.V.: Inference attacks on property-preserving encrypted databases. In: Proceedings of the 22nd ACM SIGSAC Conference on Computer and Communications Security, pp. 644–655 (2015)
2. Dwork, C., Roth, A.: The algorithmic foundations of differential privacy. Found. Trends® Theor. Comput. Sci. **9**(3–4), 211–407 (2014)
3. Zhang, Z., Yan, J., Liu, S., et al.: A face antispoofing database with diverse attacks. In: 2012 5th IAPR International Conference on Biometrics (ICB), pp. 26–31. IEEE (2012)
4. Cui, S., Song, X., Asghar, M.R., et al.: Privacy-preserving dynamic symmetric searchable encryption with controllable leakage. ACM Trans. Priv. Secur. (TOPS) **24**(3), 1–35 (2021)
5. Kadykov, V., Levina, A., Voznesensky, A.: Homomorphic encryption within lattice-based encryption system. Procedia Comput. Sci. **186**, 309–315 (2021)
6. Stefanov, E., Dijk, M., Shi, E., et al.: Path ORAM: an extremely simple oblivious RAM protocol. J. ACM (JACM) **65**(4), 1–26 (2018)
7. Chai, Q., Gong, G.: Verifiable symmetric searchable encryption for semi-honest-but-curious cloud servers. In: 2012 IEEE International Conference on Communications (ICC), pp. 917–922. IEEE (2012)
8. Popa, R.A, Li, F.H., Zeldovich, N.: An ideal-security protocol for order-preserving encoding. In: 2013 IEEE Symposium on Security and Privacy, pp. 463–477. IEEE (2013)
9. Pinto, S., Santos, N.: Demystifying arm TrustZone: a comprehensive survey. ACM Comput. Surv. (CSUR) **51**(6), 1–36 (2019)
10. Costan, V., Devadas, S.: Intel SGX explained. Cryptology ePrint Archive (2016)
11. He, X., Wei, H., Han, S., Shen, D.: Multi-party privacy-preserving record linkage method based on trusted execution environment. In: Zhao, X., Yang, S., Wang, X., Li, J. (eds.) WISA 2022. Lecture Notes in Computer Science, vol. 13579, pp. 591–602. Springer, Cham (2022). https://doi.org/10.1007/978-3-031-20309-1_52
12. Popa, R.A., Zeldovich, N., Balakrishnan, H.: CryptDB: a practical encrypted relational DBMS (2011)
13. Pappas, V., Krell, F., Vo, B., et al.: Blind seer: a scalable private DBMS. In: 2014 IEEE Symposium on Security and Privacy, pp. 359–374. IEEE (2014)
14. Fisch, B.A., Vo, B., Krell, F., et al.: Malicious-client security in blind seer: a scalable private DBMS. In: 2015 IEEE Symposium on Security and Privacy, pp. 395–410. IEEE (2015)
15. Vinayagamurthy, D., Gribov, A., Gorbunov, S.: StealthDB: a scalable encrypted database with full SQL query support. Proc. Priv. Enhancing Technol **2019**(3), 370–388 (2019)
16. Antonopoulos, P., et al.: Azure SOL database always encrypted. In: Proceedings of the 2020 ACM SIGMOD International Conference on Management of Data, pp. 1511–1525 (2020)

BertHTLG: Graph-Based Microservice Anomaly Detection Through Sentence-Bert Enhancement

Lu Chen[1,2(✉)], Qian Dang[3], Mu Chen[1,2], Biying Sun[3], Chunhui Du[3], and Ziang Lu[1,2]

[1] State Grid Smart Grid Research Institute Co., Ltd, Nanjing 210003, China
46152570@qq.com
[2] State Grid Laboratory of Power cyber-Security Protection and Monitoring Technology, Nanjing 210003, China
[3] State Grid Gansu Information & Telecommunication Company, Lanzhou, China

Abstract. Microservice systems in the industry typically comprise a large-scale distributed architecture with numerous services running on different machines. Anomalies caused by cyber attacks or other factors within such a system are often reflected in different logging systems. Existing log-based approaches for anomaly detection mainly rely on a single type of logs. To address these limitations and enhance anomaly detection, we propose BertHTLG, an approach for detecting microservice anomalies using a heterogeneous graph representation enhanced by Sentence-Bert. It leverages the heterogeneous graph representation to capture the intricate structure and heterogeneity of traces along with the embedded log events. Our approach employs RGCN based on a deep Support Vector Data Description (SVDD) model. By calculating the distances between anomalous traces and the center of the hypersphere using the trained model, we can effectively identify and distinguish anomalous traces. Evaluation on a microservice benchmark demonstrates that BertHTLG achieves remarkable precision (98.5%), recall (99.2%), and F1-Score (98.8%), surpassing state-of-the-art approaches for trace/log anomaly detection with an increase of 3.4% in F1-score. These results validate the effectiveness of BertHTLG, the contribution of the heterogeneous graph representation, and the influence pre-trained language model.

Keywords: Microservice · Anomaly Detection · Heterogeneous Graph · Sentence-Bert

1 Introduction

Microservice systems in the industry typically encompass extensive distributed systems comprising numerous services distributed across multiple machines. Operating in dynamic and unpredictable environments, these systems often experience failures stemming from infrastructure issues or application faults such

L. Yuan et al. (Eds.): WISA 2023, LNCS 14094, pp. 427–439, 2023.
https://doi.org/10.1007/978-981-99-6222-8_36

as hardware failures and configuration errors [1]. To enable engineers to respond to potential failures promptly, there is a need for automated runtime detection of anomalies within microservice systems. For microservice systems, distributed tracing plays a crucial role in profiling and monitoring executions. It encompasses both commercial systems like Dapper [2] and open-source systems such as SkyWalking [3], Zipkin [4], and Jaeger [5]. The produced trace describes the execution process (i.e., invocation chain) of a request through service instances. At the same time, log messages have also been widely used by developers to record the behaviors of each service for debugging errors. With the tremendous success of deep learning in various fields, existing researches [6–10] have also incorporated deep learning methods into microservice anomaly detection. Previous studies treated traces as sequences of service invocations, overlooking the complex invocation relationships arising from synchronous and asynchronous calls. Existing approaches consider representing traces as graphs but overlook the important log information generated within the system. Recent work considers representing traces and logs as graphs but fails to capture the heterogeneous nature between traces and logs and falls short of effectively learning the features of events, thereby affecting the graph representation. Therefore, existing work has not been able to represent traces to detect anomalous behaviors accurately. This paper introduces BertHTLG, an approach designed to detect anomalies in microservice systems. Our approach utilizes a heterogeneous graph representation to faithfully capture the intricate structure of traces and the embedded log events. To detect anomalies, we first use a large pre-trained language model SBERT [11,12] to get the feature of each node event and then employ RGCN [13] based on a deep SVDD [14] model, which enables the learning of latent representations for each trace and the creation of a compact hypersphere that encapsulates the data. We can successfully identify anomalies by calculating the distances between anomalous traces and the center of the hypersphere using the trained model. The contributions of this paper are:

- We propose a Heterogeneous Trace Log Graph which combines trace and log information to a unified representation. The events from trace and logs are distinguished by the node types to more faithfully and adequately convey the trace.
- We use a pre-trained language model Sentence-Bert and a heterogeneous graph neural network RGCN based on deep Support Vector Data Description to enhance graph representation.
- Experiments conducted on public datasets show that our method is superior to the current sequence-based and graph-based methods.

The rest of this paper is organized as follows. Section 2 introduces the related work on anomaly detection. Section 3 gives some basic background and provides the problem definition. Section 4 presents the system design. The experiments are conducted in Section 5 and Section 6 concludes the paper.

2   Related Work

2.1   Sequence-Based Anomaly Detection Methods

Traditional approaches for anomaly detection from logs or traces typically treat them as sequences. Initially, log parsing is employed to extract log events from log messages, utilizing tools such as Drain [15]. Subsequently, networks like LSTM [16] are utilized to extract relevant features. DeepLog [17] employs LSTM to learn normal execution log patterns and identifies anomalies by detecting deviations from the log data during executions. Additionally, LogAnomaly [18] combines semantic vectors generated by word embedding with numeric vectors to predict the subsequent log event using LSTM. In the context of microservice architecture, traces consist of spans. MultimodalTrace [7] represents each trace as a sequence of spans and a corresponding sequence of response times. It introduces a multimodal LSTM model for detecting anomalous traces. On the other hand, TracAnomaly [6] employs a service trace vector to represent each trace, with each possible path considered as a dimension in the vector.

Inspired by the impressive capability of large pre-trained language models to comprehend the semantic characteristics of extensive text data, recent works have diverged from the conventional sequence-based learning approach. These works [19,20] have embraced the utilization of pre-trained models to generate embedding representations of logs, enhancing the detection process.

In summary, these endeavors focus on enabling models to learn semantic information from logs or traces. However, treating them as sequences inherently disregards intricate structural information, particularly the interdependencies among spans due to synchronous and asynchronous invocations.

2.2   Graph-Based Anomaly Detection Methods

To capture the structural information inherent in traces, existing researchers have represented them as graphs and subsequently employed graph embedding policies to detect abnormalities. PUTraceAD [9] constructs a span causal graph, incorporating node features including operation name, response code, and duration time. Positive and Unlabeled algorithms are utilized to learn the parameters of the network. TraceCRL [10] considers invocation information and constructs an operation invocation graph. The graph is then trained using contrastive learning strategies. Another work [21] proposes a dual-variable graph variational autoencoder for unsupervised anomaly detection on microservice traces. To summarize, these existing works consider certain structural information of traces while overlooking other system-related details, such as logs. DeepTraLog [8] integrates trace and log information to construct a trace-event graph. The graph is then trained using gated graph neural networks [22] and Deep SVDD [14] for graph embedding representation. Although log information is considered, the heterogeneous properties of span and log nodes are neglected. Additionally, the feature representation of events utilizing the GloVe model [23] and TF-IDF [24] is insufficient for capturing semantic features.

3 Preliminaries

In this section, we first introduce several basic backgrounds, including trace and log in microservice systems and then formulate the problem of anomaly detection.

3.1 Trace and Log

The trace data model [25] can be simplified as Fig. 1, which implies the relationships between trace, span, and log. A request from users to an instantiated service instance triggers a series of service invocations, which are represented by a **trace** and each service invocation is called a **span**. Each span except the root has a parent span because of the existence of synchronous and asynchronous invocations. The execution result (successful or not) and timestamp are recorded in a span. During service invocations, unstructured **log** messages written by developers are produced by the logging system.

Example. A trace example consisting of three spans is shown in Fig. 2, the synchronous invocation Span A initiates a synchronous invocation and an asynchronous invocation corresponding to Span B and Span C respectively. Therefore, Span A is the parent of Span B and Span C. A log message such as "[AddFoodOrder] delivery info orderId is N and the foodname is bone" has a log event "[AddFoodOrder] delivery info orderId is $\langle * \rangle$ and the foodname is $\langle * \rangle$" with two parameters orderId and foodname. This log parsing procedure can be done by some approaches such as Drain [15].

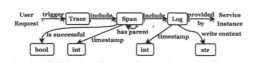

Fig. 1. Simplified trace data model.

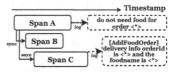

Fig. 2. Trace example.

3.2 Formulations

Trace. The trace $\mathcal{T}$ can be formulated to a set of spans $(s_0, ..., s_n)$. Each span except the root span s_i has a parent span s_p and attributes timestamp a_t and execution result a_r.

Log. The logs $\mathcal{L}$ consists of a sequence of logs $(l_0, ..., l_m)$. Each log l_i has raw log message m, attribute source s which indicates the source of span and attribute timestamp t.

Problem. Given a trace and its corresponding generated logs, we need to determine whether the trace is anomalous. Anomalies in traces can stem from execution failures or deviations, as illustrated in Fig. 2, where the food ordering system placed an order for food, which should not have occurred.

4   Methodology

4.1   Overview

High-Level Solution. As shown in Eq. 1, we attempt to find a representation method $\mathcal{R}$ that convenes trace $\mathcal{T}$ and logs $\mathcal{L}$ as a graph $\mathcal{G}$, which is later mapped to the vector space $\mathcal{I}$ using an algorithm $\mathcal{M}$ to detect anomalies.

$$\mathcal{R} : \mathcal{T} \times \mathcal{L} \rightarrow \mathcal{G}, \mathcal{M} : \mathcal{G} \rightarrow \mathcal{I} \qquad (1)$$

Workflow. To detect anomalies from trace and log, we propose BertHTLG which consists of five steps (shown in Fig. 3). First, **Data Preprocessing** extracts log events and trace events from log messages and trace messages separately from the input. Second, **Event Embedding** generates vector representations for both span and log events based on a large language model SBERT. After that, **Graph Building** creates a Heterogeneous Trace Log Graph (HTLG) for each trace to depict the relationships between its spans and log events. **Model Training** uses a heterogeneous graph neural network RGCN and deep SVDD to learn a latent representation for each HTLG and a minimized hypersphere enclosing the representations of HTLG. Finally, **Anomaly Detection** evaluates whether a trace is unusual by computing its anomaly score, which reflects the shortest distance between the latent representation of HTLG and the hypersphere.

For Data Preprocessing and Anomaly Detection steps, we follow the methods used in DeepTraLog [8]. For other steps, we will give some details next and verify their effectivenesses in the experiments.

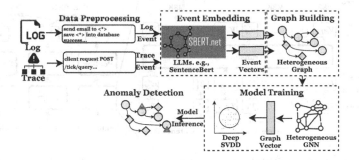

Fig. 3. Workflow of BertHTLG.

4.2   Data Preprocessing

Log Parsing. For logs, we employ Drain [15], a cutting-edge log parsing technique that offers real-time and efficient parsing capabilities while maintaining high accuracy. Each log event is associated with a trace ID and a span ID to facilitate graph construction.

Trace Parsing. For each span in a trace, we transform it into multiple span events depending on invocations types. For synchronous Client/Server spans, we generate a request event and a response event for the current span (server) and also for its parent span (client). For example, for a Client/Server span with operation name `POST service`, the current span will change to `Server Request POST service` and `Server Response POST service`, and the parent span will change to `Client Request POST service` and `Client Response POST service`. Similarly, for asynchronous Producer/Consumer spans, we generate a consumer event and a producer event for the current span (consumer) and also for its parent span (producer).

4.3 Event Embedding

After data preprocessing, each event can be seen as a sentence. Taking inspiration from the remarkable success of Pre-Trained Language Models [26], we utilize SBERT [11] for encoding the event text. This approach not only offers rich semantics but also ensures greater generality. Figure 4 shows the structure of SBERT, it can directly generate a vector of a given sentence by fine-tuning BERT [27] with siamese or triplet structures and parameters sharing rather than from scratch. It also uses large corporates and generic sentence-oriented tasks such as comparing the similarity of two sentences.

To use SBERT to encode events, we use the pre-trained model MiniLM [28], which is small but efficiently trained on over 1 billion sentence pairs. As for pooling strategies, we utilize V_{CLS} and we also explored the impacts of the other two pooling strategies on the results in our experiments.

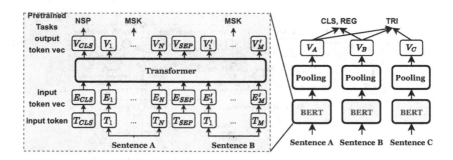

Fig. 4. Structure of SBERT who generates sentence embedding based on BERT.

4.4 Graph Building

The Heterogeneous Trace Log Graph $\mathcal{G} = (\mathcal{N}, \mathcal{E}, \mathcal{X}, \mathcal{A})$ consists of heterogeneous nodes $\mathcal{N}$, their types $\mathcal{A}$, features $\mathcal{X}$, and relationships $\mathcal{E}$ between nodes. We totally define 8 different node types shown in Table 1.

Figure 5 shows the constructed HTLG of the trace example in Fig. 2. To construct this HTLG, we first gather all the log events (type⑦) associated with

Table 1. Heterogeneous node types, Syn is synchronous and Asy is asynchronous.

Node Types	Description	Node Types	Description
①Client request	(Syn)request from client	⑤Producer	(Asy)caller
②Client response	(Syn)response from client	⑥Consumer	(Asy)receiver
③Server request	(Syn)request from server	⑦Log	Log message
④Server response	(Syn)response from server	⑧Database	Database interaction

a specific span. These log events are then arranged in chronological order based on their timestamps, establishing a sequential relationship between event logs. Next, for each span in the trace, we retrieve all the span events that belong to that particular span. Each obtained span event is inserted into the log event sequence according to the timestamp.

Additionally, we connect events and arrange types according to invocation types mentioned in Sect. 4.2. In the case of synchronous invocation, a request relationship is created between the client request event (type①) of the parent span and the server request event (type③) of the current span. Similarly, a response relationship is established between the server response event (type④) of the current span and the client response event (type②) of the parent span. For an asynchronous invocation, a request relationship is added from the producer event (type⑤) of the parent span to the consumer event (type⑥) of the current span. Invocations to the database will also generate a database event (type⑧).

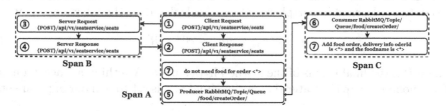

Fig. 5. Constructed heterogeneous trace log graph where the number means node types.

4.5 Model Training

Given the constructed HTLG from a trace, we train a heterogeneous graph neural network RGCN [13] combined with a one-class classification model deep SVDD [14] to detect anomalies.

Propagation. RGCN propagates features of each node according to their types, all the node feature representations are finally obtained through:

$$h_v^{(0)} = x_v, \qquad h_v^{(l+1)} = \sigma\left(\sum_{a \in \mathcal{A}} \sum_{j \in \mathcal{N}_i^a} \frac{1}{|\mathcal{N}_i^a|} W_r^{(l)} h_j^{(l)} + W_0^{(l)} h_i^{(l)}\right), \qquad (2)$$

where, $x_v \in \mathcal{X} \in \mathbb{R}^{|\mathcal{V}| \times d}$ is the inital event embedding from SBERT, d is the vector's dimension, l is the number of layers, $\mathcal{A}$ is the collection of node types, $\mathcal{N}_i^a$ is the neighbors of node i with type a, $W_r^{(l)}$ and $W_0^{(l)}$ are trainable variables.

The graph representation of a HTLG is calculated by the following equation:

$$h_g = \tanh\left(\sum_{v \in \mathcal{N}} \phi\left(f_i(h_v^{(l)}, x_v)\right) \odot \tanh\left(f_j(h_v^{(l)}, x_v)\right)\right), \quad (3)$$

where h_g is the graph vector, $\phi\left(f_i(h_v^{(l)}, x_v)\right)$ is the soft-attention mechanism, f_i and f_j are networks, $\odot$ is element-wise multiplication.

Loss Function. Motivated by works [8,14], we use a loss function as:

$$Loss = R^2 + \frac{1}{\mu N_g}\sum_{g=1}^{N_g}\max\{0, ||h_g - c||^2 - R^2\} + \frac{\lambda}{2}\sum_{l=1}^{N_l}||\theta^{(l)}||_F^2, \quad (4)$$

where c is the center of the hypersphere, R is the radius of the hypersphere, hyperparameter μ controls the trade-off between the hypersphere volume and the violations of the boundary, $\frac{\lambda}{2}\sum_{l=1}^{N_l}||\theta^{(l)}||_F^2$ is a weight decay regularizer on the RGCN parameters θ with hyperparameter λ.

4.6   Anomaly Detection

The anomaly score of a HTLG is defined as the distance from the latent representation to the learned hypersphere, which is calculated by the following equation:

$$score(h_g) = ||h_g - c||^2 - \hat{R}^2, \quad (5)$$

where $\hat{R}$ is the final radius of the learned hypersphere. We think a trace anomaly if its anomaly score is greater than a threshold γ. γ is set to 0 in our experiments.

5   Experiments

5.1   Experimental Settings

Dataset. We conduct all experiments on the dataset released in DeepTraLog [8]. It includes 132,485 traces and 7,705,050 log messages. Among the traces, 23,334 (17.6%) are anomalous ones caused by 73 faults of the 14 fault cases which are located in different services.

Training Details. We optimize RGCN parameters θ by Adam [29] with a fixed radius R in the first few epochs 10 and after every 10 epochs, we update R as the $1 - \mu$ percentile of the distances of all the graphs in the current epoch. For the hypersphere center c, we set it as the mean of all vector representations of graphs after an initial forward pass. The embedding size of each event is 384, the hidden layer of RGCN is 2 and the hidden size of each layer is 384. Other parameters and dataset split methods are the same as DeepTraLog [8].

Evaluation Metrics. To fair comparison, like DeepTraLog, we also use precision, recall, and F1-score to measure the effectiveness of anomaly detection.

Baselines. We use two sequence-based and graph-based anomaly detection approaches as the baselines.

Table 2. Experimental result compared with baselines

Category	Model	Precision(%)	Recall(%)	F1-Score(%)
Sequence-based	DeepLog [17]	60.8	94.8	74.1
	MultimodalTrace [7]	59.1	77.6	67.1
Graph-based	DeepTraLog [8]	93.0	97.8	95.4
	BertHTLG	**98.5**	**99.2**	**98.8**

Table 3. Result of ablation experiments from three aspects

Effectiveness of	Ablation Methods	Prcision(%)	Recall(%)	F1-Score(%)
Components	w/o SBERT	96.8	95.2	96.0
	w/o HTLG and RGCN	99.2	97.9	98.6
Pooling Strategies	Max Pooling	93.9	98.1	95.9
	Mean Pooling	98.8	98.4	98.6
Num_layers	num_layers = 1	98.6	97.1	97.8
	num_layers = 3	98.0	98.6	98.3
	num_layers = 4	**99.5**	95.2	97.3
	BertHTLG	98.5	**99.2**	**98.8**

- DeepLog [17] employs LSTM to learn normal execution log patterns and identifies anomalies by detecting deviations from the log data during executions.
- MultimodalTrace [7] represents each trace as a sequence of spans and a corresponding sequence of response times. It introduces a multimodal LSTM model for detecting anomalous traces.
- DeepTraLog [8] integrates trace and log information to construct a homogeneous trace-event graph. The graph is then trained using gated graph neural networks and SVDD for graph embedding representation.

5.2   Results

Overall Results. The overall results of the model comparison on the same dataset are shown in Table 2, in which the results of the baseline models are quoted from DeepTraLog [8]. All the results on the three metrics are better than other models. In particular, Graph-based anomaly detection is much better than sequence-based methods. Compared with the same graph-based method DeepTraLog, the three metrics improved by 5.5 Precision, 1.4 Recall and 3.4 F1-Score.

Ablation Study. We conducted ablation experiments from three perspectives:

- **Components.** We examined the impact of pre-trained models, as well as the design of heterogeneous graphs and algorithms, on the results. Specifically, we experimented with the combination of the pre-training model SBERT [11] and the GGNN [22] algorithm used in the DeepTraLog [8]. Additionally, we explored the utilization of event embeddings based on the GloVe [23] in conjunction with our training results obtained from heterogeneous graphs and heterogeneous graph algorithm RGCN [13].
- **Pooling Strategies.** We explored various pooling strategies for obtaining sentence embeddings, specifically focusing on the Max and Mean pooling techniques. The Max pooling method calculates the maximum value across the output vectors, while Mean pooling computes the average of all the output vectors. It is worth mentioning that we utilized the cls pooling approach, which employs the vector corresponding to the CLS-token as the final embedding.
- **Num_layers.** We investigated the impact of varying the number of layers in the heterogeneous graph algorithm RGCN [13] on the obtained results.

We systematically analyzed and evaluated the effects of varying these factors on the outcomes of our study. Table 3 shows the results of the ablation study.

Effectiveness of Two Components. We can observe that both components have improved the results. Specifically, compared to the scenario without using SBERT, incorporating only the heterogeneous graph and RGCN has resulted in a 0.6 increase in F1-score. Similarly, excluding the heterogeneous graph and algorithms and relying solely on SBERT has led to a 3.2 improvement in F1-score. This further confirms the strong semantic learning and generalization abilities of large models. It is worth noting that the improvement from the heterogeneous graph algorithm is relatively small, which could be attributed to the simplicity of the model, as it relies solely on distinguishing different neighbor node types.

Effectiveness of Pooling Strategies. Different pooling strategies also have an impact on the results, but overall, they still enhance the performance of the original model. Both cls pooling and mean pooling show significant improvements compared to max pooling. This could be attributed to the fact that max pooling only retains the maximum value, resulting in a loss of certain information.

Effectiveness of Layers. As the number of layers increases, the variation in results is minimal but still superior to those of DeepTraLog [8]. Additionally, we observe that increasing the number of layers does not necessarily lead to performance improvement, which may be attributed to over-smoothing problems [30], causing representations to become more similar.

6   Conclusion

In this paper, we propose BertHTLG, an approach for detecting microservice anomalies using a heterogeneous graph representation enhanced by Sentence-Bert. It leverages the representation to capture the intricate structure of traces along with the embedded log events within the structure. We employ a deep SVDD model based on RGCN, which learns latent representations for each trace and constructs a minimized hypersphere encapsulating the data. Anomalous traces are identified by calculating their distances to the center of the hypersphere using the trained model. Experiments conducted on a microservice benchmark demonstrate that BertHTLG significantly outperforms this existing state-of-the-art sequence-based and graph-based anomaly detection methods.

Acknowledgments. We would like to thank anonymous reviewers for their valuable comments helping us to improve this work. This work is supported by Research on Key Technologies of Power Mobile Security Protection oriented to Micro Application Architecture under Grant number 5700-202258200A-1-1-ZN.

References

1. Zhou, X., et al.: Fault analysis and debugging of microservice systems: industrial survey, benchmark system, and empirical study. IEEE Trans. Softw. Eng. **47**(2), 243–260 (2018)
2. Sigelman, B.H., et al.: Dapper, a large-scale distributed systems tracing infrastructure. Google Inc, Technical report (2010)
3. Apache skywalking. https://skywalking.apache.org/. Accessed 16 May 2023
4. Zipkin. https://zipkin.io/. Accessed 16 May 2023
5. Jaeger: open source, end-to-end distributed tracing. https://www.jaegertracing.io/. Accessed 16 May 2023
6. Liu, P., et al.: Unsupervised detection of microservice trace anomalies through service-level deep Bayesian networks. In: 2020 IEEE 31st International Symposium on Software Reliability Engineering (ISSRE), pp. 48–58 (2020)
7. Nedelkoski, S., Cardoso, J., Kao, O.: Anomaly detection from system tracing data using multimodal deep learning. In: 2019 IEEE 12th International Conference on Cloud Computing (CLOUD), pp. 179–186. IEEE (2019)
8. Zhang, C., et al.: DeepTraLog: trace-log combined microservice anomaly detection through graph-based deep learning. In: Proceedings of the 44th International Conference on Software Engineering, pp. 623–634 (2022)
9. Zhang, K., Zhang, C., Peng, X., Sha, C.: PUTraceAD: trace anomaly detection with partial labels based on GNN and PU learning. In: IEEE 33rd International Symposium on Software Reliability Engineering (ISSRE). vol. 2022, pp. 239–250. IEEE (2022)
10. Zhang, C., et al.: TraceCRL: contrastive representation learning for microservice trace analysis. In: Proceedings of the 30th ACM Joint European Software Engineering Conference and Symposium on the Foundations of Software Engineering, pp. 1221–1232 (2022)

11. Reimers, N., Gurevych, I.: Sentence-BERT: sentence embeddings using Siamese BERT-networks. In: Proceedings of the 2019 Conference on Empirical Methods in Natural Language Processing. Association for Computational Linguistics, 11 (2019)

12. Jiang, Y., Wu, D.: An integrated Chinese malicious webpages detection method based on pre-trained language models and feature fusion. In: Zhao, X., Yang, S., Wang, X., Li, J. (eds.) WISA 2022. Lecture Notes in Computer Science, vol. 13579, pp. 155–167. Springer, Cham (2022). https://doi.org/10.1007/978-3-031-20309-1_14

13. Schlichtkrull, M., Kipf, T.N., Bloem, P., van den Berg, R., Titov, I., Welling, M.: Modeling relational data with graph convolutional networks. In: Gangemi, A., et al. (eds.) ESWC 2018. LNCS, vol. 10843, pp. 593–607. Springer, Cham (2018). https://doi.org/10.1007/978-3-319-93417-4_38

14. Ruff, L., et al.: Deep one-class classification. In: International Conference on Machine Learning. PMLR, pp. 4393–4402 (2018)

15. He, P., Zhu, J., Zheng, Z., Lyu, M.R.: Drain: an online log parsing approach with fixed depth tree. In: IEEE International Conference on Web Services (ICWS), vol. 2017, pp. 33–40. IEEE (2017)

16. Hochreiter, S., Schmidhuber, J.: Long short-term memory. Neural Comput. 9(8), 1735–1780 (1997)

17. Du, M., Li, F., Zheng, G., Srikumar, V.: DeepLog: anomaly detection and diagnosis from system logs through deep learning. In: Proceedings of the 2017 ACM SIGSAC Conference on Computer and Communications Security, pp. 1285–1298 (2017)

18. Meng, W., et al.: LogAnomaly: unsupervised detection of sequential and quantitative anomalies in unstructured logs. IJCAI 19(7), 4739–4745 (2019)

19. Zhou, J., Qian, Y., Zou, Q., Liu, P., Xiang, J.: DeepsysLog: deep anomaly detection on syslog using sentence embedding and metadata. IEEE Trans. Inf. Forensics Secur. 17, 3051–3061 (2022)

20. Zhang, M., Chen, J., Liu, J., Wang, J., Shi, R., Sheng, H.: LogST: log semi-supervised anomaly detection based on sentence-BERT. In: 2022 7th International Conference on Signal and Image Processing (ICSIP), pp. 356–361. IEEE (2022)

21. Xie, Z., et al.: Unsupervised anomaly detection on microservice traces through graph VAE. Proc. ACM Web Conf. 2023, 2874–2884 (2023)

22. Li, Y., Tarlow, D., Brockschmidt, M., Zemel, R.: Gated graph sequence neural networks. arXiv preprint arXiv:1511.05493 (2015)

23. Pennington, J., Socher, R., Manning, C. D.: Glove: global vectors for word representation. In: Proceedings of the 2014 Conference on Empirical Methods in Natural Language Processing (EMNLP), pp. 1532–1543 (2014)

24. Salton, G., Buckley, C.: Term-weighting approaches in automatic text retrieval. Inf. Process. Manag. 24(5), 513–523 (1988)

25. Peng, X., Zhang, C., Zhao, Z., Isami, A., Guo, X., Cui, Y.: Trace analysis based microservice architecture measurement. In: Proceedings of the 30th ACM Joint European Software Engineering Conference and Symposium on the Foundations of Software Engineering, pp. 1589–1599 (2022)

26. Min, B., et al.: Recent advances in natural language processing via large pre-trained language models: a survey. arXiv preprint arXiv:2111.01243 (2021)

27. Devlin, J., Chang, M.-W., Lee, K., Toutanova, K.: BERT: pre-training of deep bidirectional transformers for language understanding. arXiv preprint arXiv:1810.04805 (2018)

28. all-minilm-l12-v2 language model. https://huggingface.co/sentence-transformers/all-MiniLM-L12-v2. Accessed 16 May 2023

29. Kingma, D.P., Ba, J.: Adam: a method for stochastic optimization. arXiv preprint arXiv:1412.6980 (2014)
30. Chen, D., Lin, Y., Li, W., Li, P., Zhou, J., Sun, X.: Measuring and relieving the over-smoothing problem for graph neural networks from the topological view. In: Proceedings of the AAAI Conference on Artificial Intelligence, vol. 34(04), pp. 3438–3445 (2020)

An Analysis of the Rust Programming Practice for Memory Safety Assurance

Baowen Xu[✉], Bei Chu, Hongcheng Fan, and Yang Feng

Nanjing University, Nanjing, China
bwxu@nju.edu.cn

Abstract. Memory safety is a critical concern in software development, as related issues often lead to program crashes, vulnerabilities, and security breaches, leading to severe consequences for applications and systems. This paper provides a detailed analysis of how Rust effectively addresses memory safety concerns. The paper first introduces the concepts of ownership, reference and lifetime in Rust, highlighting how they contribute to ensuring memory safety. It then delves into an examination of common memory safety issues and how they manifest in popular programming languages. Rust's solutions to these issues are compared to those of other languages, emphasizing the benefits of using Rust for enhanced memory safety. In conclusion, this paper offers a comprehensive exploration of prevalent memory safety issues in programming and demonstrates how Rust effectively addresses them. With its encompassing mechanisms and strict rules, Rust proves to be a reliable choice for developers aiming to achieve enhanced memory safety in their programming endeavors.

Keywords: Memory safety · Rust · Ownership · Reference

1 Introduction

Memory safety has always been a critical concern in software development [16]. In many programming languages, memory safety issues often result in severe consequences for applications and systems.

In the 2022 Common Weakness Enumeration (CWE) Top 25 Most Dangerous Software Weaknesses list, there are four software defects related to memory safety, including Out-of-bounds Write, Out-of-bounds Read, Use After Free, and NULL Pointer Dereference. Among these, Out-of-bounds Write holds the top position in terms of prevalence [9].

In order to address these issues, the Rust programming language emerged [7, 12]. It has garnered significant attention in recent years as a programming language known for building efficient and secure system software [3]. It offers a unique combination of low-level control over system resources, similar to languages like C and C++, while also ensuring memory and concurrency safety through mechanisms like ownership and lifetimes [4].

L. Yuan et al. (Eds.): WISA 2023, LNCS 14094, pp. 440–451, 2023.
https://doi.org/10.1007/978-981-99-6222-8_37

One of the most distinctive and innovative features of Rust is its ownership system [5]. To facilitate programming, Rust introduces the concept of borrowing [5]. Ownership can be borrowed by creating references to values, allowing multiple parts of the code to access and work with the data without transferring ownership. Besides, Rust uses the concept of lifetimes to track the relationships between references and ensure they remain valid [5].

In this paper, we aim to explain how Rust addresses issues related to memory safety. We presents an investigation into various memory safety issues commonly encountered in programming. By studying these examples, we then proceeded to showcase how Rust tackles these problems through its innovative ownership and lifetime mechanisms. Through our analysis and demonstrations, we provide insights into how Rust effectively mitigates memory safety concerns and promotes safer programming practices.

The remainder of the paper is organized as follows: Sect. 2 presents the mechanisms introduced in Rust to ensure memory safety, including ownership, references, and lifetimes. Section 3 analyzes common memory safety issues and their manifestations in popular programming languages, followed by an exploration of Rust's solutions to these issues. Finally, Sect. 4 provides a concise summary of the article's findings and highlights Rust as a dependable choice for developers seeking robust memory safety guarantees.

2   Backgrounds

In this section, we will introduce the concept of ownership and lifetime in Rust.

2.1   Ownership and Reference

Ownership. In Rust language, ownership is a fundamental concept that governs how memory is handled in Rust programs. While these rules are strictly checked in the complication, Rust programmers obtain a executable program with guaranteed memory safety [5].

With Rust's ownership model, each value has a single owner, typically represented by a variable, at any given time. The owner is responsible for the lifetime and deallocation of the value. When the owner goes out of scope, Rust automatically frees the memory associated with the value.

In contrast to many other programming languages, Rust employs move semantics as the default behavior for assignment operations. This represents a typical case of ownership transfer in Rust. Ownership transfer also occurs when variables are passed as function parameters or returned as function results, adhering to Rust's ownership principles. The ownership system in Rust ensures that memory is properly managed throughout the program's execution, enhancing memory safety and reducing the likelihood of runtime errors related to memory management.

Reference. As previously explained, ownership transfers are not limited to assignment operations. If the caller needs to retain access to the variable passed to the callee after the function call, the variable must be returned to the caller as part of the function's return value.

Undoubtedly, such code lacks conciseness and elegance for developers. Thankfully, Rust has taken this into consideration during its design and offers a feature known as *references* [5]. References allow the usage of values without transferring ownership, providing a more flexible and convenient approach in Rust programming.

In Rust, references do not own the value they point to, which ensures compliance with the ownership rules. Within Rust's ownership mechanism, the process of creating a reference is commonly known as borrowing.

Just like variables, references in Rust can be categorized into two types: immutable references and mutable references.

1. Immutable references, also referred to as shared references, provide read-only access to the referenced value.
   Rust enables multiple immutable references to coexist for the same value concurrently, promoting a shared and concurrent access model.
2. By using &mut, we can create mutable references that allow both reading and modifying values. However, Rust enforces a strict rule that only one valid mutable reference can exist for a particular value at any given time.

2.2   Lifetimes

To manage references, Rust introduces the concept of lifetimes [5]. In Rust, every reference has its own lifetime, which can be either explicitly specified by the developer or implicitly inferred by the compiler. The purpose of lifetimes is to ensure that references are valid for as long as we need them to be.

Lifetime annotations in Rust are denoted by an apostrophe (') followed by an identifier, such as 'a, 'b, 'c, and so on. These annotations describe the scope of a reference, indicating how long the reference remains valid within the program.

When writing code, it is typically necessary to follow certain rules regarding lifetimes, which include:

1. References must not outlive the values they refer to.
2. If you store a reference in a variable, the reference must remain valid for the entire duration of the variable's lifetime.
3. If there are multiple references, their lifetimes must intersect properly to satisfy validity requirements.

Developers can utilize explicit lifetime annotations or rely on the compiler's inference to ensure the validity and safety of references.

3   Common Memory Safety Issues and Solutions of Rust

In this section, we will discuss common memory safety issues and how they manifest in existing programming languages.

3.1 NULL Pointer Dereference

Description. In computing, a NULL pointer is a special value assigned to a pointer or reference to indicate that it does not point to a valid object or memory location. A NULL pointer dereference happens when an application attempts to access or manipulate data through a pointer that is expected to point to a valid memory address, but is NULL [11].

Manifestations. Below are some examples of NULL Pointer Dereference in common languages from CWE [11].

```
1   void host_lookup(char *user_supplied_addr){
2       struct hostent *hp;
3       in_addr_t *addr;
4       char hostname[64];
5       in_addr_t inet_addr(const char *cp);
6       validate_addr_form(user_supplied_addr);
7       addr = inet_addr(user_supplied_addr);
8       hp = gethostbyaddr(addr, sizeof(struct in_addr), AF_INET);
9       strcpy(hostname, hp->h_name);
10  }
```

In this example, the program accepts an IP address input from the user, validates its format, and proceeds to perform a hostname lookup. The hostname is then copied into a buffer. If an attacker supplies an apparently valid address that fails to resolve to a hostname, the `gethostbyaddr()` function would return NULL. Since the code does not verify the return value of `gethostbyaddr()`, a null pointer dereference would subsequently occur in the `strcpy()` function call.

This Android application has registered to handle a URL when sent an intent:

```
1   IntentFilter filter = new IntentFilter("com.example.URL");
2   MyReceiver receiver = new MyReceiver();
3   registerReceiver(receiver, filter);
4   public class UrlHandlerReceiver extends BroadcastReceiver {
5       @Override
6       public void onReceive(Context context, Intent intent) {
7           if("com.example.URL".equals(intent.getAction())) {
8               String URL = intent.getStringExtra("URLToOpen");
9               int length = URL.length();
10          }
11      }
12  }
```

The application assumes that the URL will always be included in the intent. However, when the URL is not present, the call to `getStringExtra()` will return null, thus causing a null pointer exception when `length()` is called.

Solution of Rust. In order to address the issue of NULL pointer dereference, a different approach is taken in Rust, compared to traditional programming languages. In safe Rust, there is no concept of a null pointer (NULL). Instead, it uses the `Option` type to represent values that may be absent.

The Option type is an enumeration with two variants: `Some` and `None`. Some wraps a concrete value, indicating its presence, while None represents the absence of a value.

In Rust, when attempting to dereference an Option type, it must be pattern matched first to determine whether it is Some or None. Only when the Option type is Some, can its value be safely dereferenced.

Below is a simple Rust code snippet demonstrating how to handle Option types:

```
1   fn main() {
2       let number: Option<i32> = Some(35);
3       match number {
4           Some(num) => {
5               println!("The number is: {}", num);
6           }
7           None => {
8               println!("There is no number.");
9           }
10      }
11  }
```

Number is an Option<i32> type assigned the value Some(42). By pattern matching, we can safely dereference the value and perform corresponding actions based on its type.

3.2 Wild Pointer

Description. Wild pointers are pointers that have not been properly initialized before their first use. Strictly speaking, in programming languages that do not enforce initialization, every pointer is considered a wild pointer initially [17]. The key issue lies in whether the pointer is initialized before its first usage.

Manifestations

```
1   int main() {
2       int *p; /* wild pointer */
3       *p = 32;
4   }
```

This is a very simple example of wild pointer in C language. We declare a pointer variable, p, without specifying its address. In this case, p becomes a wild pointer. Since p can potentially point to any address, the assignment in line 3 has a high chance of corrupting important and protected memory space, leading to program crashes.

```
1   #include <iostream>
2   int main() {
3       int *arr;
4       for(int i = 0; i < 5; i++)
5           std::cout << arr[i] << " ";
6       return 0;
7   }
```

Here is another example of wild pointers in C++ language. In the above program, a pointer arr is declared but not initialized. As a result, it is displaying the contents of random memory locations. If we compile and run this program, depending on the compiler and compilation options, it may output five numbers or result in a segmentation fault.

Solution of Rust. Due to Rust's enforcement of RAII (Resource Acquisition Is Initialization), the compiler in Rust checks whether pointers and references are initialized before their first usage.

It is worth noting that in safe Rust, dereferencing raw pointers is not allowed. Therefore, the error message mentioned above is only a part of the story.

3.3   Dangling Pointer

Description. Dangling pointers occur when an object is deleted or deallocated without modifying the value of a pointer that still points to the memory location of the deallocated object [1]. This can lead to unpredictable behavior when the dangling pointer is dereferenced, as the memory may now contain different data.

Manifestations. In many languages, when an object is deleted from memory explicitly or when the stack frame is destroyed upon return, the associated pointers are not automatically modified [17]. As a result, the pointers still point to the same memory location, even though that memory may now be used for other purposes.

```
1    {
2        char *dp = NULL;
3        {
4            char c;
5            dp = &c;
6        }
7    }
```

For example, in the above code, we declare a variable c within an inner scope and assign the address of c to the pointer dp. However, once c goes out of its scope, the memory it occupied is deallocated, leaving dp as a dangling pointer.

Another common source of dangling pointers arises from improper usage of memory allocation and deallocation functions, such as **new** and **delete** in C++, as demonstrated below.

```
1    void func() {
2        char *dp = new char[SIZE];
3        delete[] dp;
4    }
```

One frequently encountered error is returning the addresses of stack-allocated local variables. When a called function returns, the memory space for these variables is deallocated, resulting in technically undefined values or 'garbage values'.

Solution of Rust. Rust solves the problem of dangling pointers through its ownership and borrowing mechanisms. When a value is bound to a variable in Rust, the variable takes ownership of the value, and when the variable goes out of scope, Rust automatically releases the owned value.

Next comes the difference between Rust and other languages. In languages like C/C++, when we use a pointer or reference to a variable after it has gone out of scope, also known as a dangling pointer or reference, those languages do not prevent us from doing so. However, Rust performs compile-time checks and throws an error.

Rust can perform such checks because it manages references through lifetimes. As mentioned in the context, there are two crucial constraints for lifetimes. During compilation, the Rust compiler attempts to select appropriate lifetimes for references based on these constraints.

Clearly, these two constraints result in conflicting lifetime ranges. Therefore, the Rust compiler detects such situations and throws an error.

3.4 Double Free

Description. Double free errors occur when the memory deallocation function is invoked multiple times with the same memory address as an argument [1,10]. This can result in the corruption of the program's memory management data structures, potentially allowing a malicious user to write values in arbitrary memory locations.

Manifestations

```
1    char* ptr = (char*)malloc(SIZE);
2    if (abrt) {
3        free(ptr);
4    }
5    free(ptr);
```

While some double free vulnerabilities may be as straightforward as the example provided, many are scattered across hundreds of lines of code or even different files.

```
1    public final class AssetManager {
2        @Override
3        protected void finalize() throws Throwable {
4            if (mObject != 0) {
5                nativeDestroy(mObject);
6            }
7        }
8        void xmlBlockGone(int id) {
9            if (mNumRefs == 0 && mObject != 0) {
10                nativeDestroy(mObject);
11                mObject = 0;
12            }
13        }
14    }
15    final class XmlBlock {
16        private @Nullable final AssetManager mAssets;
17        private void finalize() throws Throwable {
18            if (mAssets != null) {
19                mAssets.xmlBlockGone(hashCode());
20            }
21        }
22    }
```

The above code demonstrates an example of a double free in Java [8]. Assuming we have a XmlBlock X created by a AssetManager A. After calling

A.close and both X and A are ready to be garbage collected, if A.finalize is called first (nativeDestroy), the subsequent invocation of X.finalize will trigger A.xmlBlockGone, causing the second nativeDestroy of A and resulting in a crash.

Solution of Rust. Rust addresses the issue of double free through its ownership mechanism. In Rust, when an owner goes out of scope, the value is automatically released, eliminating the need for manual resource deallocation by the developer. Rust also enforces the rule, checked at compile time, that each value has only one owner, ensuring that double free does not occur.

Rust achieves automatic resource deallocation through the use of the Drop trait, which is automatically implemented for almost all types in Rust. The Drop trait defines the behavior when a value is dropped, allowing Rust to perform necessary cleanup operations.

In the following code, we define a struct named S and implement a custom Drop trait for it. When running this code, we observe that the program outputs "Drop for S.", which demonstrates that Rust automatically invokes the drop method.

```
1    struct S;
2    impl Drop for S {
3        fn drop(&mut self) {
4            println!("Drop for S.")
5        }
6    }
7    fn main() {
8        let s = S;
9    }
```

What if we manually invoke the drop method of the struct to release the resources ahead of time? If we add the line the compiler will provide the following error message:

```
error[E0040]: explicit use of destructor method
  --> src/main.rs:9:7
   |
9  |     s.drop();
   |     --^^^^--
   |     | |
   |     | explicit destructor calls not allowed
   |     help: consider using `drop` function: `drop(s)`
```

This indicates that Rust does not allow us to explicitly call the Drop::drop method.

3.5   Buffer Overflow

Description. A buffer overflow condition occurs when a program tries to store more data in a buffer than its capacity allows or when it attempts to write data beyond the boundaries of a buffer. In this context, a buffer refers to a consecutive section of memory allocated to hold various types of data, ranging from character strings to arrays of integers [6].

Manifestations

```
1   #define BUFSIZE 256
2   int main(int argc, char **argv) {
3       char buf[BUFSIZE];
4       strcpy(buf, argv[1]);
5   }
```

In the above example, the buffer size is fixed, but there is no guarantee that the string in `argv[1]` will not exceed this size, potentially causing a buffer overflow.

```
1   #define SIZE 8
2   int main() {
3       int id[SIZE];
4       for (int i = 0; i <= SIZE; ++i) {
5           id[i] = i * 2;
6       }
7   }
```

The above code attempts to store a series of integers in an array. However, the size of the array is `SIZE`, so its indices range from 0 to `SIZE - 1`. The final assignment statement, `id[SIZE] = SIZE * 2;`, causes a buffer overflow issue.

Solution of Rust. Rust addresses the issue of buffer overflow by performing checks on buffer indices during both compile-time and runtime. For regular arrays, Rust requires specifying the size at declaration or infers it based on the code. For dynamic arrays like `Vec`, Rust stores their length and capacity on the stack.

In Rust, there are two ways to access arrays: one is the common method using the `[]` operator, and the other is through the `get` method. When using `[]`, Rust performs compile-time checks to ensure that the index does not exceed the array bounds if both are known. If the size or index is unknown at compile time, Rust inserts runtime checks. If an out-of-bounds access is detected at runtime, it triggers a panic. When accessing arrays using the `get` method, it returns an `Option` value, and it is the responsibility of the developer to manually verify its validity.

3.6   Use of Uninitialized Memory

Description. Use of uninitialized memory means reading data from a previously allocated memory region that has not been filled with initial values.

In some languages such as C and C++, stack variables are not initialized by default. They often contain random or junk data before the function was invoked [15]. In this case, the behavior of the program is unpredictable, and detecting such issues can be challenging. This type of problem is commonly referred to as a "heisenbug" [2].

Manifestations

```
1   if (isset($_POST['names'])) {
2       $nameArray = $_POST['names'];
3   }
4   echo "Hello " . $nameArray['first'];
```

The above PHP code checks if the names array in the POST request is set before assigning it to the variable $nameArray. However, if the array is not present in the POST request, $nameArray will remain uninitialized. This can result in an error when accessing the array to print the greeting message, potentially creating an opportunity for further exploitation.

The following code snippet represents a Use of Uninitialized Memory issue in the C programming language.

```
1   char *str;
2   if (i != err) {
3       str = "Hello World!";
4   }
5   printf("
```

If the value of the variable i is equal to err, then the string str in the above code will be in an uninitialized and unknown state. In such a scenario, the printf function may print junk strings.

Solution of Rust. Similar to C language, in Rust, stack variables are uninitialized by default and need to be explicitly assigned a value. However, unlike C, Rust prevents you from using variables before initializing them. Rust performs branch analysis to ensure that variables are initialized before they are used on each branch [13].

It is important to note that this check does not consider the specific values of conditions; rather, it takes into account the program's dependencies and control flow.

3.7   Data Race

Description. Data race occurs when multiple threads access and modify a shared memory region simultaneously without proper synchronization [14].

Manifestations

```
1    var counter int
2    func incrementCounter(wg *sync.WaitGroup) {
3        defer wg.Done()
4        counter++
5    }
6    func main() {
7        var wg sync.WaitGroup
8        for i := 0; i < 10; i++ {
9            wg.Add(1)
10           go incrementCounter(&wg)
11       }
12       wg.Wait()
13       fmt.Println("Counter value:", counter)
14   }
```

In this example, multiple goroutines are simultaneously modifying a shared variable called **counter**. Since there is no synchronization mechanism, such as a mutex or a semaphore, in place to coordinate access to the variable, a data race occurs. Data races can lead to inconsistent and unpredictable final values of the **counter** variable.

```
1   #include <pthread.h>
2   #include <stdio.h>
3   int counter = 0;
4   void* increment(void* arg) {
5       for (int i = 0; i < 1000000; i++) {
6           counter++;
7       }
8       return NULL;
9   }
10  int main() {
11      pthread_t thread1, thread2;
12      pthread_create(&thread1, NULL, increment, NULL);
13      pthread_create(&thread2, NULL, increment, NULL);
14      pthread_join(thread1, NULL);
15      pthread_join(thread2, NULL);
16      printf("Final value of counter:
17      return 0;
18  }
```

In this example, two threads are created, and each thread increments the shared variable **counter** in a loop. Since there is no synchronization mechanism, a data race occurs where both threads are accessing and modifying the variable simultaneously. As a result, the final value of **counter** becomes unpredictable and may vary between different runs of the program.

Solution of Rust. Rust addresses most data race issues through its ownership mechanism. As mentioned earlier, Rust distinguishes between two types of references and imposes corresponding restrictions.

Rust's distinction and restrictions on references serve the purpose of preventing data races during compile time. This concept can be likened to the rules governing readers and writers in concurrent operations. That means it is not possible to have multiple mutable references aliasing the same data simultaneously in Rust.

However, not all types adhere to inherited mutability. Certain types in Rust allow multiple aliases of a memory location while still allowing mutation. Rust addresses this by utilizing the Send and Sync traits to enforce safe concurrent behavior [13].

4   Conclusion

This article provides a comprehensive exploration of prevalent memory safety issues encountered in programming and delves into their manifestations in popular programming languages. It further elucidates how Rust effectively tackles these issues by leveraging the power of ownership, references, and other innovative mechanisms. By enforcing strict rules and guarantees, Rust empowers developers with robust memory safety, ensuring their programs are protected from a range of potential vulnerabilities. With a holistic approach encompassing various mechanisms, Rust stands as a reliable choice for developers seeking enhanced memory safety in their programming endeavors.

References

1. Caballero, J., Grieco, G., Marron, M., Nappa, A.: Undangle: early detection of dangling pointers in use-after-free and double-free vulnerabilities. In: Proceedings of the 2012 International Symposium on Software Testing and Analysis, pp. 133–143 (2012)
2. Grottke, M., Trivedi, K.S.: A classification of software faults. J. Reliab. Eng. Assoc. Jpn. **27**(7), 425–438 (2005)
3. Jiang, H., Wang, L., Tao, X., Hu, H.: RHE: relation and heterogeneousness enhanced issue participants recommendation. In: Xing, C., Fu, X., Zhang, Y., Zhang, G., Borjigin, C. (eds.) WISA 2021. LNCS, vol. 12999, pp. 605–616. Springer, Cham (2021). https://doi.org/10.1007/978-3-030-87571-8_52
4. Jung, R., Jourdan, J.H., Krebbers, R., Dreyer, D.: RustBelt: securing the foundations of the rust programming language. Proc. ACM Program. Lang. **2**(POPL), 1–34 (2017)
5. Klabnik, S., Nichols, C.: The Rust Programming Language. No Starch Press (2023)
6. Lhee, K.S., Chapin, S.J.: Buffer overflow and format string overflow vulnerabilities. Softw. Pract. Exp. **33**(5), 423–460 (2003)
7. Matsakis, N.D., Klock, F.S.: The rust language. ACM SIGAda Ada Lett. **34**(3), 103–104 (2014)
8. MITRE: CVE record | CVE. https://www.cve.org/CVERecord?id=CVE-2020-0081. Accessed 25 June 2023
9. MITRE: CWE - 2022 CWE top 25 most dangerous software weaknesses. https://cwe.mitre.org/top25/archive/2022/2022_cwe_top25.html. Accessed 24 June 2023
10. MITRE: CWE - CWE-415: Double free (4.11). https://cwe.mitre.org/data/definitions/415.html. Accessed 24 June 2023
11. MITRE: CWE - CWE-476: null pointer dereference (4.11). https://cwe.mitre.org/data/definitions/476.html. Accessed 24 June 2023
12. Rust Community: Rust programming language. https://www.rust-lang.org/. Accessed 24 June 2023
13. Rust Community: The rustonomicon. https://doc.rust-lang.org/nomicon/. Accessed 25 June 2023
14. Serebryany, K., Iskhodzhanov, T.: ThreadSanitizer: data race detection in practice. In: Proceedings of the Workshop on Binary Instrumentation and Applications, pp. 62–71 (2009)
15. Stepanov, E., Serebryany, K.: MemorySanitizer: fast detector of uninitialized memory use in C++. In: 2015 IEEE/ACM International Symposium on Code Generation and Optimization (CGO), pp. 46–55. IEEE (2015)
16. Szekeres, L., Payer, M., Wei, T., Song, D.: SoK: eternal war in memory. In: 2013 IEEE Symposium on Security and Privacy, pp. 48–62. IEEE (2013)
17. Wikipedia contributors: Dangling pointer—Wikipedia, the free encyclopedia (2023). https://en.wikipedia.org/w/index.php?title=Dangling_pointer&oldid=1155171462. Accessed 24 June 2023

Blockchain

A Blockchain Query Optimization Method Based on Hybrid Indexes

Junyan Wang, Derong Shen$^{(\boxtimes)}$, Tiezheng Nie, and Yue Kou

Northeastern University, Shenyang 110004, China
{shenderong,nietiezheng,kouyue}@cse.neu.edu.cn

Abstract. Blockchain technology possesses the characteristics of decentralization and immutability, making it widely applicable in various fields. However, existing blockchain systems display weak performance in terms of data management, typically only supporting hash-based transaction querying. Current research on query methods usually involves synchronizing data with external databases, using external query engines, or building indexes to improve query performance. In index-based research methods, most blockchain systems build indexes within blocks to enhance query efficiency, but this does not reduce the time overhead of traversing all blocks. Additionally, most indexes only support a specific query type. In this paper, we propose a highly efficient indexing structure that divides data querying into two steps: first, obtaining potential blocks from the index structure that may contain the desired data, and then searching for the specific data records within those blocks. The indexes are further divided into numerical indexes and character indexes based on the attributes being indexed. We describe the construction and use of these two index structures for efficient data querying. As invalid blocks are filtered out, system query efficiency is greatly improved. The numerical index structure further supports range queries and top-k queries on this attribute, enhancing query capabilities. We also conduct experiments to verify the performance of our indexing structure. Results demonstrate that compared with baselines, our proposed index structure improves query performance, and offers a diverse range of query types.

Keywords: Blockchain · Query · Index · Bloom Filter · B+ Tree

1 Introduction

Since the scholar Satoshi Nakamoto proposed the Bitcoin white paper [1], blockchain technology has been concerned by a large number of researchers in the computer field. Blockchain is a decentralized distributed database maintained by multiple parties [2]. It is a new application model of computer technology, which includes distributed data storage, peer-to-peer transmission protocol, consensus mechanism, and encryption algorithm [3]. Due to its own data structure limitations, there are limited ways to improve query performance in existing blockchain systems.

Since the relational database has higher query efficiency and supports multiple query methods, it is possible to break through the bottleneck of query performance by dumping

© The Author(s), under exclusive license to Springer Nature Singapore Pte Ltd. 2023
L. Yuan et al. (Eds.): WISA 2023, LNCS 14094, pp. 455–466, 2023.
https://doi.org/10.1007/978-981-99-6222-8_38

the data in the blockchain to the middleware database. However, due to the introduction of a trusted third party, it violates the original intention of the decentralization of the blockchain itself, and requires a lot of additional operations to ensure that the data in the middleware database will not be tampered with, and the verification mechanism for query results becomes more complicated.

Another popular approach is to build an index within a block to support efficient searching of blockchain data. However, most indexes only consider the performance of querying data in a block and do not address the time overhead caused by traversing all blocks. Additionally, another challenge faced by these index structures is that they only support certain specific queries.

In this paper, we address these challenges and propose a query method based on multi-level hybrid indexes that enable efficient query of blockchain data and provides multiple query modes. Specifically, we propose indexes for different attributes, which can optimize data query performance for numerical attributes and character attributes respectively. At the same time, a secondary index is introduced in the numerical index to avoid querying invalid blocks and provide multiple query methods for this attribute.

In summary, the main contributions of this paper are as follows:

- We proposed numerical index and character index respectively for different attributes, which can filter out invalid blocks. Meanwhile, the second layer of numerical index can improve query efficiency within blocks.
- Using hybrid indexes as the foundation, we propose a range of query algorithms that support various types of queries.
- Our experiments demonstrate that the proposed index structure offers excellent query performance.

The remainder of this paper is organized as follows. Section 2 introduces related work, Sect. 3 introduces the system architecture, Sect. 4 describes the construction of two index structures and the corresponding query algorithms, Sect. 5 does comparative experiments to verify the efficient query performance of the index structure. Finally, the full text is summarized in Sect. 6.

2   Related Work

To address the inefficiency of data querying in traditional blockchain, several methods and strategies have been proposed for querying genetic data, government information, and electronic certificates [4, 5]. However, these customized optimization schemes cannot be directly applied to other scenarios. Consequently, many scholars have studied more general solutions to improve data query efficiency in blockchain.

In [6], a query layer was added to the blockchain system, and the data in the blockchain was copied to external database where it was managed. The EtherQL system also offers efficient range query and top-k query methods that can be used for analyzing blockchain data. [7] proposed a verifiable query layer that extracts transactions stored in the underlying blockchain system using middleware and re-stores the data in a database to provide users with different types of query services, such as block query and transaction query. To avoid middleware from storing fake data, each built database is assigned a cryptographic hash or fingerprint. This new three-tier architecture eliminates the need to enter the block when querying, making it easier to query the middleware database. However, this scheme incurs significant update and calculation costs due to the continuous generation of new transactions and blocks.

[8] introduced ChainSQL, a new blockchain-based database system that combines blockchain with distributed databases. However, the proposed system lacks extensive experiments in evaluating the execution cost of smart contracts as well as verification and consensus costs. Although these solutions can greatly improve query efficiency, they all assume the existence of a trusted party as opposed to the blockchain itself.

[9] used built-in verifiable data structures and returned verification objects to solve the query integrity problem of blockchain databases. Huang et al. [10] proposed inverted indexes to improve the performance of the range query. But there are limitations for other types of queries.

Liu et al. [11] proposed Abstract-Trie and Operation-Record List to index the original data and operation records in the blockchain, respectively. Abstract-Trie can quickly verify whether the transaction is in the block through the hash value of the transaction. However, its disadvantage is that the search method is single, and the index only supports queries based on transaction hash values. Shlomi Linoy and their team [12] proposed the AMVSL index, which is based on the jump table and can support the query of historical data stored in the blockchain. They also proposed range queries based on AMVSL: SVRK, MVEK, and MVAK, which provide key and version query methods. Ce Zhang et al. [13] designed a gas-saving ADS to verify query results in a hybrid storage blockchain system and proposed Merkle Inverted Index and Chameleon Inverted Index to support keyword queries.

Currently, research on indexing is focused on improving the index within the block to achieve better query efficiency. However, during the querying process, the entire blockchain still needs to be traversed. Some blocks do not have the queried data, but they are still searched during the query traversal. Furthermore, due to the limitations of the proposed index structures, the supported query types are also limited.

3 System Architecture

In this section, we introduce the system architecture of this paper. As shown in Fig. 1, the architecture is divided into two layers, which are the index layer and the data layer. When the user sends a query request, the request first obtains the keys of all valid data through the index layer, and then returns all fields of the data through the keys at the data layer. Each piece of data stored in the blockchain can be described using attributes, which can be divided into two categories: numerical attributes and character attributes. For instance, consider a data record <"block", 2>, where the first attribute is a character attribute and the second attribute is a numerical attribute.

- The index layer includes numerical index and character index, and provides query interface to users.
- The data layer saves data records in the k-v database, and obtains the final query result through the key of the data found in the index layer.

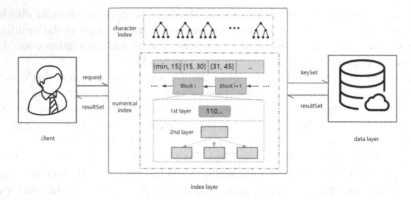

Fig. 1. The overall framework.

3.1 Numerical Index

Figure 2 in the proposed system architecture illustrates the design of the numerical index. The numerical index is created for a specific numerical attribute, and it is composed of two-level indexes that are stored in each blockchain full node.

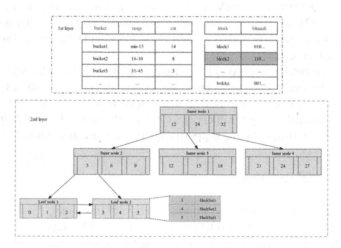

Fig. 2. The numerical index.

The First Level of Numerical Index. The first-level index contains n buckets that are divided based on the value range of the numerical attribute. Each bucket represents a specific range of attribute values, and a binary vector of length n is also maintained in each block. Each bit in the binary vector corresponds to the bucket, and when a bit is set to 1, it indicates that the block contains data within the corresponding bucket data range. Through the first-level index, invalid blocks can be quickly filtered out, where invalid blocks are those that do not contain data that meets the query conditions. Overall, the numerical index contributes to improving the efficiency of querying and retrieving data from the blockchain. For instance, if the query condition is "attr = 4", we should calculate the corresponding position based on the bucket value distribution. In Fig. 2, only block 2 meets the condition, so the other invalid blocks can be excluded based on the first index.

The Second Level of Numerical Index. The purpose of establishing a secondary index is to support data querying and retrieval. In the proposed system architecture, the secondary index is implemented based on a B+ tree. In the B+ tree, each data item's attribute value is stored in the non-leaf node, with its key value being stored in the leaf node. When we find the key value of the qualified data in the index, we only need to access the key-value database to retrieve all the corresponding data. Now that we have the only valid block, the next step is to go inside the block and search through the second level index. Each tree node is recursively searched, locating the next node through binary search within the node until the leaf node is reached. In Fig. 2, the traversal path is Inner node 1, Inner node 2 and Leaf node 2. Finally all data records are obtained in the k-v database through the result set.

3.2 Character Index

As indicated in Fig. 3, a character index is created based on a specific character attribute. It consists of several full binary trees, with each leaf node representing a block. Each node in the tree includes a Bloom filter. The Bloom filter can efficiently determine if the queried data is in the collection. The Bloom filter in the leaf node contains all the attribute values that appear in the corresponding block. For the Bloom filter in the non-leaf node, it is the union of the Bloom filters in the two child nodes. The merge operation of two Bloom filters can be completed efficiently by just combining the binary vectors.

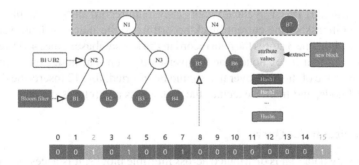

Fig. 3. The character index.

However, since the Bloom filter cannot guarantee that elements must exist, its false positive rate will become higher when there are too many elements in it. To prevent the false positive rate of the upper nodes from being too high, we divide all blocks into batches and construct a full binary tree every several blocks. When performing a query, we can quickly eliminate the invalid blocks via traversing the character index, preventing invalid searches and improving query efficiency.

4 Query Processing Method

4.1 Numerical Index Construction Method

When a new block is added to the chain, a binary vector bitmask is generated for the block based on the attribute value range of each bucket. Then, a second-level numerical index is built from the bottom up, and a bidirectional linked list is maintained.

Algorithm 1: *NumericalIndexConstruction*

Input: B: block, R: the root node of current block numerical index

```
1.  for d in B.data do
2.      SetBitMask(B.bitMask, d.attr);
3.      SaveToDatabase(d.id, d);
4.      curNode = R;
5.      while curNode.isLeaf == false
6.          insertPosition = curNode.Search(d.attr);
7.          if insertPosition >= curNode.childrenSum
8.              insertPosition --;
9.          end if
10.         curNode = curNode.children[insertPosition];
11.     end while
12.     root.InsertIntoLeafNode(curNode, d.attr, d);
13.     root = curNode.Split();
14. end for
```

In Algorithm 1, line 1 cyclically calls each data record in the new block. Line 2 updates the bitmask of the current block according to the index attribute value of the data record, while line 3 stores the data record in the k-v database. Lines 4–13 recursively traverses the current index node, line 5 traverses the non-leaf nodes on the path, lines 6–10 find the node of the next layer that should be inserted, line 12 inserts the data record into the leaf node, and line 13 executes the node splits and returns the new root node.

4.2 Numerical Index Query

When querying data that is distributed across multiple blocks, it is necessary to traverse each block and examine its bitmask to determine whether it may contain the desired data. Blocks that do not meet the criteria can then be ignored to avoid unnecessary processing and improve search efficiency.

Equality Query
In Algorithm 2, lines 1–8 recursively traverse the non-leaf nodes of the index. Lines 3–7 find the node to be searched in the next layer, while line 9 determines the position of the data record in the leaf node through binary search. Lines 11–13 loop through

each queried data record and search all the corresponding data from the k-v database according to the id of the record.

Algorithm 2: *EqualityQuery*

Input: Attr: the attribute value corresponding to the current index, R: the index root node
Output: Res: result

  1. curNode = R;
  2. **while** *curNode.isLeaf == false*
  3.   *searchPosition = curNode.Search(attr);*
  4.   **if** *searchPosition < curNode.sum*
  5.   *searchPosition --;*
  6.   **end if**
  7.   *curNode = curNode.children[searchPosition];*
  8. **end while**
  9. innerNodePosition = curNode.Search(attr);
  10. ids = curNode.entries[innerNodePosition];
  11. **for** *id* **in** *ids* **do**
  12.   *AppendToResult(res, GetFromDatabase(id));*
  13. **end for**
  14. **return** *Res;*

Range Query. For a range query (l, r), it is necessary to determine the left boundary L and the right boundary R of the query range by finding the largest node smaller than l and the smallest node larger than r, respectively. Since the data key is only stored in the leaf node, once the left and right boundary nodes are determined, it is only necessary to traverse the doubly linked list of the leaf nodes to obtain the eligible keys. After obtaining all the keys, database queries can be performed to retrieve all the data associated with those keys. This approach improves query efficiency by minimizing unnecessary processing and retrieval of irrelevant data.

Top-k Query. In some scenarios, we need to query k records with the largest or smallest attribute value. Since the leaf nodes in the B+ tree are double-linked lists with ordered keywords, it is only necessary to search sequentially from the head or tail of the linked list until all leaf nodes are traversed or the first unqualified data is traversed.

4.3 Character Index Construction

Whenever a new block arrives, a new tree node needs to be generated, and the value of this attribute of all data in this block is added to the Bloom filter. To facilitate insertion, we stipulate that each character index tree is a full binary tree and has a maximum height. When a tree exceeds the maximum height, a new index tree needs to be created. Therefore, in the entire set of character index trees, when we insert a new node, we try to merge the index trees that have not reached the maximum height until no further merging can be done.

Algorithm 3: *CharacterIndexConstruction*

Input: Data: the data record in a block, H: the current block height, Si: character index

1. *node = Si.NewLeafBloomNode(H);*
2. **for** *d* **in** *Data* **do***:*
3.   *node.bloomFilter.Put(d);*
4. **end for**
5. **if** *Si.size >= Si.capacity:*
6.   *Si.Resize();*
7. **end if**
8. *length = si.size*
9. **for** *i = length − 1;i > 0;i-- do:*
10.   *preHeight, curHeight := Si.arr[i-1].height, Si.arr[i].height;*
11.   **if** *preHeight < Si.maxHeight && preHeight == curHeight:*
12.   *parent = Si.NewBranchNode(Si.arr[i-1], Si.arr[i]);*
13.   *Si.arr[i-1] = parent;*
14.   *Si.size--;*
15.   **end if**
16.   **else**
17.   **break;**
18.   **end else**
19. **end for**

In Algorithm 3, line 1 generates a leaf node, lines 2–4 insert the attribute value of the data in the block into the Bloom filter, and lines 5–7 determine whether the entire character index set needs to be expanded. Lines 8–19 merge full binary trees, starting from the reverse order of the character index set. Line 11 checks whether the height of the current index tree is equal to the height of the previous index tree and the height of the previous index tree has not reached the maximum height limit. If the condition is satisfied, line 12 generates a new parent node, otherwise the merge process ends.

4.4 Character Index Query

When a query arrives, we need to traverse each index tree in the index collection. We start from the root node of each tree and traverse down to the leaf nodes. If a node does not contain the query data, then we need to skip the subtree rooted at this node and all the blocks corresponding to its descendant leaf nodes. By doing so, we can obtain a set of blocks containing valid data. Finally, we scan these blocks one by one to obtain a result set.

Algorithm 4: *CharacterIndexQuery*

Input: Attr: the attribute value corresponding to the current index, Si: character index
Output: Res: result

1. *blocks = [];*
2. **for** *root* **in** *Si.arr* **do**:
3.   *Append(blocks, Si.FilterInNode(root, attr));*
4. **end for**
5. **for** *b* **in** *blocks*:
6.   *AppendToResult(Res, GetFromDatabase(b));*
7. **end for**
8. **return** *Res;*

In Algorithm 4, lines 1–4 recursively traverse each index tree in the character index set to obtain a set of blocks that may contain data satisfying the query conditions. Then, lines 5–8 scan these blocks to obtain the final result to be returned.

4.5 Compound Query

For queries with multiple conditions, in order to reduce the numbers of blocks that need to be searched, we should first filter out invalid blocks using character indexed. Then, we can search through numerical indexes in these block sets, which can greatly improve query efficiency.

5 Experiment

5.1 Experiment Setup

All experiments are performed on a node with a Ryzen 5800U CPU and 16 GB RAM. The code is based on the Tendermint blockchain written in Go and tested on a local area network.

Workload. We implemented a simple application program to measure the query efficiency of our proposed index structure. To simplify the encoding, we build numerical index on the attribute 'stars_count' and character index on the attribute 'name', respectively. To evaluate the efficiency of using our index structure, we compared our method with a baseline method which does not use any index.

Dataset. The data set we used was the Github repository data set. For the purpose of simplicity, we assumed that each block contains 1000 data records. The experiment tested the total data volume of a total 200000, 500000, 800000 and 1000000 data records, respectively.

Metrics. We measured the index construction time and query time for randomly querying 50,000 pieces of data. At the same time, we also tested the space overhead of the two indexes in different experiment settings.

5.2 Numerical Index

In this subsection, we test the impact of numerical indexes on query efficiency. To further study the query efficiency of numerical indexes under different orders, we constructed three types of numerical indexes of different orders: 5th order, 9th order, and 13th order.

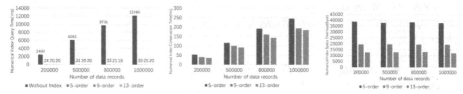

Fig. 4. Numerical index equality query time.

Fig. 5. Numerical index construction time.

Fig. 6. Numerical index space overhead.

Equality Query Time. It can be seen from Fig. 4 that under the same total amount of data, the total time for querying 50,000 pieces of data without using an index structure is two orders of magnitude higher than the total time for using a numerical index. Among all the different orders of numerical indexes, the query performance of the index structure with an order of 13 is the best, because the height of the index tree becomes smaller as the order increases.

Construction Time. In Fig. 5, we compare the construction time of numerical indexes for different orders. Under the same amount of data, the construction time for an index with an order of 13 is the shortest, while the construction time for an index with an order of 5 is the longest. This is because as the order increases, the number of nodes in the tree decreases.

Space Overhead. In our evaluation of the numerical index, we also examined its space usage within each block, as depicted in Fig. 6. With a limit of 1000 data pieces stored in each block, the space overhead of the internal secondary index is roughly equality across blocks with the same order. However, as the order of the B+ tree increases, the height of the index decreases, resulting in significant savings in space overhead for non-leaf nodes.

5.3 Character Index

In this section, we compare query performance with and without character indexes. We also divide the data volume into four levels, and set character indexes with different maximum heights. Additionally, we use a Bloom filter, which is a data structure that can quickly determine whether an element appears in a set, but has a certain error rate in determining whether an element "must exist". For the experiments presented in this article, the error rate of the Bloom filter is uniformly set to 0.1%.

Query Time. As illustrated in Fig. 7, the query time without a character index is approximately three times longer than that with one. The Bloom filter utilized in the character index helps us avoid any invalid blocks during the query process. Despite the use

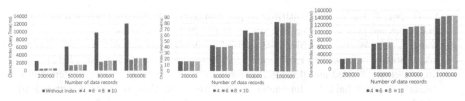

Fig. 7. Character index query time.

Fig. 8. Character index construction time.

Fig. 9. Character index space overhead.

of indexes, the query time remains higher than that for numerical attributes because comparing strings for equality is more time-consuming.

Construction Time. As depicted in Fig. 8, the construction time of a character index increases almost linearly with an increase in data volume. This is mainly due to the requirement for additional blocks to store the data, which leads to an increase in the number of nodes in the index tree. Despite this, the overall time overhead incurred by building the index remains within acceptable limits.

Space Overhead. Similarly, character indexes also require a certain amount of space overhead. As shown in Fig. 9, the space overhead of the character index is large when the amount of data is the same. As the data volume increases, the space occupied by the index grows approximately linearly with it.

6   Conclusion

In this paper, we introduce our index structure along with its corresponding query algorithm. The index is divided into a numerical index and a character index based on differing attributes, utilizing the B+ tree and Bloom filter data structures, respectively. Experimental evaluations demonstrate that our index structure can significantly improve the efficiency of data queries compared to baseline approaches. Although constructing the index incurs a certain amount of time overhead, it is negligible when compared to the performance improvement achieved by querying data with an index.

Acknowledgements. This work was supported by the National Natural Science Foundation of China (62172082, 62072084, 62072086), the Fundamental Research Funds for the central Universities (N2116008).

References

1. Nakamoto, S.: Bitcoin: a peer-to-peer electronic cash system. Decentralized Bus. Rev. **21260** (2008)
2. Zhao, X., et al.: Blockchain and distributed system. In: Wang, G., Lin, X., Hendler, J., Song, W., Xu, Z., Liu, G. (eds.) WISA 2020. LNCS, vol. 12432, pp. 629–641. Springer, Cham (2020). https://doi.org/10.1007/978-3-030-60029-7_56

3. Yu, G., Nie, T., Li, X., et al.: Distributed data management technology in blockchain system-challenges and prospects. Chin. J. Comput. 1–27 (2019). (in Chinese)
4. Wang, L., Liu, W., Han, X.: Blockchain-based government information resource sharing. In: International Conference on Parallel and Distributed Systems, pp. 804–809 (2017)
5. Xu, C., et al.: Trusted and flexible electronic certificate catalog sharing system based on consortium blockchain. In: International Conference on Computer and Communications (2019)
6. Li, Y., Zheng, K., Yan, Y., Liu, Q., Zhou, X.: EtherQL: a query layer for blockchain system. In: Candan, S., Chen, L., Pedersen, T.B., Chang, L., Hua, W. (eds.) DASFAA 2017. LNCS, vol. 10178, pp. 556–567. Springer, Cham (2017). https://doi.org/10.1007/978-3-319-55699-4_34
7. Wu, H., et al.: VQL: efficient and verifiable cloud query services for blockchain systems. IEEE Trans. Parallel Distrib. Syst. 33(6), 1393–1406 (2022)
8. Lei, Z., et al.: Transaction-based static indexing method to improve the efficiency of query on the blockchain. In: IEEE International Conference on Artificial Intelligence and Computer Applications (2021)
9. Xu, C., Zhang, C., Xu, J.: vChain: enabling verifiable Boolean range queries over blockchain databases. In: Proceedings of the 2019 International Conference on Management of Data, pp. 141–158 (2019)
10. Huang, X., Shen, D., Nie, T., Kou, Y., He, G., Xu, S.: An efficient query architecture for permissioned blockchain. In: Zhao, X., Yang, S., Wang, X., Li, J. (eds.) WISA 2022. LNCS, vol. 13579, pp. 699–711. Springer, Cham (2022). https://doi.org/10.1007/978-3-031-20309-1_61
11. Liu, M., Wang, H., Yang, F.: An efficient data query method of blockchain based on index. In: International Conference on Computer and Communications (2021)
12. Shlomi, L., Suprio, R., Natalia, S.: Authenticated multi-version index for blockchain-based range queries on historical data. In: 2022 IEEE International Conference on Blockchain (Blockchain), pp. 177–186 (2022)
13. Zhang, C., et al.: Authenticated keyword search in scalable hybrid-storage blockchains. In: 2021 IEEE 37th International Conference on Data Engineering (ICDE), pp. 996–1007. IEEE (2021)

Design Scheme of Anti-lost Functional Clothing for the Elderly Based on Blockchain

Hao Liu[1], Xueqing Zhao[1(✉)], Xin Shi[1], Guigang Zhang[2], and Yun Wang[2]

[1] Shaanxi Key Laboratory of Clothing Intelligence, School of Computer Science, Xi'an Polytechnic University, Xi'an 710048, China
zhaoxueqing@xpu.edu.cn

[2] Institute of Automation, Chinese Academy of Sciences, Beijing 100190, China

Abstract. Elderly people getting lost is a serious social problem with the increasing elderly population. This study proposes a clothing design scheme for preventing elderly people from getting lost, using blockchain and image steganography technology. Firstly, a smart contract for storing elderly identity information is designed. The elderly identity information is stored on the blockchain by interacting with the smart contract, and the smart contract returns a unique elderly identity ID corresponding to the identity information. Secondly, after performing a three-level Discrete Wavelet Transform (DWT) on the LOGO image, the elderly person's identity ID is embedded into the LL3 subband. Finally, the coefficient matrix with the embedded elderly identity ID is subjected to a three-level Inverse Discrete Wavelet Transform (IDWT) to obtain the LOGO image, which is then printed on the clothing. When seeking help after the elderly person gets lost, relevant personnel can extract the elderly identity ID from the clothing LOGO using a specific identifier and retrieve the elderly identity information from the blockchain platform. The simulation experiment involved embedding identity ID into the LL1, LL2, and LL3 subbands of the LOGO image. The results showed that embedding into the LL3 subband had the best performance, with PSNR of 30.772, SSIM of 0.968, and NC of 0.962. When the embedding strength Q was chosen as 32, the PSNR was 33.921, SSIM was 0.985, and NC was 0.913, ensuring the security and robustness of the embedded identity information without causing image distortion. This scheme not only helps lost elderly people return to their families but also effectively protects their privacy.

Keywords: Blockchain · Smart contract · Elderly prevention of getting lost · functional clothing

1 Introduction

In China, a survey study on the "search and rescue" incidents showed that as many as 40% elderly people have experienced the situation of being lost, and people over 65 years old have suffered more injuries after disappearing than we

© The Author(s), under exclusive license to Springer Nature Singapore Pte Ltd. 2023
L. Yuan et al. (Eds.): WISA 2023, LNCS 14094, pp. 467–475, 2023.
https://doi.org/10.1007/978-981-99-6222-8_39

expected. Because the decline in physical and cognitive function of the elderly can affect their ability to recognize the environment and seek help, it is very painful for the elderly and his family to go missing. With memory loss, loss of judgment, and decreased visual perception, older adults are at increased risk of being lost, and they may forget landmarks, be confused about directions, or not recognize familiar places.

With the development of science and technology, wearable devices such as "smart bracelets" can obtain the location of the elderly wearing such devices and report the safety of the elderly to the guardian, but such devices require a stable network environment and sufficient power to ensure stable operation, and long-term wear of such devices, harmful substances produced by electronic components will also cause irreversible damage to the body of the elderly. In this paper, an anti-lost clothing design scheme for the elderly is proposed. By using blockchain and image steganography technology, the elderly identity information is stored in the blockchain to get the elderly identity ID, and then the identity ID is embedded in the clothing LOGO. When the elderly need help, the relevant person can extract the identity ID from the clothing LOGO and search the elderly identity information on the blockchain to help them go home. The scheme has high information security and low cost and protects the privacy of the elderly.

2   Related Work

Image steganography is a technique that can hide secret information in images. In 2015, Singh R K et al. studied the LSB (Least Significant Bit) image steganography algorithm under different types of noise. They compared the robustness of the algorithm in hiding secret information under these noises and found that the algorithm was vulnerable to attacks and had limited robustness [1]. To further improve the robustness, Zhang Yifeng et al. proposed a two-level discrete wavelet transform multi-sub image steganography algorithm based on the particle swarm optimization algorithm in 2019. They used an improved SIFT (Scale Invariant Feature Transform) to enhance the ability to resist geometric attacks and proved that its robustness was better than that of the single-sub-image image steganography algorithm based on the two-level discrete wavelet transform [2]. In 2021, Zermi N et al. proposed an image steganography method for medical image protection. This method used blind watermarking technology and the second-level discrete wavelet transform and SVD (Singular Value Decomposition) to hide patients' medical records as secret information in medical images. The method had high quality and strong robustness, and could effectively protect the security and integrity of medical image data [3]. In all the image mentioned above steganography techniques has limitations on the amount of secret information that can be stored in the simple image and is vulnerable to various physical attacks.

Blockchain is a decentralized distributed storage technology that ensures data security and integrity through cryptographic algorithms and consensus among nodes [4]. Ethereum is a decentralized application-building platform based on blockchain technology. It uses smart contract mechanisms to develop and manage decentralized applications. Smart contracts are code protocols deployed on the

Ethereum blockchain that are automatically executed through events. They can achieve application decentralization and ensure that data is not tampered with by third parties. This paper proposes a hidden embedding scheme based on the third-level discrete wavelet transform combined with blockchain technology. This scheme is applied to the design of clothing for missing elderly people, helping them to reunite with their families quickly and safely.

3   The Proposed Scheme

3.1   Design of Smart Contract

The blockchain module proposed in this paper needs to implement two functions: uploading elderly identity information and retrieving elderly identity information. To achieve these two functions, this paper designs a smart contract algorithm based on the Solidity language, and the execution process of the smart contract algorithm is shown in Fig. 1.

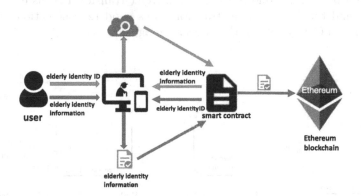

Fig. 1. Smart contract execution process

As shown in Fig. 1, the elderly identity information and the elderly identity ID are data structures mutually associated through a mapping relationship. When using this smart contract algorithm, users can upload elderly identity information through a client and trigger the smart contract to package it onto the blockchain, obtaining a unique elderly identity ID. Subsequently, when the user inputs the elderly identity ID through the client, the smart contract algorithm retrieves the corresponding elderly identity information from the blockchain based on the unique ID and returns it to the client. This algorithm provides a feasible solution for using blockchain technology to safely store and efficiently retrieve the identity information of the elderly.

3.2   Elderly Identity ID Steganography in Clothing LOGO

Performing DWT on Clothing LOGO Image. Wavelet transform decomposes a signal into a series of basic waves that are shifted and scaled by a basic wavelet. It has multiple hierarchical features of "time-frequency" and can represent the local characteristics of a signal in both the time domain and the frequency domain [5,6]. DWT can decompose the carrier image into four frequency bands: horizontal subband (LH), vertical subband (HL), diagonal subband (HH), and low-frequency subband (LL), where the main detail information of the image is concentrated in the low-frequency subband, which is the sub-image closest to the original image, while the other three subbands represent the edge detail information of the image in different directions. The low-frequency subband can be further decomposed. Since the low-frequency subband obtained by the first wavelet decomposition already contains most of the energy of the image, embedding secret information in this subband may cause image distortion. Therefore, the proposed scheme in this paper performs three-level wavelet decomposition on the carrier image to obtain a more detailed low-frequency subband (LL3) than the first decomposition, which can better hide secret information and improve its robustness [7]. The proposed scheme in this paper uses the Haar wavelet as the basic wavelet, with a time complexity of $O(n)$, where n is the length of the signal to be transformed. Therefore, the Haar wavelet transform is faster than other wavelets, and the Haar wavelet transform has good reconstruction properties, which can completely restore the original image [8].

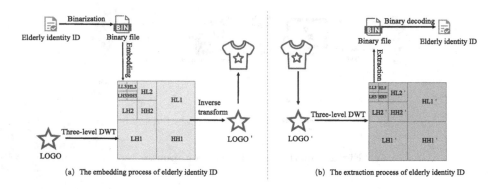

(a) The embedding process of elderly identity ID   (b) The extraction process of elderly identity ID

Fig. 2. The embedding and extraction processes of elderly identity ID

Embedding Elderly Identity ID in the LOGO. After undergoing three-level DWT, the carrier image (i.e., the clothing LOGO image) mainly concentrates its energy in the third-level low-frequency subband (LL3). In terms of robustness and human visual sensitivity, the secret information (i.e., the elderly ID) should be embedded in the most important area of the image. Therefore, this scheme chooses to embed the secret information in the LL3 subband.

After performing a three-level DWT on the carrier image, the secret information is first binarized and then scaled to the same size as the coefficient matrix of the third-level low-frequency subband (i.e., low-frequency coefficients), and then embedded in the low-frequency coefficients in sequence according to the Formula (1) shown below.

$$LL3(i,j)' = \begin{cases} LL3(i,j) + Q/4 - A & W = 0 \text{ and } A < 3Q/4 \\ LL3(i,j) + 5Q/4 - A & W = 0 \text{ and } A \geq 3Q/4 \\ LL3(i,j) + A/4 - A & W = 1 \text{ and } A < Q/4 \\ LL3(i,j) + 3Q/4 - A & W = 1 \text{ and } A \geq Q/4 \end{cases} \tag{1}$$

In Formula (1), $LL3(i,j)$ represents the low-frequency coefficient of the carrier image after undergoing three-level DWT, W is the secret information to be embedded, Q is the quantization parameter for controlling the strength of embedding, and $A = mod(LL3(i,j), Q)$. After successfully embedding the secret information into the third-level wavelet low-frequency coefficients, the inverse wavelet transform is performed on each subband to obtain the original image. The algorithmic process is shown in Fig. 2(a).

Extracting Elderly Identity ID from LOGO. After successfully embedding the secret information into the carrier image using the aforementioned steps, it is necessary to extract the secret information from the carrier image without loss when needed. To do this, the carrier image needs to undergo three-level DWT to obtain the third-level low-frequency subband (LL3). Then, Formula (2) is used to extract the embedded secret information, which is the elderly identity ID, from the LL3 subband. The algorithmic process is shown in Fig. 2(b).

$$W = \begin{cases} 0 & A < Q/2 \\ 1 & A \geq Q/2 \end{cases} \tag{2}$$

In Formula (2), $LL3(i,j)$ represents the low-frequency coefficient of the carrier image after undergoing three-level DWT, and $A = mod(LL3(i,j), Q)$.

4 Experimental Analysis

4.1 Elderly Identity Information On-Chain

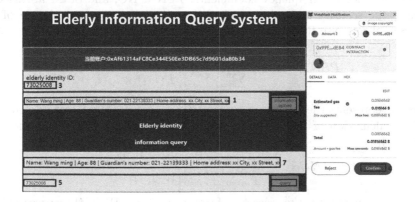

Fig. 3. Simulation results of the blockchain module

This scheme's smart contract is developed and deployed on the blockchain through Remix. Remix is an online integrated development environment (IDE) provided by Ethereum for writing smart contracts using Solidity language. Ganache is an Ethereum blockchain testing platform. Ganache provides a virtual environment that can simulate Ethereum accounts. The simulated Ethereum accounts can be imported into the Metamask wallet through a private key to facilitate the authorization and viewing of transaction details for each transaction. Metamask is a lightweight Ethereum wallet that supports the Ethereum main network as well as various Ethereum test networks. It can also be connected to a private chain through a custom RPC. The front end of this experiment is implemented based on the jQuery library. After the successful deployment of the smart contract, the simulated experimental results are shown in Fig. 3.

After the guardian has registered a blockchain account, they can enter the elderly identity information into Area 1 of Fig 3 and click on Area 2 to upload it. This will change the state variables in the smart contract, trigger an Ethereum blockchain transaction, and prompt the guardian to click on Area 4 to confirm the transaction and permanently store the elderly identity information on the blockchain. After the elderly identity information is successfully stored, the smart contract will return the corresponding elderly identity ID to the guardian, as shown in Area 3 of Fig. 3. When relevant personnel needs to access the elderly identity information, they can enter the elderly identity ID in Area 4 of Fig. 3 and click on Area 5 to search for the elderly identity information on the blockchain platform, as shown in Area 6 of Fig. 3.

4.2   Embedding and Extraction of Elderly Identity ID in LOGO

Datasets and Evaluation Metrics. This section of the simulation experiment is based on Python 3.7 and uses four datasets from the USC-SIPI Image Database, including Textures, Aerials, Miscellaneous, and Sequences, as carrier images. The elderly Identity ID is used as the secret information and embedded into the wavelet low-frequency coefficients. PSNR, NC, and SSIM are used to evaluate the security and robustness of the algorithm. The calculation of the Peak Signal-to-Noise Ratio (PSNR) is shown in Formula 3.

$$PSNR = 10\log \frac{\max\left(I_i(i,j), I_\omega(i,j)\right)}{\sum_{i=1}^{M} \sum_{j=1}^{N} I_i(i,j) \times I_\omega(i,j)} \tag{3}$$

where Ii represents the carrier image, and Iw represents the secret information.

The calculation of the Normalized correlation coefficient (NC) is shown in Formula 4.

$$NC = \frac{\sum_{i=0}^{m-1} \sum_{j=0}^{n-1} W(i,j) \times W'(i,j)}{\sum_{i=0}^{m-1} \sum_{j=0}^{n-1} (W(i,j))^2} \tag{4}$$

$W(i,j)$ represents the secret information, and $W'(i,j)$ represents the extracted secret information. As NC tends to 1, the difference between $W'(i,j)$ and $W(i,j)$ becomes smaller. An NC index of 0.75 or higher indicates good similarity performance.

The Structural Similarity Matrix(SSIM) is shown in Formula 5.

$$SSIM(M, N) = \frac{(2\mu_X\mu_Y + C_1)(2\sigma_{XY} + C_2)}{(\mu_X^2 + \mu_Y^2 + C_1)(\sigma_X^2 + \sigma_Y^2 + C_2)} \tag{5}$$

μ_X and μ_y represent the means of Ii and Iw, respectively; σ_X and σ_Y represent the variances of Ii and Iw, respectively; σ_{XY} represents the covariance between Ii and Iw; $c1$ and $c2$ are constants used to avoid division by zero.

Perceptual Quality Analysis. The experiment embedded the secret information into the low-frequency wavelet coefficients LL1, LL2, and LL3 at different levels and conducted a comparative experiment. The results are shown in Fig. 4(b). The results indicate that the secret information embedded in LL3 has a PSNR value of 30.77, the SSIM value of 0.967, and an NC value of 0.962, which is better than embedding in LL2 and LL1. Therefore, from the perspective of robustness, this study proposes embedding the elderly ID into the third-level low-frequency wavelet coefficients of the image.

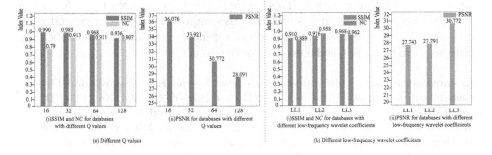

Fig. 4. Perceptual Quality Contrast

Fig. 4(a) shows the impact of different secret information embedding strengths (Q) on the embedding performance. As Q increases, the performance metrics tend to decrease. From the perspective of robustness, this study selects Q as 32. At this level, the PSNR value is 33.921 dB, SSIM is 0.985, and NC is 0.913, which can ensure the security and robustness of the secret information without causing distortion to the image.

Fig. 5. Quality of embedding and extraction of elderly identity ID

Embedding and Extracting Quality. As can be seen from Fig. 5, after the elderly identity ID is embedded by this method, the LOGO is still clear and the extracted elderly identity ID is also notably clear.

5　Conclusion

This paper proposes a new clothing design method that combines blockchain technology and DWT theory to help find lost elderly people and reunite them with their families. Firstly, we design an Ethereum-based smart contract for storing elderly identity information on the blockchain and returning the corresponding elderly identity ID to protect privacy. Next, we use a three-level DWT transform to embed the elderly identity ID into the third-level low-frequency wavelet coefficients of the elderly's clothing LOGO image and restore the LOGO through an inverse transform, which is then printed on the clothing. Finally, when an elderly person goes missing, relevant authorities can decode the elderly identity ID from the LOGO on their clothing using a specific decoder, and use this ID to query the blockchain identity authentication platform for the elderly person's identity information. This method not only ensures the information security and privacy of the elderly but also improves the efficiency and accuracy of finding the elderly. Future work could involve optimizing the embedding process, developing a specific decoder device, etc.

Acknowledgments. This work was supported by the Key Research and Development Program of Shaanxi Province in 2023 (No. 2023-YBGY-404, No. 2023-ZDLGY-48), the 2022 Research Project of Rural Public Cultural Service Research Institute, National Center for Public Culture Development, Ministry of Culture and Tourism (No.XCGGWH2022005), the 2022 Public Digital Cultural Service Project of National Center for Public Culture Development, Ministry of Culture and Tourism (No. GGSZWHFW2022-005), Shaanxi Province University Young Outstanding Talents Support Program, and Graduate Scientific Innovation Fund for Xi'an Polytechnic University(chx2023021).

References

1. Singh, R.K., Shaw, D.K., Alam, M.J.: Experimental studies of LSB watermarking with different noise. Procedia Comput. Sci. **54**, 612–620 (2015). https://doi.org/10.1016/j.procs.2015.06.071
2. Li, Y., Zhang, Y., Cheng, X., Sun, Y.: Robust watermarking algorithm based on DWT optimal multi-subgraph and SIFT geometric correction. J. Comput. Appl. Res. **36**(6), 1819–1823+1827 (2019)
3. Zermi, N., Khaldi, A., et al.: A DWT-SVD based robust digital watermarking for medical image security. Forensic Sci. Int. **320**, 110691 (2021)
4. Zhao, X., Liu, H., Hou, S., Shi, X., Wang, Y., Zhang, G.: An Ethereum-based image copyright authentication scheme. In: Zhao, X., Yang, S., Wang, X., Li, J. (eds.) Web Information Systems and Applications, WISA 2022. Lecture Notes in Computer Science, vol. 13579, pp. 724–731. Springer, Cham (2022). https://doi.org/10.1007/978-3-031-20309-1_63

5. Joshi, K., Kirola, M., et al.: Multi-focus image fusion using discrete wavelet transform method. SSRN Electron. J. (2019)
6. Diwakar, M., Tripathi, A., Joshi, K., et al.: A comparative review: medical image fusion using SWT and DWT. Mater. Today: Proc. **37**, 3411–3416 (2021). https://doi.org/10.1016/j.matpr.2020.09.278
7. Wen, B., Zhang, T., Xiong, T., Wu, C.: An image watermarking algorithm based on graph transform and DWT-SVD. Optoelectron. Laser Technol. **33**(8), 879–886 (2022). https://doi.org/10.16136/j.joel.2022.08.0840
8. Kumar, S., Kumar, R., et al.: A study of fractional Lotka-Volterra population model using Haar wavelet and Adams-Bashforth-Moulton methods. Math. Meth. Appl. Sci. **43**, 5564–5578 (2020)

Multi-User On-Chain and Off-Chain Collaborative Query Optimization Based on Consortium Blockchain

Jiali Wang[1], Yunuo Li[2], Aiping Tan[1(✉)], Zheng Gong[1], and Yan Wang[1]

[1] College of Information, Liaoning University, Shenyang, China
aipingtan@lnu.edu.cn
[2] Ming Yang Institute, Shenyang, China

Abstract. The consortium blockchain based on on-chain and off-chain data storage methods has shared cross-department business data widely. Existing research focuses on using index and other technologies to optimize a single query in the consortium blockchain, needing proper planning and scheduling for multi-user data access on-chain and off-chain. This paper fully considers the cross-department collaborative query requirements of the consortium blockchain and studies the reasonable allocation of nodes on and off the chain for multiple query tasks. The optimization model for cooperative query on and off the chain and the corresponding solution method is proposed. Firstly, the collaborative query optimization problem of on-chain and off-chain is modeled, considering the query processing capabilities of on-chain and off-chain nodes, and minimizes the query time as the optimization goal. Secondly, based on Coyote Optimization Algorithm (COA) and Genetic Algorithm (GA), a Hybrid Coyote Optimization Algorithm HCOA-GA (Hybrid COA-GA) is proposed. The experimental results show that the algorithm proposed can obtain the multi-user collaborative query optimization scheme on and off the chain and has better query time and convergence speed performance than other algorithms.

Keywords: Consortium blockchain · Multi-user · On-chain and off-chain Query · Task allocation · Optimization

1 Introduction

Distributing business data across departments has a wide range of applications. For example, in the donation system, charities and various non-profit organizations need to share information and data to facilitate the supervision of donation funds and prevent the deceit of the donates. The traditional data-sharing method will bring security risks, such as getting untrue data and a single point of data

This work is supported by the Key Science and Technology Program of the Open Competition Mechanism to Select the Best Candidates of Liaoning Province, China (Grant No.2021JH1/10400010).

storage failure. Therefore, using consortium blockchain [3] for distributed data storage has become an effective solution. However, storing a large amount of raw data from various departments on the chain is unrealistic because it will bring problems such as low throughput and high latency. The existing consortium blockchain business data-oriented to cross-department business collaboration sharing requirements usually uses on-chain and off-chain storage methods [10,18,19]. There are problems such as low data query efficiency and single query method on the consortium blockchain. Some studies usually use technologies such as built-in indexes and importing part of data into external databases to improve the query efficiency and enrich the query ability of the consortium blockchain. For example, SEBDB [22] proposed a blockchain database based on a hierarchical index, which can realize semantic query and table join operations in the consortium blockchain system. Liao [6] et al. used the external distributed database HBase as the off-chain big data storage layer of HyperledgerFabric [1] and utilized the high concurrent reading performance of the external distributed database HBase to realize efficient query of data records through the hash field on the chain. However, existing research usually assumes that there is only one data access requirement in the current consortium blockchain system. It optimizes it, that is, focuses on optimizing the data access process of a single user. When there are multiple users in practical applications, it is impossible to reasonably plan and schedule their data access on-chain and off-chain.

To solve the above problems, we propose a collaborative query model based on on-chain and off-chain storage modes and propose the corresponding solution. The main contributions of this paper are as follows:

1. The multi-user collaborative query optimization problem of on-chain and off-chain is reasonably modeled, fully considering the query processing capabilities of on-chain and off-chain nodes. It minimizes the query time as the optimization goal.
2. A hybrid heuristic algorithm HCOA-GA based on coyote and genetic algorithm is proposed to solve the proposed model.
3. The proposed algorithm is compared with the Genetic Algorithm (GA), an Improved Greedy Algorithm (Improved GA, IGA), and an Improved Simulated Annealing (Improved SA, ISA) algorithm to compare its query time and convergence speed.

The rest of this paper is organized as follows; in Sect. 2, related work is presented; in Sect. 3, the system model and related definitions are presented; in Sect. 4, the algorithm design is carried out for our proposed model; in Sect. 5, we conduct simulation experiments; and finally, in Sect. 6, we conclude this work.

2   Related Work

Blockchain [9] was initially applied to currency transactions [16]. With the development of Internet technology, the application direction of blockchain has extended to digital finance [15], Internet of Things [8], supply chain management

[12], digital asset trading, medical treatment [11] and other fields. A consortium blockchain is a special kind of blockchain. The nodes on the chain come from multiple departments or institutions, and they are responsible for maintaining the entire consortium blockchain. The traditional consortium blockchain system only supports transaction queries with block hash as an example, and its data query ability is weak. Some studies try to optimize the query performance of the consortium blockchain from two aspects.

The first method is to optimize the storage structure of the blockchain itself to improve the query efficiency or enrich the query function. This method's query processes are carried out in the blockchain network, regardless of other external databases. For example, Liu [7] et al. divided the block body into an index layer and data layer and designed an Abstract-Tire cluster index in the index layer. Abstract-Tire could obtain the original data in the block body quickly, stably, and efficiently. Liu et al. also added the index of the Operation-Record List in the index layer, which was used to connect the Operation records of the original data. SE-Chain [4] designed an efficient retrieval model. SE-Chain combined the Merkle tree and balanced binary tree in its data layer to improve the search efficiency of continuity attributes. vchain [17] proposed a lightweight verifiable blockchain framework, which used a cryptographic accumulator to add a summary to the block header to ensure that the query results returned to light nodes were not tampered, which could effectively reduce the cost of query verification.

The second method is to transfer part of the data to an external database for storage. This method is simple to implement and can enrich the query semantics. Typical research was EtherQL [5], which used MongoDB as an external database to store the output results of Ethereum client plugins. EtherQL supported range and Top-K queries in addition to existing queries on Ethereum. Literature [13] proposed a word extraction algorithm that extracted critical semantic information from the off-chain, mapping it with the on-chain data and building an inverted index according to the weight of keywords. The MST index tree was constructed using this algorithm, and the index could support range and keyword queries. Literature [21] transferred the application scenario from single chain to multi-chain based on literature [13], using a hybrid blockchain storage architecture of on-chain and off-chain to store data, and realizing cross-chain query and Top-K query.

3   On-Chain and Off-Chain System Overview

In this section, we first describe the system model and user query process of the consortium blockchain under the sharing requirements of cross-department business data (Sect. 3.1), and then specifically define and describe the multi-user collaborative query problem on and off the chain, including parameter definitions, constraints, objective functions, etc. (Section 3.2).

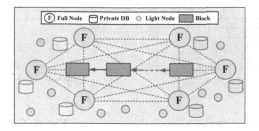

Fig. 1. System architecture

3.1   On-Chain and Off-Chain System Model

In our consortium blockchain system, as shown in Fig. 1, nodes are divided into two types: light nodes and full nodes. The smaller circles represent light nodes (i.e.users), which are used to receive query requests and verify query results. Bigger circles marked with F are full nodes. They come from multiple departments or agencies. Circles with the same color indicate that they come from the same agency.

Each full node needs to maintain two parts, an on-chain and off-chain local database, which is represented by a cylinder. In order to form a mapping between on-chain and off-chain data, we extract some key information off-chain, such as a hash of original data, data-id, keyword information, data storage location, etc., as a summary for on-chain processing. In data linking, if other full nodes find that the original data has the same copy in the local database, they will add their local database location to the transaction summary. Each block on the chain is divided into two parts: the block head and the block body. In the block header, the block number bId, the previous block hash $preHash$, the current block hash $bHash$, the timestamp tS that generated the block, and the root of the Merkle Tree $mRoot$ that generated all the transactions in the block. In the block body is stored data that can be made public and transparent, namely transactions. We define a transaction on the chain as $trans = < sp, digest >$, a set consisting of the system attribute sp and the summary $digest$. sp contains the transaction's unique identifier tId, generating the transaction timestamp tS, the transaction's signature sig and the transaction hash $tHash$. The digest contains the unique id of the original data dId, the hash value of the original data $dhash$, the keyword key extracted from the off-chain database, and the location set p stored the original data. The block structure is shown in Fig. 2.

3.2   Problem Description and Definition

In the application scenario, a full consortium blockchain node needs to process multiple query tasks in the same period. The parallel processing capability of nodes is limited. Therefore, when faced with multiple query tasks, how to allocate them to different nodes reasonably so that they can complete the task in the shortest time is the focus of this paper. The multi-user collaborative query

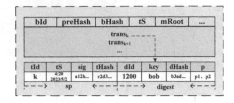

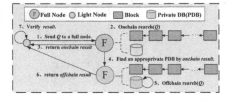

Fig. 2. Block structure   **Fig. 3.** Query process

optimization problem of on-chain and off-chain can be described as there are FN full nodes and M query tasks in the current consortium chain. For each query task, at least one node needs to be selected for on-chain and off-chain queries, and each node concurrently processes these tasks. Due to the large number of nodes distributed in different physical locations, selecting the appropriate node from many node combinations is necessary to make the shortest query time of the M tasks. To facilitate the research and ensure the scientific nature of the problem, we make the following assumptions for this problem:

1. All the full nodes stay online at all times without downtime behavior.
2. Each query task has the same priority, which ensures that multiple query tasks can be processed at the same time.
3. Each full node has a local database which is used to hold the relevant raw data.
4. There are at least two nodes in the consortium blockchain network storing the same data copies in the local database.
5. The time for a light node to issue a query request to a certain full node is the same as the time for the full node to return the query result to the light node.
6. Each partial query duration is known.

According to the actual situation and problem description, we establish the mathematical model of a multi-query task optimization problem based on on-chain and off-chain. Table 1 is the definition of some parameter variables:

As is shown in Fig. 3, in the actual application scenario of consortium blockchain, the total time of a query task is usually divided into seven parts: the request time D_{mi} from the light node where the task is located to the full node, the time P_m of the query on the chain, the time S_{ij} of the request to the off-chain database, the query time Q_m of the off-chain database, the return time D_{mi} of the query results on the chain, the return time R_{mj} of the query results on the chain, and the user verification time V_m. Among them, the on-chain query time P, the off-chain query time Q, and the user verification time V are only related to task m.

Definition 1. *For given D, P, S, Q, R, V, M, FN we can define the objective function as follows:*

$$fit = \min(\max(T_1, T_2, \cdots, T_m, \cdots, T_M)) \tag{1}$$

Table 1. Parameter definition and description.

Parameter	Description
FN	set of full nodes
LN	set of light nodes
M	set of query missions
m	number of query missions
PB	off-chain private databases
U_{on}	on-chain upper limit of processing capacity
U_{off}	off-chain upper limit of processing capacity
$g[m][i][j]$	mission m distributed by on-chain full node i and off-chain full node j
w_m	on-chain and off-chain connection property

where

$$T_m = 2 * D_{mi} + P_m + S_{ij} + Q_m + R_{mj} + V_m \tag{2}$$

$$s.t. \forall m, F(m) = l, m \in M, l \in LN \tag{3}$$

$$\forall i, \sum_{m=0}^{M-1} \sum_{j=0}^{FN-1} g[m][i][j] \leq U_{on}, i \in FN \tag{4}$$

$$\forall j, \sum_{m=0}^{M-1} \sum_{i=0}^{FN-1} g[m][i][j] \leq U_{off}, j \in FN \tag{5}$$

$$\forall m, w_m \in PB, m \in M \tag{6}$$

For Formula (1), fit is defined as our objective function, aiming at minimizing the most significant element of set T. Each element T_m of the set T can be calculated as Formula (2).Formula (3) denotes the mapping between query tasks and light nodes; that is, for each query task, the unique initiator of the query request can be found after $F(m)$. Formula (4) and Formula (5) respectively indicate that the sum of query tasks of each node in the same batch of query tasks cannot exceed the upper limit of the processing capacity of on-chain and off-chain, and Formula (6) indicates that the join attribute in the query result of on-chain must exist in the local database set.

4   HCOA-GA Algorithm Design

This subsection describes the multi-task on-chain off-chain query optimization process. The swarm intelligence optimization algorithm [2,20] achieves the optimization purpose by simulating the behavior of different populations in nature and using the information exchange and cooperation between groups. The group cultural trend proposed in the Coyote algorithm [14] increases the diversity of the population and shows excellent performance among many metaheuristic algorithms. Genetic algorithm is scalable and easy to be combined with other algorithms. In our model, affected by a variety of constraints, using the traditional genetic algorithm to perform crossover and mutation operations on randomly selected individuals may bring the problem of death of new individuals, and then affect the diversity of the population and obtain unsatisfactory task allocation results. Therefore, we used the hybrid heuristic algorithm combining the coyote algorithm and genetic algorithm (HCOA-GA) to solve the model. Specifically, the idea of population grouping and individual transition in the coyote algorithm is first cited, and then the individuals in each Wolf group are re-classified, and the crossover and mutation operations are performed according to the individual level. In addition, the population stability and individual transition standards of each Wolf group are redefined to solve our model.

4.1   Group Level Setting

HCOA-GA divides the population into multiple groups according to the coyote algorithm. The coyotes in each group are divided into three levels: α, β, and Ω, where α represents the best individual in each group, β represents the second best individual, Ω represents the remaining individuals and the worst fitness individual in each population is called ω. In addition, we use variance to replace the median in the group cultural trend in the coyote algorithm because variance is usually used to express the degree of deviation between a random variable and a mathematical expectation. The larger the variance is, the more the law of the jungle exists in this group and the more unstable the group is. Therefore, individuals in one group are more likely to die or be expelled from the original group. We use σ_p^2 to denote the degree of stability of pth group. σ_p^2 is defined as follows.

$$\sigma_p^2 = \frac{\sum_{c=0}^{C_p-1} (fit_p^c - \overline{fit_p})^2}{C_p} \tag{7}$$

where C_p denotes the number of individuals in this group, fit_p^c denotes the fitness of the current individual, and $\overline{fit_p}$ denotes the average fitness of individuals in the current group. For any group p, our criterion γ for determining coyote ω departure is defined as follows.

$$\gamma = \frac{\sigma_p^2}{\sum_{p=0}^{P-1} \sigma_p^2} \tag{8}$$

4.2   Operations of Crossover and Mutation

In our model, the overall assignment results of the nodes on and off the chain of multiple query tasks are regarded as individual, the assignment results of the nodes on and off the chain of each task are considered as genes in the individual, and the different assignment results of the same batch of tasks are denoted as a population. Before introducing crossover and mutation, we first encode the individuals. Each individual consists of M genes. Each gene represents the on-chain and off-chain node number selected by task m under the current iteration. For any gene X, we can use $X = [x^1, x^2]$, where x^1 is the number of nodes on the chain, and x^2 is the number of nodes off the chain. We use binary encoding to represent gene X.

When performing crossover and mutation operations, the uncertainty of the results after crossover and mutation will bring the problem of gene conflict and then lead to the death of new individuals. To ensure that at least one good-quality individual survives in each group, we stipulate that no operations are performed on α in the two operations of crossover and mutation. For the first operation, β and a randomly selected entity in Ω are used to perform the crossover operation, using the crossover operator based on the worst position and trying to optimize the worst genes of β. The second operation is to swap the gene in the worst position with the gene in the position randomly selected by itself in the individual ω. This step hopes to gradually improve the stability of the population from the perspective of optimizing the ω individual. Figure 4 illustrates the procedure of the two crossover operations in detail.

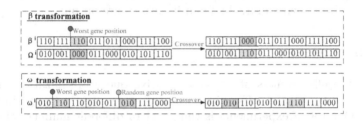

Fig. 4. Operations of crossover

For the mutation operation, we randomly select the worst-position gene of any Ω individual of the group, expecting to bring better fitness to the entity. Figure 5 shows the detailed mutation process. We dynamically specify the mutation rate ar_p^t according to the mutation operation in different stages. The dynamic mutation rate is defined as follows:

$$ar_p^t = c_1 e^{-\frac{\sigma_p^2}{t}} \tag{9}$$

Fig. 5. Operation of mutation

Among them, $c_1 \in [0.5, 1]$, t denotes the current number of iterations.

4.3 Steps of HCOA-GA

HCOA-GA is the combination and improvement of standard COA and GA algorithms. To solve the multi-user collaborative query optimization problem of on-chain and off-chain better, the specific steps are shown as follows:

Step 1. Set the coyote population N_p, the number of individuals in each group N_c, the maximum number of iterations $Iter\_max$.

Step 2. Initialize the coyote population.

Step 3. Calculate the fitness of individuals in each group.

Step 4. Define coyote α, coyote β, Ω, and coyote ω in each group and calculate the group stability of the current group.

Step 5. Make individual departure judgment according to group stability, then perform a leaving or death action.

Step 6. Update the group.

Step 7. Perform the crossover and mutation operation. Without genetic conflict, new individuals will survive or save the original.

Step 8. Update the group.

Step 9. Repeat steps 3 to 8 until the maximum number of iterations is reached. The best coyote individual and its fitness are output.

5 Experiments

To verify the effectiveness of the collaborative query optimization model for multi-user on-chain and off-chain, two sets of simulation experiments are designed. One set of experiments is the effect of different numbers of tasks on query time under different algorithms when the number of full nodes is variable. The second set of experiments compares GA, ISA, and HCOA-GA under different iterations with the same number of tasks. According to the definition of on-chain and off-chain collaborative query, an experimental data set was built. The data set was designed according to real application scenarios, and the performance comparison of different algorithms under different on-chain and off-chain query node allocation methods was considered.

In the first set of experiments, the number of full nodes FN in the consortium chain network is set to 15, 20, and 25. In the same batch of query tasks, by setting different numbers of tasks M_num, the performance of GA, IGA, ISA, and HCOA-GA were compared. The parameters in HCOA-GA are set as follows: population size $N_p = 50$, the number of individuals in each population is 10, the maximum iteration number is 300, and the four algorithms execute 20 independent experiments. The average longest query time results are shown in Fig. 6.

It can be known from Fig. 6 that the IGA performs better when the number of tasks is small because IGA seeks the optimal solution in the current state in each iteration without considering the global factors. With the increase in the number of tasks, when the nodes of the off-chain query results of a task have all reached the processing limit, there is no feasible solution now. Therefore

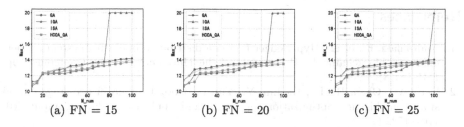

(a) FN = 15   (b) FN = 20   (c) FN = 25

Fig. 6. Query results with different number of nodes

IGA fails. GA, ISA, and HCOA-GA have strong solving performance, and the maximum time of the query task increases gently with the number of tasks. The solving performance of HCOA-GA is better than GA and ISA in the case of 300 iterations.

In the second set of experiments, we compare the convergence of GA, ISA, and HCOA-GA for solving the longest query time under different iterations. The experimental results are shown in Fig. 7. The experimental results show that HCOA-GA has faster convergence, and the solution effect is better than GA and ISA when the number of iterations is small. In practical applications, HCOA-GA can save computer resources.

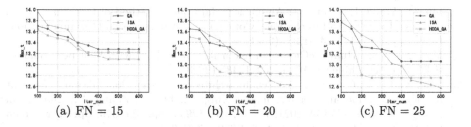

(a) FN = 15   (b) FN = 20   (c) FN = 25

Fig. 7. Query results of different number of nodes and the same number of tasks

6   Conclusion

In this paper, we first introduce the shortcomings of the existing consortium blockchain query performance and then fully consider the consortium blockchain's node selection and allocation problem based on a multi-user query. Then we construct a mathematical model for multi-user collaborative queries on and off the consortium blockchain. The purpose is to reasonably allocate multiple query tasks when there are multiple user query task requirements in the consortium chain network at a particular time, with the optimization goal of minimizing the query time and improving the query performance of the consortium chain. Finally, the proposed HCOA-GA algorithm is used to solve the model. The experimental results show that our scheme performs better in query time and convergence speed than other algorithms.

References

1. Androulaki, E., et al.: Hyperledger fabric: a distributed operating system for permissioned blockchains. In: Proceedings of the Thirteenth EuroSys Conference, pp. 1–15 (2018)
2. Deng, W., Chen, R., He, B., Liu, Y., Yin, L., Guo, J.: A novel two-stage hybrid swarm intelligence optimization algorithm and application. Soft. Comput. **16**, 1707–1722 (2012)
3. Dib, O., Brousmiche, K.L., Durand, A., Thea, E., Hamida, E.B.: Consortium blockchains: overview, applications and challenges. Int. J. Adv. Telecommun. **11**(1), 51–64 (2018)
4. Jia, D.Y., Xin, J.C., Wang, Z.Q., Lei, H., Wang, G.R.: Se-chain: a scalable storage and efficient retrieval model for blockchain. J. Comput. Sci. Technol. **36**(3), 693–706 (2021)
5. Li, Y., Zheng, K., Yan, Y., Liu, Q., Zhou, X.: EtherQL: a query layer for blockchain system. In: Candan, S., Chen, L., Pedersen, T.B., Chang, L., Hua, W. (eds.) DASFAA 2017. LNCS, vol. 10178, pp. 556–567. Springer, Cham (2017). https://doi.org/10.1007/978-3-319-55699-4_34
6. Liao, D., et al.: An efficient storage architecture based on blockchain and distributed database for public security big data. In: Hu, Z., Dychka, I., Petoukhov, S., He, M. (eds.) Advances in Computer Science for Engineering and Education, ICCSEEA 2022. Lecture Notes on Data Engineering and Communications Technologies, vol. 134, pp. 350–362. Springer, Cham (2022). https://doi.org/10.1007/978-3-031-04812-8_30
7. Liu, M., Wang, H., Yang, F.: An efficient data query method of blockchain based on index. In: 2021 7th International Conference on Computer and Communications (ICCC), pp. 1539–1544. IEEE (2021)
8. Mao, X., et al.: HuaBaseChain: an extensible blockchain with high performance. IEEE Internet Things J. **10**, 12462–12485 (2023)
9. Mao, X., Li, X., Guo, S.: A blockchain architecture design that takes into account privacy protection and regulation. In: Xing, C., Fu, X., Zhang, Y., Zhang, G., Borjigin, C. (eds.) WISA 2021. LNCS, vol. 12999, pp. 311–319. Springer, Cham (2021). https://doi.org/10.1007/978-3-030-87571-8_27
10. Miyachi, K., Mackey, T.K.: hOCBS: a privacy-preserving blockchain framework for healthcare data leveraging an on-chain and off-chain system design. Inf. Proc. Manag. **58**(3), 102535 (2021)
11. Mozumder, M.A.I., Sheeraz, M.M., Athar, A., Aich, S., Kim, H.C.: Overview: technology roadmap of the future trend of metaverse based on IoT, blockchain, AI technique, and medical domain metaverse activity. In: 2022 24th International Conference on Advanced Communication Technology (ICACT), pp. 256–261. IEEE (2022)
12. Omar, I.A., Debe, M., Jayaraman, R., Salah, K., Omar, M., Arshad, J.: Blockchain-based supply chain traceability for COVID-19 personal protective equipment. Comput. Ind. Eng. **167**, 107995 (2022)
13. Pei, Q., Zhou, E., Xiao, Y., Zhang, D., Zhao, D.: An efficient query scheme for hybrid storage blockchains based on Merkle semantic trie. In: 2020 International Symposium on Reliable Distributed Systems (SRDS), pp. 51–60. IEEE (2020)
14. Pierezan, J., Coelho, L.D.S.: Coyote optimization algorithm: a new metaheuristic for global optimization problems. In: 2018 IEEE Congress on Evolutionary Computation (CEC), pp. 1–8. IEEE (2018)

15. Tan, L., Shi, N., Yu, K., Aloqaily, M., Jararweh, Y.: A blockchain-empowered access control framework for smart devices in green internet of things. ACM Trans. Internet Technol. (TOIT) **21**(3), 1–20 (2021)
16. Urquhart, A.: The inefficiency of bitcoin. Econ. Lett. **148**, 80–82 (2016)
17. Xu, C., Zhang, C., Xu, J.: vChain: enabling verifiable Boolean range queries over blockchain databases. In: Proceedings of the 2019 International Conference on Management of Data, pp. 141–158 (2019)
18. Xu, C., Zhang, C., Xu, J., Pei, J.: SlimChain: scaling blockchain transactions through off-chain storage and parallel processing. Proc. VLDB Endowment **14**(11), 2314–2326 (2021)
19. Zhang, C., Xu, C., Wang, H., Xu, J., Choi, B.: Authenticated keyword search in scalable hybrid-storage blockchains. In: 2021 IEEE 37th International Conference on Data Engineering (ICDE), pp. 996–1007. IEEE (2021)
20. Zhang, C., Cai, Y., Hu, P., Quan, P., Song, W.: Logistics distribution route optimization using hybrid ant colony optimization algorithm. In: Zhao, X., Yang, S., Wang, X., Li, J. (eds.) Web Information Systems and Applications, WISA 2022. Lecture Notes in Computer Science, vol. 13579, pp. 510–517. Springer, Cham (2022). https://doi.org/10.1007/978-3-031-20309-1_45
21. Zhou, E., et al.: MSTDB: a hybrid storage-empowered scalable semantic blockchain database. IEEE Trans. Knowl. Data Eng. (2022)
22. Zhu, Y., Zhang, Z., Jin, C., Zhou, A., Yan, Y.: SEBDB: semantics empowered blockchain database. In: 2019 IEEE 35th International Conference on Data Engineering (ICDE), pp. 1820–1831. IEEE (2019)

BEAIV: Blockchain Empowered Accountable Integrity Verification Scheme for Cross-Chain Data

Weiwei Wei, Yuqian Zhou[✉], Dan Li, and Xina Hong

College of Computer Science and Technology, Nanjing University of Aeronautics
and Astronautics, Nanjing, China
{wwweiv,zhouyuqian}@nuaa.edu.cn

Abstract. In recent years, while blockchain technology has the potential to revolutionize various industries, the issue of data silos presents a challenge to its development. To solve this problem, cross-chain technology is used to facilitate data sharing among different blockchain systems, promoting collaboration and integration between them. However, if the sending-chain or receiving-chain tamper with data during cross-chain interaction based on their own interests, it can severely compromise the integrity of cross-chain data, thereby rendering the entire cross-chain process meaningless. Therefore, in this paper, we propose a blockchain empowered accountable integrity verification scheme for cross-chain data(BEAIV). Considering the decentralized nature of blockchain, we adopt a supervision-chain for decentralized auditing instead of relying on a centralized third-party auditor. Notably, our scheme can not only detect the corruption of data integrity but also trace responsibility for dishonest entity. Furthermore, our proposed scheme supports batch auditing to enhance the efficiency of auditing process. Security analysis and performance evaluation show that BEAIV can verify the data integrity in cross-chain interactions, enabling secure and accurate cross-chain data sharing.

Keywords: data integrity · cross chain · blockchain-enabled auditing · smart contract · privacy protection

1 Introduction

Blockchain is a collectively maintained decentralized data ledger, with many outstanding advantages like decentralization, reliability, tamper-proofing and traceability. It has been increasingly utilized in various application scenarios, such as cryptocurrency, digital supply chains, healthcare data sharing, and the Internet of Things (IoT) [9]. However, the lack of interoperability mechanisms among different blockchains has become a major bottleneck for the healthy development of blockchain technology.

In order to promote interoperability among different blockchains, several cross-chain methods have been proposed, such as the notary mechanism [5], side/relay chain [2], hash-locking [3], and distributed private key control. The utilization of cross-chain technology overcomes the limitations of isolated blockchains and facilitates the transfer of data and assets across different blockchain networks.

However, in cross-chain scenarios, the integrity of cross-chain data cannot be guaranteed as each blockchain has its own internal security mechanism and does not participate in the consensus process of other blockchains. Despite the immutability of on-chain data, the integrity of the data may be compromised during the cross-chain process, leading to inconsistencies in information between different blockchain systems. These inconsistencies can be fatal to cross-chain operations. Therefore, it is necessary to study how to verify the integrity of cross-chain data.

The research of data integrity has been extensively conducted in the context of cloud storage [1,7,8]. But these schemes brings additional computational overheads to data owner. To alleviate the heavy computational burden on the data owner, various research studies introduce third-party auditors (TPA) [10]. However, TPA has two obvious disadvantages. First, the assumption of a trusted TPA is a clear limitation since a third party can never be entirely trustworthy. Second, TPA cannot always be as stable and reliable as expected and may be subject to numerous security risks, such as hacker attacks and internal system failures [14].

To overcome the limitations of TPA, some solutions leverage decentralized blockchain technology as an alternative to centralized TPA [4,6,12,15]. Due to the decentralized nature of blockchain, it can always provide fair and accurate audit results based on the received audit proofs, enhancing the security of the system.

While data integrity auditing in cloud storage primarily involves verifying whether the data is stored correctly by the cloud storage provider, cross-chain scenarios present additional challenges, such as the potential for both sender and receiver may be malicious. Therefore, traditional cloud storage auditing schemes are not suitable for cross-chain scenarios.

In this paper, we propose a blockchain empowered accountable data integrity verification scheme for cross-chain scenarios, simply called BEAIV. The key idea of our solution is to use a supervision blockchain to perform integrity auditing of cross-chain data and automate the entire auditing process by deploying smart contracts. Our main contributions are summarized as follows:

- We take into consideration the decentralized nature of blockchain and use a supervision blockchain instead of a centralized TPA to perform cross-chain data integrity verification. The supervision-chain can not only verify the integrity of cross-chain data but also locate dishonest entities when integrity damage is detected.
- We extend our work to support batch auditing, improving auditing efficiency in the supervision-chain.

- We consider data privacy protection and adopt the technique of random masking during the auditing process to ensure that the supervision-chain cannot obtain the content of audited data blocks.

The remainder of this paper is organized as follows. Section 2 gives a brief introduction to the preliminaries covered in this paper. Section 3 defines the system model, threat model, and design goals. Section 4 presents the proposed scheme. Section 5 conducts the security analysis. Section 6 evaluates the performance of the proposed cross-chain data integrity verification scheme. Section 7 concludes this paper and points out the future work.

2   Preliminaries

2.1   Symmetric Bilinear Pairing

Let G and G_T be two multiplicative cyclic groups of large prime order q. A map function $e : G \times G \rightarrow G_T$ is a symmetric bilinear pairing only when it satisfies three properties below:

1) Bilinear: For $u, v \in G$ and $a, b \in Z_q^*$, $e\left(u^a, v^b\right) = e(u, v)^{ab}$;
2) Non-Degeneracy: $e(g, g)$ is a generator of G_T;
3) Computability: For $\forall u, v \in G$, there exists efficient algorithms to compute $e(u, v)$.

2.2   BLS-Based Homomorphic Verifiable Tag

BLS-HVT is based on the BLS signature algorithm, which can be efficiently aggregated and verified without disclosing private key.

BLS-HVT: Given a data block blk, and a cryptographic hash function $H : \{0,1\}^* \rightarrow G$. Select a random secret key $sk = a \in Z_p^*$, and compute the corresponding public key $pk = g^a$. Then the BLS-HVT for blk is $\sigma_{blk} = H(blk)^a$. For verification, the verifier will simply need to check whether $e\left(\sigma_{blk}, g\right) = e(H(blk), pk)$ holds.

3   Problem Statement

3.1   System Model

In this paper, we use a public blockchain as the supervision-chain(SVC) to regulate the integrity of cross-chain data between two blockchains, which can be consortium blockchain or private blockchain. That is to say our scheme targets the one-to-one cross-chain model. There are three different entities in system model: Sending-chain(SC), Receiving-chain(RC) and Supervision-chain(SVC).

SC: Cross-chain data D is stored on SC, it will be sent to RC as business requirements.

RC: After RC receives the data sent by SC, it first verifies whether the data and the tags are consistent. Then SC uploads the data on chain for storage after the verification is passed.

SVC: SVC checks the integrity of cross-chain data by sending audit requests to SC and RC and verifying the responded proof. SVC records verification result and audit information on chain.

3.2   Threat Models

We assume that SC and RC are untrusted since they may tamper with data according to their interests while sending or storing data. SVC is regarded as honest because anyone can join it and view information on the chain. Additionally, attacks on blockchain infrastructure and consensus mechanisms are beyond the scope of our consideration. In the cross-chain interaction environment, there are four main threats to data integrity:

1. SC may provide RC with incomplete on-chain data to save communication costs or cheat RC, attempting to mislead RC into believing that the data it received are complete.
2. RC may store the cross-chain data from SC incompletely for its own interests and try to mislead SVC into believing that the data provided by SC are incomplete.
3. SVC may extract private information from the integrity proof generated by SC and RC.
4. Cross-chain data may be corrupted due to network failure or hacking during cross-chain data transmission from SC to RC.

3.3   Design Goals

To enable responsible integrity verification for cross-chain data under the model as mentioned above, we propose the following goals:

1. **Lightweight.** Minimize additional computations on SC and RC due to cross-chain data integrity verification.
2. **Privacy protection.** SVC or off-chain adversaries cannot extract private information from the integrity proof.
3. **Efficient auditing.** Support batch auditing to improve the auditing efficiency of SVC.
4. **Accountability.** When cross-chain data integrity is corrupted, SVC needs to be able to trace responsibility for the dishonest entity.

4   The Proposed Scheme

4.1   Overview

BEAIV focuses on auditing the integrity of cross-chain data transferred by SC and stored in RC. Its workflow mainly consists of several phases, namely, (1) setup phase, (2) registry phase, (3) data sending phase, (4) data receiving phase and (5) auditing phase. The overview of integrity verification process is shown in Fig. 1, which includes six steps. Here, step 1 and step 2 belong to phase (3), step 3 belongs to phase (4), and steps 4, 5, and 6 belong to phase (5).

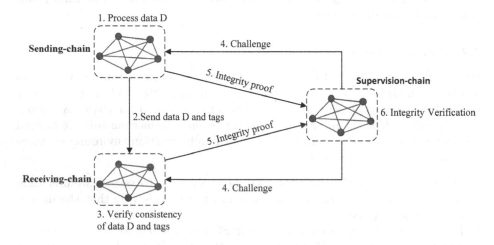

Fig. 1. Overview of the proposed scheme.

In the following subsections, we describe the smart contract used in BEAIV and the basic construction of BEAIV. Besides, to improve audit efficiency, we expanded our work to support batch auditing.

4.2 Smart Contract Used in the Scheme

In order to automate the workflow of the entire scheme, we deploy Registration smart contract and Auditing smart contract in SVC, Sending smart contract in SC, and Receiving smart contracts in RC. The functions of each smart contract are as follows:

- **Registry Smart Contract (RGC).** RGC generates public parameters for the entire system and provides registration services for regulated blockchains.
- **Sending Smart Contract (SSC).** When a cross-chain interaction occurs, SSC generates the corresponding tags for the cross-chain data.
- **Receiving Smart Contract (RSC).** When RC receives cross-chain data, RSC verifies whether the tags and data are consistent.
- **Auditing Smart Contract (ASC).** In auditing phase, ASC generates the corresponding auditing request and sends it to SC and RC respectively abd verify the integrity proof responded.

4.3   The Basic Construction

In this subsection, we present the detailed of the proposed blockchain empowered accountable cross-chain data integrity verification scheme(BEAIV). Phase (1) and phase (2) are the preparation of this scheme. During phases (3), (4), and (5), the interaction among SC, RC and SVC is depicted in Fig. 2.

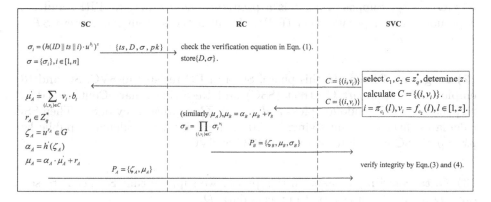

Fig. 2. The interaction among participants in BEAIV.

Before presenting the detailed scheme, we give some notations in Table 1.

Table 1. The notations in our scheme.

Notation	Description
q	A large prime order
e	A symmetric bilinear pairing
G	An elliptic curve group
G_T	A multiplicative group
Z_q	A field of residue classes modulo q
g	A generator of G
u	A random element of G
f	A pseudo-random function
π	A pseudo-random permutation
C	The challenge set generated by SVC
μ_A	The combination of masked challenged blocks of SC
μ_B	The combination of masked challenged blocks of RC
σ	The aggregated tag of challenged block generated by RC

(1) Setup Phase. First, the SVC deploys Auditing Smart Contract(ASC) and Registry Smart Contract (RGC). Then the ASC generates the system parameters for data integrity auditing. The details are as follows:

Select a large prime order q and generate a symmetric bilinear pairing e : $G \times G \to G_T$. G and G_T have the same order p. Select a field Z_q of residue classes modulo q. Pick a generator $g \in G$ and a random element $u \in G$ for cross-chain data processing in SSC of SC. Define two hash functions: $h : (0, 1)^* \to G$, $h' : G \to Z_q^*$. Suppose that f is a pseudo-random function (PRF) and π is a pseudo-random permutation (PRP). Publish the system parameters $SP = \{g, f, \pi, e, h, h', u\}$.

(2) Registry Phase. In this phase, SC and RC register in SVC. SC and RC employ Sending Smart Contract (SSC) and Receiving Smart Contract (RSC) respectively. SSC randomly picks a $sk = x \in Z_q^*$ as private key of SC. Then SSC calculates inverse inv satisfying: $x \cdot inv \equiv 1 (\mathrm{mod} p)$. SSC calculates public key $pk = g^{inv}$. SC uses pk to register in RGC of SVC.

(3) Data Sending Phase. In this phase, we suppose that SC needs to send data D to RC. SC uses SSC to process data D.

$SSC.GenTag()$. SSC splits D into n data blocks $D = \{b_1, b_2, ..., b_n\}$. It calculates a BLS-HVT for each b_i:

$$\sigma_i = (h(ID||ts||i) \cdot u^{b_i})^x. \tag{1}$$

Here ID is an unique identifier of D and ts is timestamp. Then SC sends $\{ts, D, \sigma, pk\}$ to RC, where $\sigma = \{\sigma_i\}$, $i \in [1, n]$.

(4) Data Receiving Phase. $RSC.verify()$. After receiving data, RC verifies the consistency of data D and tags σ by calculating the following equation:

$$e(\prod_{i=1}^{n} \sigma_i, pk) = e(\prod_{i=1}^{n} h_i \cdot u^{\sum_{i=1}^{n} b_i}, g). \tag{2}$$

Here $h_i = h(ID||ts||i)$.

If Eq. (2) holds, RC accepts the cross-chain data and sends cross-chain record in this format: $\{from, to, ts, h(D)\}$ to SVC. Otherwise, RC refuses to upload data on chain.

(5) Auditing Phase. When SVC decides to audit, ASC retrieves the corresponding record from the cross-chain records stored in ASC. Then ASC determines the amount of data blocks challenged z and randomly chooses two keys $c_1, c_2 \in Z_q^*$.

$ASC.GenChal()$. ASC calculates $i = \pi_{c_1}(l)$ and $v_i = f_{c_2}(l)$ for $l \in [1, z]$ to generate the challenge set $C = \{(i, v_i)\}$, where i is the index of each challenged data block and v_i is the corresponding coefficient. Then SVC sends the challenge set $C = \{(i, v_i)\}$ to SC and RC respectively.

After receiving the challenge set, SC generates the corresponding proof. The details of this procedure are as follows.

SSC.GenProof(). SSC of SC calculates the combination of the challenged blocks $\mu'_A = \sum_{(i,v_i)\in C} v_i \cdot b_i$. To prevent the content of cross-chain data from being leaked, it then selects a random element $r_A \in Z^*_q$ and computes $\zeta_A = u^{r_A} \in G$. Then it computes $\alpha_A = h'(\zeta_A)$ and $\mu_A = \alpha_A \mu'_A + r_A$. Finally, SC sends the proof $P_A = \{\zeta_A, \mu_A\}$ to SVC.

Similarly, RC performs the following operations to generate proof.

RSC.GenProof(). With the challenge set $C = \{(i, v_i)\}$, RSC of RC calculates $\mu'_B = \sum_{(i,v_i)\in C} v_i \cdot b_i$, selects randomly $r_B \in Z^*_q$ and calculates $\zeta_B = u^{r_B} \in G$. It then computes $\alpha_B = h'(\zeta_B)$ and $\mu_B = \alpha_B \mu'_B + r_B$. In addition, RSC calculates the aggregated tag $\sigma_B = \prod_{(i,v_i)\in C} \sigma_i^{v_i}$. Finally, RC sends the proof $P_B = \{\zeta_B, \mu_B, \sigma\}$ to SVC.

After receiving the proof P_A and P_B, SVC performs the verification process with ASC. The details of this process are as follows.

ASC.verify(). ASC computes $\alpha_A = h'(\zeta_A)$ and $\alpha_B = h'(\zeta_B)$ firstly. Then ASC verifies the integrity proof by calculating the following equation:

$$e(\zeta_A \cdot \zeta_B, g) \cdot e(\sigma^{\alpha_A + \alpha_B}, pk) \overset{?}{=} e((\prod_{(i,v_i)\in C} h_i^{v_i})^{\alpha_A + \alpha_B} \cdot u^{\mu_A + \mu_B}, g). \tag{3}$$

Here $h_i = h(ID\|ts\|i)$ and C is the challenged set generated by SSC.

If Eq. (3) holds, the cross-chain data is intact, and SC and RC all are honest. Otherwise, the cross-chain data integrity D is corrupted.

ASC.verify(). When the integrity of cross-chain data is corrupted. ASC verifies the following equation to identify dishonest entity:

$$e(\zeta_B, g) \cdot e(\sigma^{\alpha_B}, pk) \overset{?}{=} e((\prod_{(i,v_i)\in C} h_i^{v_i})^{\alpha_B} \cdot u^{\mu_B}, g). \tag{4}$$

If Eq. (4) holds, RC is honest. The verification process ends, SC is malicious. Otherwise, RC is malicious and SC is honest. Because when receiving cross-chain data, RC verified that the data and tags are consistent before storing it on chain. If it cannot pass the verification of SVC, then it is dishonest.

4.4   Batch Auditing

There are multiple auditing tasks that may be waiting for processing simultaneously in SVC. Batch auditing in our work supports aggregating challenges and proof to reduce pairing operation, which can reduce the computation overheads of SVC compared with single auditing.

Suppose that there are cross-chain interactions from z SCs, with t_k auditing tasks in each SC, where $k \in [1, z]$. To make the formula easy to understand, we set $\alpha = \alpha_A + \alpha_B$, $\zeta = \zeta_A \cdot \zeta_B$, $\mu = \mu_A + \mu_B$. Therefore, the batch verification equation is

$$e(\prod_{k=1}^{z}\prod_{j=1}^{t_k}\zeta_j, g) \cdot \prod_{k=1}^{z} e(\prod_{j=1}^{t_k}\sigma_j^{\alpha_j}, pk_k) \overset{?}{=} e(\prod_{k=1}^{z}\prod_{j=1}^{t_k}(\prod_{(i,v_i)\in C_j} h_i^{v_i})^{\alpha_j} \cdot u^{\sum_{k=1}^{z}\sum_{j=1}^{t_k}\mu_j}, g).$$

$$(5)$$

The detailed batch auditing process in SVC as shown in Algorithm 1.

Algorithm 1. Batch Auditing.

Input:
    The set of public key of z SCs;
    The integrity proof of z SCs, which each SC containing t_k auditing tasks;
    The integrity proof of z RCs;
Output:
    A set E that contains dishonest entities in each cross-chain record;
1: Aggregate the integrity proof of SCs and RCs;
2: Verify the Eq.(5);
3: **if** Eq.(5) is valid **then**
4:     $E = null$;
5: **else**
6:     Verify Eq.(3)and Eq.(4);
7:     Trace these dishonest entities in each cross-chain record;
8:     Record pk of dishonest entity in E;
9: **end if**
10: **return** E;

---

5 Security Analysis

In this section, we briefly evaluate the security of BEAIV. We analyze how BEAIV defends against the attacks mentioned in the threat models (Subsect. 3.2).

Theorem 1. *If SC sends the cross-chain data honestly, it can pass RC's verification.*

Proof. We prove the correctness of Eq. (2) as follow:

$$e(\prod_{i=1}^{n}\sigma_i, Y) = e(\prod_{i=1}^{n}(h_i \cdot u^{b_i})^x, g^{inv}) = e(\prod_{i=1}^{n}h_i \cdot u^{\sum_{i=1}^{n} b_i}, g^{x \cdot inv}) = e(\prod_{i=1}^{n}h_i \cdot u^{\sum_{i=1}^{n} b_i}, g)$$

Theorem 2. *If SC and RC are all honest and BEAIV is executed correctly, they can pass SVC's verification.*

Proof. We prove the correctness of BEAIV in a single auditing task.

For a single auditing task, Eq. (3) can be derived as follows:

$$LHS = e(u^{r_A} \cdot u^{r_B}, g) \cdot e((\prod_{(i,r_i) \in C} \sigma_i^{v_i})^{\alpha_A + \alpha_B}, g^{inv})$$

$$= e((\prod_{(i,v_i) \in C} h_i^{v_i})^{\alpha_A + \alpha_B} \cdot u^{(\alpha_A + \alpha_B) \sum_{(i,v_i) \in C} v_i b_i} \cdot u^{r_A + r_B}, g)$$

$$= RHS$$

Theorem 3. *If dishonest SC modifies the data and generates corresponding tags when sending cross-chain data, even if he can deceive RC, it will be checked by SVC.*

Proof. Suppose SC modifies cross-chain data D to D', and generates the correct corresponding tags σ'. Equation (3) will hold when RC verifies the data and corresponding tags. Although RC is deceived by SC, SVC will detect the malicious behavior of SC, which means the proof of SC cannot pass Eq. (4).

Theorem 4. *SVC cannot recover any information for the data blocks from the audit proofs generated by RC and SC.*

Proof. It is no doubt that the values ζ_A in proof P_A and ζ_B in proof P_B contain no information about data blocks. The challenged data blocks are hidden in σ, μ_A and μ_B. Firstly, the HVT tag σ is secure based on the CDH problem, which security had been analysed by Wang et al. in [11]. Then, for μ_A and μ_B, due to the introduction of the random value r_A and r_B respectively in them, BEAIV can resist recovery attacks as mentioned in [13]. Therefore, our scheme can preserve the privacy of the cross-chain data.

6   Performance Evaluation

In this section, we evaluate the efficiency and effectiveness of BEAIV by conducting several experiments using JAVA JDK 1.8.0 on a PC laptop which runs Windows 10 on an Intel Core i5 CPU at 2.50 GHz and 4 GB DDR4 RAM. And all algorithms are implemented using the Java Pairing-Based Cryptography Library (JPBC) version 2.0.0. In our experiments, we choose type A pairing parameters in JPBC library, which the group order is 160 bits and the base field order is 512 bits. We set the data size as 4KB and 8KB in different data size, respectively. To get more precise results, each experiment is conducted in 50 trials.

Data Preprocessing. Figure 3 shows the computation cost on the SC side during data processing. It is obvious that the time of data preprocessing increases with the size of data D. And for the same size of D, the time cost decreases as the size of the data block increases. The computation time of the tag generation algorithm grows linearly. As is shown in Fig. 4, the change trend of the size of tags is similar to that of the computation of tags.

Tags Verification. Figure 5 shows the computation cost of RC in tags verification. Thanks to less expensive operations such as exponentiation, verification cost of RC is acceptable, as we can see that it only takes about 192ms to verify cross-chain data when data size is 8M and data block size is 8KB.

Auditing Phase. Figure 6 illustrates the computation cost of SC, RC and SVC in auditing phase. It is easy to see that the proof generation cost of SC and RC increases linearly with the number of challenged blocks, while the verification cost of SVC is stable.

Batch Auditing. In Fig. 7, we present the comparison of the computation between batch auditing and single auditing of which each challenge set contains 100 data blocks. We can see that as the number of audit tasks increases, the computational time saved by batch auditing also increases. Therefore, our solution can further improve audit efficiency through batch auditing. Furthermore, we present the comparison of batch auditing with [14] under the condition that each challenge set contains 250 data blocks, and these auditing tasks are from the same SC. As shown in Fig. 8, we can see that as the number of auditing tasks increases, our average auditing efficiency improves whereas [14] stays nearly unchanged. This is because that the total time cost in BEAIV does not increase linearly when different proofs from multiple tasks are aggregated.

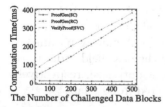

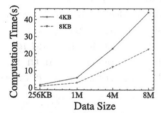

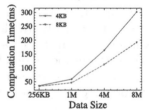

Fig. 3. The data processing computation cost of SC.

Fig. 4. The tags size generated by SC.

Fig. 5. The verification computation cost of RC.

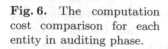

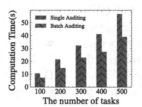

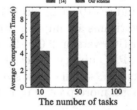

Fig. 6. The computation cost comparison for each entity in auditing phase.

Fig. 7. The comparison of the computation cost between single auditing and batch auditing.

Fig. 8. The average auditing time comparison.

7   Conclusion and Future Work

In this paper, we propose a blockchain empowered accountable integrity verification scheme(BEAIV), which introduces a supervision-chain(SVC) to audit the integrity of cross-chain data. By introducing the supervision blockchain, our scheme can automatically achieve cross-chain data integrity auditing without relying on any third-party verifier. Meanwhile, in our scheme, when cross-chain data integrity is corrupted, SVC can trace responsibility for a dishonest entity. Furthermore, BEAIV supports batch auditing to enhance the efficiency of the auditing process. The security and performance evaluation shows that our scheme is secure and available. In future work, we plan to extend our scheme to audit cross-chain data integrity for the one-to-many cross-chain model.

Acknowledgements. This work was supported by the National Key Research & Development Program of China (No. 2020YFB1005500) and Postgraduate Research & Practice Innovation Program of NUAA (No. xcxjh20221616).

References

1. Ateniese, G., et al.: Provable data possession at untrusted stores. In: Proceedings of the 14th ACM Conference on Computer and Communications Security, pp. 598–609 (2007)
2. Back, A., et al.: Enabling blockchain innovations with pegged sidechains. **72**, 201–224 (2014). www.opensciencereview.com/papers/123/enablingblockchain-innovations-with-pegged-sidechains
3. Deng, L., Chen, H., Zeng, J., Zhang, L.-J.: Research on cross-chain technology based on sidechain and hash-locking. In: Liu, S., Tekinerdogan, B., Aoyama, M., Zhang, L.-J. (eds.) EDGE 2018. LNCS, vol. 10973, pp. 144–151. Springer, Cham (2018). https://doi.org/10.1007/978-3-319-94340-4_12
4. Du, Y., Duan, H., Zhou, A., Wang, C., Au, M.H., Wang, Q.: Enabling secure and efficient decentralized storage auditing with blockchain. IEEE Trans. Dependable Secure Comput. (2021)
5. Hope-Bailie, A., Thomas, S.: Interledger: creating a standard for payments. In: Proceedings of the 25th International Conference Companion on World Wide Web, pp. 281–282 (2016)
6. Huang, P., Fan, K., Yang, H., Zhang, K., Li, H., Yang, Y.: A collaborative auditing blockchain for trustworthy data integrity in cloud storage system. IEEE Access **8**, 94780–94794 (2020)
7. Juels, A., Kaliski Jr, B.S.: PORs: proofs of retrievability for large files. In: Proceedings of the 14th ACM Conference on Computer and Communications Security, pp. 584–597 (2007)
8. Li, J., Zhang, L., Liu, J.K., Qian, H., Dong, Z.: Privacy-preserving public auditing protocol for low-performance end devices in cloud. IEEE Trans. Inf. Forensics Secur. **11**(11), 2572–2583 (2016)
9. Mao, X., Li, X., Guo, S.: A blockchain architecture design that takes into account privacy protection and regulation. In: Xing, C., Fu, X., Zhang, Y., Zhang, G., Borjigin, C. (eds.) WISA 2021. LNCS, vol. 12999, pp. 311–319. Springer, Cham (2021). https://doi.org/10.1007/978-3-030-87571-8_27

10. Ni, J., Zhang, K., Yu, Y., Yang, T.: Identity-based provable data possession from RSA assumption for secure cloud storage. IEEE Trans. Dependable Secure Comput. (2020)

11. Wang, C., Chow, S.S., Wang, Q., Ren, K., Lou, W.: Privacy-preserving public auditing for secure cloud storage. IEEE Trans. Comput. **62**(2), 362–375 (2011)

12. Wang, H., Wang, Q., He, D.: Blockchain-based private provable data possession. IEEE Trans. Dependable Secure Comput. **18**(5), 2379–2389 (2019)

13. Xu, Z., Wu, L., He, D., Khan, M.K.: Security analysis of a publicly verifiable data possession scheme for remote storage. J. Supercomput. **73**(11), 4923–4930 (2017)

14. Yu, H., Yang, Z., Sinnott, R.O.: Decentralized big data auditing for smart city environments leveraging blockchain technology. IEEE Access **7**, 6288–6296 (2018)

15. Zhang, Y., Xu, C., Lin, X., Shen, X.: Blockchain-based public integrity verification for cloud storage against procrastinating auditors. IEEE Trans. Cloud Comput. **9**(3), 923–937 (2019)

A Novel Group Signature Scheme with Time-Bound Keys for Blockchain

Dong Wang[1,2], Bingnan Zhu[1,2(✉)], Xiaoruo Li[1,2], Aoying Zheng[2], and Yanhong Liu[1,2]

[1] Henan International Joint Laboratory of Intelligent Network Theory and Key Technology, Henan University, Kaifeng 475000, China
bingnanzhu@qq.com
[2] School of Software, Henan University, Kaifeng 475000, China

Abstract. With the development of blockchain technology, there is an increasing call for privacy protection and supervision of blockchain transactions. We propose a novel privacy protection scheme for group signature. Firstly, we design a revocation algorithm based on hierarchical time tree, that can shorten the size of validity period and support the discontinuity of member time period. Secondly, we provide an anonymous certificate to protect the privacy of group members' identity. Thanks to the combination of signature scheme and blockchain, our scheme can guarantee that the identity of any user will not be revealed during the whole process. Moreover, with the features of group signature, we can also enable the tracking and revocation of risky transactions and provide security for on-chain data. Finally, our experimental evaluation of elliptic curves based on BN-382 bit prime proves that our scheme is feasible and efficient.

Keywords: Group signature · Blockchain · Revocation · Time-bound Keys

1 Introduction

Blockchain is regarded as the fifth generation of disruptive Internet technology after mobile Internet, it will lead the transformation of information Internet to the value Internet [1]. Blockchain has the characteristics of anonymity, non-tamperability and non-falsifiability. With the continuous maturity of blockchain technology, as well as the integration and cross-application with medical, financial, digital copyright [2], Internet of Things [3] and other fields, a broad application prospect is revealed [4]. However, the transparency of blockchain ledgers also brings the problem of privacy leakage to users. The most studied issue

This work was supported by the National Natural Science Foundation of China Youth Fund (12201185), Henan Province Key R&D and Promotion Special Project (232102210192) and the Key Scientific Research Projects of Henan Provincial Colleges and Universities (23A520035).

in blockchain technology is privacy leakage [5]. Blockchain privacy protection includes user anonymity and data confidentiality. The current privacy protection scheme is mainly implemented with technologies such as Zero-knowledge proof, ring signature and group signature [6]. In the process of authentication and verification of data, digital signature technology ensures the integrity, authenticity and non-repudiation of documents. The managers enforce user registration and identity verification to ensure accountability, and allow the use of pseudonyms, access control or encryption to secure data.

In the blockchain privacy protection scheme, although current group signature revocation schemes enable to revoke group members within a specified time period, no one has paid attention to the security issues caused by time period leakage. Therefore, the current imperfections in privacy protection constitute the driving force behind our work, we propose a A novel group signature scheme with time-bound keys, and the contributions are summarized as follows:

1. We construct a dynamic group signature scheme based on blockchain, it not only protects the privacy and security of user data, realizes the supervision and timely revocation of transaction users, and reduces the probability of illegal acts, but also tracks the risky transaction data.
2. We propose a new group signature revocation algorithm. It encrypts the time period and reduces the cost overhead of the time period. In particular, the algorithm supports the discontinuity of time periods.
3. We construct an anonymous certificate, and achieve identity pseudonym and encryption of personal information of group members.

2   Related Work

In current research, there are already multiple solutions for addressing blockchain regulation issues. Zheng et al. [7] proposed a linkable group signature based on blockchain, but the drawback is that if the user wants to generate a new certified public key, it will be necessary to run the registration algorithm again. Thokchom et al. [4] proposed a dynamic Cloud data shared ring group scheme, it is used to check the integrity of shared data in cloud storage, but this can be computationally costly. Mourad et al. [7] introduced a new supply chain traceability system and used ring signatures to achieve confidentiality of privacy sensitive information.

Group signature is a special kind of signature. Compared to other signature schemes, group signatures have administrator and traceability, making it possible to manage group members while guaranteeing user privacy. In 2012, chu et al. [8] innovatively introduced a group signature with time-bound keys (GS-TBK), the group members can only sign within the validity period, and it provides natural revocation (natural revocation means that a member will be automatically revoked since the valid time has passed) to reduce the total cost of revocation checking. After this, Ohara et al. [9] proposed a new signature scheme with time-limited keys. They defined a complete subtree algorithm for the time-bound keys (CS-TBK) approach and assigned a leaf node to each group member.

The signature have two zero-knowledge proofs, ensuring verification of revoked members. Moreover, the cost of the signature algorithm is constant. Emura et al. [10] also introduced the time duration method on the complete subtree and proposed the unforgeability of the signature expiration time. Manager sets a time period to associate the expiration time with each group member. The advantage of these schemes [9,10] is that the advantage of reducing signature costs, while other signature efficiencies depend on the membership time period and the size of the message. Bootle et al. [11] proposed a generalized dynamic group signature scheme with a formal security model, and it is used to discuss about the security of different group signature design paradigms. Back et al. [12] proposed a fully dynamic group signature scheme with member privacy based on signatures on equivalence classes. Li et al. [13] proposed an attribute-based hierarchical encryption scheme for data access control and introduced a regulator to regulate users' historical query records.

To the best of our knowledge, there is no scheme that combines hierarchical time trees with group signatures. In this paper, we propose a revocable group signature scheme based on hierarchical time tree and apply it in blockchain. It achieves privacy protection and audit of transaction data, and can trace the risky transaction data to ensure the safety and security of data on the chain.

3   Description of the Proposed Scheme

3.1   System Model

In this section, we propose a blockchain-based revocable time-limited group signature scheme, that aims to be efficient and scalable, and it utilizes group signature technology to design a completely anonymous user data storage protocol under blockchain architecture. In order to ensure the validity and security of user information, we introduce the time limit technique, and the group member exits the system in time after completing the signature, not only it can effectively reduce the probability of collusion attack, but also achieve the timeliness of data transactions. The system model description is shown in Fig. 1.

3.2   Time Layered Tree Algorithm

We use the Hierarchical Identity-Based En-cryption (HIBE) scheme [14] to describe the layering of time periods. Group member validity period is usually represented as some time interval, so we use hierarchical tree to define the time period of group members and further improve the signature efficiency. Figure 2 represents a hierarchical time tree (HTT) of depth 4, except for the root node, the first layer in the hierarchical tree is described as the year, the second layer is used to record the month, The third layer is described as a day, and so on. Our scheme supports more layers of time including smaller units such as minutes and seconds, we only consider the date description at the day level.

Suppose we set up an HTT of depth d and each node has at most n children, which n defined as the binary of time. The time period is defined

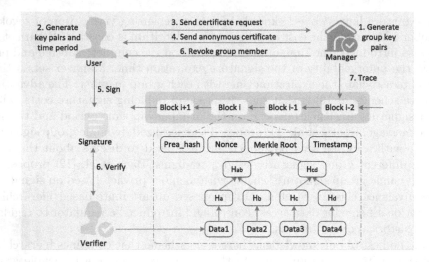

Fig. 1. System model

$\tau = \{t_1, t_2, ..., t_k\}$, where $k < d$. Each node of HTT represent a time period, if the member sets a time period that contains more than one time node and uses the minimum set-cover method to get all the number of valid time nodes, and it get the key value corresponding to the node with HTT. In our scheme, the member's validity period usually indicates a time period, including a year, a month, a day, etc. The signature proves valid only when it is covered by the whole time period. In the revocation phase, the manager generates the keys from the current day month or year using HTT and verifies whether the member is valid. If the result is invalid, the member is added to the revocation list.

For example, if the member's signature date is January 1, 2023, the validity period must be included this date, and completely covers the signature date. If the member's time period (the time period is recorded as TP) is from January 2023 to March 3, 2023, and get the following time points from HTT: "2023-Jan", "2023-Feb", "2023-Mar-1", "2023-Mar-2" and "2023-Mar-3", and we define $\tau = \{(2023, Jan), (2023, Feb), (2023, Mar, 1), (2023, Mar, 2), (2023, Mar, 3)\}$. If the current date (the current date is recorded as CD) is February 4, 2023 in revocation phase, that is $CD \in TP$, manager verified member valid, Instead, if the current date is March 4, 2023, that is $CD \notin TP$, the manager verifies invalid, and the member cannot make the signature on the message, finally member is added the revocation list.

4   Our Group Signature Scheme with Time-Bound Keys

4.1   Cryptographic Assumptions

Definition 1 (The Decisional Diffie–Hellman(DDH) assumption). *Let* $\mathbb{G}$ *be a cyclic group of prime order* p, *where* $g \leftarrow \mathbb{G}$ *such that* $\langle g \rangle = \mathbb{G}$. *Select*

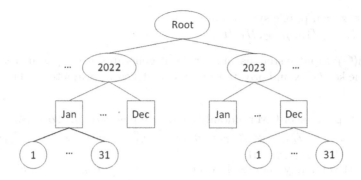

Fig. 2. Hierarchical time tree

$a, b \leftarrow \mathbb{Z}_p$, and compulte $g^a, g^b, Z \leftarrow \mathbb{G}$. We say that the DDH assumption holds if for any probabilistic polynomial time (PPT) adversary A verifies Z equals $g^{ab} \in \{\mathbb{G}, g, g^a, g^b, Z\}$. The DDH assumption is (1), where ε is denoted as a negligible function.

$$Adv_A^{DDH} =| \Pr \left[A\left(g^{ab}\right) = 1 \right] - \Pr[A(Z) = 1] |\leq \varepsilon \qquad (1)$$

Definition 2 (The Symmetric External Diffie–Hellman (SXDH) assumption). *The SXDH assumption holds in $\mathbb{G}_1$ and $\mathbb{G}_2$ if DDH is hard to compute in both $\mathbb{G}_1$ and $\mathbb{G}_2$.*

Definition 3 (The Decisional Bilinear Diffie–Hellman (DBDH) assumption). *Let $\mathbb{G}_1$ and $\mathbb{G}_2$ be cyclic groups of prime order p, where $g_1 \leftarrow \mathbb{G}_1$ and $g_2 \leftarrow \mathbb{G}_2$ such that $\langle g_1 \rangle = \mathbb{G}_1$ and $\langle g_2 \rangle = \mathbb{G}_2$. Given $g_1^a, g_2^b, g_2^c, Z \leftarrow \mathbb{G}_T$, where $a, b, c, \leftarrow \mathbb{Z}_p$. The DBDH assumption holds if for any PPT adversary A verifies Z equals $e\left(g_1, g_2\right)^{abc} \in \{\mathbb{G}_1, \mathbb{G}_2, \mathbb{G}_T, g_1, g_2^a, g_1^b, g_1^c, Z\}$. The DBDH assumption is (2), where ε is denoted as a negligible function.*

$$Adv_A^{DBDH} =| \Pr \left[e\left(g_1, g_2\right)^{abc} = 1 \right] - \Pr[A(Z) = 1] |\leq \varepsilon \qquad (2)$$

4.2 Scheme Construction

In this section, we describe the specifics of the revocable group signature scheme based on hierarchical time trees.

Setup(λ): Input a security parameter λ that satisfies the defined requirements, it outputs system parameters and a time tree.

1. Choose a bilinear group $\mathbb{G} = (\mathbb{G}_1, \mathbb{G}_2, \mathbb{G}_T, g_1, g_2)$, where $\mathbb{G}_1$, $\mathbb{G}_2$ and $\mathbb{G}_T$ are multiplicative cyclic groups, and a pairing function: $\mathbb{G}_1 \times \mathbb{G}_2 \rightarrow \mathbb{G}_T$, $\langle g_1 \rangle = \mathbb{G}_1$ and $\langle g_2 \rangle = \mathbb{G}_2$.
2. Choose two hash functions: $H_1 : \{0, 1\}^* \rightarrow \mathbb{Z}_p$ and $H_2 : \{0, 1\}^* \rightarrow \mathbb{Z}_p$.
3. Generate a time tree of depth d, time is represented as a nary string $\{1, n\}^{d-1}$, and randomly select $V_0, V_1, V_2, \ldots, V_T \leftarrow \mathbb{G}_1$ on behalf of each nodes.

4. Output global public system parameters:
$\mathbb{G} = (\mathbb{G}_1, \mathbb{G}_2, \mathbb{G}_T, g_1, g_2, H_1, H_2)$

KGen($\mathbb{G}$): Input system parameters $\mathbb{G}$, it generates group parameters of manager public key GPK and private key GSK, certificate parameter and registration Reg.

1. The group manager GM randomly chooses integers $l_1, l_2, v, \partial \leftarrow \mathbb{Z}_p$, computes $u_1 = g_2^{l_1}$, $u_2 = g_2^{l_2}$, $q_1 = g_1^v, q_2 = g_1^{\partial}$, $q_3 = g_1^{v+\partial}$, set $GSK = (l_1, l_2, \nu, \partial)$ and $GPK = (u_1, u_2, q_1, q_2, q_3)$.
2. The GM randomly chooses integers $y_1, y_2 \leftarrow \mathbb{Z}_p$, computes $NK = g_1^{y_1}$ and $Kern = g_1^{y_2}$, and sets a registration Reg.
3. Outputs the tuple $(GSK, GPK, NK, Kern)$.

Join($\mathbb{G}, GPK, NK, Kern$): The user requests to join the group and generates the public key UPK and private key USK by running join algorithm, and sets the survival period by HTT, and finally sends Qcert to GM.

1. The user randomly chooses $n_1, n_2 \leftarrow \mathbb{Z}_p$, computes $upk_1 = g_1^{n_1}$, $upk_2 = g_1^{n_2}$, set $usk_1 = n_1$, $usk_2 = n_2$, $USK = (usk_1, usk_2)$ and $UPK = (upk_1, upk_2)$.
2. The user chooses a time T, implies that the user is valid for a period in group. The $\mathbb{T}$ is a set-cover used to describe T, and it consists of several time elements $\tau = (t_1 t_2 t_3 \ldots t_{k_\tau}) \leftarrow \{1\,n\}^{k_\tau}$, where any $\tau \leftarrow \mathbb{T}$ satisfies $k_\tau < d$, it randomly selects an integer $\rho_i \leftarrow \mathbb{Z}_p$, and computes:

$$D_0 = g_2^{\rho_i} \tag{3}$$

$$D_{i,\tau} = g_1^{n_1}(V_0 \prod_{j=1}^{t_{k_\tau}} V_j^{t_{k_j}})^{\rho_i}, \tau \in \mathbb{T} \tag{4}$$

$$D_{j,\tau} = V_j^{\rho_i}, j \in \{k_{\tau+1}, \ldots, d\}, \tau \in \mathbb{T}. \tag{5}$$

3. The user randomly choose integers $\mu_1, \mu_2 \leftarrow \mathbb{Z}_p$, computes identity of the certificate $Una = Nk^{\mu_1} \cdot kern^{\mu_2}$, and sets user identity properties $feas = \{fea_1, fea_2, fea_3, \ldots fea_n\}$, it computes challenge value $c = H_1(Una \parallel UPK \parallel NK \parallel Kern)$, computes $w = Nk^{n_1} \cdot Kern^{n_2}$, $R_1 = \mu_1 - c \cdot n_1, R_2 = \mu_2 - c \cdot n_2$
4. Finally, the user sends a certificate request $Qcert = (Una, UPK, feas, w, R_1, R_2)$ to the GM.

Issue($\mathbb{G}, GPK, NK, Kern, Qcert$): In order to assist users in joining the group, we need communication between the manager and the user to verify the certificates, then GM verifies whether Qcert is hold, generates user's certificate ϖ based on user's Una and feas, and sends it to user. ϖ is taken as personal certificate after verifying successfully by user, so far the user becomes the member of the group.

1. After the GM receives a certificate request $Qcert = (Una, UPK, feas, w, R_1, R_2)$ from the user, starts to verify its correctness, computes $\bar{w} = NK^{R_1} \cdot Kern^{R_2}$ and $\tilde{w} = \bar{w} \cdot w^c$, then it generates $\tilde{c} = H_1(\tilde{w} \| UPK \| NK \| Kern)$, verifies whether the user's challenge values c and $\tilde{c}$ are equal. If the verification result holds, the user's certificate request is valid, and it adds the user's Una, feas, T and to the Reg, otherwise, the user's certificate request is invalid and GM refuses to grant the certificate for the user.

2. The GM generates user's credit $Creds = H_1(UPK, fea_1, fea_2, \dots fea_n)$, randomly chooses an integer $f \leftarrow \mathbb{Z}_p$, computes $P_1 = \left(upk_1^{l_1} \cdot upk_2^{l_2}\right)$, $P_2 = (upk_1^{l_1} \cdot upk_2^{l_2})^f$ and $B_1 = g_2^f$. Then it generates user's certificate $\varpi = (P_1, P_2, Una, Creds, B_1)$ to send user.

3. After the user receives the certificate ϖ from GM. It computes $\delta_1 = e(upk_1, u_1) \cdot e(upk_2, u_2)$ and $\delta_2 = e(P_1, B_1)$, verifies whether $\delta_1 = e(P_1, g_2)$ and $\delta_2 = e(P_2, g_2)$, if hold, the ϖ is taken as the user's certificate of identity, otherwise return $\perp$.

Revoke $(\mathbb{G}, T, \varpi_i, D_0, D_{i,\tau}, D_{j,\tau})$: The GM periodically checks members time period and updates the revocation list. GM gets the encrypted time based on the HTT by the current time. Then it verifies if the member is expired and sets the revocation mark to each expired member to help verification. In the end GM adds expired members to the revocation list RL.

1. The GM express the current time to $\tau_c = (t_1, t_2, \dots, t_k) \in \{1, n\}^k$.

2. The GM computes $D'_{i,\tau_c} = \left(V_0 \prod_{j=1}^{t_k} V_j^{t_j}\right)^v$ and verifies $e(D_{i,\tau_c}, g_2^v) = e(D'_{i,\tau_c}, D_0) \cdot e(upk_1, g_2^v)$. If the result doesn't hold, GM searches for the prefix $\bar{\tau}_c$ of τ_c from its set-cover, if $\bar{\tau}_c$ exists, it gets $\bar{\tau}_c = (t_{1'}, t_{2'}, \dots t_{k'})$ and $D_{i,\bar{\tau}_c}$, where $k' < k$, and computes $D'_{i,\bar{\tau}_c} = D_{i,\bar{\tau}_c} \prod_{j=k'+1}^{k} D_{j,\bar{\tau}_c}^{t_j}$.

3. The GM verifies whether $e\left(D'_{i,\bar{\tau}_c}, g_2^v\right) = e\left(D'_{i,\tau_c}, D_0\right) \cdot e(upk_1, g_2^v)$. If the result holds, the member is valid, otherwise it computes mark $r_g = q_2^{n_1}$ and add member's r_g to RL.

Sign $(\mathbb{G}, \mathbf{M}, \boldsymbol{USK}, \boldsymbol{GPK}, \varpi_i)$: Input message M, the member runs the signature algorithm to sign the message. Finally, the signature σ is outputted.

1. Send message $M \in \{0,1\}^*$ to member.
2. The member randomly selects $\zeta_1, \alpha \leftarrow \mathbb{Z}_p$, and computes the tuple $(\psi_1, \psi_2, \psi_3, \psi_4, \psi_5)$:

$$\psi_1 = upk_1 \cdot q_3^\alpha \tag{6}$$

$$\psi_2 = g_1^\alpha \tag{7}$$

$$\psi_3 = e(g_2, q_2)^{n_1 \zeta_1} \tag{8}$$

$$\psi_4 = g_2^{\zeta_1} \tag{9}$$

$$\psi_5 = g_1^{n_2 \zeta_1} \tag{10}$$

3. The member starts to compute the signature of knowledge (SPK) [15]: $V = SPK : \{(n_1, \alpha, \zeta_1) : \psi_1 = spk_1 \cdot q_3^\alpha \wedge \psi_2 = g_1^\alpha \wedge \psi_3 = e(g_1, u_1)^{n_1 \alpha} \wedge \psi_4 = g_2^\alpha \wedge \psi_6 = g_1^{n_2} \zeta_1\}(M)$. The details about the SPK calculation are as follows. It sets $\zeta_2 = n_1 \cdot \zeta_1$ and $\zeta_3 = n_2 \cdot \zeta_1$, and randomly selects $d_1, d_2, d_3, d_4, d_5 \leftarrow \mathbb{Z}_p$, computes:

$$\theta_1 = e(g_1, g_2)^{d_1} \cdot e(q_3, g_2)^{d_2} \tag{11}$$

$$\theta_2 = g_1^{d_2} \tag{12}$$

$$\theta_3 = e(g_2, q_2)^{d_3} \tag{13}$$

$$\theta_4 = g_2^{d_4} \tag{14}$$

$$\theta_5 = g_1^{d_5} \tag{15}$$

4. The member generates challenge values (16), and computes the following values by φ:

$$\varphi = H_2(\psi_1, \psi_2, \psi_3, \psi_4, \psi_5, \theta_1, \theta_2, \theta_3, \theta_4, \theta_5, M) \tag{16}$$

$$S_1 = d_1 - \varphi n_1 \tag{17}$$

$$S_2 = d_2 - \varphi \alpha \tag{18}$$

$$S_3 = d_3 - \varphi \zeta_2 \tag{19}$$

$$S_4 = d_4 - \varphi \zeta_1 \tag{20}$$

$$S_5 = d_5 - \varphi \zeta_3 \tag{21}$$

5. Output the group signature $\sigma = (\psi_1, \psi_2, \psi_3, \psi_4, \psi_5, S_1, S_2, S_3, S_4, S_5, \varphi)$.

$Verify(\sigma, M, GPK, UPK, RL)$: Inputs the signature σ and the message M, the verifier runs the verification algorithm to verify σ is valid.

1. Send σ and M to the verifier.
2. The verifier chooses $\{r_{i,g}\} \in RL$, computes $\psi_3' = e(r_g, \psi_4)$. Then it verifies whether the values ψ_3' and ψ_3 are equal, if the verification result holds, the verifier continues. Otherwise, it outputs $\bot$.
3. The verifier computes as following:

$$\theta_1' = e(g_1, g_2)^{S_1} \cdot e(q_3, g_2)^{S_2} \cdot e(\psi_1, g_2)^\varphi \tag{22}$$

$$\theta_2' = g_1^{S_2} \cdot \psi_2^\varphi \tag{23}$$

$$\theta_3' = e(g_2, q_2)^{S_3} \cdot \psi_3^\varphi \tag{24}$$

$$\theta_4' = g_2^{S_4} \cdot \psi_4^\varphi \tag{25}$$

$$\theta_5' = g_1^{S_5} \cdot \psi_5^\varphi \tag{26}$$

From the above results, we calculate the challenge value (27). Finally, it checks whether the challenge φ' and φ are equal, if the result holds, the signature is valid. Otherwise, the algorithm outputs $\bot$.

$$\varphi' = H_2(\psi_1, \psi_2, \psi_3, \psi_4, \psi_5, \theta_1', \theta_2', \theta_3', \theta_4', \theta_5', M) \tag{27}$$

Trace(σ, GSK, Reg): After the user sends the request, the GM traces the signed member and gets the real information of the signed member.

1. The user launches trace request to MG.
2. The GM verifies member's identity. If the identity is correct then it gets tracking parameters (ψ_1, ψ_2) from σ and computes $K = \psi_1/\psi_2^{v+\partial}$. If K is found in the registration Reg, MG outputs member's information {Una, feas, T}, otherwise return $\perp$.

5   Security Analysis

5.1   Correctness

In this section, the correctness of the proposed ring signature algorithm is verified.

Step 1: In the Revoke phase, if the proof time period is valid, it must satisfy $e\left(D_{i,\tau_c}, g_2^v\right) = e\left(D'_{i,\tau_c}, D_0\right) \cdot e\left(upk_1, g_2^v\right)$, proved in part as follows:

$$
\begin{aligned}
e\left(D_{i,\tau_c}, g_2^v\right) &= e\left(D'_{i,\tau_c}, D_0\right) \cdot e\left(upk_1, g_2^v\right) \\
&= e\left(\left(V_0 \textstyle\prod_{j=1}^{t_k} V_j^{t_j}\right)^{\rho_i}, g_2^v\right) \cdot e\left(g_1^{n_1}, g_2^v\right) \\
&= e\left(g_1^{n_1}\left(V_0 \textstyle\prod_{j=1}^{t_k} V_j^{t_j}\right)^{\rho_i}, g_2^v\right) \\
&= e\left(D_{i,\tau_c}, g_2^v\right)
\end{aligned}
\tag{28}
$$

Step 2: In the Vertify phase, if the σ is valid, it must satisfy $\varphi = \varphi'$, proved in part as follows:

$$
\begin{aligned}
\theta_1' &= e(g_1, g_2)^{S_1} \cdot e(q_3, g_2)^{S_2} \cdot e(\psi_1, g_2)^{\varphi} \\
&= e(g_1, g_2)^{d_1 - \varphi n_1} \cdot e(q_3, g_2)^{d_2 - \varphi\alpha} \cdot e(upk_1 \cdot q_3^{\alpha}, g_2)^{\varphi} \\
&= \frac{e(g_1,g_2)^{d_1} \cdot e(q_3,g_2)^{d_2} \cdot e\left(g_1^{n_1}, g_2\right)^{\varphi} \cdot e(q_3^{\alpha}, g_2)^{\varphi}}{e(g_1,g_2)^{\varphi n_1} \cdot e(q_3,g_2)^{\varphi\alpha}} \\
&= e(g_1, g_2)^{d_1} \cdot e(q_3, g_2)^{d_2} = \theta_1
\end{aligned}
\tag{29}
$$

5.2   Anonymity

Anonymity in group signatures generally refers to the inability to find the real identity of the signer without the private key of the group members, i.e., the inability to identify the group signature. In other words, the attack on anonymity is also an attack on the signature ciphertext. The anonymity of our group signature scheme can be reduced to the reliability of zero-knowledge proofs and the CCA security of linear encryption schemes. For our SPK signature $\sigma = (\psi_1, \psi_2, \psi_3, \psi_4, \psi_5, S_1, S_2, S_3, S_4, S_5, \varphi)$, it is composed of multiple zero-knowledge proofs and the signer's key is contained in the $\psi_1, \psi_2, ..., \psi_5$. Yet again, $\psi_1, \psi_2, ..., \psi_5$ are all random values, thus ensuring the anonymity of the signer's identity. Under the assumption of DDH and DBDH puzzles, only the adversary can solve the puzzles with non-negligible probability, then it can also successfully attack our scheme, but this is contrary to the fact that the adversary cannot obtain any valid information from the signatures, so our scheme satisfies anonymity.

5.3   Non-frameability

Non-frameability means that no adversary can generate a forged signature of an honest member that can be traced. In our scheme, the group signature consists of the signature keys and knowledge proofs, and the unforgeability of the group signature ensures that all valid signatures can be traced. If adversary generates a forged signature without the keys, it can break the untraceability of the proposed group signature. However, the advantage of the adversary winning is negligible. Therefore, our scheme satisfies unframeability.

5.4   Traceability

The traceability attack is the ability of an adversary to forge an signature of the group member. The following two types of forgery signature in our scheme: Forging the signature of a current group member and forging the signature of a revoked group member. For the first type of forgery, it can be attributed to the reliability of the EUF-CMA security and knowledge proof. The second type of forgery is essentially the forgery of a signature after the time period of a group member has expired. Since the forged signature does not exist in the RL, the forgery attack is attributed to EUF-CMA security. It is the EUF-CMA security and the reliability of the SPK proof that ensures that our signature scheme satisfy traceability.

6   Performance

In this section, the theoretical overhead cost of our scheme is evaluated and the performance is compared with the scheme [9, 10, 16, 17]. Finally, our scheme is proved to be efficient based on the experimental results.

We use the BN curve over a 382-bit prime field (BN382) to ensure the 128-bit security of our solution. Then we implemented the implementation of our solution on the desktop by JAVA (JDK8) and implementation environment is as follows: CPU: Intel i5-12500, OS: Windows 10, Compilation environment: IntelliJ IDEA 2021.1.3 and Cryptolibrary: the JPBC library.

Table 1 summarizes the benchmarks of the pairwise operations on the elliptic curve of BN382 in our simulation experiments. The $\text{Mul}(\mathbb{G}_1)$, $\text{Mul}(\mathbb{G}_2)$, $\text{Exp}(\mathbb{G}_T)$ are scalar multiplication on G_1, G_2, and exponentiation on G_T, respectively. P is the time to perform one pairing operation. Figure 3 compares our Sign and Verify costs with other schemes [10, 16, 17], and our scheme has the least time in the signature phase compared to others, and also has excellent performance in the verification phase. Figure 4 summarizes the evaluation of the signature size of our scheme compared to the other schemes [9,10,16], since our simulation experiments are based on the BN382 elliptic curve, the sizes of a scalar value in $\mathbb{Z}_p$, an element in $\mathbb{G}_1$, an element in $\mathbb{G}_2$, and an element in $\mathbb{G}_T$ are 48 bytes, 49 bytes, 97 bytes, and 384 bytes, respectively. So our signature size is 722 Bytes in this case, it reduced the signature size by 20.3%, 7.5% and 7.2%, respectively, compared to other schemes [9,10,16], so it has better performance.

Table 1. Benchmarks of group operations

Operations	Time (us)
Mul $(\mathbb{G}_1)$	297.42
Mul $(\mathbb{G}_2)$	339.37
Mul $(\mathbb{G}_T)$	853.72
P	1689.56

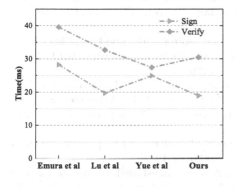

Fig. 3. Computation Cost

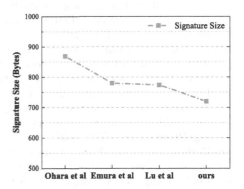

Fig. 4. Signature size

7 Conclusion

In view of the difficulty of handling risky transactions in blockchain, we propose a revocable group signature scheme based on hierarchical time tree. First, we design an anonymous certificate to protect the privacy of group member identities, Second, we design a hierarchical time tree to achieve timely revocation of group members while ensuring that the validity period is encrypted. Finally, the experimental evaluation proves the feasibility and security. Since group signatures possess the characteristics of administrators, although it can facilitate us to achieve the supervision of transactions, it can also lead to over-centralization of our scheme, so this will be our next research and improvement direction.

References

1. Nofer, M., et al.: Blockchain. Bus. Inf. Syst. Eng. **59**, 183–187 (2017). https://doi. org/10.1007/s12599-017-0467-3
2. Treleaven, P., Brown, R.G., Yang, D.: Blockchain technology in finance. Computer **50**(9), 14–17 (2017)
3. Fernández-Caramés, T.M., Fraga-Lamas, P.: A review on the use of blockchain for the Internet of Things. IEEE Access **6**, 32979–33001 (2018)
4. Tang, F., Feng, Z., et al.: Privacy-preserving scheme in the blockchain based on group signature with multiple managers. Secur. Commun. Netw. **2021**, 1–8 (2021)
5. Shao, W., et al.: AttriChain: decentralized traceable anonymous identities in privacy-preserving permissioned blockchain. Comput. Secur. **99**, 102069 (2020)
6. Li, X., Mei, Y., et al.: A blockchain privacy protection scheme based on ring signature. IEEE Access **8**, 76765–76772 (2020)
7. Maouchi, M.E., et al.: DECOUPLES: a decentralized, unlinkable and privacy-preserving traceability system for the supply chain, pp. 364–373. ACM (2019)
8. Chu, C.-K., Liu, J.K., Huang, X., Zhou, J.: Verifier-local revocation group signatures with time-bound keys, p. 26. ACM Press (2012)
9. Ohara, K., Emura, K., Hanaoka, G., Ishida, A., Ohta, K., Sakai, Y.: Shortening the Libert-Peters-Yung revocable group signature scheme by using the random oracle methodology. IEICE Trans. Fundam. Electron. Commun. Comput. Sci. **E102.A**(9), 1101–1117 (2019)
10. Emura, K., Hayashi, T., Ishida, A.: Group signatures with time-bound keys revisited: a new model, an efficient construction, and its implementation. IEEE Trans. Dependable Secure Comput. **17**(2), 292–305 (2017)
11. Bootle, J., Cerulli, A., Chaidos, P., Ghadafi, E., Groth, J.: Foundations of fully dynamic group signatures. In: Manulis, M., Sadeghi, A.-R., Schneider, S. (eds.) ACNS 2016. LNCS, vol. 9696, pp. 117–136. Springer, Cham (2016). https://doi. org/10.1007/978-3-319-39555-5_7
12. Backes, M., Hanzlik, L., Schneider-Bensch, J.: Membership privacy for fully dynamic group signatures. pp. 2181–2198. ACM (2019)
13. Li, X., Dong, X., Xu, X., He, G., Xu, S.: A blockchain-based scheme for efficient medical data sharing with attribute-based hierarchical encryption. In: Zhao, X., Yang, S., Wang, X., Li, J. (eds.) WISA 2022. LNCS, vol. 13579, pp. 661–673. Springer, Cham (2022). https://doi.org/10.1007/978-3-031-20309-1_58
14. Liu, J.K., Yuen, T.H., Zhang, P., Liang, K.: Time-based direct revocable ciphertext-policy attribute-based encryption with short revocation list. In: Preneel, B., Vercauteren, F. (eds.) ACNS 2018. LNCS, vol. 10892, pp. 516–534. Springer, Cham (2018). https://doi.org/10.1007/978-3-319-93387-0_27
15. Nakanishi, T., Funabiki, N.: Verifier-local revocation group signature schemes with backward unlinkability from bilinear maps. In: Roy, B. (ed.) ASIACRYPT 2005. LNCS, vol. 3788, pp. 533–548. Springer, Heidelberg (2005). https://doi.org/10. 1007/11593447_29
16. Lu, J., Shen, J., Vijayakumar, P., Gupta, B.B.: Blockchain-based secure data storage protocol for sensors in the industrial Internet of Things. IEEE Trans. Ind. Inform. **18**(8), 5422–5431 (2021)
17. Yue, X., Xi, M., et al.: A revocable group signatures scheme to provide privacy-preserving authentications. Mob. Netw. Appl. **26**(4), 1412–1429 (2021). https:// doi.org/10.1007/s11036-019-01459-5

An Efficient Storage Optimization Scheme for Blockchain Based on Hash Slot

Jiquan Wang, Tiezheng Nie$^{(\boxtimes)}$, Derong Shen, and Yue Kou

Northeastern University, Shenyang 110169, China
`nietiezheng@cse.neu.edu.cn`

Abstract. Blockchain has developed tremendously in recent years with its own features such as anonymity and immutability. The advent of the big data era has caused the size of data on blockchain systems to increase exponentially, which leads to a great challenge to the storage capacity of nodes. Therefore, this paper proposes a storage optimization scheme based on hash slot algorithm to reduce the storage volume of nodes. We propose a model to evaluate the liveness of each block and adopt different storage strategies for blocks of different liveness to reduce the storage overhead of each node and the additional communication cost when querying ledger data. With this scheme, the ledger data can be distributed more reasonably on each node according to the storage capacity of each node by allocating the hash slots using the hash slot allocation algorithm we proposed. And experiment results show that our scheme can reduce the storage volume overhead of nodes by 78% and cause negligible additional time when accessing the ledger data.

Keywords: Blockchain · Storage optimization · Hash slot

1 Introduction

The emergence of Bitcoin [1] brought blockchain into the limelight. Blockchain mainly adopts cryptographic algorithm, hash algorithm and consensus algorithm to ensure the secure transfer of value on the chain, traceability and tamper-evident of data, as well as the consistency of the blockchain data stored by each node. These techniques make the blockchain decentralized, anonymous, and secure. The introduction of smart contract technology into blockchain for the first time by Ethernet [2] has led to the rapid development of blockchain technology, which has now been applied to various fields such as law [3], copyright protection [4, 5], privacy protection [6], assets [7], education [8], and medical care [9].

However, current blockchain systems adopt a highly redundant strategy for storing ledger data, which requires each node in the system to store a complete copy of ledger data to ensure the security and reliability of the data. For most blockchain systems, the storage system of ledger data mainly consists of two parts: file system and storage system based on key-value database (e.g., LevelDB). The file system is used to store the actual block data, while the key-value database is used to store the index information of each block and transaction in the file system for easy access. However, with the advent of the big data era, blockchain technology is also being applied to a variety of scenarios, making the size of its ledger data expand rapidly. Taking the current mainstream public blockchain systems as examples, until now, the data size of Bitcoin full node has exceeded 470 GB, and the data size of Ethereum full node is close to 1 TB. Such a huge storage overhead not only makes more and more full nodes unaffordable and turn to run as a light node, but also makes the initial synchronization time required for new nodes very long, which makes the running of the blockchain system depends on a small number of full nodes and the security of the system cannot be guaranteed.

Many of the current research has directly used the same copy number optimization method for each block to reduce storage overhead, but each block is accessed at different frequencies in the system, and using the same number copy allocation scheme for blocks accessed at high frequencies as for blocks accessed at low frequencies will result in significant cross-node communication overhead. In addition, some schemes do not consider the division of whole ledger data according to the actual storage capacity of each node, which makes the distribution of ledger data not balanced and reasonable. Therefore, this paper proposes a storage optimization scheme for blockchain based on hash slot algorithm, which can effectively reduce the node storage overhead and allocate the data stored in each node more reasonably, and also improve the access efficiency to different types of blocks. The main contributions of this paper are as follows:

1. We propose a storage optimization scheme based on hash slot algorithm for blockchain, which can significantly reduce the storage overhead of each node. We also propose an algorithm to allot the hash slots according to the storage capacity of each node, which can make the distribution of ledger data more reasonable.
2. We propose a model to evaluate the liveness of each block and adopt different storage modes to store each type of block according to its liveness, which minimizes the additional time when accessing the ledger data.
3. We evaluate the proposed scheme and the experimental results show that the proposed scheme is effective in reducing the storage overhead of the nodes and causes a negligible increase on the ledger data query time.

The remainder of this paper is organized as follows. Section 2 will introduce the related works, and the relevant model and algorithm will be given in Sect. 3. The strategy of block management will be introduced in Sect. 4, while Sect. 5 will describe the experiments and evaluations, and finally the conclusion will be given in Sect. 6.

2 Related Work

Current work on blockchain storage optimization can be mainly divided into schemes based on data sharding, block encoding, and data pruning.

Elastico [10] is the first to propose the idea of slicing on a public blockchain. It divides nodes into multiple slices, each of which can process different transactions in parallel to improve the throughput of the system, yet each node still needs to store the whole ledger data. Zamani et al. [11] proposed RapidChain to solve the above problem. Each slice in RapidChain stores only part of the blockchain data and ensures the security of cross-slice transactions by using periodic inter-slice verification.

BFT-Store [12] is a storage optimization scheme implemented by erasure code in consortium blockchain. The scheme uses the RS algorithm to encode blocks and Byzantine fault tolerance is achieved by combining the BFT consensus algorithm. Dai et al. [13] proposed a network coding-based storage optimization scheme NC-DS, which blocks are encoded into n packets for propagation in the network, and each node only needs to store about 1/n of the data volume of a block.

SSChain [14] reduces the storage overhead of nodes by using checkpoints to summarize the UTXOs of the current system and pruning the blocks before the checkpoints. SCC [15] reduces the storage overhead of light nodes by introducing compressed blocks, which contain the hashes of the latest blocks in the system, so that light nodes only need to store the latest blocks and compressed blocks of the system instead of the whole ledger data.

The hash slot algorithm used in Redis Cluster is a distributed data storage algorithm. Each data in the system is mapped into a slot and each slot is stored on a specific node. Through this double-layer mapping relationship, we can easily store and query a piece of data in the system, so we choose this algorithm for block storage management.

3 Model Definition and Algorithm

3.1 The Model of Block Liveness

Depending on various usage scenarios, blocks in a blockchain may contain payment records between users, instructions about the creation and invocation of smart contracts, and private data between organizations, respectively. Each block has a different access frequency depending on the content it contains and how long it has been mined out. For example, under the blockchain with UTXO model, if the number of UTXO in a block is higher, then the probability that it will be accessed at a future stage is also higher. And the newer a block is (i.e., the shorter the time when the block was mined from the current time), the higher the probability that it will be accessed at a future stage, because the information contained in the block is also time sensitive.

We denote the block at height i as B_i and give the definition of the liveness of block B_i as:

$$L(B_i) = W_1 * U_i + W_2 * \left(1 - e^{-AN_i}\right) + W_3 * e^{-\left(\frac{T_{now}-T_i}{BLRIT} - 1\right)} \quad (1)$$

where $L(B_i)$ is the liveness of B_i; U_i is the ratio of the number of unspent transaction outputs to the number of all transaction outputs in B_i; $BLRIT$ represents the block liveness refresh interval time. It indicates how often the liveness of a block is updated and it means that after each block is mined, the liveness of this block is recalculated every $BLRIT$ time; AN_i represents the number B_i has been accessed in the last $BLRIT$ time; T_{now} represents the current time; T_i represents the moment when the B_i was mined out; W_1, W_2 and W_3 are the weights of each parameter.

The formula given above shows that the more UTXO a block contains, the more times it has been accessed in the last $BLRIT$ time and the newer it is, the higher its liveness is, and vice versa.

We set a threshold T_h for the block liveness to distinguish between high-liveness block (HLB) and low-liveness block (LLB). When the liveness of a block is higher than T_h, the block is marked as a HLB and vice versa as a LLB. The size of T_h can be chosen flexibly according to the actual situation of the blockchain system.

3.2 Hash Slot Allocation Algorithm

The hash slot algorithm is designed for distributed data storage and querying. In the hash slot algorithm, there are a lot of slots in the system, and the data to be stored will be placed into a specified slot according to certain rules, and then each slot is bound to a node in the system through certain rules, and the node is responsible for the storage of the data in the slot. Each node in the system stores a routing table, which records the nodes corresponding to each slot. When storing or querying required data, the number of the slot where the target data is located is first calculated according to the rules, and then the target node is got by querying the routing table, and finally the data will be stored or queried on the target node. Figure 1 shows an example of hash slot algorithm.

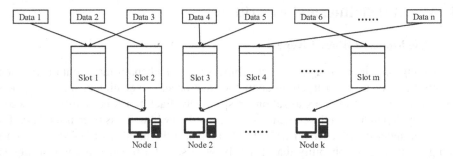

Fig. 1. The example of hash slot algorithm

In Redis, a record of data will be stored on a single node determined by *CRC16* algorithm and routing table and each node in the system is responsible for storing the same number of sequential slots. So, there will be some problems if this scheme is used directly in blockchain scenarios for block data storage. Firstly, a record of data is only stored on a single node, if the node goes down, it will cause the loss of the data stored on it, which is not allowed in the blockchain system, and the solution to this problem will be introduced later. Another problem is that the allocation of hash slots using the average allocation strategy does not take the actual storage capacity of each node into account. Each node is allocated the same number of slots, and for a node with small storage capacity, it may not have enough space to store the allocated slot data, while for a node with large storage capacity, it may still have a very large remaining space, which will result in unreasonable data allocation.

To solve the above problem, we improve the hash slot allocation algorithm to make it more reasonable. We denote the storage space that node j can use for block storage as C_j, the total number of hash slots in the system as N_{slots}, and the total number of nodes in the network as N_{nodes}, then the number of slots allocated to each node j is:

$$SN(j) = \frac{C_j}{\sum_{k=1}^{N_{nodes}} C_k} * N_{slots} \tag{2}$$

In this way each node will be assigned the corresponding number of hash slots based on its storage capacity. For the specific slot allocation, each slot is allocated to each node according to the slot number. If the number of slots currently allocated to node j reaches $SN(j)$, it indicates that it has allocated enough slots and no more new slots will be allocated to it, and the remaining nodes will continue to be allocated the remaining slots until all slots are allocated.

Algorithm 1: Hash Slot Allocation

       Input: node list *Nodes*, the total slot number N_{slots}

       Output: routing table *RTable*

1.   total = 0; remain = {}; RTable = []
2.   **for** node in Nodes **do**
3.   total += node.capacity
4.   **for** node in Nodes **do**
5.   remain[node] = node.capacity / total * N_{slots}
6.   **for** i = 0 to N_{slots}-1 **do**
7.   **while** remain[Nodes[p]] == 0 **do** p = (p+1)%Nodes.length
8.   RTable.add(Nodes[p])
9.   remain[Nodes[p]] -= 1
10.   p = (p+1)%Nodes.length
11.   **return** RTable

In Algorithm 1, the number of slots allocated to each node is calculated in line 2–5, while the allocation of hash slots and the generation of routing table are done in line 6–10.

4 Strategy of Block Management

4.1 Block Storage Process

We assign a node ID to each node in the network by hash algorithm. The input to the hash function is the IP address and port number of the node, and the output is a binary node ID. The predecessor and successor nodes of each node can be determined by sorting the node IDs. A master node needs to be elected to perform the block liveness calculation and hash slot allocation. The selection method of the master node can be determined according to the actual situation and is not the focus of this paper.

According to the model of block liveness proposed above, in this paper, we use different storage strategies for blocks of different liveness. For high-liveness blocks, we use the storage strategy of the original blockchain system to store them on all nodes, while for low-liveness blocks, we use hash slot algorithm to store them on a few nodes in the network, and the rest of the nodes do not need to store them.

As illustrated in Fig. 2, after each new block is mined, it is stored on all nodes in the first *BLRIT* time, because each block is accessed most frequently when it is first mined and then decreases significantly, so storing new mined blocks on all nodes will not cause additional cross-node communication cost when accessing them. When the survival time of a block reaches *BLRIT* time, the master node collects the access numbers of this block on each node during the first *BLRIT* time and performs a block liveness calculation on it using the model of block liveness proposed above. If the result is a high-liveness block, the block will continue to be stored on all nodes until the end of the next *BLRIT* time, at which point the liveness calculation will be performed on it again and the corresponding action will be executed based on the result. If the result is a low-liveness block, a hash slot algorithm is applied to it for distributed storage. First the master node calculates $CRC32(block\_hash)\%N_{slots}$ for the block to get the slot number where the block should be placed, then queries the routing table to get the target node responsible for storing the slot data, and in order to achieve r copies of redundant storage, the r-1 successor nodes of the target node also need to store the block and they are marked as target node too. Then the master node broadcasts the block slot storage message to the network which contains the block hash, slot number, target node list and the signature of master node. When the other nodes in the network receive the message, they will calculate whether the slot number for the block and the assignment of the target nodes responsible for storing the block are correct. After receiving the message and verifying it, the target storage nodes will broadcast confirmation storage message to the network, and the rest of the nodes in the network can delete their locally stored block body of the block to reduce their storage overhead after receiving the confirmation storage message from the target storage nodes. The block header is reserved for query verification.

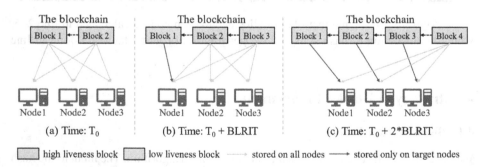

Fig. 2. The process of block storage

4.2 Data Query and Verification

When a node needs to query a block or transaction data, it first looks it up in its local storage, and if it has stored the block data locally, it can get the query result directly. Otherwise, it first calculates $CRC32(block\_hash)\%N_{slots}$ to get the slot number of the target block, and then queries the routing table to find the storage node corresponding to the slot and initiates a query request to that node. If the node responds to the request and returns result, the query ends. If the node does not respond, it continues to send requests to its r-1 successor nodes until the query result is obtained. The whole process of data querying is illustrated in Fig. 3.

The returned result contains the block or transaction data and its verification object. For the transaction, the verification object is the merkle proof path of the transaction in the block, and SPV is used for verification.

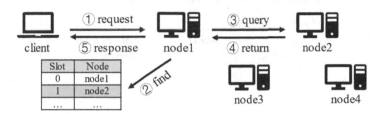

Fig. 3. Data query process

4.3 Node Join in and Quit

During the running of the system, new nodes may join the network or existing nodes may drop out, when this happens, the hash slot reallocation and data migration is required.

When a new node joins in the network, it first requests all current high-liveness blocks and all block headers from other nodes and store them. After completion, it reports its available storage space for block storage to the master node. When the master node receives the message, it runs the hash slot allocation algorithm to update the routing table and broadcasts the routing table update message to the network. After each node receives the message, each node request and store the new slots data which it needs to store from the other node that storing these slots data in the network. At the same time, each node will identify the slots it no longer needs to store and simply removes these slots data from its local storage. Once all the nodes have done this, the data migration is complete.

When a node quits the network, the master node also reruns the hash slot allocation algorithm and updates the routing table, and then each node simply updates its stored slot data according to the above process.

5 Experiments

5.1 Experiment Setup

We have conducted a series of comparative experiments and evaluations of our proposed storage scheme. We compared the original storage strategy of blockchain with our proposed strategy. The simulation experiments were done on a computer with a 2.9 Ghz

CPU and 16G RAM, and the parameters of the experiments are shown in Table 1. We build a network with 8 virtual nodes and set the block copy number to 2 according to the total number of nodes in the network. N_{slots} was set to 2048 to ensure that the blocks can be distributed evenly into each slot. We set the *BLRIT* to the time of generating 100 blocks which ensures the liveness of a block can be updated in a timely manner.

Table 1. Experimental parameters

Parameter	Value	Parameter	Value
N_{nodes}	8	r	2
N_{slots}	2048	*BLRIT*	time of generating 100 blocks

5.2 Comparison of Storage Overhead

First, we compare the storage overhead of our proposed storage scheme with that of the original storage scheme in the blockchain (baseline), i.e., every node will store the whole ledger data. The node6 and node7 are set to have half the storage capacity of the other nodes, and 2000 new blocks are added to the system. Both the total storage cost of the system and the storage cost of each node are evaluated, and we also evaluate the storage cost of our scheme without using the model of block liveness (MBL), i.e., all blocks will be directly stored only on the target nodes based on the hash slot algorithm when it was mined out. The experimental results are as follows.

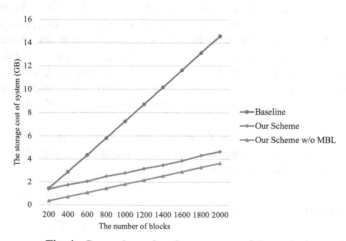

Fig. 4. Comparison of total storage cost of the system

Figure 4 shows the comparison of the total storage cost of the system under the 3 schemes. After adding 2000 blocks, the total storage cost of baseline reaches about

14.57 GB, while our proposed scheme is less than 4.65 GB, which is 68.1% lower and saves a lot of space. And if the model of block liveness is not considered, the occupied space can be further reduced by about 74.9%. As the number of blocks keeps increasing in the running system, the reduction of storage cost brought by our scheme will be more and more.

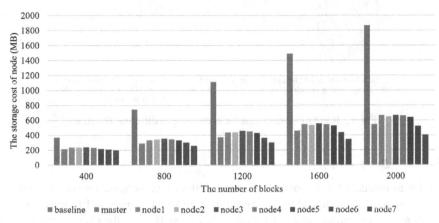

Fig. 5. Comparison of the storage cost of each node

Figure 5 shows the storage cost of each node for different number of blocks in the system. Compared to baseline, the storage cost of each node in our scheme is significantly reduced, up to 78.3% when the number of blocks is 2000, which proves the effectiveness of our storage scheme in reducing the storage overhead.

5.3   Evaluation of Data Distribution

In order to evaluate the effectiveness of the hash slot allocation algorithm in this paper, the ratio of storage cost for each node in different situation was analyzed. In situation a, each node has the same storage capacity, while in situation b, the storage capacity of node 6 and node 7 are the half of other nodes, and these are used to evaluate whether the size of the ledger data that each node is assigned to store matches its storage capacity. The results are shown in Fig. 6 after 2000 blocks were adding to the system.

It can be concluded from Fig. 6 that in situation a, when the storage capacity of each node is same, the size of the ledger data that each node is assigned to store is also almost the same. While in situation b, when the storage capacity of node6 and node7 are the half of the other nodes, the size of the ledger data that node6 and node7 are assigned to store are also the least among all nodes, which is consistent with their storage capacities. And for the other six nodes, the size of the data assigned to each node is also comparatively average. So, these proves the effectiveness of the hash slot allocation algorithm proposed in this paper and the ledger data can be distributed more reasonable on the nodes.

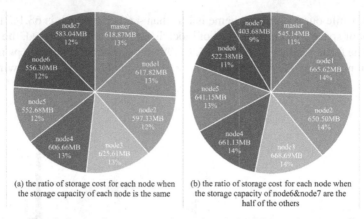

(a) the ratio of storage cost for each node when the storage capacity of each node is the same

(b) the ratio of storage cost for each node when the storage capacity of node6&node7 are the half of the others

Fig. 6. Ratio of storage cost for each node in different situations

5.4 Data Query Time

In order to evaluate the impact of our proposed storage scheme on the efficiency of ledger data access, we tested the query time of different numbers of block data under the full node, our scheme, and our scheme without MBL after the experiment 5.2 were completed. We queried the latest specific number of block data in the system, and the results are shown in Fig. 7.

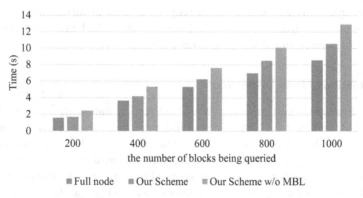

Fig. 7. Comparison of block data query time

It can be seen that the full node takes the shortest query time because it stores the complete ledger data so all query requests can be done by reading the block data directly from its local storage without the need to communication and data transmission with other nodes in the network. Compared to not applying the model of block liveness, the HLBs are stored on every node in our scheme, so these blocks are available directly from the local storage on each node without any additional communication cost when being queried, which makes the query time reduces greatly. Since the HLBs are also the blocks that are frequently accessed, most of the query requests can be done locally in our scheme and this will cause a negligible increase on query time, which proves the effectiveness of the model of block liveness in minimizing the additional query time. As can be seen, compared to the substantial reduction in storage overhead brought by our proposed scheme, the increase in query time is negligible under the current good network communication environment, which proves the rationality and effectiveness of our proposed storage scheme.

6 Conclusion

The advent of the big data era has led to an exponential increase in the size of data stored in blockchain systems, which has caused more and more nodes to become overwhelmed and drop out of the system.

In order to reduce the storage overhead for nodes, this paper proposes a storage optimization scheme based on hash slot algorithm. We first propose a model to evaluate the liveness of a block so that different storage strategies are applied to different type of blocks. Then we propose a more reasonable hash slot allocation algorithm that takes the actual storage capacity of each node into account. Based on the above proposed model and algorithm, we describe the block storage process, the data query and verification scheme, and the data migration methods when nodes join and exit the network in detail. The experimental results show that our scheme can reduce the storage overhead of nodes by about 78%, while the data allocation becomes more reasonable on each node and the additional communication cost brought by our scheme when querying data is negligible.

The proposed scheme in this paper still has room for improvement in the future. We will further evaluate the scalability and security of our proposed scheme in a network consisting of a large number of nodes, and improve our scheme based on the evaluation results in the future.

Acknowledgements. This work was supported by the National Natural Science Foundation of China (62072086, 62172082, 62072084), the Fundamental Research Funds for the central Universities (N2116008).

References

1. Nakamoto, s.: Bitcoin: A peer-to-peer electronic cash system (2008). https://bitcoin.org/bit coin.pdf
2. Wood, G.: Ethereum: a secure decentralised generalised transaction ledger. Ethereum Project Yellow Paper, vol. 151, pp. 1–32 (2014)
3. Li, M., Lal, C., Conti, M., Hu, D.: LEChain: a blockchain-based lawful evidence management scheme for digital forensics. Future Gener. Comput. Syst. **115**, 406–420 (2021)
4. Wang, B., Jiawei, S., Wang, W., Zhao, P.: Image copyright protection based on blockchain and zero-watermark. IEEE Trans. Netw. Sci. Eng. **9**(4), 2188–2199 (2022)
5. Zhao, X., Liu, H., Hou, S., Shi, X., Wang, Y., Zhang, G.: An ethereum-based image copyright authentication scheme. In: Zhao, X., Yang, S., Wang, X., Li, J. (eds.) WISA 2022. LNCS, vol. 13579, pp. 724–731. Springer, Cham (2022). https://doi.org/10.1007/978-3-031-20309-1_63

6. Mao, X., Li, X., Guo, S.: A blockchain architecture design that takes into account privacy protection and regulation. In: Xing, C., Fu, X., Zhang, Y., Zhang, G., Borjigin, C. (eds.) WISA 2021. LNCS, vol. 12999, pp. 311–319. Springer, Cham (2021). https://doi.org/10.1007/978-3-030-87571-8_27

7. Papi, F.G., Hübner, J.F., de Brito, M.: A Blockchain integration to support transactions of assets in multi-agent systems. Eng. Appl. Artif. Intell. **107**, 104534 (2022)

8. Chen, X., Zou, D., Cheng, G., Xie, H., Jong, M.: Blockchain in smart education: contributors, collaborations, applications and research topics. Educ. Inf. Technol. **28**(4), 4597–4627 (2023)

9. Xu, G., et al.: A privacy-preserving medical data sharing scheme based on blockchain. IEEE J. Biomed. Health Inform. **27**(2), 698–709 (2022)

10. Luu, L., Narayanan, V., Zheng, C., Baweja, K., Gilbert, S., Saxena, P.: A secure sharding protocol for open blockchains. In: Proceedings of the 2016 ACM SIGSAC Conference on Computer and Communications Security, pp. 17–30 (2016)

11. Zamani, M., Movahedi, M., Raykova, M.: RapidChain: scaling blockchain via full sharding. In: Proceedings of the 2018 ACM SIGSAC Conference on Computer and Communications Security, pp. 931–948 (2018)

12. Qi, X., Zhang, Z., Jin, C., Zhou, A.: A reliable storage partition for permissioned blockchain. IEEE Trans. Knowl. Data Eng. **33**(1), 14–27 (2020)

13. Dai, M., Zhang, S., Wang, H., Jin, S.: A low storage room requirement framework for distributed ledger in blockchain. IEEE Access **6**, 22970–22975 (2018)

14. Chen, H., Wang, Y.: SSChain: a full sharding protocol for public blockchain without data migration overhead. Pervasive Mob. Comput. **59**, 101055 (2019)

15. Kim, T., Noh, J., Cho, S.: SCC: storage compression consensus for blockchain in lightweight IoT network. In: 2019 IEEE International Conference on Consumer Electronics (ICCE), pp. 1–4. IEEE (2019)

Parallel and Distributed Systems

Parallel and Distributed Systems

An IoT Service Development Framework Driven by Business Event Description

Junhua Li[1,2](✉), Guiling Wang[1,2], Jianhang Hu[1,2], Hai Wang[1,2], and Haoran Zhang[1,2]

[1] Beijing Key Laboratory on Integration and Analysis of Large-Scale Stream Data, Beijing, China
juniorslee@163.com

[2] School of Information Science and Technology, North China University of Technology, Beijing, China

Abstract. In order to enable the integration between Business Process Management (BPM) and the Internet of Things (IoT), it is imperative to create an IoT service capable of encapsulating IoT device resources, converting raw data generated by IoT devices into high-level business events and transmitting these events to BPM. This would serve as a means of augmenting the intelligence and automation of business processes. Custom development of parsing and computation modules is required for different streaming data, and IoT services are built based on descriptions of business events. However, this process often involves a lot of repetitive work, and these valuable tasks are difficult to meet the requirements of changing data demands in different IoT scenarios. To solve this problem, this study proposes a code generation development framework for real-time stream data processing application, driven by business event descriptions in IoT scenarios. The framework can automatically generate real-time stream data processing applications that meet data processing needs based on domain data requirements described by business personnel in the IoT scenario. Through relevant experiments, the designed development framework demonstrates its potential to help non-software developers quickly configure Flink-based stream computing applications for scenarios and has a certain flexibility to adapt to real-time stream data processing requirements in common IoT scenarios.

Keywords: IoT service · Stream processing · Business event · Code generation

1 Introduction

In recent years, with the continuous deepening of the level of informationization, automation, and intelligence in modern technological society, the Internet of Things (IoT) industry has become an indispensable part of the new generation of information technology, and has penetrated into various industries in modern

L. Yuan et al. (Eds.): WISA 2023, LNCS 14094, pp. 527–538, 2023.
https://doi.org/10.1007/978-981-99-6222-8_44

society [1, 2]. Stream data has characteristics such as infinity, real-time, volatility, and disorderliness. Stream data is generated in real-time by data sources, and its value is short-lived. Real-time processing is required, and generally, calculations are performed directly in memory after the data arrives. Real-time stream computing is suitable for scenes with relatively loose calculation accuracy requirements and high response time requirements. Therefore, through the deployment of well-configured IoT sensors, massive data in the business environment can be collected in real-time and processed accordingly. The massive raw data collected by IoT sensors can be transformed into interesting business events and output events of IoT services through stream processing.

However, in practical IoT applications, how to model IoT services is the most critical step in the design and development of IoT systems. In modeling IoT services, several dimensions need to be considered to ensure the reliability and validity of the IoT services. First, the input and output of the IoT service need to be determined to ensure that the IoT service can be invoked properly and provide the required output. Also, to ensure the scalability and adaptability of the IoT service, it is necessary to provide the ability to publish and subscribe to event streams so that the output business event stream results of the IoT service are available to external BPM systems. In the process of IoT service modeling, it is also necessary to provide the operation of filtering, aggregation and analysis of event streams so that business users can add processing policies for IoT services to meet business requirements.

This article proposes a Flink-based, domain-specific, data-driven framework for real-time streaming data processing applications, which provides domain data requirement description tools for business personnel who are not familiar with big data components, and automatically generates Flink real-time data processing applications that meet data processing requirements.

2   Related Work

With the development of the Internet of Things (IoT), two key challenges have emerged: communicating with devices and managing them [3]. The most relevant challenge is coordinating tasks between different devices, which currently involves a single IoT device solution. However, IoT components from different platforms typically lack unified protocols and standards in terms of data formats, communication protocols, and interface definitions. The service paradigm is a key means of overcoming these challenges by abstractly converting IoT devices into IoT services, which through service mapping are viewed as unified IoT objects [4]. In short, services operate at a higher level of abstraction than raw IoT data, enabling the transformation of data into actionable objects and making data more useful.

IoT Service Model. In order to address the aforementioned problems, the service-oriented architecture (SOA) has been proven to be an appropriate approach [5]. SOA provides a standardized way of describing IoT services and provides service description information to callers. It utilizes a unified service-based pattern to describe IoT devices and implements the description of different components at the service level [6]. However, SOA still needs to solve some problems when modeling IoT services, such as the fact that most IoT services are provided by devices with limited resources, thus requiring a more lightweight service.

Subsequent work has explored the use of traditional service models by extending existing service models (OWL-S and DPWS) [7,8]. For example, in [9], the authors analyzed the characteristics of IoT services, designed the core model of IoT using the OWL language, and analyzed the relationships among various concepts in the custom model (IoT core concepts: entities, devices, resources, and services). However, the protocol stack of SOAP/WSDL services is primarily designed for resource-rich infrastructure and is not suitable for resource-limited IoT services. Additionally, the model does not involve modeling means for heterogeneous devices in complex scenarios.

Later, some work attempted to extend IoT services on newer RESTful services by using lightweight protocols and data formats to reduce service overhead. However, these methods usually establish their own private service description models, which limits their application in the IoT field and weakens their ability to describe IoT devices and objects.

IoT Service Development. IoT services are services that can be extended to the physical environment, which perceives entities in the physical environment through sensor networks and acts on them. Unlike traditional web services, the characteristics of IoT services are influenced by factors such as time limits, resource limits, and device failure probability in the physical environment. Streaming data generated by IoT is characterized by infinity, real-time, volatility and disorder. Streaming data is generated in real time by the data source, the timeliness of the data value is short, and the real-time requirement of processing is high, in which, for example, to optimize the efficiency of query processing [10] is also a point we consider. The response speed, service energy consumption, and fault tolerance of IoT services have important effects on the overall characteristics of IoT systems.

For IoT service development frameworks, tools, and platforms, Ci et al. propose a virtual IoT device construction method based on edge services, which abstracts heterogeneous IoT devices into a unified virtual device and supports application construction based on virtual IoT devices. Mikkonen et al. address the IoT sensing data access Mikkonen et al. propose a unified homogeneous IoT software development architecture to address the serious heterogeneity and incompatibility problems of software development and deployment methods [11]. Harbin Institute of Technology (HIT) proposed an intention-aware interactive IoT service [12]. Therefore, comprehensive modeling of IoT services is necessary in the development process of IoT systems to facilitate in-depth analysis and description of services. These works are essential for ensuring the reliability and efficiency of IoT systems.

3 Business Event Model and Framework Design

3.1 Business Event Model

Business events refer to specific events or actions that occur in the business process, often also referred to as business process events. Business events are a key concept that is closely related to business process modeling and management. It contains information about the business process and can be used to describe the state, changes, and behavior of the business process. These events may be automatically triggered by the business system, or manually triggered by users or other systems. Business events can also be used for analysis and decision-making. For example, if a large number of orders fail during the payment process, an investigation may be necessary to identify the cause and take appropriate action. These business events can be used to describe changes in user behavior or trigger actions in subsequent processes.

In IoT systems, business events typically refer to IoT data being transformed into events that are meaningful to the business. In this process, IoT raw data needs to be analyzed and processed to extract information relevant to the business, ultimately resulting in multiple events with business meaning. The business event model, as the upper layer of IoT objects, plays the most fundamental role in describing the model of this transformation process. The conversion and output of business events are in the process of working with IoT services, as shown in the Fig. 1.

In business event modeling, the model usually includes the following aspects:

Firstly, data source access. The raw data stream of IoT devices has been abstracted and integrated by IoT objects. The addresses of different IoT device data sources are stored in the IoT objects. The business event model is based on these data sources in the IoT objects.

Secondly, data format and data constraints. The data format module uses the data identifiers and various data items defined in the IoT devices, and the data constraint module is the constraint and limitation on these raw data items.

Thirdly, data calculation requirements and business event conversion. Select the IoT devices involved in the IoT object, and define the calculation requirements based on specific IoT devices. This article standardizes several commonly used calculation requirements. For example, judgment of threshold for data collection of IoT devices, calculation of average, accumulation, variance, etc.

Fourthly, business event output. This article focuses on combining IoT services with business processes, so business events are processed as the processing results of IoT services. They will be sent to the business process for real-time processing of process-related tasks.

In addition to defining the process of converting raw data of IoT devices into business events, the role of the business event model is to automatically generate IoT service servers based on the business event model. Based on pre-defined data constraints and data calculation requirement files, IoT service servers can be easily generated and deployed for operation.

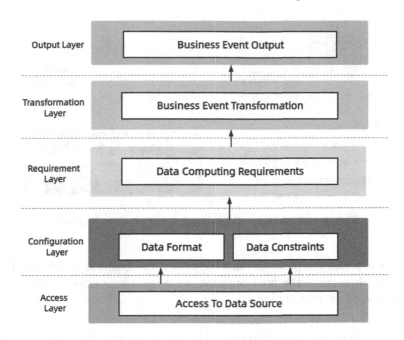

Fig. 1. IoT Service Work Process

3.2   Basic Development Framework Design

The overall design concept of the development framework for this project mainly revolves around modeling efficient and easy-to-use IoT services based on high-demand requirements, as shown in Fig. 2, which depicts the entire process of constructing an IoT service. The overall purpose of the IoT service modeling module is to build suitable IoT services based on various types of IoT devices.

Developing business event applications in the context of IoT typically requires modeling the business event rules defined in the modeling components of the interested business events based on the raw streaming data from different IoT objects. This involves accessing the different IoT data sources configured by the user through a unified template file and directing them to the messaging middleware as raw data streams for processing in the big data stream processing system being developed. Therefore, this framework includes a data processing flow, which requires designing a data source access template file, preprocessing the raw continuous data obtained from the messaging middleware, and designing two files, IoT-DF (IoT-Data Format) and IoT-DC (IoT-Data Constraint), based on the user-defined data format and range constraints. These two files can parse multi-level structured raw data into single-level data items that can be processed by the system and check the value range of the data items while processing abnormal values. Next, the data needs to be matched with the corresponding business event rules for real-time processing and calculation, in order to output the business events of interest to the user. The most essential file in

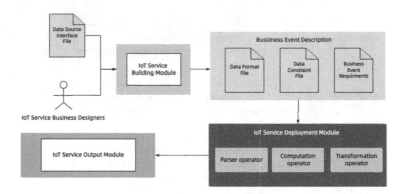

Fig. 2. IoT service build execution full process diagram

this process is the template file that describes the business events in the IoT service, which outputs the final business events to the configured sink address for further analysis and mining.

Data Specification Description. The data analysis for specific business events requires IoT data formats and related constraints. Therefore, it is necessary to describe the basic semantic information and normative requirements of the data, so that IoT service development frameworks can adapt to the automatic parsing, filtering, and analysis of different business data.

The IoT Data Constraint (IoT-DC) is a description of the valid range of values for an item, including the minimum and maximum values of the data item and a list of exception values. The constraints imposed by the CR configuration file are the initial processing of IoT devices, and the pre-processing part directly discards data that does not meet the constraints or is abnormal, thereby improving the conversion efficiency of the overall business event. The definition of data constraints is as follows, and the Fig. 3 including only the reference of the device with ID 123456 and the data constraints. If there are multiple devices, only change "DeviceID" to the corresponding device ID. The normal range of measurement values is 0–100. If it exceeds this range, it is considered an exceptional value. For the open/closed state, since there are only two states, the allowed values can be directly specified by adding the "validValues" attribute, and all values other than true or false are considered exceptional values.

Business Event Description. In the Flink stream processing framework, the data calculation requirements define the way in which data is exchanged and consumed between the various computing units (operators) in the Flink topology structure. The data exchange method in Flink is also known as data stream grouping.

In an IoT system, the IoT service function model refers to the model that describes the features and attributes of IoT services, as well as the generation

```
<IoTDeviceConstraint>
 <DeviceID>123456</DeviceID>
 <DataConstraint>
  <Constraint dataKey="" min_measurement="0" max_measurement="100" />
  <Constraint dataKey="OpenCloseState" validValues="true,false"/>
 </DataConstraint>
</IoTDeviceConstraint>
```

Fig. 3. Design model for data constraint requirements

and provision of IoT services. The IoT service function model is an important component of the IoT system, which defines the behavior and functions of IoT services and generates and provides services through the server.

In IoT, service description can be used to describe the features and attributes of IoT services, such as service type, device ID, and sensor data format. Through the service description, users can have a clear understanding of the specific function and usage of the service, thereby better using and managing IoT services. Below is a simple XML description format example for a temperature IoT service (Fig. 4):

```
<BusinessEvent>
 <Name>Temperature monitoring alarm IoT BusinessEvent</Name>
 <Description>Abnormal temperature alarm detected</Description>
 <Input>
  <parameter name="threshold" type="double" unit="°C">Define temperature threshold</parameter>
  <parameter name="SensorData" type="string">Define the actual temperature value collected by the sensor</parameter>
 </Input>
 <Operators>
        <ThresholdOperator>True</ThresholdOperator>
        <SumOperator>False</SumOperator>
 <Operators>
 <BusinessEventOutput>
  <SinkAddress name="Kafka">192.168.10.102</SinkAddress>
  <Port>9092</Port>
  <Topic>tempevent</Topic>
  <parameter name="alert" type="string">Alarm Event</parameter>
 <IoTDataFormat>
  <Description>Timestamp,DeviceId,EventName</Description>
 </IoTDataFormat>
 </BusinessEventOutput>
</BusinessEvent>
```

Fig. 4. Business event description file for IoT service

This XML description file defines an IoT service named "TemperatureSensor", which describes the service's inputs, outputs, and available functions. The input parameter of the service is a floating-point number named "threshold", which represents the temperature threshold. When the temperature exceeds this threshold, an alert event will be triggered. The output of the service includes a business event "alert", indicating a temperature alert event.

4   Implementation of IoT Service Development Framework

4.1   Overall Implementation Strategy

First, users generate three types of configuration files: IoT-DF, IoT-DC, and IoT-CR based on specific business events' data specification and data calculation requirements. Second, data parsing and calculation modules are generated from the configuration files, and a merge calculation module is selectively generated. Finally, by using an appropriate data flow grouping strategy, the three modules are assembled into the basic structure of Streaming Dataflows and formed into an executable Java project, which is packaged and deployed to run on the Flink cluster.

4.2   Generate Data Parsing Code

In order to support the parsing of data with multi-level structures, a template-based data parsing module (ParseOperator) and a data calculation module (CalculateOperator) have been introduced into the development framework. At the same time, in consideration of the performance of data parsing, a data parsing code based on DF and CR definitions is automatically generated as hard coding.

Data Source Access Template Generation. For the user-defined IoT-DF file used in IoT devices, it defines the data identification and various data items, while the CR (Constraint Rules) file imposes constraints and limitations on the original data.

The user-defined IoT-DF file includes the data source file, and the auxiliary framework uses the StringTemplate backend template engine to generate the corresponding access data source file based on the preset configuration parameters in order to obtain the user-configured information such as KafkaBrokers, group_id, and Topic. Below is a generated code template Fig. 5.

```
// set kafka connection information
Properties props = new Properties();
props.setProperty("boostrap.servers", $kafka_brokers$ );
props.setProperty("group.id", $group_id$ );

// creating a Flink streaming environment
final StreamExecutionEnvironment env = StreamExecutionEnvironment.getExecutionEnvironment();
// add kafka data source
FlinkKafkaConsumer<String> kafkaConsumer = new FlinkKafkaConsumer<String>(
        $topic_name$,
        new SimpleStringSchema(),
        props);
```

Fig. 5. Access data source code template

Data Analysis Module Generation. The generation of the data parsing module is mainly used for preprocessing and filtering the read data. The data preprocessing template file have two types of templates defined for data source processing. The first type processes general values collected, and the parameter <device_id> that needs to be replaced in this template specifies which IoT device's parameters are to be processed. MIN_MEASUREMENT and MAX_MEASUREMENT represent the parameter values read from the IoT data constraint file. The second type processes the collected status values, and the parameter VALID_STATUSES that needs to be replaced represents the status list that the sensor read from the data constraint file has.

4.3   Generate Data Calculation Code

Actual business requirements have different calculation requirements for different types of data, but common calculation needs have certain similarities. For example, measuring certain data metrics over a period of time involves relatively simple algorithms such as cumulative values and averages. Therefore, it is necessary to first define common basic algorithms. Secondly, design a data calculation template that is applicable to different calculation requirements. Finally, generate data calculation modules based on the above calculation template and calculation needs.

Data Calculation and Business Event Processing Template Generation. The data calculation requirements include the transformation of business events. For real-time data processing, each business event is matched according to the configured rules, and each business event may require processing for several calculation requirements. This module provides a series of built-in transformation functions and operators, and also supports user-defined functions, which can be flexibly customized according to specific business needs.

Therefore, for the calculation of the same data item in different statistical time periods, a longest window (maxWindowSize) needs to be maintained for the data item. As shown in Fig. 6, the WindowSize time window variable can be read from the CR configuration file to replace the parameters in the template file, and the boolean value in each operator is used to determine whether the calculation of the business event requires the corresponding operator. This allows for flexible configuration of different operators, using SumOperator and AvgOperator as examples in this template.

After meticulously configuring the data source specification files and business event specification files required by the framework proposed in this article according to established standards, non-developers can generate IoT services based on business event descriptions. By utilizing the entry program we provide in Fig. 5, users need only provide the correct path information for the configuration files. The program then executes the SubmitFlinkJob() function, which reads and parses the corresponding XML files based on the given paths to retrieve the necessary configuration parameters, and configures data source integration and operator selection accordingly. Finally, user-defined Flink job programs are generated using the backend template engine, and Flink jobs are automatically submitted through preinstalled scripts. Task output addresses are directed to the corresponding sink as stipulated in the business event configuration file. This process requires no additional coding from developers, but only necessitates the proper configuration of the files corresponding to the business event descriptions in line with established standards for generating the corresponding IoT service programs in the IoT scenario.

```
boolean SumOperator = $SummerOperator$;
boolean AvgOperator = $AvgOperator$;
if (SumOperator == true) {
    input.keyBy(SensorData::getDeviceId) KeyedStream<String, Object>
            .timeWindow(Time.seconds($windowSize$)) WindowedStream<String, Object, TimeWindow>
            .apply(new WindowFunction<SensorData, Double, String, TimeWindow>() {
                @Override
                public void apply(String deviceId, TimeWindow window, Iterable<SensorData> input, Collector<Double> out) throws Exception {
                    double sum = 0;
                    for (SensorData data : input) {
                        sum += data.getValue();
                    }
                    out.collect(sum);
                }
            });
}
if (AvgOperator == true) {
    input.keyBy(SensorData::getDeviceId) KeyedStream<String, Object>
            .timeWindow(Time.seconds($windowSize$)) WindowedStream<String, Object, TimeWindow>
            .apply(new WindowFunction<SensorData, Double, String, TimeWindow>() {
                @Override
                public void apply(String deviceId, TimeWindow window, Iterable<SensorData> input, Collector<Double> out) throws Exception {
                    double sum = 0;
                    int count = 0;
                    for (SensorData data : input) {
                        sum += data.getValue();
                        count++;
                    }
                    double avg = sum / count;
                    out.collect(avg);
                }
            });
}
```

Fig. 6. Business event processing operator template

4.4 Experimental Results and Evaluation

Experimental content: Given data samples and processing requirements, 2 professional Flink developers wrote Flink application code, and 2 non-Flink developers generated applications through the framework.

Data sample:
[1615242327], sensor, {bohai: {fog:134.34, light:on, buzzer:off}}
[1615242476], sensor, {bohai: {fog:156.62, light:off, buzzer:on}}
[1615242582], sensor, {bohai: {fog:183.23, light:off, buzzer:off}}
[1615242747], sensor, {bohai: {fog:204.72, light:off, buzzer:on}}

Processing Requirements: This article will use a maritime transportation scenario as an experimental background. Specifically, we take the example of a transport ship named Boyuan operating in foggy weather conditions for maritime transportation. To ensure safety, when the ship sends a notification request, the regulatory agency will receive the request and start monitoring the transportation operations. The regulatory agency will use uninterrupted border activity subscription IoT services to monitor the ship's foggy environment, navigation light regulations, and foghorn regulations while sailing in foggy weather. This real data contains various IoT data from the ship, including timestamps, device IDs, locations, fog concentrations, navigation light status, and foghorn status at different times. The fog concentration is in decimal form. These services will monitor the foggy environment in which the ship is located and activate navigation light and foghorn status monitoring services when severe foggy weather is detected, outputting corresponding events.

Experiment results: Two developers familiar with Flink programming and Four non-Flink developers completed Flink application development that met the requirements within two hours and successfully ran it with simulated data. Two developers generated two Flink applications based on processing requirements, but in fact, only one was needed. Another Flink developer failed to complete the task within the specified time due to incorrect parameter configuration, resulting in incorrect calculation results.

In the above experiment, All non-Flink developers were able to configure a runnable stream computing application within two hours without any programming technology help. Therefore, we believe that Flink application development frameworks and related tools can provide effective development support. We hope to continuously maintain and enhance the auxiliary development framework in application practice, in order to adapt to more types of data demands.

5   Conclusion

This article targets business requirements personnel and designs and implements a data-driven Flink application development framework. By modeling and describing data specification and computation requirements, and by designing template-based Flink programming units, an auxiliary development framework for Flink applications is implemented, shortening the development cycle of real-time processing applications for streaming data. The Flink application development framework and related tools can provide effective development support for non-Flink developers, significantly improving development efficiency and shortening application debugging time in the face of constantly changing requirements. In the future, we hope to continue to maintain and improve the auxiliary development framework in practical applications, making it adaptable to more types of data requirements and providing more comprehensive and user-friendly development support for Flink application development.

Acknowledgment. This work is supported by the International Cooperation and Exchange Program of National Natural Science Foundation of China (No. 62061136006) and the Key Program of National Natural Science Foundation of China (No. 61832004).

References

1. Gruhn, V., et al.: BRIBOT: towards a service-based methodology for bridging business processes and IoT big data. In: Hacid, H., Kao, O., Mecella, M., Moha, N., Paik, H. (eds.) ICSOC 2021. LNCS, vol. 13121, pp. 597–611. Springer, Cham (2021). https://doi.org/10.1007/978-3-030-91431-8_37
2. Kirikkayis, Y., Gallik, F., Reichert, M.: Towards a comprehensive BPMN extension for modeling IoT-aware processes in business process models. In: Guizzardi, R., Ralyté, J., Franch, X. (eds.) RCIS 2022. LNBIP, vol. 446, pp. 711–718. Springer, Cham (2022). https://doi.org/10.1007/978-3-031-05760-1_47
3. Sheng, M., Qin, Y., Yao, L., Benatallah, B.: Managing the Web of Things: Linking the Real World to the Web. Morgan Kaufmann (2017)
4. Bouguettaya, A., et al.: A service computing manifesto: the next 10 years. Commun. ACM **60**(4), 64–72 (2017)
5. Gama, K., Touseau, L., Donsez, D.: Combining heterogeneous service technologies for building an Internet of Things middleware. Comput. Commun. **35**(4), 405–417 (2012)
6. Fanjiang, Y.-Y., Syu, Y., Ma, S.-P., Kuo, J.-Y.: An overview and classification of service description approaches in automated service composition research. IEEE Trans. Serv. Comput. **10**(2), 176–189 (2015)
7. Issarny, V., Bouloukakis, G., Georgantas, N., Billet, B.: Revisiting service-oriented architecture for the IoT: a middleware perspective. In: Sheng, Q.Z., Stroulia, E., Tata, S., Bhiri, S. (eds.) ICSOC 2016. LNCS, vol. 9936, pp. 3–17. Springer, Cham (2016). https://doi.org/10.1007/978-3-319-46295-0_1
8. Guinard, D., Trifa, V., Karnouskos, S., Spiess, P., Savio, D.: Interacting with the SOA-based Internet of Things: discovery, query, selection, and on-demand provisioning of web services. IEEE Trans. Serv. Comput. **3**(3), 223–235 (2010)
9. De, S., Barnaghi, P., Bauer, M., Meissner, S.: Service modelling for the Internet of Things. In: 2011 Federated Conference on Computer Science and Information Systems (FedCSIS), pp. 949–955. IEEE (2011)
10. Qi, X., Wang, M., Wen, Y., Zhang, H., Yuan, X.: Weighted cost model for optimized query processing. In: Zhao, X., Yang, S., Wang, X., Li, J. (eds.) WISA 2022. LNCS, vol. 13579, pp. 473–484. Springer, Cham (2022). https://doi.org/10.1007/978-3-031-20309-1_42
11. Mikkonen, T., Pautasso, C., Taivalsaari, A.: Isomorphic Internet of Things architectures with web technologies. Computer **54**(7), 69–78 (2021)
12. Huang, C., et al.: Intent-aware interactive Internet of Things for enhanced collaborative ambient intelligence. IEEE Internet Comput. **26**(5), 68–75 (2022)

Slew-Driven Layer Assignment for Advanced Non-default-rule Wires

Ren Lu[1,2], Wei Zhang[1,2], Lieqiu Jiang[1,2], and Genggeng Liu[1,2(✉)]

[1] College of Computer and Data Science, Fuzhou University, Fuzhou 350116, China
liu_genggeng@126.com
[2] Key Laboratory of Network Computing and Intelligent Information Processing, Fujian Province, Fuzhou 350116, China

Abstract. With the rapid increase in circuit density in Very Large Scale Integration, the proportion of interconnect delay in circuit timing is also increasing. This makes the importance of layer assignment algorithms increasingly prominent in circuit design. However, most previous layer assignment algorithms prioritize optimizing timing exclusively from the perspective of interconnect delay, thereby disregarding the impact of slew violations on circuits. Therefore, this paper proposes a slew-driven layer assignment algorithm, which considers the timing of different routing layers and introduces non-default-rule wires to design a layer assignment algorithm that can significantly optimize delay, congestion, and slew violations. This algorithm mainly includes three key technologies: 1) Introducing non-default-rule wires technology to optimize timing and adopting a negotiation based approach to ensure that the final routing scheme does not have overflow; 2) Proposing a slew prioritization strategy that comprehensively considers slew and interconnect delay during the routing process; 3) Proposing a timing critical awareness strategy to further optimize the slew and interconnect delay without worsening the overflow. The experimental results show that the proposed algorithm has significant effects on optimizing delay and reducing slew violations.

Keywords: Interconnect Delay · Layer Assignment Algorithms · Slew Violations · Non-default-rule Wires

1 Introduction

With the continuous progress of chip manufacturing technology, chip size continues to shrink and integration level continues to improve [1, 2]. This trend leads to an increasing routing density per unit area of chips, which in turn increases the proportion of interconnect delay in circuit timing, becoming a bottleneck limiting circuit performance improvement [3, 4]. During routing phase of the physical design, [5, 6] indicate the X-architecture Steiner Minimum Tree (XSMT) is the best connection model for multi-terminal nets in global routing algorithms under non-Manhattan structures. In the entire physical process, the consideration of interconnect delay in the early stages of interconnect synthesis (such as global routing) is particularly important.

This work was supported in part by the Fujian Natural Science Funds under Grant 2023J06017.

At this stage, layer assignment, as one of the key factors affecting interconnect delay, determines the layer of wires and via [7]. Specifically, the layer assignment process is a critical step in 2D global routing, where each segment of the 2D net is assigned to an appropriate routing layer and connected to different layers through via, thereby reducing overall delay. Since interconnect delay is primarily composed of via delay and wire delay, most existing layer assignment algorithms focus on minimizing the number of via required [8–11]. On the premise that it has been proved that the layer assignment problem with the minimum number of via is NP complete [8], layer assignment algorithms based on integer linear programming method [9], dynamic programming method [10] and negotiation-based [11] have been proposed successively.

While these algorithms are effective in minimizing the number of via, they do not fully optimized the timing from the perspective of interconnect delay, resulting in insufficient optimization efforts for delay. Consequently, researchers have proposed multiple delay optimization methods to address this limitation [12–14]. [12, 13] have focused on time-aware dynamic programming algorithms to handle single-net layer assignments without considering congestion. [14] extended the algorithm to include both timing and congestion considerations, leading to more comprehensive and effective interconnect optimization.

In the domain of layer assignment, non-default rule (NDR) wires have been applied to advanced manufacturing processes [15], which are divided into two types: parallel wires and wide wires. Although the previous default rule wires only occupies one track, it has a high delay. Figure 1a gives an example of using default-width wires. Figure 1b and c show examples of using parallel and wide wires to reduce delay, respectively. To accurately estimate the coupling capacitance of the net, [16] incorporated the estimated value of the coupling capacitance into the delay calculation of the net, thereby avoiding the assignment of nets that exceed the upper layer's wire carrying capacity and optimizing overall delay.

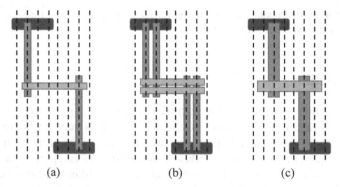

Fig. 1. Examples of (a) default-width wires, (b) parallel wires, and (c) wide wires.

In addition, it is important to consider the impact of potential slew violations on the matching between global and detail routing in the net. To address this issue, [17] introduces slew constraints into the construction of routing trees. However, the introduction of too many buffers can result in the underutilization of buffer resources, thus further constraining the already scarce routing resources available. In response to this challenge, [12, 18–20] have further optimized the timing of the routing tree by introducing buffering technology while minimizing the amount of resources occupied by the buffer.

Therefore, this paper proposes a slew-driven layer assignment algorithm. This paper's contribution is summarized as follows:

(1) Several effective comprehensive optimization strategies have been proposed using a negotiation-based framework, including the slew prioritization strategy and the timing criticality perception strategy.
(2) Slew prioritization strategy adjusts the routing order and comprehensively considers slew and delay to obtain more reasonable routing results.
(3) Timing critical network perception strategy dynamically adjusts the weights for each net and selects solutions with fewer slew violations. Further optimize slew and delay without worsening the overflow.

2 Problem Formulation

2.1 Delay Model

This paper employs the *Elmore* delay model to estimate the delay with utmost precision and accuracy [21]. The *delay(s)* follows this model which can be determined by the following expression:

$$delay(s) = R(s) \times (C(s)/2 + C_{down}(s)) \tag{1}$$

where $R(s)$ and $C(s)$ represent the resistance of the s segment and capacitance of the s segment, respectively. Additionally, $C_{down}(s)$ represents the downstream capacitance of the s segment. By calculating the delay of each segment, the accumulated delay of any path from the source to the sink can be computed:

$$delay(si) = \sum_{s \in path(si)} delay(s) \tag{2}$$

where *path* represents the path. A net often comprises multiple paths. In such cases, each path can be assigned a weight by the user to account for its relative importance. The total *delay(N)* can be determined as the weighted sum of the delay of each path in the net:

$$delay(N) = \sum_{si \in S(N)} a_{si} \times delay(si) \tag{3}$$

where $S(N)$ denotes the set of N paths, and a_{si} represents the weight of path si. In previous studies, it was assumed that the delay ratio for each path was the same.

2.2 Slew Model

In addition to delay, our algorithm also considers slew to reduce potential slew violations. To accurately quantify the number of slew violations, we utilize the *PERI* model based on the *Elmore* model, which has been demonstrated to yield an error rate less than 1% [22]. In a 2D grid diagram, each grid is represented as a tree structure with one source and one or more sinks. For each segment, the pins near the source are counted as upstream pins *pu*, while the pins near the sink are counted as downstream pins *pd*. Utilizing the *PERI* model, we calculate the slew value for segment s with the following formula:

$$slew_{pd}(s) = \sqrt{slew_{pu}(s)^2 + slew_{step}(s)^2} \tag{4}$$

where $slew_{pu}(s)$ and $slew_{pd}(s)$ represent the slew values of segment s input and output, respectively, and $slew_{step}(s)$ is the path of the slew. To calculate the path of the slew, we use the *Bakoglu's metric*. The value of $slew_{step}(s)$ can be calculated as follows:

$$slew_{step}(s) = \ln 9 \times delay(s) \tag{5}$$

where the segment *delay(s)* can be calculated from Eq. 1. In addition, the calculation method for the via slew is as follows:

$$slew_{vd}(s) = \sqrt{slew_{vu}(s)^2 + slew_{step}(vd, vn)^2} \tag{6}$$

where v_u and v_d refer to the slew values of the via input and via output and $slew_{vu}(s)$ is the slew value of the upstream pin. If the number of slews for the sink go over the stated slew constraint, we assume that the violation occurs.

2.3   Problem Model

The grid graph model is a widely used method for modeling layer assignment problems. As illustrated in Fig. 2a, each routing layer is evenly divided into a series of global tiles. Moreover, this model is represented through the k-layer three-dimensional grid graph $G^k(V^k, E^k)$, as shown in Fig. 2b. A via is represented as an edge connecting two different layers. Adjacent units within the same layer are connected by 3D mesh edges ($e \in E^k$) to establish layout wires. Each track can be placed using default width routing segments.

The layer assignment problem is a critical issue in designing electronic circuits. It involves assigning the routing of wires to different layers within a three-dimensional grid graph. This problem can be represented mathematically as follows:

Given a triplet (G^k, G, S), $G(V, E)$ represents the 2D grid diagram compacted from G^k. The 2D global routing solution can be denoted as the variable "S". The primary objective of the layer assignment algorithm is to assign segments on the 2D grid graph to the 3D grid graph to obtain the 3D global routing result S^k. Importantly, this process must guarantee good distributability and reduce slew violations.

Figure 2c displays the 2D global routing outcomes, while Fig. 2d illustrates that the segments associated with the 2D routing of n_1 are consecutively appointed to the 3D grid graph. Among them, solid and dashed lines represent segments and via, respectively. In addition, if the number of different segments on the 3D edge exceeds its capacity, it can be considered that overflow has occurred at that edge. Every net that passes through the edge and overflows is defined as an illegal net.

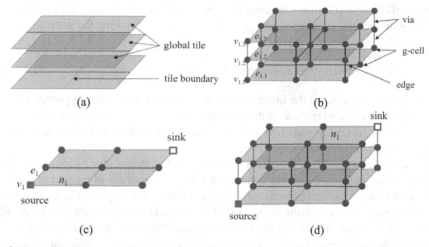

Fig. 2. Examples of (a) multi-layer routing, (b) three-dimensional grid graph, (c) 2D grid graph, and (d) compacted 3D grid graph from (c).

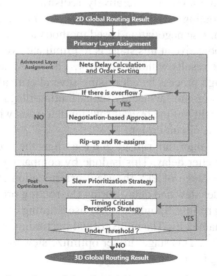

Fig. 3. The flow chart of the slew-driver layer assignment algorithm.

3   The Proposed Algorithm

Figure 3 depicts the proposed algorithm. It consists of three stages: primary layer assignment (PLA), advanced layer assignment (ALA), and post optimization (PO). To achieve efficient and optimized layer assignment, the algorithm explores all possible 3D routing solutions using dynamic programming. This comprehensive approach ensures that the optimal solution is selected, thereby improving the performance, reliability, and manufacturability of the circuit design. For each stage, the algorithm searches through all possible 3D routing solution, calculates the generation value for each scheme, and selects the one with the best generation value. The formula for computing the generation value is given as:

$$cost(n) = \alpha \times delay(n) + \beta \times via(n) + \lambda \times cong(n) \tag{7}$$

where $delay(n)$ represents the interconnect delay of net n and $via(n)$ represents the via count. Besides, $cong(n)$ represents the congestion cost. To control the proportion of each component of the cost, parameters α, β, and λ are user-defined and adjusted according to the different optimization objectives.

In the initial stage, PLA, the goal is to obtain an ideal delay layer assignment scheme while keeping the delay and via count in a comparable order of magnitude. This is achieved by assigning the optimal layer to each net using the single net layer assignment algorithm, without considering the use of NDR wires. Due to the main goal of the PLA stage being to obtain an ideal delay layer assignment scheme, and considering both to obtain a better solution, α and β set to 10 and 2 respectively. In addition, in order to alleviate the pressure of handling congestion in the future, the value of λ is set to 0.5. These data are based on multiple experiments and empirical observations.

In the ALA stage, the algorithm iteratively redistributes illegal nets and reduces net delay by utilizing routing resources effectively. At this stage, the nets delay will be calculated and sorted. Use negotiation-based methods to eliminate all overflows. As the main optimization objective at this stage is to eliminate overflow, we let λ increase to 1 to further refine the results without overflow. The ALA stage plays an important role in the physical design process by addressing several critical issues that impact the performance of the final circuit layout. By redistributing illegal nets and reducing net delay, it enhances the quality of the overall routing outcomes while also improving the timing performance of the circuit design.

In the final stage, the algorithm rips up and reassigns the network in sequence, and selects a solution with better delay. This is done by comparing the obtained nets with the original nets. In the stage of PO, the slew prioritization strategy is first used to comprehensively consider slew and delay during the routing process, in order to obtain more reasonable routing results. Simultaneously use a timing critical perception strategy to set the value of α dynamically. The more critical the timing, the greater the value, the larger the net size α value. This stage further optimize slew violation and interconnect delay without worsening the overflow.

3.1 Negotiation-Based Approach

In our approach, a negotiation-based method [23] was proposed to iteratively increase congestion costs and avoid overflow. The congestion cost $cong(n)$ can be computed as follows:

$$cong(n) = p_n \times h_n \tag{8}$$

$$p_n = \max\{0, b \times (u(n) - cap(n))\} \tag{9}$$

where p_n represents the penalty for the current overflow, and h_n represents the historical cost in the current iteration. Besides, $cap(n)$ and $u(n)$ represent the capacity and current usage of net n, respectively. To control the trade-off between overflow and congestion during the routing process, a user-defined constant b is introduced. If the results of the second stage still violate congestion constraints, historical costs are used to increase congestion costs. Specifically, the historical cost of net n in the $(i + 1)$-th iteration is expressed as h_n^{i+1}, which is calculated based on its historical cost in the i-th iteration h_n^i. The iteration count i is determined as follows:

$$h_n^{i+1} = \begin{cases} h_n^i + \rho \times 2^i, & \text{if } n \text{ has overflow} \\ 0, & \text{otherwise} \end{cases} \tag{10}$$

where ρ is a user-defined parameter, which can be used to control the growth rate of historical costs.

3.2 Slew Prioritization Strategy

Slew violation is an important design constraint in timing, and excessive slew violations can compromise the degree of matching between overall and detailed routing. Therefore, on the basis of designing delay optimization algorithms, it is necessary to further design effective strategies to optimize the number of slew violations and ensure that the chip has good performance.

The routing sequence also affects the routing results. Although the optimization of net delay is closely related to the optimization of slew to a certain extent, there are certain differences in their focus. For example, delay optimization focuses on the selection of routing layers for downstream segments, while slew optimization requires simultaneous consideration of routing layer selection for both upstream and downstream segments. Based on this, we propose a slew prioritization strategy that comprehensively considers slew and delay to obtain more reasonable routing results.

To achieve a good balance between interconnect delay and slew violations, we consider the slew as factors in evaluating routing priority. The routing priority is defined as follows:

$$sort(i) = d \times sl(i) + f \times del(i) \tag{11}$$

where $sl(i)$ represents the default number of slew, $del(i)$ represents interconnect delay, and d and f are user-defined parameters.

3.3 Timing Critical Perception Strategy

The selection of appropriate routing layers for different net segments is crucial in determining the timing criticality of each segment. This is particularly important in multi-layer structures where upper layer routing resources are limited, leading to competition among multiple net segments. To address this issue, it is essential to prioritize the selection of routing layers based on their timing criticality. An optimal routing layer assignment aims to minimize both the interconnect delay and the slew violations. To achieve this, an effective strategy must be designed to select the most suitable routing layers for each net segment, and then optimize the interconnect delay and the slew violations of routing layer assignment results.

The algorithm mainly consists of statistics, rip-up, and re-assigns. Lines 1 to 6 are for the statistical stage. At this stage, our strategy involves traversing the 3D routing net to count the number of existing slew violations and delay. Lines 7 to 14 are the rip-up stage. This stage will discuss the classification of time critical and non critical routing nets. For time critical nets, the timing critical perception strategy will dynamically adjust the value α of for each nets. The calculation formula is as follows:

$$\alpha = (k + 1) \times (limit - i + 1) \times \alpha \tag{12}$$

where *limit* represents the number of layers in the critical network. From the formula, we can see that the nets with more critical timing will have a larger α value.

Algorithm 1 Timing critical perception strategy

Input: net set N
1: **for** $k<2$ **do**
2:      sort each net from begin to the end
3:    **for** each net **do**
4:        record each net of old one
5:        record delay of each net
6:        record slew count of each net
7:        rip up every nets
8:        **if** the net is crucial in timing **then**
9:              increase *Lambda* each time until it to 0
10:             let *Beta* be 0.01
11:             dynamic programming routing
12:      **else**
13:             dynamic programming routing
14:        **end if**
15:        **if** the new net's delay is bigger **or** slew count is bigger **then**
16:              rip up the nets and recover the new one
17:              reassign the nets with the old one
18:        **end if**
19:    **end for**
20: **end for**

Lines 15 to 20 constitute the re-assign stage, which is a critical stage in the routing process. During this phase, illegal nets are identified and subsequently reassigned to valid routing layers. To determine the optimal routing scheme, both the delay and the number of slew violations in the nets before and after rewinding are compared. The key objective during the re-assign phase is to minimize the number of slew violations while simultaneously minimizing overall delay. This requires a rigorous assessment of the timing constraints and requirements of each net segment, as well as careful mapping onto the available routing layers. If the reassigned routing scheme exhibits fewer slew violations or lower delay than the existing solution, it will be retained. On the other hand, if the reassigned routing scheme yields more slew violations or increased delay, it will be replaced by the previous last solution.

4 Experimental Results

The proposed algorithm was implemented using the C++ programming language and executed on a Linux workstation equipped with an Intel Xeon processor of 3.5G and 128 GB memory to carry out all experiments. There are a total of 10 benchmarks circuits in the experimental samples. The benchmark circuits set is the DAC12 benchmarks [24]. It provides a set of benchmarks representing modern industrial design. The layout solution of the benchmark was obtained by NTUplace4 [25]. To generate appropriate slew constraints, we followed the approach presented in [20]. Additionally, the 3D global routing results generated by NCTU-GR 2.0 [26] were compressed into 2D global routing results to evaluate the performance of our algorithm.

Table 1. Experimental comparisons of DLA and SYX.

Benchmark	DLA						SYX					
	TD	MD	0.50%	1.00%	5.00%	#sv	TD	MD	0.50%	1.00%	5.00%	#sv
sp2	2222970	503.9	145.0	116.2	51.5	4934	2267690	581.4	156.4	122.8	53.3	6012
sp3	775357	535.1	89.1	66.2	23.2	38743	770432	587.3	90.1	66.6	23.0	45625
sp6	654807	253.7	57.3	44.0	16.9	42125	638596	262.6	55.2	42.1	16.3	46473
sp7	624896	233.2	43.1	31.4	10.9	160274	607791	204.6	42.0	30.3	10.3	153597
sp9	413723	212.5	46.0	34.2	13.1	81698	405126	170.7	44.9	33.1	12.7	79359
sp11	750541	1602.3	109.7	68.2	19.8	27072	749705	1710.9	109.1	67.9	19.6	24988
sp12	542782	581.5	40.2	29.9	10.1	193072	521266	604.0	38.0	27.8	9.5	175935
sp14	407862	205.0	56.0	41.9	15.2	52747	392748	256.3	52.2	38.9	14.4	50658
sp16	544414	186.3	54.4	42.9	18.9	30786	522838	206.8	51.6	40.5	18.0	29875
sp19	193844	304.5	24.6	19.0	8.7	74544	190711	304.3	23.5	8.2	5.0	75143
Ratio	1.000	1.000	1.000	1.000	1.000	1.000	0.981	1.043	0.980	0.918	0.931	1.020

In order to validate the efficacy of the slew prioritization strategy, this paper conducted two experiments to compare the performance of our proposed algorithm with and without this strategy. The experimental results are presented in Table 1, where "DLA" [15] represents "SYX-SGG" algorithm without any proposed key strategies, while "SYX" represents "DLA" algorithm that adopt the slew prioritization strategy. To evaluate the performance of the algorithms, utilize the DAC12 benchmarks, which are commonly used for industrial design evaluations. The evaluation criteria included TD (total delay), MD (maximum delay), and #sv (number of slew violations). Additionally, this strategy evaluated the average delay of the critical nets at various percentages including 0.5%, 1%, and 5%. All delay results were measured in picoseconds. The ratio of each DLA parameter is set to 1.000. For all SYX parameters, the ratio of TD as an example is calculated as $\dfrac{\sum_{i=1}^{n} \frac{SYX\_TD_i}{DLA\_TD_i}}{n}$, where n is the number of benchmark, SYX_TD_i and

DLA_TD$_i$ represent the TD values of algorithm "SYX" and algorithm "DLA" on the *i*th benchmark, respectively.

From Table 1, it can be seen that the performance of the proposed algorithm improved significantly with the adoption of the slew prioritization strategy. The "SYX" exhibited lower TD values compared to the "DLA"', reducing by 1.9%, indicating a reduction in delay. Moreover, the average delay of the critical nets was considerably lower for the "SYX", particularly for networks with high priorities. For instance, for top 0.5% and top 1% delay, the SYX outperformed the DLA by reducing 2% and 8.2%, respectively. For the top 5% delay, the "SYX" still achieved a noteworthy improvement, reducing by 6.9%. And the problem of deterioration is solved in the final algorithm of fusing other strategies.

To verify the effectiveness of the timing critical perception strategy, this algorithm compared the complete version of the "SYX-SGG" algorithm that integrates multiple strategies. That is, an algorithm that adds a timing critical perception strategy on the basis of the slew prioritization strategy.

Table 2. Experimental comparisons of DLA and SYX-SGG.

Benchmark	DLA						SYX-SGG					
	TD	MD	0.50%	1.00%	5.00%	#sv	TD	MD	0.50%	1.00%	5.00%	#sv
sp2	2222970	503.9	145.0	116.2	51.5	4934	2190190	581.4	144.5	115.3	50.9	4594
sp3	775357	535.1	89.1	66.2	23.2	38743	757607	567.6	89.2	66.2	22.7	42899
sp6	654807	253.7	57.3	44.0	16.9	42125	622646	259.4	54.4	41.5	15.9	43858
sp7	624896	233.2	43.1	31.4	10.9	160274	589615	200.1	41.7	29.9	10.1	143183
sp9	413723	212.5	46.0	34.2	13.1	81698	393794	181.2	44.7	32.6	12.3	80708
sp11	750541	1602.3	109.7	68.2	19.8	27072	725343	1605.4	108.3	66.8	19.1	24123
sp12	542782	581.5	40.2	29.9	10.1	193072	511797	546.7	38.5	28.1	9.5	175287
sp14	407862	205.0	56.0	41.9	15.2	52747	379826	215.7	51.7	38.2	13.9	49492
sp16	544414	186.3	54.4	42.9	18.9	30786	515139	210.5	51.7	40.5	17.7	30241
sp19	193844	304.5	24.6	19.0	8.7	74544	181123	272.4	22.2	17.2	7.8	74067
Ratio	1.000	1.000	1.000	1.000	1.000	1.000	0.953	0.996	0.963	0.956	0.947	0.967

The comparative analysis of the proposed "SYX-SGG" algorithm with the previous layer assignment algorithm "DLA" [12] is presented in Table 2. The results indicate that the "SYX-SGG" algorithm has effectively reduced the number of slew violations compared to "DLA".

In terms of delay optimization, the "SYX-SGG" algorithm has shown consistent improvements across all metrics. The total delay, maximum delay, top 0.5% delay, top 1% delay, and top 5% delay have decreased by 4.7%, 0.4%, 3.7%, 4.4%, and 5.3%, respectively. These results demonstrate the effectiveness of the proposed algorithm in improving the timing performance of global routing algorithms. Furthermore, this algorithm has demonstrated excellent performance in reducing the number of slew violations by 3.3%. This indicates a significant improvement in signal integrity, which is crucial for ensuring reliable circuit operation.

5 Conclusion

This paper proposes a slew-driven layer assignment for advanced non-default-rule wires layer assignment algorithm that can use NDR wires to simultaneously optimize overflow, interconnect delay and slew violation. First, a negotiation based approach is adopted to iteratively increase congestion costs and avoid overflow. Second, the slew prioritization strategy adjusts the routing order and comprehensively considers slew and delay to obtain more reasonable routing results. Third, the timing critical network perception strategy dynamically adjusts the weights for each network and selects solutions with fewer slew violations, effectively achieving the overall optimization of the slew. The results demonstrate that this algorithm significantly reduces interconnect delays while simultaneously reducing the slew violations.

References

1. Saxena, P., Menezes, N., Cocchini, P., Kirkpatrick, D.A.: Repeater scaling and its impact on CAD. IEEE Trans. Comput.-Aided Des. Integr. Circ. Syst. **23**(4), 451–463 (2004)
2. Kim, Q.M., Ahn, B., Kim, J., Lee, B., Chong, J.: Thermal aware timing budget for buffer insertion in early stage of physical design. In: 2012 IEEE International Symposium on Circuits and Systems, Seoul, Korea, pp. 357–360 (2012)
3. Liu, G., et al.: Timing-aware layer assignment for advanced process technologies considering via pillars. IEEE Trans. Comput.-Aided Des. Integr. Circ. Syst. **41**(6), 1957–1970 (2022)
4. Jiang, L., et al.: LA-SVR: a high-performance layer assignment algorithm with slew violations reduction. In: 30th International Conference on Very Large Scale Integration, Patras, Greece, pp. 1–6 (2022)
5. Zhou, R., Liu, G., Guo, W., Wang, X.: An X-architecture SMT algorithm based on competitive swarm optimizer. In: Xing, C., Fu, X., Zhang, Y., Zhang, G., Borjigin, C. (eds.) WISA 2021. LNCS, vol. 12999, pp. 39–404. Springer, Cham (2021). https://doi.org/10.1007/978-3-030-87571-8_34
6. Chen, X., Zhou, R., Liu, G., Wang, X.: SLPSO-based X-architecture steiner minimum tree construction. In: Wang, G., Lin, X., Hendler, J., Song, W., Xu, Z., Liu, G. (eds.) WISA 2020. LNCS, vol. 12432, pp. 131–142. Springer, Cham (2020). https://doi.org/10.1007/978-3-030-60029-7_12
7. Zhang, X., et al.: MiniDelay: multi-strategy timing-aware layer assignment for advanced technology nodes. In: 2020 Design, Automation & Test in Europe Conference & Exhibition, Grenoble, France, pp. 586–591 (2020)
8. Naclerio, N.J., Masude, S., Nakajima, K.: The via minimization problem is NP-complete. IEEE Trans. Comput. **38**(11), 1604–1608 (1989)
9. Cho, M., Pan, D.Z.: BoxRouter: a new global router based on box expansion and progressive ILP. IEEE Trans. Comput.-Aided Des. Integr. Circ. Syst. **26**(12), 2130–2143 (2007)
10. Lee, T.H., Wang, T.C.: Congestion-constrained layer assignment for via minimization in global routing. IEEE Trans. Comput.-Aided Des. Integr. Circ. Syst. **27**(9), 1643–1656 (2008)
11. Liu, W.H., Li, Y.L.: Negotiation-based layer assignment for via count and via overflow minimization. In: 16th Proceedings of Asia and South Pacific Design Automation Conference, Yokohama, Japan, pp. 539–544 (2011)
12. Li, Z., Alpert, C.J., Hu, S., Muhmud, T., Quay, S.T., Villarrubia, P.G.: Fast interconnect synthesis with layer assignment. In: Proceedings of the 2009 International Symposium on Physical Design, New York, NY, USA, pp. 71–77. Association for Computing Machinery (2008)

13. Hu, S., Li, Z., Alpert, C.J.: A faster approximation scheme for timing driven minimum cost layer assignment. In: Proceedings of International Symposium on Physical Design, San Diego, California, USA, pp. 167–174 (2009)
14. Ao, J., Dong, S., Chen, S., Goto, S.: Delay-driven layer assignment in global routing under multi-tier interconnect structure. In: Proceedings of International Symposium on Physical Design, Stateline Nevada, USA, pp. 101–107 (2013)
15. Han, S.Y., Liu, W.H., Ewetz, R., Koh, C.K., Chao, K.Y., Wang, T.C.: Delay-driven layer assignment for advanced technology nodes. In: Proceedings of the 2017 Asia and South Pacific Design Automation Conference, Chiba, Japan, pp. 456–462. IEEE Computer Society Press, Los Alamitos (2017)
16. Ewetz, R., Liu, W.H., Chao, K.Y., Wang, T.C., Koh, C.K.: A study on the use of parallel wiring techniques for sub-20 nm designs. In: Proceedings of the 24th Edition of the Great Lakes Symposium on VLSI, New York, NY, USA, pp. 129–134 (2014)
17. Huang, T., Young, E.F.Y.: Construction of rectilinear Steiner minimum trees with slew constraints over obstacles. In: Proceedings ACM International Conference on Computer-Aided Design, New York, NY, USA, pp. 144–151 (2012)
18. Hu, S., Li, Z., Alpert, C.J.: A fully polynomial time approximation scheme for timing driven minimum cost buffer insertion. In: Proceedings of Design Automation Conference, San Francisco, CA, USA, pp. 424–429 (2009)
19. Hu, S., et al.: Fast algorithms for slew-constrained minimum cost buffering. IEEE Trans. Comput.-Aided Des. Integr. Circ. Syst. 26(11), 2009–2022 (2007)
20. Liu, D., Yu, B., Chowdhury, S., Pan, D.Z.: TILA-S: timing-driven incremental layer assignment avoiding slew violations. IEEE Trans. Comput.-Aided Des. Integr. Circ. Syst. 37(1), 231–244 (2017)
21. Elmore, W.C.: The transient response of damped linear networks with particular regard to wideband amplifier. J. Appl. Phys. 19(1), 55–63 (1948)
22. Kashyap, C., Alpert, C., Liu, F., Devgan, A.: Closed-form expressions for extending step delay and slew metrics to ramp inputs for RC trees. IEEE Trans. Comput.-Aided Des. Integr. Circ. Syst. 23(4), 509–516 (2004)
23. McMurchie, L., Ebeling, C.: PathFinder: a negotiation-based performance-driven router for FPGAs. In: Reconfigurable Computing, Napa Valley, CA, USA, pp. 365–381 (2008)
24. Viswanathan, N., Alpert, C., Sze, C., Li, Z., Wei, Y.: The DAC 2012 routability-driven placement contest and benchmark suite. In: Design Automation Conference 2012, San Francisco, CA, USA, pp. 774–782. IEEE (2012)
25. Hsu, M.K., et al.: NTUplace4h: a novel routability-driven placement algorithm for hierarchical mixed-size circuit designs. IEEE Trans. Comput.-Aided Des. Integr. Circ. Syst. 33(12), 1914–1927 (2014)
26. Liu, W.H., Kao, W.C., Li, Y.L., Chao, K.Y.: NCTU-GR 2.0: multithreaded collision-aware global routing with bounded-length maze routing. IEEE Trans. Comput.-Aided Des. Integr. Circ. Syst. 32(5), 709–722 (2013)

Parallelize Accelerated Triangle Counting Using Bit-Wise on GPU

Li Lin, Dian Ouyang$^{(\boxtimes)}$, Zhipeng He, and Chengqian Li

School of Computer Science and Network Engineering, Guangzhou University,
Guangzhou, China
dian.ouyang@gzhu.edu.cn

Abstract. Triangle counting is a graph algorithm that calculates the
number of triangles in a graph, the number of triangles is a key metric for
a large number of graph algorithms. Traditional triangle counting algorithms are divided into vertex-iterator and edge-iterator when traversing
the graph. As the scale of graph data grows, the use of CPU with other
architectural platforms for triangle counting has become mainstream.
Our accelerating method proposes an algorithm for triangle counting on
a single machine GPU, and performs a two-dimensional partition algorithm for large-scale graph data, in order to ensure that large graph data
can be correctly loaded into GPU memory and the independence of each
partition to obtain the right result. According to the high concurrency of
GPU, a bit-wise operation intersection algorithm is proposed. We experiment with our algorithm to verify that our method effectively speeds up
triangle counting algorithm on a single-machine GPU.

Keywords: Triangle Counting · GPU · Bit-Wise

1 Introduction

Triangle counting algorithm is the basic algorithm of other graph algorithms,
such as K-Truss [1], Clustering Coefficient [2]. Traditional triangle counting algorithm can be divided into vertex-iterator and edge-iterator [3]. Vertex-iterator
is to detect the presence of an edge between all neighbors of each vertex. Edge-
iterator iterates through each edge in the graph to find the common neighbor
vertices of the neighbor lists of two points. Depending on how to intersect lists,
there are algorithms including merge-based, binary-search hashing-based and
bitmap [3] algorithms, where bitmap is a special form of hashing.

CPU have very powerful and complex computing units and control capabilities [4], many works have proposed multicore CPUs as well as heterogeneous
triangle counting algorithms, including multi-core CPUs [5], as well as external memory devices [6], and in-memory devices [7]. Our work focuses on the
implementation of triangle counting algorithms on GPU. The existing GPU-
implemented algorithm has achieved great results. [8] uses the basic triangle
counting algorithm of the GPU. [9] allocates the number of threads according to

L. Yuan et al. (Eds.): WISA 2023, LNCS 14094, pp. 551–558, 2023.
https://doi.org/10.1007/978-981-99-6222-8_46

the workload of each edge. [3] proposes the strategy of using bitmap as a list for fast lookup for processing triangle counting, [10] proposed a triangle counting system tricore, a distributed GPU processing strategy, [11] proposed TRUST, a triangle counting method based on hashing and vertex-centric using distributed GPUs, [12] proposed the most basic parallel strategy using one thread to process one edge. [13] based on map-reduce, [9] used binary search, [12] used the merge-based intersection algorithm, [10] proves that the intersection algorithm based on binary search is superior to the merge-based algorithm on the GPU architecture. [14] uses a similar triangle counting method like [10], besides with a system of a single external memory. [15] instead of proposing a new triangle counting algorithm, but designed a preprocessing strategy before calculating triangles.

Unlike previous lists intersection algorithm to perform triangle counting algorithm based on GPU, we propose a triangle counting algorithm on a GPU-based heterogeneous platform, which uses bit-wise on GPU, and adopts a two-dimensional partition algorithm for large-scale graphs. As well as implementing a preprocessing strategy to optimize the direction of edges between vertices, the purpose is to homogenize the balance of workload between thread blocks.

We experimented with our algorithm and others on a single GPU RTX A6000 based running platform, and for different graphs, our triangle counting algorithm is $1.3\times$ to $21.3\times$ faster than the-state-of-the-art binary-search intersection algorithm implemented on GPU [10] and $1.1\times$ to $1.8\times$ faster than merge-based intersection algorithm implemented on GPU [12].

Generally, the contributions of this paper can be summarized as follows.

1) This project is an algorithm for triangle counting on GPU-based heterogeneous platforms. After partition of a large-scale graph, handling the triangle counting subtask of each independent partition, we propose a new triangle counting algorithm of bit-wise intersection algorithm implemented on GPU.
2) Real-world network graphs are very sparse, their vertex distribution is skewed degree distribution. Therefore, we implement a way to change the direction of each edge during preprocessing to solve the problem of workload imbalance of each thread in warps on GPU, avoid thread diversity on GPU.
3) In order to efficiently perform bit-wise operation, we optimize the distribution of values for each row in the compressed sparse row (CSR). In addition, the data is reasonably stored in GPU memory, which reduces frequent data interaction between CPU main memory and GPU.

2   Background

In this section, we analyze the processing task on GPU, and the graph format commonly used and previous work on the triangle counting algorithm.

2.1   GPU Architecture

The threads on GPU are executed according to the SIMD (Single Instruction Multiple Data) style. Each thread on GPU processes an edge, and the adjacency table length of each vertex determines the workload of the threads. The running time of a warp is the longest running time of the thread, if the distribution of adjacency tables of the vertices in the graph is extremely unbalanced, it will lead to an imbalance in thread workload.

2.2 Graph Format

CSR is a mainstream graph data compression format we adopt for data compression, suppose each element in adjacency list is represented as $A[i]$ and each element in begin position array is represented as $B[j]$. The adjacency list stores all edges for each vertex, and the begin position array stores the location of the first element of each vertex adjacency list. That is, the neighbor vertices of vertex v starts at $A[B[v]]$ and ends at $A[B[v+1]] - 1$.

2.3 Existing Solution

Our method is BFS-based edge-parallelism-iterator, and it's intersection algorithm is related to the intersection algorithm implemented by matrix-multiplication method, assuming that matrix A represents the adjacency matrix of the undirected graph $G(V, E)$, the value of $A^2[i][j]$ represents the vertex i, j has how many paths of length 2-hop. When $A[i][j] = 1$ and $A^2[i][j] = 1$, it means i, j have both 1-hop path and 2-hop paths to form a triangle. Our method uses bit-wise to optimize matrix-multiplication method, it avoids excessive spatial overhead of the latter and simplifies computational complexity.

3 Method

In this section, our goal is to use bit-wise to execute the triangle counting algorithm on the GPU. Figure 1 shows overview of our method.

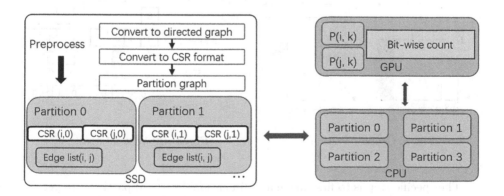

Fig. 1. Bit-wise triangle counting overview.

3.1   Preprocessing

Duplicate edges and looped edges will be removed, and the edge data structure of the original undirected graph is converted into the edge list representation of the directed graph. We use the CSR data format to save all vertex information, and the amount of space used occupies $|V| + |E|$, where $|V|$ is the number of vertices in the graph, $|E|$ is the number of edges in the graph. In the process of converting to a directed graph, it is common practice to use the rank-by-degree strategy to set the direction of the edges, but this usually brings about the imbalance of thread workload.

Therefore, we implemented a preprocessing strategy to balance out-degree of vertices. We first collect the degree $d(u)$ of each vertex u and calculate the average degree $\overline{d} = (|E|)/(|V|)$, when $d(u) \geq \overline{d}$, we treat vertex u as a center vertex, denoted by v_c, otherwise a non-center vertex, denoted by v_n. When two vertices of an edge are v_c, connected to v_n, we set the direction of the edge is $v_n \rightarrow v_c$, when both two vertices of an edge are v_c, or both are v_n, the pointing of the edge can be set arbitrarily. In this way, the degree of v_c is reduced and the degree of v_n is increased. When the two connected vertices are v_c, or both are v_n, the direction of the edge can be set arbitrarily, because changing either of the two will result in the same. Detailed proof is in [15].

3.2   Graph Partition

To avoid running out of memory, the graph data must be partitioned. First, we use a 2-dimensional graph partition algorithm to divide the task of triangle counting into multiple subtasks, each subtask needs to have two corresponding CSR partitions data, and an edge list partition provides data support. Figure 2 shows the process of creating a 2-dimension CSR and 2-dimension edge list.

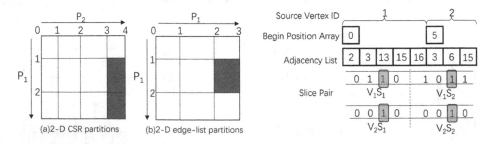

(a)2-D CSR partitions   (b)2-D edge-list partitions

Fig. 2. Graph partition.   **Fig. 3.** Bit-wise intersection.

The specific step is to first make a 1-dimensional vertical partition to get P_2, so that each P_2 partition has the same size as possible. Then take the same strategy, make a horizontal partition, take the smallest horizontal partition size as the size of the horizontal vertex range partition, and get partition P_1. Figure 2(a) shows the size of the CSR after partitioning as $P_1 \times P_2$, Fig. 2(b) is the corresponding edge list partition as $P_1 \times P_1$. For example, edge list partition is (i, j), we need CSR partitions (i, k) and (j, k), the value of k is 0 to 3, $i = 1$, $j = 2$, that means we need to take the CSR partitions data 4 times to obtain the triangles in edge list partition $(1, 2)$.

3.3   Bit-Wise Triangle Counting Algorithm

Adjacency matrix $A[i][j]$ refers to whether vertex i, j is connected by an edge, and $A^2[i][j]$ indicates that vertex i, j has a connection of 2-hop path. And the number of triangles in graph G is the number of non-zero elements in matrix A AND matrix A^2. As the values of $A[i][j]$ can only be 1 or 0, the operation of matrix multiplication can be converted to bit-wise operation of AND.

   To get the bit slices, for edge (u, v), we will get their corresponding neighbor lists u_list, v_list, respectively, then each time get m(slice fixed size) values $t \in [k, k + m - 1]$ of u_list and v_list. If $u\_list[t] = 1$, we set the corresponding bit of a slice to 1. After the data is read, the algorithm of intersecting the neighboring vertices of two vertices is calculated by parallel bit-wise operation to calculate the number of triangles. For the retrieved CSR partition data, we convert the data into bit slices data with a fixed slice size. The pseudocode is shown in the Algorithm 1, the neighbor nodes of vertex u are from the col array and row array of CSR, and using bit-wise to intersect the k-th slice of vertex u and v. The INTERSECT phase is shown in the Fig. 3, assuming that $m = 4$, and the jth slice of each vertex i is represented as $V_i S_j$.

Algorithm 1. Triangle counting with bit-wise

1: **Input** : Graph CSR row[], col[], edgeId
2: **Output** : The number of triangles in Graph
3: TCnum=0
4: upos=col[edgeId]
5: u=row[upos]
6: adjList[u]=col[u+1]-col[u]
7: **foreach** v $\in$ adjList[u] **do**
8:     **foreach** slice pair $(\mathbf{V_u S_k}, \mathbf{V_v S_k})$ **do**
9:         TCnum+=INTERSECT$(\mathbf{V_u S_k}, \mathbf{V_v S_k})$
10:     **end for**
11: **end for**
12: **return** TCnum

Vertex 1 is connected to vertex 2, requiring the number of triangles composed by edge $(1, 2)$, starting from the index position $idx = 0$, corresponding vertex is 2, starting to fill $V_1 S_1 = 0110$ with slice size 4, and construct the corresponding slice $V_2 S_1 = 0010$ of vertex 2, and perform AND bit operation on the two slices, and the result is 1. Consecutive values of 0 at vertex 1 are skipped. At the next time, $idx = 3$, corresponding vertex is 13, fill the slice $V_1S_2 = 1011$ of vertex 1, and the slice $V_2S_2 = 0010$ of vertex 2, and the AND bit operation is also performed, and the result is 1, the total number of triangle is 2.

Although bit operations are performed on a row and column of the CSR, the CSR is stored in the GPU memory, frequent data interaction between the CPU main memory and the GPU is reduced, and the use of storage space in the GPU is not particularly high, the cores of the GPU can be fully utilized.

4   Evaluation

In this section, we will introduce the implemented experimental environment and the graph datasets and then evaluate the results of the experiment.

4.1   Experimental Environment

We implemented the method on GPU RTX A6000, Intel(R) Xeon(R) Silver 4216 CPU, CUDA toolkit is 7.0, including nvcc, with gcc version 9.3.0 and the operating system was ubuntu20.04, we set the compilation flag to $-O2$.

The datasets we use are from Stanford Network Analysis Project, they are all real-world social network graphs data. All of them will first be transformed from the original data format in order to fit our method.

Table 1. Running Time of Different Triangle Counting Methods on GPU (in seconds).

Dataset	Size	Nodes	Edges	Triangles	Tricore	Adam	Tcbw (CPU)	Tcbw (GPU)
roadNet-PA	44M	1.04M	1.47M	65.57K	0.23	**0.18**	1.64	**0.18**
roadNet-CA	84M	1.87M	2.64M	0.12M	**0.19**	0.27	5	0.25
com-Amazon	13M	0.32M	0.88M	0.64M	0.27	0.24	0.33	**0.14**
facebook-cb	836K	4K	0.08M	1.54M	0.26	0.21	**0.05**	0.12
com-DBLP	14M	0.3M	1M	2.12M	0.27	0.18	0.39	**0.14**
com-Youtube	37M	1.08M	2.85M	2.91M	0.41	0.25	1.38	**0.18**
cit-Patents	268M	3.6M	15.75M	7.17M	0.97	**0.47**	35.5	0.65
as-Skitter	143M	1.62M	10.58M	27.44M	0.74	0.41	7.59	**0.37**
com-lj	479M	3.81M	33.07M	169.6M	0.91	**0.83**	63.6	1.07
twitter-cb	43M	79K	1.69M	12.48M	3.4	0.2	0.68	**0.16**
com-orkut	3.2G	2.93M	0.11G	0.58G	11.65	**2.47**	316	5.56
com-Friendster	31G	62.57M	1.68G	3.88G	**46.84**	*	-	262.07

* means that can not run successfully on the current device.
- means that result is more than 30 min.

4.2   Evaluation of Experimental Results

We focus on the comparison of triangle counting algorithms on GPU, we compare the following algorithms implemented on GPU:

Tricore [10] is the state-of-the-art triangle counting using binary search based GPU.

Adam et al. [12] is the state-of-the-art triangle counting using merge-based on GPU.

From Table 1, the result shows the running time of different GPU triangle counting algorithms, we can know Adam et al. uses vertex sorting by the size of vertex degree. However, let the small degree vertex point to the larger degree vertex is not suitable for the graph of skewed degree distribution. The preprocessing time is proportional to the number of sides, and the time to calculate the number of triangles is proportional to the number of triangles. Adam et al. does not have the ability to handle large-scale graph and does not design partitioning algorithms. Tricore and our method apply partitioning algorithms that doubles the size of the graph, but has the ability to handle large-scale graphs. Tricore doesn't designed vertex average degree algorithms, which will lead to a greater impact on the speed of graph running for vertex degree distributions that vary greatly, such as com-Amazon, facebook-combined, twitter-combined, our algorithm is nearly 1.3× to 21.3× faster than Tricore. On large graph, Tricore is better than our method, because the slices for bit-wise is regenerated each time, and that's what we're working on in future, Adam et al. performs best at some small graphs, because it uses a simple graph format and does not implement the graph partition algorithm, but our method still has efficiency gains on some graphs, 1.1× to 1.8× faster than Adam et al. And we also implement our method on CPU, using Intel(R) Xeon(R) Silver 4216 CPU, which shows in the column of Tcbw (CPU). Our GPU method is almost 4.3× to 59.4× faster than the CPU method, the latter does not have the consumption of data transfer between the GPU and the CPU, it has an advantage in graphs with high cache hit ratios like facebook-combined, but the degree of parallelism is not as high as the former. The default running time is measured in seconds.

5   Conclusion

From the above experimental results and analysis, we propose a method that can calculate the number of triangles correctly, and accelerate the algorithm computational efficiency. We implement a preprocessing algorithm that optimizes threads balancing on GPU and partitioning for large-scale graph data to avoid running out of GPU memory. The results show that our proposed method consumes fewer running time than existing work on skewed degree distribution graphs, and that optimized the efficiency of triangle counting algorithm.

Acknowledgments. The authors are very grateful to the anonymous reviewers for their valuable comments.

References

1. Chen, P.-L., Chou, C.-K., Chen, M.-S.: Distributed algorithms for k-truss decomposition. In: 2014 IEEE International Conference on Big Data (Big Data), pp. 471–480. IEEE (2014)

2. Li, X., Chang, L., Zheng, K., Huang, Z., Zhou, X.: Ranking weighted clustering coefficient in large dynamic graphs. World Wide Web **20**, 855–883 (2017). https://doi.org/10.1007/s11280-016-0420-2
3. Bisson, M., Fatica, M.: High performance exact triangle counting on GPUs. IEEE Trans. Parallel Distrib. Syst. **28**(12), 3501–3510 (2017)
4. Qi, X., Wang, M., Wen, Y., Zhang, H., Yuan, X.: Weighted cost model for optimized query processing. In: Zhao, X., Yang, S., Wang, X., Li, J. (eds.) WISA 2022. LNCS, vol. 13579, pp. 473–484. Springer International Publishing, Cham (2022). https://doi.org/10.1007/978-3-031-20309-1_42
5. Shun, J., Tangwongsan, K.: Multicore triangle computations without tuning. In: 2015 IEEE 31st International Conference on Data Engineering, pp. 149–160. IEEE (2015)
6. Giechaskiel, I., Panagopoulos, G., Yoneki, E.: PDTL: parallel and distributed triangle listing for massive graphs. In: 2015 44th International Conference on Parallel Processing, pp. 370–379. IEEE (2015)
7. Wang, X., et al.: Triangle counting accelerations: from algorithm to in-memory computing architecture. IEEE Trans. Comput. **71**(10), 2462–2472 (2021)
8. Wang, L., Wang, Y., Yang, C., Owens, J.D.: A comparative study on exact triangle counting algorithms on the GPU. In: Proceedings of the ACM Workshop on High Performance Graph Processing, pp. 1–8 (2016)
9. Green, O., et al.: Logarithmic radix binning and vectorized triangle counting. In: 2018 IEEE High Performance Extreme Computing Conference (HPEC), pp. 1–7. IEEE (2018)
10. Hu, Y., Liu, H., Huang, H.H.: TriCore: parallel triangle counting on GPUs. In: SC 2018: International Conference for High Performance Computing, Networking, Storage and Analysis, pp. 171–182. IEEE (2018)
11. Pandey, S., et al.: TRUST: triangle counting reloaded on GPUs. IEEE Trans. Parallel Distrib. Syst. **32**(11), 2646–2660 (2021)
12. Polak, A.: Counting triangles in large graphs on GPU. In: 2016 IEEE International Parallel and Distributed Processing Symposium Workshops (IPDPSW), pp. 740–746. IEEE (2016)
13. Kolda, T.G., Pinar, A., Plantenga, T., Seshadhri, C., Task, C.: Counting triangles in massive graphs with MapReduce. SIAM J. Sci. Comput. **36**(5), S48–S77 (2014)
14. Huang, J., Wang, H., Fei, X., Wang, X., Chen, W.: TC-Stream: large-scale graph triangle counting on a single machine using GPUs. IEEE Trans. Parallel Distrib. Syst. **33**(11), 3067–3078 (2022)
15. Hu, L., Zou, L., Liu, Y.: Accelerating triangle counting on GPU. In: Li, G., Li, Z., Idreos, S., Srivastava, D. (eds.) SIGMOD 2021: International Conference on Management of Data, Virtual Event, China, 20–25 June 2021, pp. 736–748. ACM (2021)

RL-Based CEP Operator Placement Method on Edge Networks Using Response Time Feedback

Yuyou Wang[1], Hao Hu[1(✉)], Hongyu Kuang[1], Chenyou Fan[2], Liang Wang[1], and Xianping Tao[1]

[1] State Key Lab for Novel Software Technology, Nanjing University, Nanjing, Jiangsu, China
wangyy@smail.nju.edu.cn, {myou,khy,wl,txp}@nju.edu.cn
[2] South China Normal University, Guangzhou, Guangdong, China
fanchenyou@scnu.edu.cn

Abstract. The placement of operators in Complex Event Processing (CEP) services, handling real-time data with DAGs, faces challenges due to the NP-hard nature and edge environment complexity. Prior research by Cai et al. used a predictive and greedy approach to minimize delay during placement, but it degrades with increased node count or event rate fluctuations. We propose a novel approach for CEP operator placement using response time feedback to adapt to the dynamic edge environment. We formulate the problem as a Markov Decision Process and use reinforcement learning for optimal policy learning. Our objective is to minimize total response time in edge environments. Extensive simulations evaluate our approach, which outperforms the greedy method with a 25% average reduction in response time.

Keywords: reinforcement learning · complex event processing · edge computing · operator placement · response time

1 Introduction

Complex event processing (CEP) services can process large amounts of real-time data by using a set of operators organized as a directed acyclic graph called *operator graph*. CEP services were commonly deployed on cloud servers and consumers could access them through the network. However, the bottleneck of network communications often limits the quality and user experiences of CEP services. Recently, there is a growing research interest on deploying CEP services on edge networks to improve the quality of service (QoS) [3,13]. The reason of this trend is that moving CEP services from the remote cloud data center to the edge network can significantly reduce the data transmission from cloud servers to the end users [12] and thus help improve the QoS of redeployed services. However, edge devices have limited computing resources compared to servers in data centers. Thus, traditional operator placement approaches [9,10] are not suitable for edge networks.

© The Author(s), under exclusive license to Springer Nature Singapore Pte Ltd. 2023
L. Yuan et al. (Eds.): WISA 2023, LNCS 14094, pp. 559–571, 2023.
https://doi.org/10.1007/978-981-99-6222-8_47

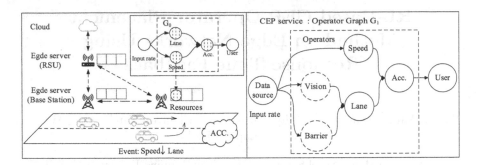

Fig. 1. Operator Placement of a CEP Service on an Edge Network (left) and Operator Graph of Accident CEP Service (right). Edge resources are represented by square boxes.

Researchers further proposed several strategies. These strategies choose to either deploy operators on the nearest edge node [13] in the network, or deploy operators evenly distributed in the edge network [4]. However, these two strategies cannot effectively improve the QoS of CEP services deployed on the edge network due to the easily caused congestion, or not fully using the computing resources of nearby edge nodes, respectively.

Recently, Cai et al. [2] proposed a combined model describing both CEP services and edge network. The benefits of adopting Reinforcement Learning (RL) are listed in the literature [20]. Unlike heuristic and metaheuristic methods that may get stuck in local optima and have difficult-to-adjust hyperparameters, RL methods are self-learning and self-adaptive, making them suitable for modeling complex high-dimensional scenarios in the real world. The complexity and uncertainty of placement problem make traditional optimization methods less efficient, whereas RL are suitable options as they can learn the optimal offloading policy by interacting directly with the environment without prior knowledge. They identified that the response time is a critical QoS metric and thus proposed an approach to reduce the average response time of CEP services on the edge by greedily deploying operators of CEP services with current minimal delay based on predicted response time for each CEP service. They assumed that the event input rate remains unchanged when making a placement decision and predicted the response time with fixed monitoring intervals. However, in a real-world edge environment, their assumption is often violated, and their approach deteriorates rapidly when either the input rate or the number of operators increase while the edge node capacity remains unchanged or even decreases. Therefore, a major challenge is how to deal with the changing edge environment when making deployment decisions.

This paper proposes an approach based on reinforcement learning (RL) to solve the problem. The approach considers three important aspects of the problem: the size of operator graphs, the computing resources of edge nodes, and the events input rate. Before diving into the technical details, we first explain our general idea with an example.

Figure 1(left) illustrates a scenario where complex events are employed to assist in traffic management. In this scenario, autonomous vehicles are present, and an edge network is deployed on base stations and road side units (RSUs) to provide assistance services for drivers. The objective is to notify drivers of potential car accidents by analyzing speed, lane position, vision, and obstacles. The CEP service involves a simplified operator graph with three operators, but it can become more complex with additional operators like speed, lane, vision, obstacle, and accident. Note that the size of the operator graph (Ω), a.k.a, the number of operators in graph, indicates the complexity of the task of the CEP service. Theoretically, the number of feasible placement candidate of CEP services is exponential to the number of operators in Ω. In addition, because each edge node in the edge network has limited resources, it is infeasible to place an entire CEP service in a single edge node. Moreover, the event input rate (denoted as *rate*) of operator graph G_0 and G_1 changes with the traffic from idle to busy. The input rate may increase because of traffic jam or change randomly following some distribution, e.g. Poisson distribution. We argue that the input rate should be taken into consideration combining operator graphs and computing resources when making a placement decision.

In this paper, we propose a RL-based approach using response time feedback to adapt the dynamic edge environment. The main contributions of this paper include:

- We model the operator placement as a Markov decision process, and add input rate (denoted as *rate*) as a variable in our model.
- We propose an RL-based CEP placement approach on edge networks with ϵ-greedy heuristic to select optimal placement.
- We conduct a simulation based evaluation on Omnet++, and the simulation results show that our approach can reduce the response time by 25% on average.

The rest of the paper is structured as follows. Section 2 reviews background. Section 3 describes the problem formulation and presents our RL-based algorithm. Section 4 reports the evaluation results, Sect. 5 presents some related work, and finally, Sect. 6 concludes the paper.

2   Background

2.1   Edge Network

Edge nodes are represented as vertexes V_{edge} [2]. The network connections between edge nodes are represented as E_{edge}. An edge network can be represented as $G_{edge} = (V_{edge}, E_{edge})$. Every node v_i in the edge network has an attribute: $c(v_i)$, the amount of resources available in the edge node v_i. When a user sends a query to the nearest edge node for computing resources, the edge node manages the placement of the operator graph. This edge node is called as *manage edge node*. Events are sent to the operator graph for processing. The input rate of events is represented as *rate*.

2.2   Operator Placement Problem

When the manage edge node receives a CEP service query, the operator graph is pushed into an operator-graphs-queue $Q = \{G_{cep}(1), \ldots, G_{cep}(N_{cep})\}$. Optimal operator placement problem in edge computing consists in determining a suitable mapping between operators Ω and edge nodes V_{edge} to minimize the average response time of operator graphs.

Response time is the time of responding along the slowest path from data source to data consumer. The response time of an operator graph G_{cep} can be calculated as (Table 1):

$$T_r(G_{cep}) = \max_{p \in \pi_{G_{cep}}} T_p. \tag{1}$$

where $\max\limits_{p \in \pi_{G_{cep}}} T_p$ means the worst end-to-end delay from a data source to the consumer. T_p denotes the delay of a path in the operator graph G_{cep}. The path p can be represented as $(\omega_{p_1}, \omega_{p_2}, \ldots, \omega_{p_{n_p}})$, where n_p denotes the number of operators in p. These operators are deployed on the edge nodes $(v_{p_1}, v_{p_2}, \ldots, v_{p_{n_p}})$.

Table 1. Main notation adopted in the paper

Symbol	Description
$G_{edge} = (V_{edge}, E_{edge})$	The edge network G_{edge} consisting of the edge nodes V_{edge} and the connections between edge nodes E_{edge}
v_i	The ith edge node in V_{edge}
(v_i, v_j)	The connection between the edge nodes v_i and v_j
$c(v_i)$	The resources capacity of the edge node v_i
$G_{cep} = (\Omega, L)$	An operator graph consisting of the operators Ω and the event streams L
$Q = \{G_{cep}(1), \ldots, G_{cep}(N_{cep})\}$	The set of operator graphs Q to be deployed on G_{edge}
Ω	The set of operators of G_{cep}
$w(v_i, v_j)$	The transmission rate between edge nodes v_i and v_j
$T_r(X)$	The response time of the operator graph according to a placement X
$X(\omega_i) = v_u$	Deploy the operator ω_i on the edge node v_u
Δ_t	The time interval of replacement judgement
$rate$	The input rate (events per second)

3 RL-Based Operator Placement

3.1 Problem Formulation

As Fig. 2 shows, there are three different types of operator graphs (i.e., G_1, G_2, G_3), each of which may has many *operator graph instances* consisting of multiple operators that need to be placed on edge nodes. In experiment, each time of making a placement decision, we need to deploy operator graph instances including three operator graphs. All the operators from operator graph instances are put in virtual operator candidate pool. We formulate the operator placement problem on edge networks as an RL problem, which can be solved through iteratively learning the placement policy during training procedures. We formulate the edge environment and deployment situation as the environment in RL, and we assume a virtual "decision maker" as the agent who is making placement decisions by following the placement policy. The agent can observe the state s of the environment and select corresponding actions a according to current state. This action will have an impact on the environment and make the state of the environment change with a certain probability called the state transition probability. The feedback of a certain placement is measured by a pre-defined reward function. Markov Decision Process (MDP) is a mathematically idealized form of the reinforcement learning problem for which precise theoretical statements can be made [14]. In particular, our MDP is characterized by a 4-tuple (S, A, P, R), detailed as below:

State S: is a finite state space. A state that an agent can observe is represented by a set of vectors, including: The resource capacity of each edge node: $\langle c(v_1), c(v_2), ..., c(v_i), ..., c(v_n) \rangle$, where v_i is the ith edge node in V_{edge} and n is the number of edge nodes. The transmission rate between edge nodes v_i and v_j: $w(v_i, v_j)$, which means there is a connection between the edge nodes v_i and v_j. The rate of input event *rate*. In particular, considering the state space explosion

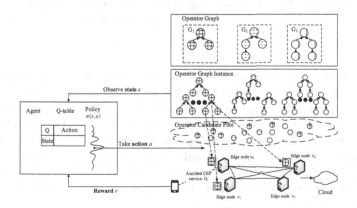

Fig. 2. Markov Decision Process for operator placement problem; MDP is characterized by a 4-tuple (S, A, P, R).

problem, we divide the input rate range into k discrete intervals according to the upper and lower limits.

As mentioned above, $S = \{c, w, rate\}$. Specifically, state $s_{nk} \in S$ represents the edge network $G_{edge} = (V_{edge}, E_{edge})$ has n edge nodes, each node has capacity c, the input rate of events for operator graph is divided into k discrete intervals and the current state of input rate for CEP is $rate$.

Action A: is a finite action space. An action w.r.t. a state is defined as placing a certain operator on a certain edge node. Specifically, action $a_{iu}^{cep} \in A$ represents placing an available operator w_i, $w_i \in \Omega$ from operator graph G_{cep} on edge node v_u at current state, where Ω is the set of operators of G_{cep}. Formally, an action a represents a mapping from operator w_i to edge node v_u as $X(w_i) = v_u$.

Probability P: is the state transition probability. In our model, after taking an action, the current state will transfer to next state and the state transition probability is determined by the resources required for action a and the environment. Thus, the state transition probability is hard to formalize, which is also known as model-free reinforcement learning. In addition, We use ϵ-greedy policy to balance exploration and exploitation of different placement strategies [5], with a gradually decreasing ϵ defined as:

$$\epsilon = \begin{cases} \epsilon_0 - \frac{count}{N_r} & \text{if } \epsilon_0 - \frac{count}{N_r} > \epsilon_e, \\ \epsilon_e & \text{if } \epsilon_0 - \frac{count}{N_r} < \epsilon_e. \end{cases} \quad (2)$$

Table 2. Reward

Action	Reward	Description
$X(w_i) = v_u (i < |\Omega|)$	0	Legal action, but operator graph deployment has not been completed
$X(w_i) = v_u (i = |\Omega|)$	$-T_r(G_{cep})$	Legal action, and operator graph deployment completed

Reward R: is a reward function to measure the quality of an action taken in the current state, which we define as $R(s, a)$. As mentioned above, the goal of our problem is to reduce the response time $T_r(G_{cep})$, which is influenced by the input rate according to the formula of response time [2]. Therefore, we design the reward value to be inversely proportional to the response time observed after applying an action. If an action is illegal $(c(v_u) < c(w_i)$, the resources required by placing operator w_i exceed those provided by the edge node v_u), this action will not be taken and receive no reward. The reward function is defined in Table 2.

Value Function: The goal of reinforcement learning is to learn an optimal policy of selecting the optimal action conditioned on the current state, which is defined as the mapping from state to possible action $\pi : s \rightarrow a$. In order to find the optimal policy, immediate reward is used as the signal of the quality of the selected action. In addition, the agents will interact with the environment through selected actions and thus the ultimate goal is to maximize the total reward in the long run. Therefore, an optimal strategy should take into account the long-term situation and balance immediate reward and long-term reward. A common practice is to set the value of a state as the total amount of reward an agent can expect to accumulate over the future, starting from that state. Thus, the state-action value function (Q-function) $Q_\pi(s_t, a_t)$ and state value function $V_\pi(s_t)$ are defined as below, where γ is the discount factor of future reward.

$$Q_\pi(s_t, a_t) = E_{s_{t+1}, a_{t+1}, \ldots}[\sum_{l=0}^{\infty} \gamma^l r_{t+l}]. \tag{3}$$

$$V_\pi(s_t) = E_{a_t, s_{t+1}, \ldots}[\sum_{l=0}^{\infty} \gamma^l r_{t+l}]. \tag{4}$$

We now describe our valued-based RL approach to solve the operator placement problem. To search for the optimal policy, we learn an optimal state-action value function which estimates the action probability distribution given a state. In the operator placement scenario, the action is to place an operator on an edge node, so that the action space is finite and discrete.

The main idea of our algorithm is that the agent observes the reward obtained after applying the action and updates the Q-value function accordingly. If the action selected by the agent gets a lower response time (higher reward), the reward will gradually accumulate. We will use "RLG" to refer to our algorithm in the following sections.

RLG is based on the classic model-free reinforcement learning method *Temporal-Difference Learning, TD* [16]. RLG is an on-policy TD algorithm. On-policy means the agent learns about policy π from experience sampled from π. *Temporal-Difference Learning* is proved to be optimal under certain conditions. RLG use value function in Eq. (3) to estimate the long-term reward of a state-action combination. All learned values from this function constitute the *Q-matrix* as shown in Fig. 2. The optimal action-value function is defined as:

$$Q^*(s, a) = E[R(s, a) + \gamma \max_{a' \in A_{s'}} Q^*(s', a')]. \tag{5}$$

In the beginning, RLG initialize the *Q-matrix* with all 0 entries. In learning process, the *Q-matrix* updates following the rule:

$$Q(s, a) \leftarrow Q(s.a) + \alpha[r + \gamma Q(s', a') - Q(s, a)]. \tag{6}$$

Although the value function $Q(s, a)$ is different from the optimal $Q^*(s, a)$ at the beginning, the RLG algorithm makes $Q(s, a) \to Q^*(s, a)$ after enough iterations. In training, we set learning step $\alpha = 0.5$. We tried different discount rate and finally we set $\gamma = 0.9$ and the initial value of greedy exploration probability is 0.5. Thus, at the beginning, it is more likely to explore new strategies.

In RLG, we use ϵ-greedy method to solve the exploration-exploitation dilemma [5] in reinforcement learning. Specifically, we select action randomly with probability ϵ, or we choose the action that get the highest expectation of $Q(s, a)$ with probability $1 - \epsilon$. At the same time, in order to speed up the convergence of the algorithm, we use a gradually decreased ϵ. When the training reaches enough rounds, the probability of exploration is gradually decreased to ϵ_e, where ϵ_e is usually very small. This ϵ-greedy exploration method tries to exploit actions with high known rewards with high probability to take advantage of the learned experiences, and also makes sure that the agent makes sufficient explorations.

4   Evaluation

4.1   Experiment Settings

We compare our algorithm with three baseline algorithms: **Greedy Algorithm (GA)** [13] deploys operators on the edge node closest to the input streams of the operators except the edge nodes which do not have sufficient resources. **Load Balance Algorithm (LB)** [4] calculates the average response time of every edge node, and then reduces the load of the overloaded edge nodes by redirecting some of the operators to the underloaded edge nodes. **Response Time Aware Algorithm (RTA)** [2] first calculates the response time of different paths in an operator graph, and then improves the placement of the path with the largest end-to-end delay.

Because both GA and LB do not consider the application structure, we formulate the following rule for both algorithms: *the order of placing operators is determined by the shortest logical distance from the data source.* We conduct a series of experiments to study our proposed approach on three aspects of our problem: the size of operator graphs, the computing resources of edge nodes and the input rate as mentioned in Sect. 1:

Scalability to Different Operator Graphs and Edge Networks. We first evaluate algorithms in a network where different operator graphs are deployed as well as in different edge networks where the same operator graphs are deployed. Then we evaluate different edge networks having different capacity to simulate edge networks with different computing resources. We also evaluate different size of networks and simulate a more complicated edge network.

Robustness to Different Input Rate. We focus on the difference between RLG and RTA when the input rate is gradually increased in range of 600 eps– 2800 eps. We change the operator graph and edge networks in the same way in scalability (different number of operators or different node capacity).

Influences of Different Monitor Intervals. We measure the response time of two different input rates *rate* which in range of 600 eps–2800 eps. We also compare the performance of RTA at different monitoring intervals with our approach. We set $\Delta_t = 1, 5, 10\,\text{s}$, respectively.

4.2   Simulations

We use Omnet++ [18] to simulate MCEP [8] and internet of vehicles. Also we use the open source traffic simulation environment SUMO [1] to produce a traffic environment with more than 2500 vehicles running on real geolocations which we obtained from OpenStreetMap (OSM) (www.openstreetmap.org). We extract the road number, position and acceleration as supporting data in order to perform CEP service operator placement.

Edge Networks. In the scalability experiment, we run three sets of simulations. For first simulation, we run the algorithms in a network where different operator graphs are deployed. The edge network G_{edge} has 10 nodes and 15 random connections. For second simulation, we run our algorithm and baseline algorithm in the different edge networks where the same operator graphs are deployed. For third simulation, we run our algorithm and baseline algorithm in the different edge networks with different size. In each simulation, the transmission rate $w(v_i, v_j)$ is generated randomly between 10 to 20 Mbps following the previous work.

Operator Graph. We randomly generate three different types of operator graphs, each type of operator graph contains 3, 4, and 5 operators as shown in Fig. 2.

Input Rate. In the input rate and monitor intervals experiments, we gradually increase the input rate *rate* of the operator graphs in range of 600 eps–2800 eps to simulate a system from idle to busy. We divide the total range into eight segments, i.e., $[600, 875), [875, 1150)..., [2525, 2800)$, each with uniform range without overlapping.

4.3   Experiment Results

Scalability to Different Operator Graphs and Edge Networks. Figure 3 shows the total response time of RTA and RLG when the input rate follows the Poisson distribution. Overall, RLG outperforms the RTA. Next, we will analyze the detailed differences between RLG and baseline methods from the following three aspects.

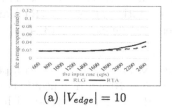
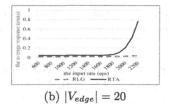

(a) $|V_{edge}| = 10$ (b) $|V_{edge}| = 20$

Fig. 3. The response time difference with different size of edge networks.

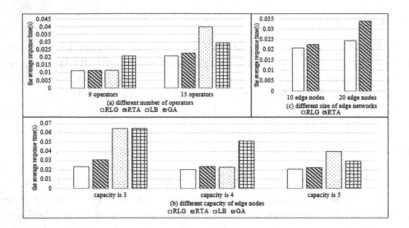

Fig. 4. The response time comparison with different operator graphs and edge networks between RLG, GA, LB and RTA.

Number of operators: As Fig. 4(a) shows, the more operators needed to be deployed, the higher the response time. Compared with all other baseline algorithms, RLG method increases less, indicating that RLG is more scalable when deploying different operator graphs with different number of operators. **Node Capacity:** As Fig. 4(b) shows, when the edge networks provide less computing resources, the response time of all methods is significantly increase. In all cases, RLG achieves better results compared with all other baseline algorithms. **Size of edge networks:** As Fig. 4(c) shows, when the edge networks get more complicated (the edge networks contain more edge nodes), the response time of both methods significant increases. Comparing the two methods in the same edge network (20 edge nodes), RLG gets about 25% improvement compared with RTA.

Robustness to Different Input Rate. As Fig. 5(a), (b), (c) shows, when the operator graph is relatively simple, both methods are almost the same. When number of operators increase, RLG is clearly much more robust than RTA. For example, as shown in Fig. 5(c), when $\Omega = 5$ the input rate is 2800 eps, the response time of RTA is 0.12 s, while that of RLG is 0.04 s. As shown in Fig. 6, with increase of input rate, RLG consistently outperforms RTA and yields smaller response time with different choices of node capacity (3, 4, 5).

Influence of Different Monitor Intervals. Figure 7(a) compares the response time of RLG with RTA of $\Delta_t = 1, 5, 10$, with continuous changing input rate from 800 to 2800 eps. We observed that RTA method with different monitoring intervals yielded quite different results, and all three RTA variants performed less effectively than RLG. One reason is that RTA method is quite sensitive to the choice of monitoring interval, and when input rate is high while monitoring interval is large, the predicted response time becomes totally deviated

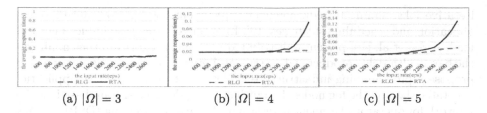

(a) $|\Omega| = 3$ (b) $|\Omega| = 4$ (c) $|\Omega| = 5$

Fig. 5. The response time with different operator graphs when the input rate gradually increases. In each simulation we place three operator graphs in total.

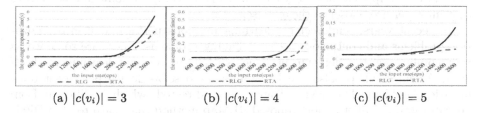

(a) $|c(v_i)| = 3$ (b) $|c(v_i)| = 4$ (c) $|c(v_i)| = 5$

Fig. 6. The response time with different node capacity when the input rate gradually increases. In each simulation we place three operator graphs in total.

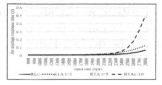
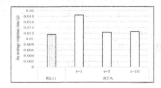

(a) Input rate increases gradually

(b) Input rate follows the Poisson distribution

Fig. 7. Comparison of response time between RLG and RTA under different input rates.

from the actual response time. RLG does not suffer from this issue and becomes quite robust against different input rates. Figure 7(b) shows the input rate follows the Poisson distribution. Interestingly, a larger the monitoring interval gives a lower response time of placement. This is because RTA method makes more accurate predictions of response time after monitoring longer time to predict the subsequent input.

5   Related Work

Tziritas et al. [17] propose that place operator graph in network to help to reduce the network traffic. They propose the basic form of *operator placement problem*. Several algorithms [9,10,17] have been proposed to solve operator placement problem. These algorithms are characterized by different assumptions and optimization goals. Yue et al. proposed a method to predict the runtime of distributed iterative jobs, aiming to reduce cluster deployment cost and optimize resource allocation and scheduling strategies [19]. In order to reduce the induced network load, Rizou et al. [10] present a distributed placement algorithm that minimizes the bandwidth-delay product of data streams between operators. Saurez et al. [13] propose a programming infrastructure Foglets to store and retrieve application generated data on the fog nodes. Pietzuch et al. [9] design a stream-based overlay network for operator placement in distributed stream-processing (DSP) systems. The earliest work to deploy CEP on edge is proposed by Hong et al. [3]. They think deploying CEP on edge can satisfy the latency requirement of mobile situation awareness applications. Edge computing is the computational infrastructures that make services closer to the end users [12].

In recent years, reinforcement learning methods perform well in optimization problems and continuous decision-making problems. There are some works that combine service deployment issues with reinforcement learning methods [6,7]. Azalia et al. [7] also use RL approach to solve device placement optimization problem. The device diagram is undirected, which is different from operator graph. Liu et al. [6] proposed RLMap method that learn to map DFGs onto spatially-programmed CGRAs directly from experiences. Recently, several papers related to our work have been included in two surveys [11,20] from the ACM Computing Survey journal. This not only indicates that the placement problem continues to receive significant attention but also highlights the increasing importance of reinforcement learning techniques as a crucial means of addressing this issue. However, their services to be deployed are different from operator graph, and the hardware devices that provide computing resources are not dynamic.

6   Conclusion

In this paper, we addressed the CEP operator placement problem on the edge by proposing an RL-based algorithm that leverages response time feedback. Our approach was evaluated through experiments, yielding the following key findings: 1) It outperforms all baseline algorithms (GA, LB, RTA) by reducing congestion and response time in most cases, 2) RLG exhibits higher robustness compared to RTA, and 3) Our approach demonstrates scalability to different operator graphs and edge networks, considering varying operator numbers, node capacities, and network sizes.

References

1. Behrisch, M., Bieker, L., Erdmann, J., Krajzewicz, D.: SUMO-simulation of urban mobility: an overview. In: Proceedings of SIMUL 2011, The Third International Conference on Advances in System Simulation. ThinkMind (2011)
2. Cai, X., Kuang, H., Hu, H., Song, W., Lü, J.: Response time aware operator placement for complex event processing in edge computing. In: Pahl, C., Vukovic, M., Yin, J., Yu, Q. (eds.) ICSOC 2018. LNCS, vol. 11236, pp. 264–278. Springer, Cham (2018). https://doi.org/10.1007/978-3-030-03596-9_18

3. Hong, K., Lillethun, D.J., Ramachandran, U., Ottenwalder, B., Koldehofe, B.: Opportunistic spatio-temporal event processing for mobile situation awareness. In: Proceedings of the 7th ACM International Conference on Distributed Event-Based Systems, DEBS 2013, pp. 195–206 (2013)

4. Jia, M., Liang, W., Xu, Z., Huang, M.: Cloudlet load balancing in wireless metropolitan area networks. In: IEEE INFOCOM 2016 - The 35th Annual IEEE International Conference on Computer Communications, pp. 1–9 (2016)

5. Jiang, J., Lu, Z.: Generative exploration and exploitation. CoRR abs/1904.09605 (2019). http://arxiv.org/abs/1904.09605

6. Liu, D., et al.: Data-flow graph mapping optimization for CGRA with deep reinforcement learning. IEEE Trans. CAD Integr. Circ. Syst. **38**(12), 2271–2283 (2019)

7. Mirhoseini, A., et al.: Device placement optimization with reinforcement learning. In: Proceedings of the 34th International Conference on Machine Learning, ICML 2017, Sydney, NSW, Australia, 6–11 August 2017, pp. 2430–2439 (2017). http://proceedings.mlr.press/v70/mirhoseini17a.html

8. Ottenwalder, B., Koldehofe, B., Rothermel, K., Hong, K., Lillethun, D.J., Ramachandran, U.: MCEP: a mobility-aware complex event processing system. ACM Trans. Internet Technol. **14**(1), 6 (2014)

9. Pietzuch, P.R., Ledlie, J., Shneidman, J., Roussopoulos, M., Welsh, M., Seltzer, M.I.: Network-aware operator placement for stream-processing systems. In: 22nd International Conference on Data Engineering (ICDE 2006), p. 49 (2006)

10. Rizou, S., Durr, F., Rothermel, K.: Solving the multi-operator placement problem in large-scale operator networks. In: 2010 Proceedings of 19th International Conference on Computer Communications and Networks, pp. 1–6 (2010)

11. Salaht, F.A., Desprez, F., Lebre, A.: An overview of service placement problem in fog and edge computing. ACM Comput. Surv. **53**(3), 1–35 (2020)

12. Satyanarayanan, M.: The emergence of edge computing. Computer **50**(1), 30–39 (2017)

13. Saurez, E., Hong, K., Lillethun, D., Ramachandran, U., Ottenwalder, B.: Incremental deployment and migration of geo-distributed situation awareness applications in the fog. In: Proceedings of the 10th ACM International Conference on Distributed and Event-Based Systems, DEBS 2016, pp. 258–269. ACM (2016)

14. Sutton, R.S., Barto, A.G.: Reinforcement Learning - An Introduction. Adaptive Computation and Machine Learning. MIT Press, Cambridge (1998)

15. Tesauro, G.: Temporal difference learning and TD-Gammon. Commun. ACM **38**(3), 58–68 (1995)

16. Tsitsiklis, J.N., Van Roy, B.: An analysis of temporal-difference learning with function approximation. IEEE Trans. Autom. Control (2002)

17. Tziritas, N., Loukopoulos, T., Khan, S.U., Xu, C.Z., Zomaya, A.Y.: On improving constrained single and group operator placement using evictions in big data environments. IEEE Trans. Serv. Comput. **9**(5), 818–831 (2016)

18. Varga, A., Hornig, R.: An overview of the OMNeT++ simulation environment, p. 60 (2008)

19. Yue, X., Shi, L., Zhao, Y., Ji, H., Wang, G.: Online runtime prediction method for distributed iterative jobs. In: Xing, C., Fu, X., Zhang, Y., Zhang, G., Borjigin, C. (eds.) WISA 2021. LNCS, vol. 12999, pp. 156–168. Springer, Cham (2021). https://doi.org/10.1007/978-3-030-87571-8_14

20. Zabihi, Z., Moghadam, A.M.E., Rezvani, M.H.: Reinforcement learning methods for computing offloading: a systematic review. ACM Comput. Surv. (2023)

Database for Artificial Intelligence

A Suitability Assessment Framework for Medical Cell Images in Chromosome Analysis

Zefeng Mo[1], Chengchuang Lin[3], Hanbiao Chen[2], Zhihao Hou[1], Zhuangwei Li[1], Gansen Zhao[1(✉)], and Aihua Yin[2(✉)]

[1] School of Computer Science, South China Normal University, Guangzhou 510631, China
{mozef,2021023238,zhuangwei.li,gzhao}@m.scnu.edu.cn
[2] Guangdong Women and Children Hospital, Guangzhou 511400, China
chenhanbiao2000@126.com, yinaiwa@vip.126.com
[3] Guangdong Planning and Designing Institute of Telecommunications Co., Ltd., Guangzhou 510630, China
chengchuang.lin@m.scnu.edu.cn

Abstract. The process of chromosome karyotype analysis is a highly time-consuming and error-prone task heavily relying on the experience of the cytogeneticists and influenced by factors such as fatigue and decrease of attention. Many efforts have dedicated to automatic chromosome karyotype analysis using various computer vision techniques based on geometric morphology and deep learning. However, few of them have paid attention to selections of high-suitability medical cell images for chromosome karyotype analysis. High-suitability cell images not only can significantly decrease the difficulty of manual chromosome karyotype analysis, but also can boost the analysis performance of automatic chromosome karyotype analysis algorithms. This paper proposes a suitability assessment framework for evaluating the suitabilities of cell images to address the issue of selecting high-suitability medical cell images for the inputs of chromosome karyotype analysis. The quantitative experimental results show that using the proposed suitability assessment framework to select suitable inputs can significantly boost chromosome segmentation performance by **5.06** percentage points of mAP, **2.4** percentage points of AP^{50}, and **3.58** percentage points of AP^{75}. The qualitative experiments with a group of cell images show that the corresponding suitability results evaluated by the proposed framework are highly in accordance with results evaluated by the experienced analysts, demonstrating the effectiveness of the proposed method to address the selection issue of suitable medical cell images.

Keywords: Medical Image Suitability Evaluation · Chromosome Karyotype Analysis · Chromosome Instance Segmentation · Quality Assessment

© The Author(s), under exclusive license to Springer Nature Singapore Pte Ltd. 2023
L. Yuan et al. (Eds.): WISA 2023, LNCS 14094, pp. 575–586, 2023.
https://doi.org/10.1007/978-981-99-6222-8_48

1   Introduction

Chromosomal analysis provides valuable information about a person's genetic constitution because normal chromosomes have normal morphologies and sizes. Chromosomal anomalies result in numerous genetic diseases and are responsible for gestational losses, implantation failures, and congenital malformations. Chromosome karyotype analysis, a most common and vital approach in chromosomal analysis, refers to segmenting chromosome instances from selected cell images and arranging them into corresponding chromosome karyotypes according to ISCN (international system for human cytogenetic nomenclature) criteria. Figure 1 illustrates the procedure of chromosome karyotype analysis, which consists of three major stages: medical image selections, chromosome instance segmentation, and chromosome classification. Chromosome karyotype analysis is a time-consuming, costly, and error-prone process, considering the experience of the experts and influencing factors such as fatigue and attention. Accordingly, many efforts have been dedicated to developing computer-assisted algorithms or deep learning-based models for automatic or partial-automatic chromosome analysis. These efforts focus majorly on automatic chromosome instance segmentation [1,2] (Stage 2) or chromosome classification [3,4] (Stage 3). Few of them pay attention to medical image selections (Stage 1), which also plays an important role in having excellent performance of karyotype analysis.

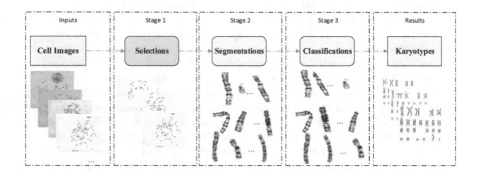

Fig. 1. The clinical procedure of chromosome karyotype analysis.

The medical cell image selection is vital for accurate chromosomal analysis, as low-suitable inputs are usually responsible for failures of automatic chromosomal analysis. Moreover, high-suitable inputs may significantly boost the performance of chromosomal analysis algorithms. However, the hypothesis that high-quality cell images have been selected by analysts is adopted by existing methods can not be hold in most cases. These cases face the challenges of how to convert medical cell image selection experiences of analysts and define explicit criteria to select high-suitable medical cell images.

This paper converts the medical cell image selection problem (Stage 1 in Fig. 1) to assess the suitabilities of cell images and then rates and ranks them according to their suitabilities. Therefore, this work proposes a suitability assessment framework for medical cell images based on analyzing multiple vital factors related to chromosome analysis. The proposed framework assesses suitabilities of given cell images by first considering various dimensions in factors including image quality, noises, impurities, and behaviors of chromosomes in given cell images, and then build up a comprehensive assessment based on linear weighting method.

Extensive experiments have been conducted to demonstrate the effectiveness of the proposed suitability assessment framework. The quantitative evaluating experiment results of the chromosome segmentation task on a hold-out medical cell image dataset show that using the proposed framework to select high-suitability cell images can significantly boost chromosome segmentation performance by **5.06** percentage points of mAP, **2.4** percentage points of AP^{50}, **3.58** percentage points of AP^{75} without modifying chromosome analysis algorithms. The qualitative experiments with a group of cell images show that the corresponding suitability results evaluated by the proposed framework are highly coordinated with results evaluated by the experienced analysts.

The contributions of this paper can be concluded as follows:

- This work identifies the unique role of medical cell imaging selection in chromosome analysis and the critical challenges in the current clinical application of medical imaging selections.
- This work explores a way to tackle the selection problem of medical cell images and proposes a novel metric termed *suitability* to indicate suitable degrees for chromosome analysis of given cell images of the same tested person.
- This work contributes a suitability assessment framework to select suitable medical cell images among all medical cell images of the same person for performing chromosome analysis. Besides, the proposed suitability assessment introduce the expertise of cytogeneticists to achieve a complete automatic selection process.
- This work achieved an improvement in segmentation performance through a series of quantitative experiments using the proposed framework to select highly suitable cell images compared to unselected cell images as test data and demonstrated the effectiveness of the proposed framework through the qualitative experiments compared with the expertise of experienced analysts.

2   Related Works

This section reviews two research aspects, including automatic chromosome analysis methods trapped by insufficient-suitability cell image inputs and works introduced to dedicate the issue of image suitability assessment.

Nirmala Madian and K.B. Jayanthi [5] proposed different algorithms to separate the touching chromosomes and segment overlapping chromosomes from G-Band metaphase images. Their methods works well on separating touching chromosome clusters or overlapping clusters but fails to separate those complicated clusters in complicated cell images. Shervin Minaee et al. [6] proposed a fully automatic segmentation algorithm based on chromosomes' geometric features to segment chromosome instances from medical cell images. Their algorithm were tested on 62 chromosome clusters. It was found that 8.9% complicated chromosome clusters were incorrectly separated due to too many chromosomes overlapping with each other. However, this method still requires manual interventions for complicated cell images.

Many deep learning-based chromosome segmentation methods [2,7,8] are also facing the challenge of failing in complex and low-suitability medical cell images. Although the improvement of existing geometry-based algorithms and deep learning-based models may successfully analyze more complicated medical cell images, selecting high-suitability image inputs is a much more straightforward and effective way to facilitate existing algorithms and models for better performance.

Many applications analyze suitabilities of images by assessing image qualities. However, image quality assessment is a vital but not easy task in various image processing applications [9,10]. Traditionally, image quality has been evaluated by human subjects, which is expensive and too slow for real-world applications [11]. Recently, many objective methods were proposed to address the issue of image quality assessment. These methods mainly include full-reference image quality assessment (FRIQA) methods, no-reference image quality assessment (NRIQA) methods, and reduced-reference image quality assessment (RRIQA) methods. The objective of the above methods is to evaluate the general qualities of the image, such as signal-to-noise ratio, sharpness, and contrast. However, other factors are vital for chromosome analysis but not taken into consideration in existing image quality assessment methods, such as cell impurities, the number of chromosome clusters, and the number of non-cluster chromosome instances.

Accordingly, motivated by the urgent necessity for chromosome analysis and limitations of existing image quality assessment methods, this paper proposes a suitability assessment framework for medical cell images by a linear summation of four measurements: image quality metric, noises and cell impurities, chromosome clusters, and non-cluster chromosomes metrics.

3   Method

This section first analyzes the factors that affect chromosome analysis and proposes a framework for evaluating medical cell imaging suitability by assessing corresponding factors.

3.1   Quality Factors Analysis for Medical Cell Images

In manual chromosomal analysis, decisions on either selecting or excluding medical cell images based on their image suitabilities are made by chromosomal analysts dependent on their experiences. The factors that affect the suitabilities of medical cell images include the sharpness, cell impurities, noises, geometries of chromosome instances, etc. Figure 2 shows an example of medical cell images that indicates various factors affecting the suitability of the given cell image.

In the given example of Fig. 2, red rectangular frames represent cell impurities or noises that are not related to chromosome analysis in medical images. However, these cell impurities or noises may jeopardize chromosome segmentation precisions of algorithms, which means more impurities or noises in the given cell image indicating lower cell suitability. Green rectangular frames show examples of chromosome clusters that are the main obstacles in chromosome instance segmentation tasks. Hence, a given cell image with fewer chromosome clusters is better for chromosome analysis. Chromosomes in the blue rectangular frames mean they are non-cluster chromosome instances that do not require to be further segmented or separated by chromosome instance segmentation algorithms or models. Accordingly, a medical cell image with more non cluster chromosome instances is more suitable for analysis.

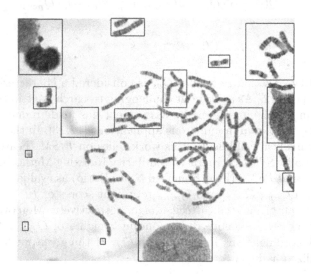

Fig. 2. An example of medical cell images indicating various suitability factors. Red rectangles indicate cell impurities or noises, while blue rectangles refer to chromosome instances that do not belong to the given cell. Green rectangles represent chromosome clusters. (Color figure online)

3.2   Suitability Assessment Framework

Based on the analysis of factors affecting chromosome analysis, this paper proposes an effective suitability assessment framework for medical cell images by defining a unified score termed by suitability score Q, whose value ranging from 0 to 100 represents the suitability of a given cell image where higher Q means more suitable for analysis. The suitability score Q is obtained by a weighted linear summation of four metrics: image quality score (Q_{qs}), image noise score (Q_{in}), chromosome cluster score (Q_{cs}), and non-cluster chromosome instance score (Q_{nc}). The objective of medical cell image suitability assessment is to output a unified score Q_I that represents the suitability of given image I. A suitability assessment framework $\mathcal{Q}(\cdot)$ accepts a cell image and outputs the score of the input image. Accordingly, the suitability assessment of a given image I can be formalized as Eq. 1.

$$Q_I = \mathcal{Q}(I) \tag{1}$$

The proposed suitability assessment framework $\mathcal{Q}(\cdot)$ assesses the suitability score Q of the given sample by four different scores and corresponding weights: Q_{qs}, Q_{in}, Q_{cs}, and Q_{nc}. The above scores Q_{qs}, Q_{in}, Q_{cs} and Q_{nc} range from 0 to 100. The total suitability assessment framework $\mathcal{Q}(\cdot)$ is defined as Eq. 2, where α, β, γ, and ζ are corresponding weights.

$$\mathcal{Q}(\cdot) = \alpha \cdot Q_{qs} + \beta \cdot Q_{in} + \gamma \cdot Q_{cs} + \zeta \cdot Q_{nc} \tag{2}$$

$$\alpha + \beta + \gamma + \zeta = 1 \tag{3}$$

Image Quality. The image quality Q_{qs} is considered a characteristic property of a general image [12]. As a vital but challenging research issue in the image processing domain [9], the image quality assessment has made remarkable progress and extensively employed in numerous applications [12,13]. In the image quality assessment of medical cell images, this work relies on *FISM* (Feature Similarity Indexing Method), *SSIM* (Structured Similarity Indexing Method), *MSE* (Mean Square Error), and *PSNR* (Peak Signal to Noise Ratio) as evaluated metrics [12].

Let Q_{fism}, Q_{ssim}, Q_{mse}, and Q_{psnr} represent scores of *FISM*, *SSIM*, *MSE*, and *PSNR* of the given medical cell image, respectively. Moreover, let λ_{fism}, λ_{ssim}, λ_{mse}, and λ_{psnr} denote corresponding weights of Q_{fism}, Q_{ssim}, Q_{mse}, and Q_{psnr}. Accordingly, the image quality Q_{qs} of the given medical cell image can be formalized as Eq. 4.

$$\begin{aligned} Q_{qs} = \lambda_{fism} \cdot Q_{fism} + \lambda_{ssim} \cdot Q_{ssim} \\ + \lambda_{mse} \cdot Q_{mse} + \lambda_{psnr} \cdot Q_{psnr} \end{aligned} \tag{4}$$

Image Noise Score. In clinical chromosome analysis, all medical cell images of the same person are taken on the same cell culture slide by a professional microscope. Besides the target metaphase cell, many impurities of other cells and noises may fall into the view field of the microscope, where these impurities and noises may cover some chromosome instances and lead to running failures of many chromosome instance segmentation algorithms.

In the proposed cell image suitability assessment framework, Q_{in} is introduced to denote the metric of cell image noise whose value ranges from 0 to 100, where a higher score means fewer cell impurities or noises. After that, let S_{snc} denote the penalized score caused by a cell impurity or noise and use N_{in} to represent the total number of cell impurities and noises. Accordingly, the calculation of Q_{in} can be defined as Eq. 5.

$$Q_{in} = \max(0, 100 - S_{snc} \cdot N_{in}) \tag{5}$$

Chromosome Cluster Score. Given a medical cell image, more chromosome clusters in the given image means more difficult to segment and separate chromosome instances. Accordingly, this work proposes S_{cs} to denote penalized score caused by a chromosome cluster. Afterward, let N_{cs} to represent the total number of chromosome clusters. Finally, let Q_{cs} represent the evaluated metric of chromosome cluster, and its calculation can be formalized as Eq. 6.

$$Q_{cs} = \max(0, 100 - S_{cs} \cdot N_{cs}) \tag{6}$$

Non-cluster Chromosome Instance Score. Non-cluster chromosome instances are perfect for clinical chromosome analysis. A given medical cell image with more non-cluster chromosome instances means fewer chromosome instances overlapping or touching with each other. Therefore, when a given medical cell image has more non-cluster chromosome instances, the given image has a higher non-cluster chromosome instance score.

Let Q_{nc} represent non-cluster chromosome instance score, S_{ncc} denotes the penalized score caused by lacking one non-cluster chromosome instance, and N_{ncc} indicates the total number of non-cluster chromosome instances. Moreover, given a medical cell image of a normal human has a total of 46 chromosome instances. Consequently, the calculation of Q_{nc} can be defined as Eq. 7.

$$Q_{nc} = \max(0, 100 - S_{ncc} \cdot (46 - N_{ncc})) \tag{7}$$

3.3 Implementations

This work develops a prototype to implement the proposed framework. First, the developed prototype employs a image quality assessment toolkit[1] to calculate Q_{fism}, Q_{ssim}, Q_{mse}, and Q_{psnr} for the image quality score Q_{qs}. Second, the developed prototype applies a geometric connectivity algorithm implemented in the scikit-image toolkit to separate a given medical cell image into multiple image slices. Finally, the developed prototype distinguishes noise slices, cell impurity slices, cluster slices, and non-cluster chromosome instance slices relying on method [14].

4 Experiment

4.1 Experimental Objectives

The experimental objectives consist of three aspects. The first objective is to demonstrate the importance of medical image selections to downstream chromosome analysis performance, while the second one is to evaluate the effectiveness to boost downstream chromosome analysis performance of the proposed

---

[1] https://github.com/ocampor/image-quality.

framework. Besides, the last objective is to verify selection consistencies by the proposed framework with experienced analysts.

4.2 Experimental Settings

The experiment workstation running the proposed framework is a MacBook Pro laptop whose CPU is i5-8259U and RAM capacity is 8 GB. The dataset used in this work were medical cell images collected anonymously from Guangdong Women and Children Hospital which consists of a total of 1,149 cell images from over 60 individuals, with approximately 15 to 20 cell images of varying image quality per individual. First, 50 images were selected randomly and ranked according to their suitabilities by a professional chromosome analysis expert. This suitability rank analyzed by the expert is taken as the gold standard. Second, all correspond parameters of proposed framework were initially set to mean weights i.i.e $\alpha = \beta = \gamma = \zeta = 0.25$, $\lambda_{fism} = \lambda_{ssim} = \lambda_{mse} = \lambda_{psnr} = 0.25$, and $S_{snc} = S_{cs} = S_{ncc} = 5$. Multiple rounds of human-computer interactions were introduced to determine parameters of the proposed framework. Finally, all parameters were finely adjusted to approximate the proposed framework's rank to the gold standard. The corresponding weights were adjusted to $\alpha = \beta = 0.2$, $\gamma = \zeta = 0.3$, $\lambda_{fism} = \lambda_{ssim} = \lambda_{mse} = \lambda_{psnr} = 0.25$, and $S_{snc} = S_{cs} = S_{ncc} = 2$.

4.3 Quantitative Experimental Evaluations

This paper demonstrates the importance of the proposed cell image assessment framework by performing the downstream chromosome analysis tasks using medical cell images with different suitability ranges to demonstrate further the importance of cell image quality assessment to chromosome analysis. First, this paper trained a chromosome instance segmentation model utilizing Mask R-CNN instance segmentation model whose super-parameters set to $\{lr = 0.005,$ $epochs = 500,$ $momentum = 0.9,$ $weightdecay = 0.0005,$ $stepsize = 3,$ $gamma = 0.9,$ $batchsize = 32\}$. Second, this work prepared a hold-out dataset of cell images whose suitabilities were evaluated by the proposed framework. Third, this work utilizes the trained chromosome segmentation model to evaluate the whole hold-out dataset and partial cell images with various suitability score ranges, respectively.

Table 1 summarizes the experimental results of the same chromosome instance segmentation model evaluated on medical cell images with various suitability score ranges. Specifically, the chromosome instance segmentation model was firstly evaluated on the whole hold-out dataset and obtained the instance segmentation performance with 70.54% mAP, 93.23% AP^{50}, and 88.52% AP^{75}. When evaluated on cell images in the hold-out dataset whose suitability scores are lower than 80, the segmentation performance of the same chromosome instance segmentation model dropped down to 66.86% mAP, 91.64% AP^{50}, and 88.48% AP^{75}. However, when evaluated on those cell images in the hold-out dataset whose suitability scores are between 80 and 90, the segmentation performance obtained 71.99% mAP, 93.84% AP^{50}, and 90.13% AP^{75}. Finally,

Table 1. Chromosome segmentation performance of cell images with different suitability score intervals.

quality score	mAP	AP^{50}	AP^{75}
whole	70.54%	93.23%	88.52%
$\leq$80	66.86%	91.64%	84.88%
	(-3.68%↓)	(-1.59%↓)	(-3.64%↓)
80–90	71.99%	93.84%	90.13%
	(1.45%↑)	(0.61%↑)	(1.61%↑)
$\geq$90	75.60%	95.63%	92.10%
	(5.06%↑)	(2.4%↑)	(3.58%↑)

when evaluated on those cell images in the hold-out dataset whose suitability scores are higher than 90, the corresponding segmentation performance of the same model hit 75.60% mAP, 95.63% AP^{50}, and 92.10% AP^{75}. Comparing with evaluating on the whole hold-out dataset, when evaluating those samples whose suitability scores are higher than 90, the corresponding experimental results of mAP improves by **5.06** percentage points, AP^{50} raises **2.4** percentage points, and AP^{75} increases **3.58** percentage points. These improvements boosted by suitable inputs are significant without modifying the given chromosome instance segmentation model.

According to the above quantitative experimental results, this work obtains the following conclusions. First, the suitabilities of given cell images are vital for chromosome analysis. Using the same algorithm, given cell images with higher suitability scores have more accurate chromosome analysis performance. Second, selecting cell images with high suitability scores for chromosome analysis can significantly boost analysis performance on the condition of no modification or improvement of the given chromosome analysis algorithms. Third, the proposed cell suitability assessment framework can effectively distinguish medical cell images suitable for downstream chromosome analysis tasks by assessing their qualities.

4.4 Qualitative Experimental Evaluations

Table 2 shows a group of medical cell images and their corresponding suitability scores assessed by the proposed framework. These given cases are sorted according to their suitability scores, from low to high. The image suitability of each case is finally evaluated into a final quantitative score by the proposed framework. Meanwhile, four corresponding measurement metrics are given in Table 2.

According to the results shown in Table 2, there are many cell impurities and noises in the first case whose suitability score is 58.9. Meanwhile, numerous chromosome instances are overlapping and touching with each other. Therefore, it is not an ideal medical cell image for further chromosome analysis. These impurities and clusters may cause automatic chromosome analysis algorithm failures and require chromosomal analysts to consume many efforts to remove impurities and separate chromosome instances. Inversely, the last example, whose suitability score is 92.1, has few impurities or noises. Most of chromosome instances of this case are separated, which is an ideal chromosome analysis input. The suit-

Table 2. Examples of medical cell images and their corresponding suitability scores assessed by the proposed framework. This *rank* is given by a professional chromosome analysis expert, where the smaller the value, the more suitable for chromosome analysis.

Images				
Q_{qs}	63.3	73.9	77.9	52.1
Q_{in}	90	95	85	95
Q_{cs}	60	52	64	64
Q_{nc}	34	42	48	62
Q	**58.9**	**62.9**	**66.2**	**68.3**
rank	12	11	10	9
Images				
Q_{qs}	71.1	69.8	70.4	69.9
Q_{in}	95	90	100	100
Q_{cs}	68	72	76	84
Q_{nc}	54	66	62	72
Q	**70.8**	**73.4**	**75.5**	**80.8**
rank	8	6	7	5
Images				
Q_{qs}	69.8	78.2	71.2	78.7
Q_{in}	95	100	100	100
Q_{cs}	76	88	92	96
Q_{nc}	94	74	88	92
Q	**83.9**	**84.2**	**88.2**	**92.1**
rank	4	3	1	2

ability rank of these examples shown in Table 2 is highly in accordance with the rank given by an experienced clinical chromosome analysis expert, demonstrating the effectiveness of the proposed framework for assessing the suitabilities of medical cell images.

5   Conclusion

High-suitability medical cell images are an indispensable guarantee for the accuracy of chromosome analysis and diagnosis of genetic diseases. This paper proposed an effective suitability assessment framework for medical cell images to convert the experiential knowledge of clinical chromosome analysis experts into quantifiable suitability scores to select high-suitability medical cell images. The quantitative experimental results demonstrate that the proposed framework can effectively distinguish medical cell images suitable for downstream chromosome analysis to boost chromosome analysis performance significantly. Moreover, the qualitative experimental results confirm that the proposed framework's cell image selection results are highly in accordance with the gold standard. According to our best knowledge, this article is the first work to propose a quantifiable quality assessment framework for medical cell images. This work studied the significances of medical cell imaging selection in chromosome analysis and identified critical challenges in the current clinical application of medical imaging selections. The quantitative experimental results of downstream chromosome analysis with improvements of **5.06** percentage points of mAP, **2.4** percentage points of AP^{50}, **3.58** percentage points of AP^{75} have demonstrated the effectiveness and significances of this work.

This work is not without limitations. Accordingly, the work can only be evaluated by downstream chromosome analysis tasks due to the lack of well-accepted evaluated metric or system for suitability assessment algorithms of medical cell images. Therefore, this work quantitatively evaluates performance improvements of the chromosome instance segmentation task boosted by medical input selections with the proposed framework. In the future, this work will explore more reasonable and straightforward evaluation metrics or systems for medical image suitability assessment algorithms.

References

1. Nikolaou, A., Papakostas, G.A.: Exploiting deep learning for overlapping chromosome segmentation. In: Shukla, P.K., Singh, K.P., Tripathi, A.K., Engelbrecht, A. (eds.) CVR 2022. AIS, pp. 309–329. Springer, Singapore (2023). https://doi.org/10.1007/978-981-19-7892-0_24
2. Huang, K., Lin, C., Huang, R., et al.: A novel chromosome instance segmentation method based on geometry and deep learning. In: 2021 International Joint Conference on Neural Networks (IJCNN), pp. 1–8. IEEE (2021)
3. Lin, C., Chen, H., Huang, J., et al.: ChromosomeNet: a massive dataset enabling benchmarking and building basedlines of clinical chromosome classification. Comput. Biol. Chem. **100**, 107731 (2022)

4. Gong, Z., Peng, B., Shen, A., et al.: Attention-based densely connected convolutional network for chromosome classification. In: Proceedings of the 8th International Conference on Computing and Artificial Intelligence, pp. 534–540 (2022)
5. Madian, N., Jayanthi, K.B.: Overlapped chromosome segmentation and separation of touching chromosome for automated chromosome classification. In: 2012 Annual International Conference of the IEEE Engineering in Medicine and Biology Society, pp. 5392–5395. IEEE (2012)
6. Minaee, S., Fotouhi, M., Khalaj, B.H.: A geometric approach to fully automatic chromosome segmentation. In: 2014 IEEE Signal Processing in Medicine and Biology Symposium (SPMB), pp. 1–6. IEEE (2014)
7. Lin, C., Zhao, G., Yin, A., Ding, B., Guo, L., Chen, H.: A multi-stages chromosome segmentation and mixed classification method for chromosome automatic karyotyping. In: Wang, G., Lin, X., Hendler, J., Song, W., Xu, Z., Liu, G. (eds.) WISA 2020. LNCS, vol. 12432, pp. 365–376. Springer, Cham (2020). https://doi.org/10.1007/978-3-030-60029-7_34
8. Feng, T., Chen, B., Zhang, Y.: Chromosome segmentation framework based on improved mask region-based convolutional neural network. J. Comput. Appl. **40**(11), 3332–3339 (2020)
9. Wang, Z., Bovik, A.C., Lu, L.: Why is image quality assessment so difficult? In: 2002 IEEE International Conference on Acoustics, Speech, and Signal Processing, vol. 4, p. IV-3313. IEEE (2002)
10. Ji, Y., Li, J., Huang, Z., Xie, W., Zhao, D.: A data dimensionality reduction method based on mRMR and genetic algorithm for high-dimensional small sample data. In: Zhao, X., Yang, S., Wang, X., Li, J. (eds.) WISA 2022. LNCS, vol. 13579, pp. 485–496. Springer, Cham (2022). https://doi.org/10.1007/978-3-031-20309-1_43
11. Wang, Z., Bovik, A.C.: Modern Image Quality Assessment. Synthesis Lectures on Image, Video, and Multimedia Processing, vol. 2, no. 1, pp. 1–156 (2006)
12. Sara, U., Akter, M., Uddin, M.S.: Image quality assessment through FSIM, SSIM, MSE and PSNR-a comparative study. J. Comput. Commun. **7**(3), 8–18 (2019)
13. Li, X.: Blind image quality assessment. In: Proceedings of the International Conference on Image Processing, vol. 1, p. I-I. IEEE (2002)
14. Lin, C., Yin, A., Wu, Q., et al.: Chromosome cluster identification framework based on geometric features and machine learning algorithms. In: 2020 IEEE International Conference on Bioinformatics and Biomedicine (BIBM), pp. 2357–2363. IEEE (2020)

Design Innovation and Application Practice Based on Automatic Thrombolysis After Ischemic Stroke

Jiayi Cai[1] (✉) ⓘ and Jialiang Cai[2] ⓘ

[1] Faculty of Humanities and Arts, Macau University of Science and Technology, Macau, China
Cai849027924@qq.com
[2] Macao Institute of Materials Science and Engineering, Macau University of Science and Technology, Macau, China

Abstract. In response to the global trend of aging and changing lifestyles, the researchers concluded, based on clinical practice, that 30 percent of patients treated with intravenous thrombolytic therapy for ischemic stroke are not able to receive thrombolytic injections well enough to target the treatment algorithm after the onset of symptoms. The research team jointly developed and designed an innovative automated thrombolysis care process that eliminates the time-consuming process of going to the hospital at the onset of symptoms, with an optimal treatment time of less than four hours from onset of symptoms to treatment. The hospital will be based on the patient's various medical histories. Antithrombotic therapy for control patients, testing thrombolytic drug dispensing, thrombolytic effectiveness testing (customized program for patient history), and research on the ratio of intentionally home-treated patients to intentionally hospital-treated patients. Automated thrombolytic drug safety testing (immunogenicity), total drug dosing (customized protocols for patient history), drug testing using organs that enter the body's metabolism, p-value assessment, and drug half-life for a total of 317 validated patients to provide real data. Based on the assistance of a researcher specializing in mechanical engineering the design includes four areas of innovation. Performance, detection methods, intelligent, synchronized connection to the mobile terminal, in the daily design and operation of the work to provide the most valuable technical solutions, and data through the app in real-time view. In the research to address the patient mechanical thrombolysis treatment process is also a large number of healthcare resources needed also with human resources. The research team based on research, development, testing, research, innovation repeated testing of the research to come up with innovative automatic thrombolysis of the experimental conclusions.

Keywords: Ischemic stroke · automatic thrombolysis · design innovation · application practice

L. Yuan et al. (Eds.): WISA 2023, LNCS 14094, pp. 587–600, 2023.
https://doi.org/10.1007/978-981-99-6222-8_49

1 Introduction

It has been found that more than 80% of elderly people over the age of 65 have a chance of having a stroke, and with the global trend of aging and changes in people's lifestyles, the number of elderly patients having a stroke is expected to continue to increase in the future [1]. Because of the poor prognosis of activity and emergency treatment in the elderly, social attention and enhanced post-stroke emergency care are needed more than in younger patients, and this study has some research value for automatic thrombolytic injection in ischemic stroke for emergency treatment of the elderly, as well as the ability to perform automatic thrombolytic intravenous injection in case of emergency. It is also a new research direction and a more intelligent treatment for the elderly to distinguish the traditional machine embolization [2]. Also machine embolization has a poor prognosis for elderly patients over 65 years of age in emergency situations [3]. An analysis of a report on elderly patients over 65 years of age indicated that the mortality rate was higher than in younger people [4]. The intravenous regimen used in ischemic stroke is the best care for patients 2.5 h after the onset of symptoms [2]. The size of the intravenous dose is inversely proportional to the duration of treatment for the physical condition of the patient after the stroke and needs to be administered within 4 h [5]. It has been found that the high incidence of ischemic stroke is early in the morning, which causes the symptoms of ischemic stroke in most elderly people without being fully awake [6–9], resulting in missing the best time for treatment. At the same time, we focused the research analysis on the population due to the state of the patients, similar to some patients with insomnia, patients with loss of speech, patients with cognitive impairment, and we required this group to account for 35% of the research, as well as a single-arm injection study of 62 cases of automatic thrombolysis in ischemic stroke. [10] is a team approved for testing at the Centers for Medicare, primarily for the therapeutic care of mechanical intravenous injections in ischemic stroke. The use of rtPA studies has a high benefit when compared to younger patients less than 25 years old, with a 57% frequency of ischemic stroke symptoms between the ages of 85 and 65 years [11–16]. However, at this stage of research there is no well-developed innovative system for automated thrombolysis techniques for ischemic stroke, and there is a lack of adequate understanding of the effectiveness and safety of physical examination and dispensing of medications based on physical condition for older adults [12, 17–24]. Moreover, the FDA (Food and Drug Ad ministration) is the internationally approved thrombolytic agent for the treatment of ischemic stroke with auto-thrombolytic injections for ALS [25–27]. This protocol is also widely used in the treatment process of ALS patients [28].

2 Materials and Methods

Through patient selection and location screening our study was officially approved by the Macau University of Science and Technology Hospital - KU Hospital and the question-naire survey was conducted in the Macau Peninsula - Iao Hon Community, the project study period was from July 10, 2018 - November 02, 2021, to close the project. During this period, the study visited and researched a total of 317 valid social questionnaires of the target patients and centrally integrated them into the electronic medical record backup for the elderly aged between 25 and 85 years old. The researchers were divided into three age groups for the study: 25–45 years, 45–65 years, and 65–85 years. The study evaluation technique was divided into three phases, the first phase was to categorize the patients into male and female and age class screening; the second phase was to determine the physical parameters of the patients, including previous illnesses and antithrombotic therapy, and knowledge of thrombolytic efficacy tests. The third stage was to screen for a history of ischemic stroke with mechanical intravenous therapy.

Clinical testing of medications was performed and the final p-value for evaluation was calculated. We used the WUS Magnetic Resonance Imaging (MRI) machine equipment, CT Body Intelligence Scanner, and Mechanical Venous Thrombolysis tPA equipment within the Macao University of Science and Technology Hospital-KUH. Combined with the clinical trial control study, we experimented with the specific operation procedure of mechanical thrombolysis through the data parameters of the control group. And in advance to test the various functions of the patient's body, and its corresponding dispensing, while the occurrence of stroke can be timely intelligent feedback patients can be timely injection operation. Our research is based on ischemic stroke innovative automatic intravenous thrombolysis drug delivery. The steps saved to go to the hospital to make an appointment, registration, processing, consultation, and prescribing drugs. Mechanical thrombolysis and other operations of time.

The research team has jointly developed and designed an automated thrombolysis care process to reduce the time spent on hospital visits at the onset of the disease (the optimal treatment time for a patient from the onset of the disease to treatment is less than 4 h). The current methodology focuses on the use of structured data [29] and this study effectively improved thrombolysis outcomes in patients with ischemic stroke based on detecting the patient's medical history in advance and administering medication in advance, and automating the process of administering medication in the event of an emergency (DNT). This study is innovative and deserves to be widely used in subsequent emergency care.

In response to the global trend of aging and changing lifestyles, researchers have concluded, based on clinical practice, that 30% of patients with ischemic stroke do not respond well to the treatment algorithm for thrombolytic injections after the onset of symptoms (the optimal time from onset of symptoms to treatment is less than 4 h). The dosage and injections were not well detected, and the study concluded that the above patients did not have a good prognosis.

This study was conducted from July 10, 2018 - November 02, 2021, the end of the project, the selected site of the Macau University of Science and Technology - the research group of the School of Design of the University Hospital and the Faculty of Pharmacy of the University of Science and Technology (MUST) has jointly developed and designed an innovative automated thrombolytic care process to avoid the time spent on the various examination processes at the onset of the disease to avoid the time spent on the various check-up processes in the hospital (the optimal time of the patient from the onset of the disease to the treatment is less than 4 h). The hospital will be based on the patient's various medical histories. Antithrombotic therapy for control patients, testing thrombolytic drug dispensing, thrombolytic effectiveness testing (customized protocols for patient history), and research on the ratio of intentionally home-treated patients to intentionally hospital-treated patients. Automated thrombolytic drug safety testing (immunogenicity), total drug dosage (customized protocols for patient history), drug testing using organs that enter the body's metabolism, p-value assessment, and drug half-life for a total of 317 validated patients to provide real data.

Based on the assistance of a researcher specializing in Mechanical Engineering the design includes four areas of innovation. a) In terms of performance, independently program a DDS chip with the added advantages of small size, high detection accuracy, and good endurance. b) In terms of performance, independently program a DDS chip with the added advantages of small size, high detection accuracy, and good endurance. The programmable system modeling scheme can prioritize the solution to detect abnormal waveforms and frequencies in the human body. b) In terms of detection, innovative automatic triggering of system testing through the system methodology apparatus for the pulse period and center frequency. Meanwhile, the innovative integrated design better serves ischemic stroke patients. In patients who have not yet received mechanical thrombolysis compared to automatic thrombolysis can better directly detect and serve patients (patients from the onset of the disease to the treatment of the best time in less than 4 h). c) Intelligent, through the internal trigger chip research data support, to better ensure that patients are more in place to detect and deliver drugs for automatic intravenous thrombolysis treatment. d) Synchronized connection to cell phones to provide the most valuable technical solutions in the day-to-day design of the operational work, data through the app. e) The system is designed to provide the most efficient and effective way to detect and treat ischemic stroke patients. Value of the technical program, data through the app in real-time view.

3   Clinical Study Groups and Technology Collection Evaluation

This study focused on social questionnaire research for patients, the study protocol was discussed with the researchers of the Faculty of Pharmacy and granted approval. The duration of our study was from July 10, 2018-November 02, 2021 to close the project. A total of 317 patients with stroke cerebral infarction admitted from July 10, 2018 to November 02, 2021 were selected as the control group for this study based on questionnaire survey and hospital clinical trial in order to verify the innovative experimental points of this study through research analysis. The 317 cases in this study included 184 males and 133 females; 87 males and 47 females were younger than 65 years old; 97 males and 86 females were older than 65 years old; the number of patients younger than 25 years old was 35, accounting for 11%; the number of patients younger than 45 years old was 92, accounting for 29%; and the number of patients younger than 65 years old was 152, accounting for 15%. We also opened a research team to analyze the historical cases of 317 patients during the period from March 02, 2021 to November 07, 2021, and analyzed the important aspects of the whole process through a detailed on-site questionnaire survey. And recorded. The optimal treatment time is less than 4 h, so the research team redesigned and studied how to establish an innovative, efficient and simple treatment process and testing procedure. A standard, intravenous, refined therapeutic care process for thrombolytic therapy was also achieved.

As shown in Fig. 1 below, the evaluation index was divided into three innovative and important phases: a) Medical history observation phase, which was analyzed by statistical criteria, and seven medical histories with high impact on patients' thrombolysis were counted, namely, cardiac infarction, hypertension, diabetes, hyperlipidemia, liver and kidney diseases, and other disease symptoms. Based on the above, our team added the DNT process required by the patients. The results of the study integration showed that 33 patients (10.7%) were between 25 and 45 years old, 82 patients (26.9%) were between 45 and 65 years old, and 62 patients (19.5%) were between 65 and 85 years old, and we performed safety tests (immunity tests) for these patients and came up with the best drug for the treatment Fibrin Specific (++++); the total dose of the drug was tested at $2.25 - 6.25 \times 10$ IU; the new organ of metabolism of the drug after entering the patient's body was Hepatic, and the experimental study showed that this drug was safe for use.

The safety test (immunity test) was conducted for this group of patients and the best drug to be used was Fibrin Specific (++); the total dose of the drug was tested as MI: 50–100 mg, IS: 0.9 mg/kg, PE: 100 mg; the organ of metabolism of the drug into the patient's body is Renal, and the experimental study concluded that this drug is safe for use.

The safety test (immunity test) was conducted for these patients and the best drug for treatment was Nom-fibrin. Specific; the total dose of the drug was $2.25 - 6.25 \times 10$ IU; the new organ of metabolism of the drug after entering the patient's body was Renal, and the experimental study showed that this drug is safe for use.

The safety test (immunity test) was performed on 143 patients aged 25–45 years (45.1%), 69 patients aged 45–65 years (21.7%) and 74 patients aged 65–85 years (23.3%). The total dose of the drug was $2.25 - 6.25 \times 10$ IU, and the new organ of metabolism of the drug after entering the patient's body was Hepatic, which was found to be safe through experimental studies.

For patients with a history of liver and kidney disease, 98 patients (30.9%) were between 25 and 45 years of age, 34 patients (10.7%) were between 45 and 65 years of age, and 64 patients (20.1%) were between 65 and 85 years of age, and for these patients, we conducted safety tests (immunity tests) and concluded that the best drug for treatment was Fibrin Specific (++); the total dose of the drug was tested at 20 IU, 30 mg–50 mg; the new organ of metabolism of the drug after entering the patient's body was Renal, and the experimental study showed that this drug was safe for use.

Other disease histories were recorded in 87 cases (27.4%) between the ages of 25 and 45, 41 cases (12.9%) between the ages of 45 and 65, and 46 cases (14.5%) between the ages of 65 and 85, and we conducted safety tests (immunity tests) on these patients and concluded that the best drug for treatment was Nom-fibrin Specific. The total dose of the drug was tested as MI: 50–100 mg, IS: 0.9 mg/kg, PE: 100mg; the organ of metabolism of the drug into the patient's body was Fisher, and the experimental study concluded that the drug was safe for use.

The second innovative and important phase of this evaluation index: b) controlled observational experimental phase, analyzed by statistical criteria, was conducted for patients with ischemic stroke for antithrombotic therapy, testing the optimal ratio of thrombolytic dosing, and automatic intravenous thrombolytic drug effectiveness testing. The focus of the study was to compare the traditional way of drug delivery process, and the research team explored the time points delayed by emergency thrombolysis in the event of stroke, and to enhance the concept of automated thrombolytic care. In this study, 32 patients (10.1%) in the 25–45 age group, 6 (1.89%) in the 45–65 age group, and 26 (8.2%) in the 65–85 age group were treated with antithrombotic therapy. The best drug for the treatment of these patients is Fibrin Specific (++++); the total dose of the drug is 30 mg–50 mg; the organ of metabolism of the drug after entering the patient's body is Hepatic and renal, and the experimental research shows that this drug is safe for use.

The third innovative and important stage of this evaluation index: c) adopted a statistical research approach, by combining the first two paragraphs of innovative research methods, we used spss 11.0 research statistical scheme, through computerized data measurement analysis to p value < 0.05-p < 0.01 as the evaluation criteria derived, indicating that the smaller the value of the difference is statistically significant, the more his research can be established value the higher the value of his study's validity (Fig. 3A).

As shown in the results in Fig. 1 below, patients with ischemic stroke were tested for antithrombotic therapy, optimal thrombolytic dosing ratio, and automatic intravenous thrombolytic drug effectiveness. The p value values analyzed for all three tests were between 0.2–0.3. Through continuous comparison between multiple departments and multiple study groups it was finally determined that the innovative study data comparison concluded that the health care effectiveness of mechanical and automated thrombolysis was consistent.

Category 1 / Category 2	Questionnaire Total number of people: 317	Number of people <65 years old	Number of people ≥65 years old	Safety test (immunological)	Total drug dose	Entering human metabolism	Drug test use	p value	Drug half-life	Drug test use	Regimen
Total number of men and women in the survey	Males 184 / Female 133	Male 87 / Female 47	Male 97 / Female 86	Non-fibrin Specific / Fibrin Specific(+++)	20 IU / 30mg–50mg	Renal	J	< 0.0037	15 / 11–20	J	Infusion
Percentage of men and women	<25 years old 35 (11%)	<45 years old 92 (29%)	<65 years old 152 (48%)	<85 years old 38 (12%)							
Basic symptom study	Between 25–45 years old	Between 45–65 years old	Between 65–85 years old	Safety test (immunological)	Total drug dose	Entering human metabolism	Drug test use	p value	Drug half-life	abbreviation	Regimen
Cardiac infarction	33 (10.7%)	82 (26.9%)	62 (19.5%)	Fibrin Specific(+++)	2.25–6.25×10 IU	Hepatic	Fisher	0.5	20	sk	infusion
Hypertension	116 (36.5%)	64 (20.1%)	52 (16.4%)	Fibrin Specific(++)	M:50–100mg IS:0.9mg/kg PE:100mg	Renal	Fisher	0.6	4–8	uPA	infusion
Diabetes	235 (74.1%)	149 (47.0%)	86 (21.1%)	Non-fibrin Specific	2.25–6.25×10 IU	Renal	Fisher	0.9	15	rtPA	infusion
High blood fat	143 (45.1%)	69 (21.7%)	74 (23.3%)	Non-fibrin Specific Immunogenic	2.25–6.25×10 IU	Hepatic	Fisher	0.7	20	rPA	bolus
Liver and kidney diseases	98 (30.9%)	34 (10.7%)	64 (20.1%)	Fibrin Specific(++)	20 IU / 30mg–50mg	Renal	Fisher	0.5	4–8	sk	Double bolus
Other Disease Symptoms	87 (27.4%)	41 (12.9%)	46 (14.5%)	Non-fibrin Specific	M:50–100mg IS:0.9mg/kg PE:100mg	Renal	Fisher	0.4	15	uPA	infusion
Anti-thrombotic therapy	32 (10.1%)	6 (1.89%)	26 (8.2%)	Fibrin Specific(+++)	30mg–50mg	Hepatic Renal	Fisher	0.2	11–20	sk	Double bolus
Testing Thrombolytic Dosing	Between 25–45 years old	Between 45–65 years old	Between 65–85 years old	Safety test (immunological)	Total drug dose	Entering human metabolism	Drug test use	p value	Drug half-life	rtPA	infusion
Thrombolytic Efficacy Testing	261 (82.3%)	129 (40.6%)	13 (41.6%)	Intravenous arm	M:50–100mg IS:0.9mg/kg PE:100mg	Renal	Fisher	0.3	20	rtPA	infusion
Intentional Home Treatment Patients	211 (66.5%)	119 (27.5%)	92 (29.0%)	Automated thrombolytic therapy	2.25–6.25×10 IU	Renal	Fisher	0.2	4–8	rPA	Double bolus
Intended Hospital Patients	137 (43.2%)	56 (17.6%)	81 (25.5%)	Intravenous arm	20 IU / 30mg–50mg	Renal	Fisher	0.3	4–8	sk	bolus

Fig. 1. 317 patients were studied and the evaluation indicators were divided into three important stages of innovation. (A) The group aimed at redesigning and studying how to establish innovative, efficient and concise treatment processes and testing procedures. By realizing the integration study of indicators for thrombolytic therapy, the fine therapeutic care flow chart for intravenous injection.

4   Characterization of Research Design Innovations and Processes

In the initial research integration we found. The use of analytical computing research program can better compare the results of the sample of 317 studies. At the same time combined with the mRS classification method, the valid data to integrate again, so as to calculate the best advantage of the sample and compared to the sample of the research is not enough, this study through the research of 65–85 years of age of the elderly occurrence of ischemic stroke occurs in a higher incidence of ischemic stroke, and this type of population is rarely in less than four hours in a timely manner to the hospital for treatment, more than the optimal time for treatment will be the emergence of symptoms of large infarcts, some of which are very difficult to detect ischemic stroke symptoms in time.

As shown in Fig. 2 below, analyze and study 317 cases of stroke cerebral infarction patient comparison, by parsing the patient began to appear face twitching, speech impediment slowness, unilateral limb movement delay and obstacles, based on one or more of the above is judged to be the emergence of the symptoms began to the hospital time, for the process of wishing to process, the process of diagnosis, testing process, the process of treatment through the collection of the two compared to the cycle time (from the discovery of the optimal) treatment time was less than four hours).

In terms of the safety analysis study, the researchers listed in detail the flow of the treatment plan for patients traditional machine thrombus removal (MT) is a residual stroke due to large vessel occlusion (Fig. 2A). Through the conclusion of the study it was found that the higher the age increase the longer the time to discover the pathology to the hospital visit has a greater impact on the degree of deterioration of ALS patients. This is also valid for younger young patients aged 25–45 years. Our study also has more research significance and value by studying the post morbidity process and subtracting the series of processes such as routine registration and booking of appointment, emergency channel visit, priority visit for the elderly (green channel), 4 h golden rescue time, arrival of ischemic stroke patients at the rescue room, monitoring of ECG and blood glucose and various indicators, major venous thrombolysis, dispensing of western medicines to assist in the treatment and real time ECG monitoring, our innovative study reduces this series of processes. Innovative study reduces this process.

The patient only needs to perform the automated thrombolysis process at home. All that is needed is to test the patient's body functions in advance and dispense medication accordingly, while the occurrence of a stroke can provide timely and intelligent feedback to the patient so that he or she can be injected in time for the operation. Our research based on ischemic stroke innovative automatic intravenous thrombolysis drug delivery (Fig. 2B). Based on the research of 317 research samples to refine the split variable data index analysis (Fig. 2C).

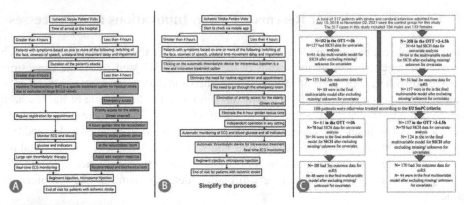

Fig. 2. Based on one or more of the above, we determined the time to the hospital from the onset of symptoms, the process of processing as desired, the consultation process, the testing process, and the treatment process by collecting and comparing the two-cycle time diagrams. (A) For the safety analysis study, we detailed the flow chart of the treatment plan for patients with residual stroke due to large vessel occlusion by conventional machine embolization (MT). (B) A comparison of the findings of the study shows that the higher the age of the patient, the longer the time to the hospital for the detection of the pathology and the greater the impact on the deterioration of the ALS patient. (C) Refined split-variable data indicator analysis based on the 317 study samples studied.

5   Automatic Thrombolysis System Overall Construction Application Design

With the increase in age and the change of lifestyle, researchers found that 30% of ischemic stroke patients could not be given corresponding treatment in time after the onset (the best time from onset to treatment was no more than 4 h). According to the survey, the prognosis of these patients is not very good. On this basis, a new type of automatic thrombolysis driving system with independent intellectual property rights is proposed, and the overall block diagram and automatic drug delivery process (DNT) design are carried out, so as to provide patients with mechanical thrombolytic therapy A new way of thinking is provided.

The research team has obtained innovative self-thrombolysis test results in research, development, experimentation, research, and innovative trial-and-error studies. Finally, a series of experiments were carried out on the device, which proved its good adaptability, and everyone can tailor a suitable application plan for it and give specific guidance. Increase their logical diversity at the multi-stakeholder group and system level to help them formulate and execute strategies.

This research is innovative research on the automatic thrombolysis drive system for patients with ischemic stroke. Based on a DDS chip designed with the assistance of graduate students majoring in mechanical engineering, the programmable system model solution can give priority to the detection of abnormal waveforms in the human body. And frequency, the system test is automatically triggered by the system method for the pulse period and center frequency. At the same time, the innovative integrated design better serves patients with ischemic stroke.

As shown in Fig. 3 below, shows the overall framework of the automatic thrombolytic drive system. The research requires the use of a special DDS chip and computer programming, using a programmable gate array solution (ZYNQ-7Z020) to input various parameters at the FEP input terminal (DDS signal source control transmitter), including the waveform maximum range and waveform range value, pulse cycle and emergency dispatch physical button settings.

Send the SSD1306 signal and DAC signal through the (ZYNQ-7Z020) system to output various detection indicators of multiple functions. If the system programming setting is exceeded, the power amplifier will start and automatically trigger the thrombolytic system (3A). The system research is based on the FPGA phase accumulator for programming research, the frequency sent by the human body is the signal port, and the FEP (ZYNQ-7Z020) is the input port. Transform into a waveform detection indicator system. Possessed programmable functions can make the range of data more adjustable by programming setting 2B. Convert the waveform data and waveform frequency of the signal source into visualized data. The Waveform data range is 0.9 MHz–1.9 MHz; the waveform frequency data range is 35 Vpp–55 Vpp (3B).

The data detection waveforms used in this study are two indicators of sine wave pulse center frequency detection and sine wave pulse period detection. According to the data of 317 patients, it can be concluded that the best detection data range of the sine wave pulse center frequency is 1–1000; the best repetition range of the sine wave pulse cycle detection data is 1 Hz–2000 Hz. The system automatically activates an alarm (3C) if the specification program exceeds the index through our design and programming. Research and development of automatic thrombolysis drive system chip structure test diagram (3D).

Appearance humanized design and appearance system design (3E) for the automatic thrombolytic drive system. Show the research framework (3F) for the structure of this study. Models, spare parts, quantity, fenestration size, diameter parameters, chip layout, and innovative structure wireframe of the automated thrombolytic drive system for patients with ischemic stroke.

The research is aimed at innovative research on the automatic thrombolysis drive system for patients with ischemic stroke. Based on a DDS chip designed with the assistance of graduate students majoring in mechanical engineering, the programmable system model solution can give priority to the detection of human abnormalities. Frequency, through the system method automatically triggers the system test for the pulse period and center frequency.

At the same time, the innovative integrated design better serves patients with ischemic stroke. Compared with patients who have not yet received mechanical thrombolysis, automatic thrombolysis can better directly detect and serve patients, and provide research data support to better ensure that patients are more properly detected and delivered drugs for automatic intravenous thrombolysis treatment.

In this study, a combination of questionnaire survey and hospital clinical trials was used to select 317 patients with stroke who were hospitalized during the period from July 10, 2018, to November 2, 2021, as controls, and analyzed them to verify the validity of this study. Innovation. Among the 317 subjects, there were 184 males and 133 females; among the population under 65 years old, there were 87 males and 47 females; among the population over 65 years old, there were 97 males and 86 females; Arranging and dividing into four age groups: 35 people under the age of 25, accounting for 11%; 92 people under the age of 45, accounting for 29%; 152 people under the age of 65, accounting for 100% 48 out of 100; 38 people under the age of 85, accounting for 12%.

At the same time, we set up a research team to analyze the historical cases of these 317 patients during the period from March 2, 2021, to November 7, 2021, based on the above age group ratios and conducted detailed on-site questionnaire surveys. Link analysis and record diagram (3G). The survey questionnaire outline is based on the user's characteristic area (city center, suburban junction, suburb, county, etc.), gender, age, education level, marriage, several children (whether living with children); working status, occupation, working hours and age Income; leisure life and values, hobbies, leisure life, and life circle. User needs are based on purchase motivation (why buy), purchase type (first purchase/additional purchase), budget, purchase influencing factors, and purchase channels (recommendation, online shopping, physical store, etc.); the development form is based on appearance, function and configuration, user experience and Brand and price research.

The result of the comprehensive bar chart sequence is how to calculate the average comprehensive score. The five detailed data average 5 points for the comprehensive score. The wearing method is 4.61; whether it is portable is 4.58; the size is 3.33; the appearance is 2.44; the color is 2.14; and the others are 0.11. Based on the sense of technology, the proportion is 8.33%, the sense of fashion is 8.33%, the sense of sport is 0%, the sense of cuteness is 50%, and the sense of simplicity is 33.33% (3H).

Design based on the assistance of fellows in mechanical engineering includes four areas of innovation. In terms of performance, independently program a DDS chip, which has additional advantages such as small size, high detection accuracy, and good battery life. The programmable system model solution can give priority to solving the waveform and frequency of abnormal detection of the human body. Using the system amplifier to conduct statistical analysis on the collected data, the abnormal number of people, abnormal rate, and other important indicators can be detected. Data statistics can be viewed through the mobile terminal.

In the detection method, the innovative automatic trigger system test is carried out for the pulse period and center frequency through the system amplifier. At the same time, the innovative integrated design better serves patients with ischemic stroke. Compared with patients who have not yet received mechanical thrombolysis, automatic thrombolysis can better directly detect and serve patients (the optimal time from patient onset to treatment is within 4 h).

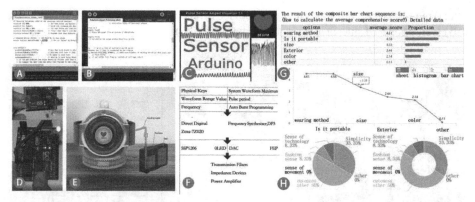

Fig. 3. The overall framework diagram of the automatic thrombolytic drive system. (A) (ZYNQ-7Z020) The system sends SSD1306 signal and DAC signal to output multi-functional detection indicators. (B) Based on FEP (ZYNQ-7Z020) as the input terminal. Convert it into a waveform detection index system diagram. (C) The set specification scheme of the design programming exceeds the index and the system automatically activates the alarm system. (D) Schematic diagram of the chip structure test for developing an automated thrombolysis-driven system. (E) Humanized appearance design and appearance system design for the automatic thrombolytic drive system. (F) Research framework showing the structure of this study. (G) On-site questionnaire survey record map. (H) The result of the comprehensive bar graph sequence is how to calculate the average comprehensive score.

6 Automatic Thrombolytic Injection Nursing Process Design for Ischemic Stroke

This study is based on methodological research, assessment of clinical study population and technology collection, analysis of study design innovation and process characteristics, design of structural applications based on automatic trigger thrombolysis function, and study of the general framework diagram of the automatic thrombolysis drive system. We again designed the data integration based on the technical process and interface UI. By reviewing the nature of this study, the researchers found that there are still deficiencies in the product interface.

As shown in Fig. 4 below, the data integration design was carried out for the interface for visualization upgrade, system internal programming diagram and cell phone page UI. The implementation of the program red color for endangerment warning, green color warm reminder, yellow indicators to indicate the safety of the three for a simple visualization interface. The elderly can also better observe the physical condition when wearing. Watch information timely feedback to the user (4A).

Set circuit amplifier can view the data comprehensively. Through data analysis and integration of different time periods of heart rate collection system map, this study is based on arduino development layout and processing display software to assist in programming and development of innovative design. The indicators of this device are: board diameter: 16mm, board thickness: 1.6 mm (innovative pcb board thickness), led peak wavelength: 515 nm, power supply voltage: 3.3 v–5.5 v, output signal type: analog signal, output signal size: 0 v–3.3 v (3.3v power supply or 3.3 v–5.5 V power supply), current size: 0 ma–4 ma (>5 v). Diagram of each device index parameter and programming data (4B) Cell phone UI interface flow design diagram (4C).

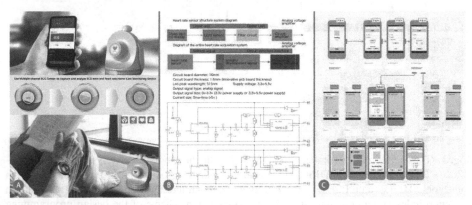

Fig. 4. Design diagram for visualization of interface upgrade, system internal programming diagram and cell phone page UI for data integration. (A) Implementation plan red color for endangerment warning, green color for warm reminder, yellow indicator for safety three for simple visualization interface. The elderly can better observe the physical condition when wearing. Watch information timely feedback to the user. (B) Set circuit amplifier to view data more comprehensively. Through data analysis and integration of different time periods of heart rate collection system map, this study based on arduino development layout and processing display software to assist in programming and development of innovative design map. (C) Mobile phone UI interface flow design diagram.

7 Conclusion and Future Work

The shortcomings of this study are the lack of more control group of patients with a total of 317 cases of stroke cerebral infarction admitted from July 10, 2018, to November 02, 2021, which always has a small amount of bias in attempting to study the conclusions. And there is a lack of more valid patients to provide real data to support the study data. And from the technical point of view, the methodology is limited in its innovation. Lack of specific analysis. At the same time, based on our innovative automatic thrombolysis and clinical trials, mechanical engineering researchers assisted in the design of a DDS chip, with a programmable system model solution that can be prioritized to address the detection of abnormal waveforms and frequencies in the human body, through the system methodology for the pulse period and the center frequency of the system automatically triggered by the system test. At the same time, the innovative integrated design better serves ischemic stroke patients. Compared to patients who have not yet received mechanical thrombolysis, automated thrombolysis provides better direct detection and service to patients, and provides research data to support and better ensure that patients are more on point for detection and delivery of medications for automated intravenous thrombolysis. Thrombolytic injections were administered following the treatment algorithm, and both dosage and injections were well detected, with good prognostic value for the patient.

References

1. Chen, R.L., Balami, J.S., Esiri, M.M., Chen, L.K., Buchan, A.M.: Ischemic stroke in the elderly: an overview of evidence. Nat. Rev. Neurol. **6**, 256–265 (2010)
2. Powers, W.J., Rabinstein, A.A., Ackerson, T., et al.: Guidelines for the early management of patients with acute ischemic stroke: 2019 update to the 2018 guidelines for the early management of acute ischemic stroke: a guideline for healthcare professionals from the American Heart Association/American Stroke Association. Stroke **50**, e344–e418 (2019)
3. Alawieh, A., Starke, R.M., Chatterjee, A.R., et al.: Outcomes of endovascular thrombectomy in the elderly: a 'real-world' multicenter study. J. Neurointerv. Surg. **11**, 545–553 (2019)
4. Zhao, W., Ma, P., Zhang, P., Yue, X.: Mechanical thrombectomy for acute ischemic stroke in octogenarians: a systematic review and meta-analysis. Front. Neurol. **10**, 1355 (2020)
5. Lees, K.R., Bluhmki, E., von Kummer, R., et al.: Time to treatment with intravenous alteplase and outcome in stroke: an updated pooled analysis of ecass, atlantis, ninds, and epithet trials. Lancet **375**, 1695–1703 (2010)
6. Elliott, W.J.: Circadian variation in the timing of stroke onset: a metaanalysis. Stroke **29**, 992–996 (1998)
7. Casetta, I., Granieri, E., Fallica, E., la Cecilia, O., Paolino, E., Manfredini, R.: Patient demographic and clinical features and circadian variation in onset of ischemic stroke. Arch. Neurol. **59**, 48–53 (2002)
8. Lago, A., Geffner, D., Tembl, J., Landete, L., Valero, C., Baquero, M.: Circadian variation in acute ischemic stroke: a hospital-based study. Stroke **29**, 1873–1875 (1998)
9. Marler, J.R., Price, T.R., Clark, G.L., et al.: Morning increase in onset of ischemic stroke. Stroke **20**, 473–476 (1989)
10. Adeoye, O., Hornung, R., Khatri, P., Kleindorfer, D.: Recombinant tissuetype plasminogen activator use for ischemic stroke in the united states: a doubling of treatment rates over the course of 5 years. Stroke **42**, 1952–1955 (2011)
11. Bray, B.D., Campbell, J., Hoffman, A., et al.: Stroke thrombolysis in England: an age stratified analysis of practice and outcome. Age Ageing **42**, 240–245 (2013)
12. Mateen, F.J., Buchan, A.M., Hill, M.D.: Outcomes of thrombolysis for acute ischemic stroke in octogenarians versus nonagenarians. Stroke **41**, 1833–1835 (2010)
13. Mateen, F.J., Nasser, M., Spencer, B.R., et al.: Outcomes of intravenous tissue plasminogen activator for acute ischemic stroke in patients aged 90 years or older. Mayo. Clin. Proc. **84**, 334–338 (2009)
14. Sun, M.C., Lai, T.B., Jeng, J.S., et al.: Safety of intravenous thrombolysis for ischaemic stroke in Asian octogenarians and nonagenarians. Age Ageing **44**, 158–161 (2015)
15. Sarikaya, H., Arnold, M., Engelter, S.T., et al.: Intravenous thrombolysis in nonagenarians with ischemic stroke. Stroke **42**, 1967–1970 (2011)
16. Ford, G.A., Ahmed, N., Azevedo, E., et al.: Intravenous alteplase for stroke in those older than 80 years old. Stroke **41**, 2568–2574 (2010)
17. Derex, L., Nighoghossian, N.: Thrombolysis, stroke-unit admission and early rehabilitation in elderly patients. Nat.. Rev Neurol. **5**, 506–511 (2009)
18. Chen, R.L., Balami, J.S., Esiri, M.M., et al.: Ischemic stroke in the elderly: an overview of evidence. Nat. Rev. Neurol. **6**, 2562–2565 (2010)
19. Russo, T., Felzani, G., Marini, C.: Stroke in the very old: a systematic review of studies on incidence, outcome, and resource use. J. Aging Res. **2011**, 108785 (2011)
20. Arora, R., Salamon, E., Katz, J.M., et al.: Use and outcomes of intravenous thrombolysis for acute ischemic stroke in patients ≥90 years of age. Stroke **47**, 2347–2354 (2016)
21. Mutgi, S.A., Behrouz,:. Practice diversity among US stroke physicians with respect to the 3- to 4.5-h window. Int. J. Stroke **10**, E24–E25 (2015)

22. Mishra, N.K., Ahmed, N., Andersen, G., et al.: Thrombolysis in very elderly people: controlled comparison of SITS international stroke thrombolysis registry and virtual international stroke trials archive. BMJ **341**, c6046 (2010)
23. The National Institute of Neurological Disorders and Stroke rt-PA Stroke Study Group. Tissue plasminogen activator for acute ischemic stroke. N Engl. J. Med. **333**, 1581–1587 (1995)
24. Sandercock, P., Wardlaw, J.M., Lindley, R.I., et al.: The benefits and harms of intravenous thrombolysis with recombinant tissue plasminogen activator within 6 h of acute ischaemic stroke (the Third International Stroke Trial [IST-3]): a randomised controlled trial. Lancet **379**, 2352–2363 (2012)
25. Wardlaw, J.M., et al.: Recombinant tissue plasminogen activator for acute ischaemic stroke: an updated systematic review and meta-analysis. Lancet **379**, 2364–2372 (2012)
26. Polk, S., Stafford, C., Adkins, A., Efird, J., Colello, M., Nathaniel, I.T.: Contraindications with recombinant tissue plasminogen activator (rt-PA) in acute ischemic stroke population. Neurol. Psychiatry Brain Res. **27**, 6–11 (2018)
27. Fredwall, M., Sternberg, S., Blackhurst, D., Lee, A., Leacock, R., Nathaniel, T.I.: Gender differences in exclusion criteria for recombinant tissue-type plasminogen activator. J. Stroke Cerebrovasc. Dis. **25**, 2569–2574 (2016)
28. Wapshott, T., Blum, B., Williams, K., Nathaniel, I.T.: Investigation of gender differences and exclusive criteria in a diabetic acute ischemic stroke population treated with recombinant tissue-type plasminogen activator (rtPA). J. Vasc. Interv. Neurol. **9**, 26–32 (2017)
29. Wang, R.: A multi-modal knowledge graph platform based on medical data lake. In: Zhao, X., Yang, S., Wang, X., Li, J. (eds). Web Information Systems and Applications. WISA 2022. LNCS, vol. 13579. Springer, Cham (2022). https://doi.org/10.1007/978-3-031-20309-1_2

Friction-Based Nanotransparent Fibers for Electronic Skin for Medical Applications

Jiayi Cai[1](✉) (iD) and Jialiang Cai[2] (iD)

[1] Faculty of Humanities and Arts, Macau University of Science and Technology, Macau, China
Cai849027924@qq.com
[2] Macao Institute of Materials Science and Engineering, Macau University of Science and Technology, Macau, China

Abstract. Based on the three-dimensional deformation of scoliosis and transverse misalignment, the study investigates the clinical application and computational method evaluation of 4D printed scoliosis rehabilitation devices, based on RHINO software with morphology software, mechanical analysis software, magisso magnetic field training software, and 3D imaging equipment for clinical profiling and in-depth analysis of models. The original national design patent of the study has future marketable advantages such as recyclable, renewable, green, safe, and low-cost for mass production. The experimentalists conducted a clinical study based on 49 patients with 3D deformation of scoliosis and transverse misalignment (11 to 17 years old spinal deformation) during November 2020 - June 2021, respectively based on the clinical application of the current filming records, innovative The SOSORT/SRS assessment criteria and the scoliosis rehabilitation equipment calculation method were studied in the field and proved to be better for the assessment and physical treatment of patients with scoliosis and transverse subluxation in 3D deformities. Significant benefits to patient quality of life and rehabilitation prognosis and validated the accuracy of the model.

Keywords: Scoliosis rehabilitation devices · 4D printing · Clinical applications · Learning computing system development

1 Introduction

Systemic disorders of the human body based on three-dimensional deformities of scoliosis and transverse subluxation frequently result in deformities of the body in (3D) different dimensions. Most 3D angles are mostly greater than 10° or more [1, 2]. It occurs internally and is difficult to detect in external life, and is more frequent in the age range of 11 to 17 years old. In severe cases, it can cause significant pain and developmental problems in the heart, lungs, and bones of the body [3]. The current treatment modalities and therapeutic equipment are not very effective because of the long period and expensive equipment. The existing traditional research method is for computerized tomography and 3D scan data calculation to obtain 3D data reconstruction and view quantification [4]. However, a large number of studies have shown [5, 6] that there is not much improvement and cure for patients' quality of life and rehabilitation.

© The Author(s), under exclusive license to Springer Nature Singapore Pte Ltd. 2023
L. Yuan et al. (Eds.): WISA 2023, LNCS 14094, pp. 601–612, 2023.
https://doi.org/10.1007/978-981-99-6222-8_50

This study aims to innovate research findings and patient field records, through the 4D printing of scoliosis rehabilitation equipment based on the clinical application of the current situation of the whole record, carbon principle construction steel Q235A for simulation of the introduction of the real analysis and evaluation, and through the RHINO software with morphology software, mechanical analysis software, Magisto magnetic field training software classification of the components studied on the human body has a substantial contribution to the human body.

4D printing scoliosis rehabilitation equipment statistical testing and quantitative integration of analytical charts, 4D printing scoliosis rehabilitation equipment calculation methodology research conclusions show that the 4D printing scoliosis rehabilitation equipment can enhance the patient's skeletal muscles on both sides of the muscle and exercise strength, correct the body posture, and at the same time, enhance the stability and flexibility of the spine. Improving spinal blood circulation and relieving the nervous system has an important role [7], and improving lung capacity [8] and reducing low back pain has a significant effect [9]. Combined with the International Scoliosis Orthopaedic and Rehabilitation Therapy Society's proposed rehabilitation data and corrective program index evaluation [10]. Thus, we can achieve the internal adjustment of the body and our innovative research 4D printing scoliosis rehabilitation equipment and external equipment for better rehabilitation and conditioning, to achieve the purpose of symptom reduction, rehabilitation, and prognosis.

2   Materials and Methods

The study was based on the advantageous properties of 4D printed shape-memory polymer material with 350% recovery performance, along with good stability performance to the manipulation environment, traditional therapeutic exercise studies, and a good prognosis for spinal patients [11].

Based on a clinical study of 49 patients (11 to 17 years of age with spinal deformity) with scoliosis and transverse misalignment in three dimensions during November 2020-June 2021, the investigators demonstrated better assessment and physical therapy based on the clinical application of current filming records, innovative SOSORT/SRS assessment criteria, and scoliosis rehabilitation device calculation method study field operations, respectively. Patients with three-dimensional deformities of scoliosis and transverse misalignment.

The study analysis was based on the carbon principle construction steel Q235A for the simulated introduction of realistic analysis and evaluation. Forty-nine patients available for spinal radiographs were recruited for a 7-month follow-up study. Three specific types of scoliosis (ALS), scoliosis (JIS), and anterior spinal convexity (retroflection misalignment) were recorded. The study also followed the patient's daily life from the first week of wear to the tenth week for a total of 70 days and provided important future references.

The scoliosis rehabilitation device can be adapted to different participants through the application testing the body heat distribution of the respondents. Scoliosis rehabilitation equipment improves load support and mobility and prevents work-related musculoskeletal disorders. It can reduce musculoskeletal problems and improve productivity.

Its performance and shape (angle/length) can be adjusted for body size, worker strength, and working conditions, and it is light enough (approximately 1.6 kg) to be used in most situations. t-tests and K-S tests for three-dimensional forces and torques at the spine level in patients with scoliosis and transverse misalignment were recorded in the field for a two-week follow-up period from one hour of wear per day to three hours of wear per day, yielding P values, and mean plots of TDA [12, 13]. Significant benefits on patients' quality of life and rehabilitation prognosis and validated the accuracy of the model.

3   4D Printed Scoliosis Rehabilitation Device Clinical Trial Data

The study is based on the current state of clinical application of 4D printed scoliosis rehabilitation devices photographed. As shown in Fig. 1 below, the study is based on the progression of ALS curve research in young people and the experimenters based on voluntary participation and clinical study through 49 patients with scoliosis with transverse misalignment in 3D deformation between November 2020 and June 2021, this study is based on the (COBB) to assess the way to avoid inherent errors [6, 14, 15].

This study tested a scoliosis rehabilitation device to help patients with adolescent idiopathic scoliosis (spinal deformities between the ages of 11 and 17 years). The analysis of this study was based on the carbon principle construction steel Q235A for the simulated introduction of real-world analysis evaluation. Forty-nine patients available for spinal radiographs were recruited for a 7-month follow-up study. Three images were taken for three specific types of scoliosis (ALS), scoliosis (JIS), and anterior spinal convexity (retroversion misalignment). As shown in Fig. 1 below, the study's status quo photogram was based on lateral S, T, L, and H profiles; and the offset dimension data of the lateral side was clinically profiled by a 3D imaging device (1A).

The study status shot is based on the patient's lateral lying S, T, L, and H profiles; and the offset dimensional data of the lateral side is clinically profiled by the 3D imaging device (1B). The study status shot is based on the patient standing S, T, L, and H profile; and the offset dimensional data of the lateral side is clinically profiled by a 3D imaging device (1C). The body block analysis was determined by fixing the body block in four zones, respectively, H for the hip position to the pelvis and the lumbar spine toward the ankle (LEV); L for the lumbar spine region from the lumbar portion of the subject to the upper lumbar spine (UEV) and the lower middle (LEV); T for the upper thoracic vertebral region (UEV) to the lower thoracic vertebral region (LEV); S for the upper scapular muscle group (UEV) to the neck upper-end vertebral region (LEV) (1D). Three shots were taken for three specific types of scoliosis (ALS), scoliosis (JIS), and anterior spinal convexity (retroversion misalignment) (1E). In traditional scoliosis rehabilitation equipment terminals, the patient rehabilitation process can not be monitored in real-time, and lack of cloud data storage, a lack of feedback to professional doctors, can not fit the skin appropriately, space occupation, the operation is cumbersome, and other social status research conclusions. (1F).

The four categories of record research were replaced with simplified geometric trapezoids. Researchers based on four important to be overcome basis: innovative 4D printing model, affected by human body temperature model will shrink to the appropriate body shape. More ergonomic, the product is scientifically comfortable and efficient in correcting human posture and gradually recovering, helping adolescents with idio-pathic scoliosis lightweight, soft, and fashionable; digital recording imaging customized patient parameters, digitally designed and recorded; doctors online remote monitoring and tracking patient data records, digital monitoring of progress and conditions; inno-vative patents, 4D printed scoliosis rehabilitation equipment multi-structural innovation Patented design, soft material with parametric breathable material (1G).

This study is based on 49 patients with 3D deformation of scoliosis and transverse misalignment for clinical research integration chart, and through RHINO software with morphology software, mechanical analysis software, Magisto magnetic field training software to classify the studied components with substantial contribution to the human body. The heat distribution contribution map based on the monitoring of DLAs was mainly focused on the part of the region above the lumbar spine and the part of the region below the scapula. The training concentration demonstration integration yielded significant heat concentration and rehabilitation effects of the clinical application of scoliosis rehabilitation equipment. Based on the whole spine CT rays, MRI screening, and electrophysiological examinations are available as alternatives to the X-ray data collection library (1H).

4   4D Printed Scoliosis Rehabilitation Equipment Testing

Based on the existing hot study map of conventional 3D printed scoliosis rehabilitation devices, we conducted a clinical study based on voluntary participation of 49 patients with scoliosis and transverse misalignment in 3D deformation between November 2020 and June 2021, this study tested scoliosis rehabilitation devices to help patients with adolescent idiopathic scoliosis (spinal deformity between 11 and 17 years old). The study was divided into 9 groups and the characteristic points that emerged from the study have validated the innovative 4D printed scoliosis rehabilitation device as having a significant clinical rehabilitation effect. And based on existing SOSORT/SRS rubrics [16, 17] and distributed collaborative planning processes [18].

The study also followed up the patients' daily life from the first week of wear to the tenth week for a total of 70 days to document the study and provide an important future reference. This is shown in Fig. 2 below. A graph of the fixed-point recording data based on the innovative 4D printed scoliosis rehabilitation device developed by our team. The strongest intensity is the orange color indicator, the medium intensity is the orange color indicator, and the weaker intensity is the yellow color indicator (2A).

This study is based on the statistical test and quantitative integration analysis graph of the 4D printed scoliosis rehabilitation device. It is shown in Fig. 2 below. The first category of patients were number 1 for scoliosis/psychology, number 2 for scolio-sis/rehabilitation, number 3 for scoliosis/treatment, number 4 for quality of life/support, and number 5 for lumbar spine/surgery in the form of a group of clinical patients who met the five criteria. The spinal impact of scoliosis was higher in the first and seventh week of treatment, and in scoliosis/psychology patients had some psychological stress in the tenth week. In addition quality of life/support this study has new features for future clinical studies [19, 20].The second category of patients were number 6 for external spinal fusion/method, number 7 for scoliosis/surgery, number 8 for spinal fusion/device,

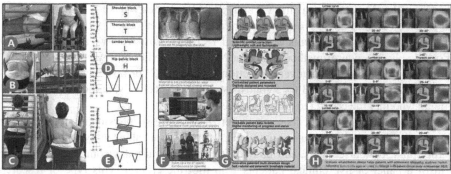

Fig. 1. Current state shot diagram of the clinical application of the 4D-printed scoliosis-based rehabilitation device. (A) The study status shot diagram is based on lateral S, T, L, and H profiles; and the offset dimensional data diagram of the lateral side was clinically profiled by 3D imaging equipment. (B) The study status shot is based on a lateral S, T, L, and H profile of the patient; and the offset dimensional data of the lateral side was clinically profiled by a 3D imaging device. (C) The study status shot is based on the patient standing S, T, L, and H profiles; and the offset dimension data of the lateral side was clinically dissected by a 3D imaging device. (D) H is the hip position to the pelvis and the lumbar spine toward the ankle (LEV); L is the lumbar region from the lumbar portion of the subject to the upper lumbar spine (UEV) with the lower middle (LEV); T is the upper thoracic vertebral region (UEV) to the lower thoracic vertebral region (LEV); and S is the upper scapular muscle group (UEV) to the upper cervical vertebral region (LEV). (E) Three specific types of scoliosis (ALS), scoliosis (JIS), and anterior spinal convexity (retroflexion misalignment) were photographed. (F) Traditional scoliosis rehabilitation equipment terminal, the patient rehabilitation process can not be monitored in real time, and lack of cloud data storage, lack of feedback to professional doctors, can not fit the skin properly, space occupied more, operation is cumbersome and other social status research conclusions. (G) 4D printing scoliosis rehabilitation equipment multi-structure innovation patent design, soft material and parameterized breathable material. (H) Based on the whole spine CT ray, MRI screening, electrophysiological examination can replace the X-ray data collection library.

and number 9 for internal fixator/posture, respectively, and the four clinical patients who met the criteria constituted a group format. And this group study was tested around external assistive devices, while at the same time, the existing rehabilitation treatment modalities were transferred from traditional surgery to innovative studies of external devices [21–23].

The third group of patients constituted a panel format of five clinical patients who met the criteria for scoliosis/congenital for number 10, scoliosis/epidemiology for number 11, scoliosis/diagnosis for number 12, spine/deformity for number 13, and lumbar spine/diagnostic imaging for number 14, respectively. As idiosyncratic diagnostic indicators, clinical studies with 3D projection analysis help subsequent studies of scoliosis with unknown benchmarks in medical history studies and provide a better understanding of the patient's body deformation status, leading to more accurate data criteria.

The fourth category of patients are number 15 for kyphosis/diagnostic imaging, number 16 for spine/diagnostic imaging, and number 17 for 3D imaging/methods. With the continuous development of existing VR and AR technologies, our study also investigated and analyzed a multi-angle group approach to imaging methods affecting diagnosis.

The fifth category of patients was number 18 for scoliosis/classification, number 19 for orthopedic procedures/methods, number 20 for orthopedic surgery/adverse reactions, number 21 for spinal fusion/adverse reactions, number 22 for postoperative complications/etiology, and number 23 for kyphosis/surgery, respectively, and other six clinical patients who met the criteria to constitute the panel format.

The sixth category of patients was number 24 for scoliosis/physiopathology, number 25 for scoliosis/pathology/etiology, number 26 for postural balance/physiology, number 27 for spinal fusion/scoliosis, and number 28 for cerebral palsy/complications in the form of a panel of five clinical patients who met the criteria. This type of research integration is related to human genetics and brain genetic mechanisms, and it was found that this type is accompanied by a high distribution of symptoms from week 1 to week 7 in patients with idiopathic cerebral palsy/complications has very important hotspots and implications for our study of patients with scoliosis.

The seventh category is number 29 for lumbar spondylosis/diagnostic imaging, number 30 for internal fixator/posture, and number 31 for scoliosis/congenital and other three clinical patients who meet the criteria to form a group format. Based on the congenital spine history is more predominantly genetic idiopathic. The need for internal fixators and postural adjustment studies was high.

The eighth category of patients is number 32 for scoliosis/epidemiology, number 33 for stereoscopic imaging, 3D/methodology, number 34 for scoliosis/classification, number 35 for scoliosis/treatment, and number 36 for scoliosis/congenital and other five clinical patients who meet the criteria form the group format. This study was based on the asymmetrical body posture associated with the later stages of the body and the important role of scoliosis dislocation and hip position dislocation caused by this (2B). Based on the growth brace perspective presentation diagram, it evolves with the patient's treatment while observing and stimulating the patient to effectively and comfortably treat scoliosis (2C).

The scoliosis rehabilitation device is controlled by an integrated flexible pressure sensor that collects data and provides information to the specialist and the patient. It also suggests adjustments to the patient and the specialist, with a live view of the respondent (2D). The scoliosis rehabilitation device can be adapted to different participants by testing the respondent's body heat distribution map through an application. Scoliosis rehabilitation equipment can improve load support and mobility and prevent work-related musculoskeletal disorders. It can reduce musculoskeletal problems and improve productivity. Its performance and shape (angle/length) can be adjusted for body size, worker strength and working conditions, and it is light enough (about 1.6 kg) to be used in most situations (2E).

Based on the growth support perspective display diagram, our researchers are living digitally by iterating modern equipment second generation. We digitally designed to share more time and space with our self-real world counterparts. We believe that the future of wearable computing for scoliosis rehabilitation sensing devices will create opportunities to integrate our digital and physical lives into a more cohesive and meaningful whole. Scoliosis rehabilitation sensing devices and parametric custom surfaces are the future of wearable form and function. However, early forms of wearable devices are still finding use and value (2F).

Research and establishment of 4D printed scoliosis rehabilitation equipment for clinical application and calculation methods to study four key scientific issues: innovative 4D printing model, under the action of human body temperature, the model will shrink to a suitable shape. Based on RHINO software with morphological software, mechanical analysis software, Magisto magnetic field training software, and three-dimensional imaging equipment, the clinical application and calculation method evaluation of clinical anatomy and depth analysis models are carried out. More in line with ergonomics, the product is scientific and comfortable, and can effectively correct the posture of the body, making spontaneous scoliosis lighter and more fashionable for young people; the idiopathic scoliosis correction device is based on the patient's It is customized according to the needs, and it is digital, and it is designed and recorded; doctors can remotely monitor and track patients' medical records and records through the network, and realize digital monitoring of diseases; multi-layer structure, multi-layer structure, flexibility, and Breathable material, multi-layer structure patent. For idiopathic scoliosis and spinal deformity, it should also be treated at the end of development. At the same time, corresponding countermeasures will be taken to prevent the occurrence of diseases.

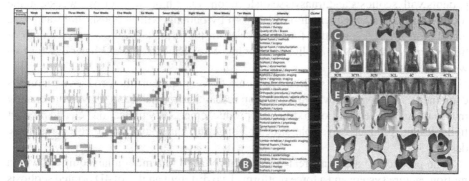

Fig. 2. 4D printed scoliosis rehabilitation device testing diagram. (A) Graph of hotspot studies based on existing conventional 3D printed scoliosis rehabilitation devices. (B) Graph of statistical testing and quantitative integration analysis based on 4D printed scoliosis rehabilitation devices. (C) Graph based on the growth support perspective presentation. (D) Respondents' real-world photography. (E) Body heat distribution of respondents in the application test. (F) View based on growth brace.

5   Study on the Calculation Method of 4D Printed Scoliosis Rehabilitation Device

Research to establish the 4 key scientific issues of scoliosis shaping machine: the innovative 4d printing model will shrink to a form suitable for body temperature action. The product is scientifically comfortable and can effectively correct body posture and better mitigate adolescent idiopathic scoliosis. A study of computational methods based on 4D printed scoliosis rehabilitation devices.

This is shown in Fig. 3 below. We configured 3 radiographic machines based in the laboratory to measure the degree of scoliosis of each test subject by (Vicon, USA) and embedded 3 x-ray independent force platforms (AMTI.UK) in the laboratory to monitor the range of reaction forces and parameter numerical maps (3A) of the product. On this basis a further (COBB) quantitative progression risk numerical analysis assessment is required [2, 24]. The researchers based the study on voluntary participation and clinical study through 49 patients with three-dimensional deformities of scoliosis with transverse misalignment during November 2020-June 2021, this study tested that scoliosis rehabilitation devices can help patients with adolescent idiopathic scoliosis (spinal deformities between 11 and 17 years of age).

The five phases of the study were: Phase 1, where patients with scoliosis and transverse subluxation wore the device for an average of 1 h per day for two weeks; Phase 2, where patients with scoliosis and transverse subluxation wore the device for an average of 1.5 h per day for two weeks; Phase 3, where patients with scoliosis and transverse subluxation wore the device for an average of 1.5 h per day for two weeks; and Phase 4, where patients with scoliosis and transverse subluxation wore the device for an average of 1 h per day for two weeks. In the third stage, the average length of time worn by patients with scoliosis and transverse dislocation was 2 h per day for two weeks; in the fourth stage, the average length of time worn by patients with scoliosis and transverse dislocation was 2.5 h per day for two weeks; in the fifth stage, the average length of time worn by patients with scoliosis and transverse dislocation was 3 h per day for two weeks. (3B).

Based on 4D printing scoliosis rehabilitation equipment, the product structure is explained as follows: innovative UI system research; breathable and recyclable coating material; innovative detail knob structure; product emergency indication belt and product light design; product recyclable structure material; innovative parametric breathable design (3C).

The product is based on a real-world picture taken in a motion lab equipped with a motion capture system and a force platform. Two different angles of 4D printed scoliosis rehabilitation equipment from three quarters of the sides (3D). Front and back view of 4D printed scoliosis rehabilitation equipment (3E). To perform the data collection procedure based on computer and X-ray data, 49 patients with 3D deformation of scoliosis with transverse misalignment (35 males, mean (SD)-age: 12.7 years (± 0.9) (range: 11–16 years), weight: 42.5 kg (± 3.9), height: 1.58 m (± 1.5) and 14 TDA as control group (8 males)), mean (SD) age: 15.3 years (± 2.1) (range: 11–16 years), weight: 39.8 kg (± 9), height. 1.61 m (± 0.25), were recruited to this study. The AIS group consisted of 6 thoracic, 6 lumbar/thoracolumbar and 3 hyperbolic types, with Cobb adjustment angles between 10 and 35 degrees (mean ($\pm$SD): 26 (± 7)) versus between 15 and 45 degrees (mean ($\pm$SD): 29 (± 5)). Inclusion criteria: two categories of AIS participants with Cobb angles of 10 degrees but less than 35 degrees and Cobb angles of 15 degrees but less than 45 degrees were included. Patients with AIS with an angle less than 50 degrees were recruited.

Healthy participants with less than 5 degrees of trunk asymmetry in healthy individuals were tested by the Scoliosis (AIS) system to monitor the p-values of t-tests and K-S tests for 3D forces and torques at the spine level for 4D printed scoliosis rehabilitation devices in patients with 3D deformations of scoliosis with transverse misalignment in the patient's two-week state of use, and mean plots of TDA. The study based on the 4D printed scoliosis rehabilitation device calculation method tested whether the variance of intervertebral effort at the vertex level was equal after testing whether the variance of intervertebral effort at the vertex level of the 4D printed scoliosis rehabilitation device was equal. The means of the two study groups within a gait cycle were compared by applying a t-test. Variables with p-values less than 0.0 and p-values less than 0.05 for the applied t-test variables were those that were significantly different between the two groups.

4D Printed Scoliosis Rehabilitation Equipment The position of the major curve on the spine affects the major curve on the spine affects the trunk kinematic parameters and 4D Printed Scoliosis Rehabilitation Equipment leads to asymmetric parameters with leads to asymmetric trunk movements thus affecting their gait. So for patients with 3D deformation of scoliosis with transverse misalignment, the study showed that the mean is not representative of the whole one cycle time and all types of scoliosis, t-test and K-S test p-values for 3D force and torque at the spine level for patients with 3D deformation of scoliosis with transverse misalignment, and mean plot of TDA (3F).

Research to establish the 4 key scientific issues of scoliosis shaping machine: the innovative 4d printing model will shrink to a form suitable for body temperature action. The product is scientifically comfortable and effective in correcting body posture to make adolescent idiopathic scoliosis lighter and drier. 4d printed scoliosis rehabilitation equipment is digitally designed and documented, made to the patient's specifications. The doctor observes and tracks the patient's medical history and records online at a distance to digitally observe the condition. Invention patent, multiple structure patent for idiopathic scoliosis bridge machine, continuous load material, counted breathable material. Prevention and response. The initially developed idiopathic scoliosis bridging device assists in compressing the spine and increasing lung capacity. It also relieves patient tension and is helpful in the treatment of idiopathic scoliosis and spinal deformities. Other original studies have evaluated patients who received only traction by using idiopathic scoliosis orthopedic correction for a period of 43 cycles, but the two methods have no difference in timeliness, but less experienced physicians have a more effective and safer method for patients and angle 4d printed scoliosis rehabilitation devices. Rehabilitation and correction can be done in a controlled manner. The risks of surgery and osteotomy are reduced. At the same time, the study documented in detail from visits to the patients' daily lives wearing the 4d printed scoliosis rehabilitation device for the first week 10 for a total of 70 days in the study, the period of time the patients wore the rehabilitation program at home, and the number of times the patients were able to adhere to the rehabilitation training at home to achieve a cure worked more significantly.

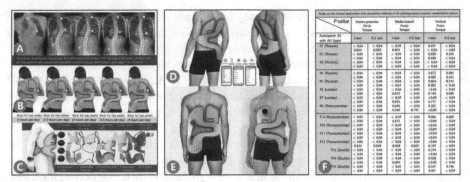

Fig. 3. Diagram of the computational method study based on 4D printed scoliosis rehabilitation device. (A) Force platform (AMTI.UK) monitoring the reaction force range and parameter values of the product. (B) 3D deformation of scoliosis with transverse misalignment in a patient rehabilitation real picture. (C) 4D print-based construction of a scoliosis rehabilitation device. (D) A three-quarter side view of 4D printed scoliosis rehabilitation equipment from two different angles. (E) Front and back view of 4D printed scoliosis rehabilitation equipment. (F) P-values of t-test and K-S test for 3D forces and torques at the spine level for patients with 3D deformation of scoliosis and transverse misalignment, and the mean plot of TDA.

6   Conclusion and Future Work

This study is based on 4D printed scoliosis rehabilitation device, RHINO software with morphology software, mechanical analysis software, magisso magnetic field training software, and 3D imaging equipment for clinical application and computational method evaluation of clinical profiling and depth analysis models. The original national design patent of the study has future marketable advantages such as recyclable, renewable, green, safe, and low-cost for mass production. The experimentalists conducted a clinical study based on 49 patients (11 to 17 years old with spinal deformation) with 3D deformation of scoliosis and transverse misalignment during November 2020 - June 2021, based on clinical application of current filming records, innovative The SOSORT/SRS assessment criteria and the scoliosis rehabilitation device calculation method were studied in the field and proved to be better for the assessment and physical treatment of patients with scoliosis and transverse subluxation in 3D. This study lacks personalized research studies and tests on the customization of 4D printed scoliosis rehabilitation devices, although 4D printing has a wide range of applications and areas based on customization are the focus of future research needed. Future applications of 4D printing for medical applications will meet different innovative needs, while providing better flexibility in the design and research of medical models. In the future, this research will improve the efficiency indicators and safety performance, and this emerging field will create more clinical effectiveness.

References

1. SOSORT guideline committee, Weiss, H.R., Negrini, S., Rigo, M., Kotwicki, T., Hawes, M.C., et al.: Indications for conservative management of scoliosis (guidelines). Scoliosis 1, 5 (2006)
2. Mohanty, S.P., Pai Kanhangad, M., Gullia, A.: Curve severity and apical vertebral rotation and their association with curve flexibility in adolescent idiopathic scoliosis. Musculoskelet. Surg. **105**, 303–308 (2021)
3. Hefti, F.: Pathogenesis and biomechanics of Adolescent Idiopathic Scoliosis (AIS). J. Child. Orthop. **7**, 17–24 (2013)
4. Ma, Q., et al.: Correlation between spinal coronal balance and static baropodometry in children with adolescent idiopathic scoliosis. Gait Posture **75**, 93–97 (2020)
5. Hattori, T., Sakaura, H., Iwasaki, M., Nagamoto, Y., Yoshikawa, H., Sugamoto, K.: In vivo three-dimensional segmental analysis of adolescent idiopathic scoliosis. Eur. Spine J. **20**, 1745–1750 (2011)
6. Kuklo, T.R., Potter, B.K., Schroeder, T.M., O'Brien, M.F.: Comparison of manual and digital measurements in adolescent idiopathic scoliosis. Spine **31**, 1240–1246 (2006)
7. Stark, L., Roberts, L., Wheaton, W., Acham, A., Boothby, N., Ager, A.: Measuring violence against women amidst war and displacement in northern Uganda using the "neighbourhood method." J. Epidemiol. Community Health **64**(12), 1056–1061 (2010)
8. Park, S.Y., Shim, J.H.: Effect of 8 weeks of Schroth exercise (three-dimensional convergence exercise) on pulmonary function, Cobb's angle, and erector spinae muscle activity in idiopathic scoliosis. J. Korea Converg. Soc. **5**, 61–68 (2014)
9. Lebel, A., Lebel, V.A.: Severe progressive scoliosis in an adult female possibly secondary thoracic surgery in childhood treated with scoliosis specific Schroth physiotherapy: case presentation. Scoliosis Spinal Disord. **11**(Suppl 2), 41 (2016)
10. Negrini, S., et al.: 2016 SOSORT Guidelines: orthopaedic and rehabilitation treatment of idiopathic scoliosis during growth. Scoliosis Spinal Disord. **13**, 3 (2018)
11. Kuru, T., Yeldan, İ, Dereli, E.E., Özdinçler, A.R., Dikici, F., Çolak, İ: The efficacy of three-dimensional Schroth exercises in adolescent idiopathic scoliosis: a randomised controlled clinical trial. Clin. Rehabil. **30**, 181–190 (2016)
12. Atkinson, G., Nevill, A.M.: Selected issues in the design and analysis of sport performance research. J. Sports Sci. **19**, 811–827 (2001)
13. Hopkins, W.G.: Measures of reliability in sports medicine and science. Sport. Med. **30**, 1–15 (2000)
14. Ricart, P.A., Andres, T.M., Apazidis, A., Errico, T.J., Trobisch, P.D.: Validity of cobb angle measurements using digitally photographed radiographs. Spine J. **11**, 942–946 (2011)
15. Srinivasalu, S., Modi, H.N., SMehta, S., Suh, S.-W., Chen, T., Murun, T.: Cobb angle measurement of scoliosis using computer measurement of digitally acquired radiographs-intraobserver and interobserver variability. Asian Spine J. **2**, 90 (2008)
16. Pasquini, G., Cecchi, F., Bini, C., et al.: The outcome of a modified version of the Cheneau brace in adolescent idiopathic scoliosis (AIS) based on SRS and SOSORT criteria: a retrospective study. Eur. J. Phys. Rehabil. Med. **52**, 618–629 (2016)
17. Ersen, O., Bilgic, S., Koca, K., et al.: Difference between Spinecor brace and Thoracolumbosacral orthosis for deformity correction and quality of life in adolescent idiopathic scoliosis. Acta Orthop. Belg. **82**, 710–714 (2016)
18. Liu, X., Wang, K., Fan, X., Xiong, Z., Tong, X., Dong, B.: A route planning method of UAV based on multi-aircraft cooperation. In: Zhao, X., Yang, S., Wang, X., Li, J. (eds). Web Information Systems and Applications. WISA 2022. Lecture Notes in Computer Science, vol. 13579. Springer, Cham (2022). https://doi.org/10.1007/978-3-031-20309-1_38

19. Shi, B., Mao, S., Xu, L., et al.: Factors favoring regain of the lost vertical spinal height through posterior spinal fusion in adolescent idiopathic scoliosis. Sci. Rep. **6**, 29115 (2016)
20. Wessels, M., Homminga, J.J., Hekman, E.E., et al.: A novel anchoring system for use in a nonfusion scoliosis correction device. Spine J. **14**, 2740–2747 (2014)
21. Kraff, O., Bitz, A.K., Kruszona, S., et al.: An eight-channel phased array RF coil for spine MR imaging at 7 T. Invest Radiol. **44**, 734–740 (2009)
22. Huang, Y., Feng, G., Song, Y., et al.: Efficacy and safety of one-stage posterior hemivertebral resection for unbalanced multiple hemivertebrae: a more than 2-year follow-up. Clin. Neurol Neurosurg. **160**, 130–136 (2017)
23. Kato, S., Lewis, S.J.: Temporary iliac fixation to salvage an acute L4 chance fracture: following pedicle screw fixation for adolescent idiopathic scoliosis. Spine (Phila Pa 1976) **42**, E313–E316 (2017)
24. Eijgenraam, S.M., et al.: Development and assessment of a digital X-ray software tool to determine vertebral rotation in adolescent idiopathic scoliosis. Spine J. **17**, 260–265 (2017)

Self-powered Flexible Electronic Skin Based on Ultra-stretchable Frictional Nano-integration

Jiayi Cai[1]([✉]) [iD] and Jialiang Cai[2] [iD]

[1] Faculty of Humanities and Arts, Macau University of Science and Technology, Macau, China
Cai849027924@qq.com
[2] Macao Institute of Materials Science and Engineering, Macau University of Science and Technology, Macau, China

Abstract. The prospect of self-powered, highly sensitive electronic skin in areas such as human health and smart electronics is key to the future and sustainable direction. In this work, we investigate and demonstrate an ultra-stretchable frictional nano-integrated self-powered flexible e-skin that is innovatively designed with a unique patented configuration for pressure value detection and tensile strain. The contents that should be available for several key sensor functions such as pressure, temperature, humidity, flow, and material in the flexible e-skin are discussed, which can realize low-cost small-volume manufacturing under multifunctional conditions from multiple angles and can enhance the comprehensive advantages such as small size, customizability, and experimental efficiency of the ultra-stretchable friction nano-integrated self-powered flexible e-skin. The edge-applied force of the e-skin was observed in hand dynamics experiments. The bending angle increases with increasing output voltage and the process is controlled by frequency. 0%–100% stretchability allows for greater tailoring to any part of the body and can be cut and applied according to the volume of the module to be customized. The aim is to consolidate and improve the clinical rehabilitation of patients used in medicine, which is a promising method.

Keywords: Friction nanogenerators · Transparent antibacterial e-skin · Healthcare · Development of clinical applications

1 Introduction

The human skin is an important and natural system that receives real-time information about the temperature, heat, stimulation, and humidity influenced by the surrounding environment. It is converted into chemical and physical signals and interconverted through the nervous system [1]. Electronic skin can truly imitate the feedback of human skin [2], in addition to meeting the requirements of high accuracy, low power consumption, high sensitivity, and manufacturing cost [3]. Partial progress has been made in stretchable flexible-based pressure-sensitive electronic skin [4] and can improve the accuracy and sensitivity of pressure sensor transmission. Multiple structures are usually constructed on nano-integrated flexible substrates [5].

© The Author(s), under exclusive license to Springer Nature Singapore Pte Ltd. 2023
L. Yuan et al. (Eds.): WISA 2023, LNCS 14094, pp. 613–624, 2023.
https://doi.org/10.1007/978-981-99-6222-8_51

In this study, the innovation of a self-powered flexible electronic skin system based on super-stretchable frictional nano-integrated, silk has the advantages of being non-toxic, breathable, anti-inflammatory, and easily degradable, which is a highly promising biomaterial for application. And based on it, a large number of electrical sensor components are integrated and store them. It can work even when it is not exercising or sleeping. Currently, electronic skins have disadvantages such as short service life, large size, and weight [6].

This study is based on three research frameworks clinical experimental data of self-powered flexible e-skin based on super-stretchable friction nano-integration, sensing and stretchable capability of self-powered flexible e-skin based on super-stretchable friction nano-integration, and application study and evaluation of self-powered flexible e-skin based on super-stretchable friction nano-integration. And in the study conducted on it, its excellent performance on the function and stretchability of the electronic skin is shown. In addition, in the skin system, two electrodes are configured in the friction system, and there will be two electrodes moving between the contact points of the subject's skin and the device when the frictional movement occurs. This method does not require an external power source to convert mechanical and kinetic energy into electrical energy. Based on this, the research on multi-sensor electronic skin and scale reduction will greatly improve the interaction and design realism in a virtual reality environment.

The new frictional self-powered transparent antibacterial e-skin in our research and development field is a non-toxic, breathable, non-inflammatory response, and degradable natural biomaterial. Despite these contributions, there are still obstacles in the development process for the conductive aspect of the charge transfer properties. On this basis, the development of light transmission coefficient >95, ductility >300%, high-temperature resistance of 120 °C, conductivity <2%, and controlled degradation properties. In this project, we propose to use the Ag NFs network as a heat source, heat its surface at 60 °C, and solidly load it on the SFCM surface to change its properties and cool it into a film, thus significantly enhancing its conductivity, degradation, and flexibility. E-skin is a new type of skin that integrates multiple senses such as heat, force, and sensation. At the same time, its temperature changes when it comes in contact with the skin. E-skin has both conductive and convective effects. Mechanical pressure can be sensed and monitored by temperature control.

2 Materials and Methods

Research-based on super stretchable friction nano-integrated self-powered flexible electronic skin system, through questionnaires and clinical research studies, and in the future, electronic skin will become more flexible and stretchable. This report is true because it is getting smaller and smaller, so it is technological level and power consumption will increase dramatically. The requirements for technology are greatly increased [7]. Also, the studied energy storage and harvesting schemes are suitable for e-skin applications. In daily operation, electronic skin should have two basic criteria self-powered and portable.

This study is based on three research frameworks clinical experimental data of self-powered flexible e-skin based on ultra-stretchable friction nano-integration, sensing and stretchability of self-powered flexible e-skin based on ultra-stretchable friction nano-integration, and application study and evaluation of self-powered flexible e-skin based on ultra-stretchable friction nano-integration. Based on this, the preparation process of frictional nano self-powered antimicrobial agent is investigated.

This project proposes to compound SF with SFCM to regulate the properties of SFCM, reduce the film-forming temperature of SFCM, and improve its electrical conductivity, degradability, and flexibility by increasing the SFCM pores and porosity in SFCM. A set of 35 by 47-inch inductive elements is divided into groups to be used as temperature sensors. Then they are transferred to the film. Finally, an array of 70x94 sensing units was integrated with hardware using an AG printing process.

In the section on body validation of a self-powered flexible e-skin system based on ultra-stretchable friction nano-integration, it is presented that the e-skin system can be stretched from 0% to 100% stretchability allowing greater tailoring to any part of the body and can be cut and applied according to the volume of the module to be customized.

The study is based on the motion-driven haptic perception evaluation of a self-powered flexible e-skin with ultra-stretchable frictional nano-integration, and the edge-applied force of the e-skin is observed in hand dynamics experiments. The bending angle increases as the output voltage becomes larger, and the process is controlled by frequency. The piezoelectric voltage is 0.034 V, 0.139 V, 0.234 V, 0.634 V, and 0.749 V for bending angles of 5°, 10°, 35°, 45°, 70°, and 90°, respectively. Based on this, clinical and experimental studies were conducted to preserve and optimize its advantages. The project team has developed a stretchable and retractable dual-mode flexible electronic skin, which has been successfully used in clinical trials. This makes it easier to operate and apply in practice. Meanwhile, its novel modular sensing structure can effectively maintain the capacitance value of the device and process the feedback electrical signal using its internal micro-nano chip.

3 Clinical Trial Data of Self-powered Flexible Electronic Skin Based on Ultra-stretchable Frictional Nano-integration

As a new generation of stretchable electronic skin [8, 9], a self-powered flexible electronic skin based on super stretchable frictional nano-integration is shown in Fig. 1 below. The material chosen is silk, which is a promising biomaterial with the advantages of non-toxicity, breathability, anti-inflammation, and easy degradation. However, it also has certain limitations, such as ductility and durability.

In this project, an ultra-thin stretchable electronic skin for medical and smart prosthesis is proposed to adopt the six innovative processes of "cicada pupa collection - item debonding - material dialysis - property modification - cooling into film - hardware introduction". That are: "cicada nymph collection" – "object debonding" – "material dialysis" – "property modification" – "cooling into a film" – "hardware introduction". "Cooling into a film" – "Hardware introduction". On this basis, a light transmission coefficient > 95, ductility coefficient > 300%, high-temperature resistance of 120°C, conductivity coefficient < 2%, and inhibit degradation are developed.

In this project, an ultra-thin stretchable electronic skin for medical and smart prosthesis is proposed to take this new silver nanowire as the research object and modify two different nanofibers (NFs) structures on its surface to realize the harvesting of human body temperature and the conversion of electrical energy. The other metal nanofiber structure based on NFs allows real-time monitoring of temperature.

Based on this, a frictional nano self-driven antibacterial antimicrobial material with 35 × 47 sensor arrays is used to achieve body temperature measurement on human skin. The self-powered flexible electronic skin based on ultra-stretchable frictional nano-integration employs a nickel nanosized organic polymer material with a physical structure. Based on this, this project proposes a novel antibacterial electronic skin system that can be used for friction nano self-powering. It is also capable of detecting and collecting air pressure values, various temperature indicators of the human body, and equipment.

Through the research of ultra-thin stretchable electronic skin for a medical and smart prosthesis in this project, it is confirmed that this novel frictional self-powered transparent antimicrobial electronic skin is a non-toxic, breathable, non-inflammatory reaction, and degradable natural biomaterial. It can be directly covered on the human skin for preliminary data collection. Based on this, a new wearable antibacterial material is developed. The cicada pupae were collected and then left to dry. The degumming was mainly to remove the silk glue from the cocoons and then to retain the fibers in the cocoons, which is a key part of the subsequent film formation (1A). The preparation process of frictional nano-self-driven antimicrobial e-skin of this novel antimicrobial material was analyzed, and the heating and temperature monitoring was carried out.

In this project an ultra-thin stretchable e-skin for medical and smart prosthesis is proposed to co-blend SF with SFCM, introduce it into SFCM to enhance the pores and porosity of SFCM, and combine it with SFCM to change its properties and cool it to form a film with good electrical conductivity, degradability, and flexibility. A 35 × 47 array of sensing units is divided into a temperature sensor. It is then transferred to a thin film. Finally, an array of 70 × 94 sensor cells was integrated into the hardware using AG printing technology (1B). In the studied case, the STENG-based e-skin is highly stretchable and consists of a flexible polymer with PDMS and Eco-flex [12], a material that can cause irritation and discomfort in some cases during long-term use. On the other hand, the new super-stretchable frictional nano-integrated self-powered flexible e-skin in our research and development field is a non-toxic, breathable, non-inflammatory, and degradable natural biomaterial.

Despite these contributions, there are still obstacles in the development process for the conductive aspects of the charge transfer properties. On the other hand, silver nanowires AgNWs are an ideal conductive layer for self-powered flexible electronic skins based on super-stretchable friction nano-integration. It is because the material possesses controlled electrical and mechanical properties [10].

However, there are still obstacles in the development of ideal conducting layers, to solve such unstable structures and to enhance the interaction properties between the material and silver nanowire AgNWs. The study incorporates a new nanomaterial in the conductive network of silver nanowires AgNWs. It also facilitates the introduction of conductive networks in the study. The thickness of the electronic skin can be reduced and the stretchability of the conductive layer can be improved.

In this work, we prepared a self-powered flexible e-skin with ultra-stretchable friction nano-integration, using silver nanowires AgNWs and rGO synergistic conductive electrode modules. And based on strong stretchability and conductivity for the preparation of stretchable electrodes, TPU was selected as the protective layer with a frictional electric layer (1C). To study the evaluation of the motion-driven haptic perception of self-powered flexible e-skin based on super-stretchable frictional nano-integration, the edge-applied force of the e-skin was observed in hand dynamics experiments. The bending angle increases as the output voltage becomes larger, and the process is controlled by frequency. The piezoelectric voltages were 0.034 V, 0.139 V, 0.234 V, 0.634 V, and 0.749 V for bending angles of 5°, 10°, 35°, 45°, 70°, and 90°, respectively, and the values of the piezoelectric voltages were constant for bending activities on the human finger and at the joints. The output voltage of the self-powered flexible electronic skin based on ultra-stretchable friction nano-integration is determined by the bending frequency. The study experimented with creating sensor arrays for bending activities on human fingers and joints, and the front view showed that the sensor 70 × 94 array was also able to conformally adhere to the grooves of the bent structure parts of the human body (1D). The side view shows that the 70 × 94 array of sensors can also conformally adhere to the grooves of the bent structure of the human body at 5°, 10°, 35°, 45°, 70°, and 90° respectively (1E). The three-quarter side view of the bent 5°, 10°, 35°, 45°, 70°, and 90° sensor 70 × 94 arrays are also able to conformally adhere to the grooves of the curved structural parts of the body (1F). The study is based on the fact that the 70 × 94-pixel sensor can be arrayed to fit completely on various parts of the human hand, including the back of the hand, the back of the fingers, the wrist, and the palm.

It can be concluded that the self-powered flexible electronic skin sensor based on ultra-stretchable friction nano-integration can fit well on natural skin and maintain the performance of the product without any discomfort as the human skin moves during operation and use. The study is based on the 70 × 94-pixel sensor that can be arrayed to fit completely on the human wrist (1G). Self-powered flexible electronic skin based on super stretchable friction nano-integration is attached to the fingers and knuckles respectively. And make a variety of gesture changes can be made with the movement to maintain a good fit on the skin. It does not cause any skin irritation or discomfort to the user (1H). Self-powered flexible electronic skin based on super stretchable friction nano-integration is attached to the palm respectively (1I). E-skin-based has greatly contributed to the contemporary lifestyle and self-powered sensors have been used for motion sensing [11], and motion monitoring [12].

4 Sensing and Stretchability of Self-powered Flexible Electronic Skin Based on Ultra-stretchable Friction Nano-integration

The study considered that the frictional electric layer of a self-powered flexible electronic skin based on ultra-stretchable frictional nano-integration is crucial for the sensing capability of the system. Therefore, a material with ecological material, biocompatibility, and good wear resistance was chosen [13]. However, several experiments are required to obtain a more complex frictional nano self-powered antimicrobial material.

This is shown in Fig. 2 below. In this project, we propose to use metal nanofibers, silver nanowires, PTFE, carbon nanotube nylon, and porous nanofibers (STENG) as

618   J. Cai and J. Cai

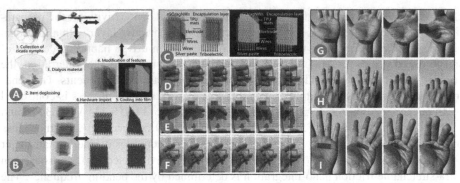

Fig. 1. Clinical trial data of self-powered flexible e-skin based on ultra-stretchable frictional nano-integration. (A) This project is an ultra-thin stretchable e-skin for a medical and smart prosthesis with six innovative processes of "cicada pupa collection - article debonding - material dialysis - property modification - cooling into film - hardware introduction". (B) This project is a kind of ultra-thin stretchable electronic skin for medical and smart prosthesis. (C) To study the incorporation of a new nanomaterial into the conductive network of AgNWs with silver nanowires. (D) The front view shows that the sensor 70 × 94 array is also able to conformally adhere to the grooves of the curved structure of the human body. (E) The side view shows that the sensor 70 × 94 array can also conformally adhere to the grooves of the body's curved structure at 5°, 10°, 35°, 45°, 70°, and 90° respectively. (F) Three-quarter side views of the bent 5°, 10°, 35°, 45°, 70°, and 90° sensor 70 × 94 arrays were also able to conformally adhere to the grooves of the curved structural parts of the body. (G) Study of 70 × 94 pixel-based sensors that can be arrayed to fit completely on the human wrist. (H) Self-powered flexible electronic skin based on super-stretchable friction nano-integration attached to fingers and knuckles, respectively. (I) Self-powered flexible electronic skin based on ultra-stretchable friction nano-integration attached to the palm, respectively.

electrode materials. In this process, air permeability, light permeability, electrical properties, thermal properties, and biocompatibility are to be ensured. The technology is a stretchability based on ultra-stretchable friction nano-integration with stretchability from 0% to 100%, enabling a greater range of tailoring and use for any part of the human body.

Tests yielded a mechanical hysteresis range of 0%–3% for the self-powered flexible e-skin based on ultra-extendable friction nano-integration. It is also reversible after 10, 20, 40, 60, 80, and 100 cycles of continuous stretching. Thus the material acts as a skin-friendly biosensor that can be adsorbed on various parts of the body. And the real-time signal is continuously monitored during the operator activity. The self-powered flexible electronic skin based on ultra-stretchable friction nano-integration can actively detect the activity state of the skin and shows the current activity of the human skin under stretching, squeezing, fist clenching, bending, and twisting deformation, respectively. It can be seen that with different ranges of activity and activity intensity, the voltage and current output generated by the self-powered flexible electronic skin based on ultra-stretchable friction nano-integration is varied. To test the output and input piezoelectric current of the actual electronic skin, it can be seen that in stretching 9 nA, while having high stability, and also verifies the conclusion that the skin has active monitoring and stable output piezoelectric current.

In this research work, a flexible and self-powered, and highly efficient electronic skin based on super stretchable friction nano-integrated self-powered flexible electronic skin is prepared, and the output and input voltage of the electronic skin can be significantly affected by different external applied forces. The current activity of human skin under stretching, squeezing, fist clenching, bending, and twisting deformation. The piezoelectric-photocatalytic process of the electronic skin provides a new research field and direction for the development of self-powered.

The self-powered flexible electronic skin based on ultra-stretchable frictional nano-integration can be deformed by body movements and output precise piezoelectric voltage by the piezoelectricity of PVDF polymer. An external programming system is applied to the super-stretchable friction nano-integrated self-powered flexible electronic skin and the bending angle and frequency are recorded separately, and all tests are conducted at room temperature of 30 degrees Celsius (2A).

To better investigate the antibacterial properties of the super-stretchable friction nano-integrated self-powered flexible e-skin, we tested the effect of the e-skin on the bacterial reproduction environment and bactericidal effect under a natural static environment and verified the high piezoelectric photocatalytic activity of the super-stretchable friction nano-integrated self-powered flexible e-skin under different UV/ultrasonic irradiation conditions. Catalytic activity. The kinetic profile of MB can also be degraded rapidly [14, 15].

As shown in Fig. 2 below. The study is based on the photographs of the sample solution of self-powered flexible e-skin based on super-stretchable friction nano-integration during the degradation process. The study MB degradation rate was 98%, congo red degradation rate was 90%, and phenol red degradation rate was 73%, and the experimental results verified that the self-powered flexible e-skin based on super-stretchable friction nano-integration has a high piezoelectric photocatalytic activity. It is capable of self-cleaning repair and degradation of a variety of organic pollutants in life. The self-cleaning capability was also verified in a 24h experiment, where the e-skin was first subjected to MB contamination and continuously irradiated with natural light UV for more than 6 h, and the test results verified the ability to degrade a variety of organic pollutants in life (2B). The experiments further tested the antimicrobial properties of the super-stretchable friction nano-integrated self-powered flexible e-skin. The researchers used the effect of Gram-positive bacteria strain propagation/sterilization [16].It was deduced that the super stretchable friction nano-integrated self-powered flexible e-skin was effective in inhibiting the growth of bacteria and resisting the bacterial population and inhibiting the growth of bacteria in the natural environment with natural growth, UV environment, and ultrasonic irradiation environment for more than 6 h of continuous exposure to natural light UV. The electronic skin is found to deepen in color and oxidation reaction (2C) in the plate under UV irradiation.

It has the advantages of being non-toxic, breathable, anti-inflammatory, and easily degradable, which is a promising biomaterial for application. It can detect the current activity under various stretching, squeezing, fist clenching, bending, and twisting deformations in motion in real-time. The piezoelectric photocatalytic activity of super-stretchable frictional nano-integrated self-powered flexible electronic skin has significant self-cleaning advantages. Simultaneous experiments were conducted to assign eight different ultrasonic powers (0 W, 50 W, 100 W, 150 W, 200 W, 250 W, 300 W, 350 W, 400 W) to the ultra-stretchable friction nano-integrated self-powered flexible e-skin to catalyze the photodegradation of MB solution recorded by the e-skin. For the experiments, the UV light was kept at a constant power of 100W. It was concluded that the piezoelectric photocatalytic activity of the self-powered flexible e-skin with super-stretchable friction nano-integration increased with increasing ultrasonic radiation power. Ultrasonic power above 300 W leads to a large deformation of the self-powered flexible e-skin system with super-stretchable friction nano-integration. The most suitable ultrasound power is 200 W and UV power is 100 W. Electronic skin is a new type of skin that integrates multiple senses such as heat, force, and sensation.

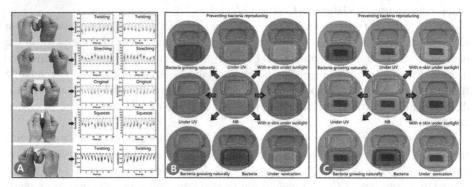

Fig. 2. Diagram of sensing and stretchability capabilities of a self-powered flexible electronic skin based on ultra-stretchable friction nano-integration. (A) Precise piezoelectric voltage output via piezoelectricity of PVDF polymer. An external programming system was simultaneously applied to the super-stretchable friction nano-integrated self-powered flexible electronic skin, and the bending angle and frequency were recorded separately, all tests were performed at room temperature of 30 degrees C. Experimental plots. (B) The degradation rates were 98% for MB, 90% for congo red and, 73% for phenol red, and the experimental results verified that the self-powered flexible electronic skin based on super-stretchable friction nano-integration has high piezoelectric photocatalytic activity. (C) The system can effectively inhibit the growth of bacteria and is lower than the number of bacteria and inhibit the growth of bacteria in the natural environment of natural growth, UV environment, and ultrasonic irradiation environment.

5 Research and Evaluation of Self-powered Flexible Electronic Skin Applications Based on Ultra-stretchable Frictional Nano-Integration

A self-powered flexible electronic skin based on ultra-stretchable friction nano-integration is different from human skin which uses organic tissues and nerve impulses, etc. to achieve sensing functions. And consider user emotional factors [17].

This project will work on two aspects: first, a bionic skin, consisting of pressure, chemical, electrochemical, and temperature sensors that are matched to each other. All sensors are arranged in a cascade on the same plane. Their sensing properties should match those of human skin; secondly, some physiological signals of the body are monitored. The method is to rub two electrodes on the skin so that the two electrodes are in contact with the skin and the device generates frictional motion between the two electrodes. This method does not require an external power source to convert mechanical and kinetic energy into electrical energy. In this case, the perceptual features of the electronic skin must be covered by the physiological information of the body to achieve the conversion of energy and mechanical energy to electrical energy. The system works by electrostatic and systemic induction [18].

This is shown in Fig. 3 below. In contrast to a single ultra-stretchable friction nano-integrated self-powered flexible e-skin sensor, the proximity sensor can identify the temperature, shape, and location of external objects through mapping data, while providing an alert in case of contact with a foreign object. An ultra-thin stretchable electronic skin for medical and smart prosthetics facilitates improved safety precautions and enhances the sensing and stretchability of the skin for robotic sensing control. The study integrates an array of 70 × 94 sensor units into the hardware. The spatial sensing and localization capability of the sensor array in planar stationary and deformation activity states. Study of 70 × 94 pixel-based sensors that can be arrayed for complete fit on the human wrist. Electrical results of a super-stretchable friction nano-integrated self-powered flexible e-skin planar sensor array on the wrist when the finger approaches the right side of the wrist and stops until the distance is around 1.5 cm. The corresponding super-stretchable friction nano-integrated self-powered flexible e-skin current map variation also clearly shows the position of the finger, and these results all verify that the sensor can identify and localize the position on the hand in three dimensions as well (3A).

The self-powered flexible e-skin based on ultra-stretchable friction nano-integration was attached to the finger and knuckle, respectively. The response rates used to monitor the fingers and joints were 37% and 49%, respectively. (3B). Self-powered flexible electronic skin based on super-stretchable friction nano-integration was attached to the palm, respectively, and all these results verified the ability of the sensor to identify and localize position on the hand in three dimensions as well (3C).

The human skin acts as a positive-level frictional electric material and the e-skin acts as a negative-level frictional electric material, and when the human skin is separated from the e-skin, then a small amount of electrical signal electrons are generated due to electrostatic induction. This generates electrical signals that can provide a self-powered flexible e-skin system with ultra-stretchable friction nano integration (3D).

The friction nano self-powered transparent antibacterial e-skin system is highly applicable, stretchable, scalable, and biocompatible. Based on this, light transmission coefficient >95, ductility >300%, high-temperature resistance 120 °C, conductivity <2%, and controlled degradation properties are developed. This project proposes to use the Ag NFs network as the heat source, heat its surface at 60 °C, and solidly load it on the SFCM surface to change its properties and cool it to form a film, thus significantly enhancing its electrical conductivity, degradability, and flexibility. Silk is an ideal biomedical material because of its natural, non-toxic, breathable, non-inflammatory, and degradable properties.

The super-stretchable frictional nano-integrated self-powered flexible electronic skin we studied can be successfully used as a wearable skin that sensitively detects subtle external stresses and has superior potential in detecting body part activity (3E). In summary, this paper takes the function of sensors in human skin as an entry point, explores what the functions of several key sensors such as pressure, temperature, humidity, flow, and material should be in a flexible electronic skin, and examines how the mechanical functions of the device sensors can be implemented (3F).

The innovative design of this topic allows for low-cost, low-volume manufacturing under multi-angle multifunctional conditions, and enhances small size, customizability, and test efficiency. In particular, the frictional nano-self-electrogenic transparent antibacterial e-skin can realize nano-self-electroluminescence with increased difficulty in signal modulation and coupling of various physical parameters due to the increased structural and process complexity of the components. Through the research of this project, we have developed a novel super-stretchable frictional nano-integrated self-powered flexible e-skin. Electronic skin is a new type of intelligent skin that integrates multiple sensors such as heat and force. At the same time, its temperature changes when it is in contact with the skin. Electronic skin has both conductive and convective effects. The mechanical pressure can be sensed and monitored by temperature control. In this way, the function of the device is ensured without putting additional burden on the skin. This is prepared to better mimic human skin and to provide multiple protection and interaction for amputees and patients (3G).

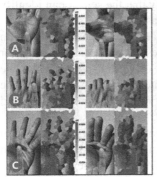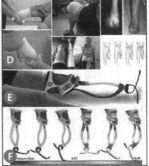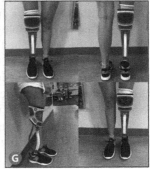

Fig. 3. Application study and evaluation of self-powered flexible e-skin based on ultra-stretchable friction nano-integration. (A) Changes in the current map of the super-stretchable friction nano-integrated self-powered flexible e-skin also clearly show the position of the finger. (B) Response rates of 37% and 49% for monitoring fingers and joints, respectively. (C) Self-powered flexible electronic skin based on super-stretchable friction nano-integration is attached to the palm. (D) Electrostatic induction generates a small number of electrons for electrical signals. (E) The study can be attached to complex surfaces with different bending, folding bending, twisting, and coupling bending. (F) The implementation of the mechanical function of the device sensor was investigated and tested. (G) The self-powered flexible electronic skin with super-stretchable frictional nano-integration ensures the functionality.

6 Conclusion and Future Work

Our research and development area of this new super-stretchable friction nano-integrated self-powered flexible electronic skin is a non-toxic, breathable, non-inflammatory response, and degradable natural biomaterial. Despite these contributions, there are still obstacles in the development process for the conductive aspects of the charge transfer properties. On this basis, the development of light transmission coefficient >95, ductility >300%, high-temperature resistance of 120 °C, conductivity <2%, and controlled degradation properties. In this project, we propose to use the Ag NFs network as a heat source, heat its surface at 60 °C, and solidly load it on the SFCM surface to change its properties and cool it into a film, thus significantly enhancing its conductivity, degradability, and flexibility. Light transmission, breathability, biocompatibility, stretchability, and thermal and electrical properties are also ensured. We have developed a novel ultra-stretchable friction nano-integrated self-powered flexible electronic skin system that can be stretched from 0% to 100% stretchability allowing greater tailoring to any part of the body and can be cut and applied according to the volume of the module to be customized. The module is also stable in terms of signal and sensor parameters based on test results. It also generates pressure to keep the temperature of the product constant. Also, this innovative research method allows for independent replacement of the parts after they are damaged. And through experiments, the research proves that it can better reduce costs and achieve mass production. It improves the patient's daily self-care ability and quality of life. The main shortcomings of the current project are the lack of personnel available for repeat testing and the lack of sufficient sample size, which needs to be further expanded in the future. Verify the effectiveness and stability of this research super stretchable friction nano-integrated self-powered flexible e-skin system for medical application of e-skin development clinically.

References

1. Kim, J., et al.: Stretchable silicon nanoribbon electronics for skin prosthesis. Nat. Commun. **5**, 5747–5757 (2014)
2. Lee, J.H., et al.: Micropatterned P(VDF-TrFE) film-based piezoelectric nanogenerators for highly sensitive self-powered pressure sensors. Adv. Funct. Mater. **25**, 3203–3209 (2015)
3. Darabi, M.A., et al.: Gum sensor: a stretchable, wearable, and foldable sensor based on carbon nanotube/chewing gum membrane. ACS Appl. Mater. Inter. **7**, 26195–26205 (2015)
4. Kaltenbrunner, M., et al.: An ultra-lightweight design for imperceptible plastic electronics. Nature **499**, 458–463 (2013)
5. Park, J., et al.: Fingertip skin-inspired microstructured ferroelectric skins discriminate static/dynamic pressure and temperature stimuli. Sci. Adv. **1**, e1500661 (2015)
6. Armand, M., Tarascon, J.M.: Building better batteries. Nature **451**, 652 (2008)
7. Ho, D.H., et al.: Stretchable and multimodal all graphene electronic skin. Adv. Mater. **28**, 2601–2608 (2016)
8. Qiu, H., et al.: Long, Nano Energy **58**, 536–542 (2019)
9. Parida, K., et al.: Nat. Commun. **10**, 2158 (2019)
10. Lin, L., et al.: ACS Nano **7**, 8266–8274 (2013)
11. Pu, X., et al.: Sci. Adv. **3**, e1700015 (2017)

12. Shi, Y., et al.: Integrated all-fiber electronic skin toward self-powered sensing sports systems. ACS Appl. Mater. Interfaces **13**, 50329–50337 (2021)
13. Jiang, Y., et al.: UV-protective, self-cleaning, and antibacterial nanofiber-based triboelectric nanogenerators for self-powered human motion monitoring. ACS Appl. Mater. Interfaces **13**, 11205–11214 (2021)
14. Tan, C., et al.: A high performance wearable strain sensor with advanced thermal management for motion monitoring. Nat. Commun. **11**, 3530 (2020)
15. Xue, X.Y., et al.: Nano Energy **13**, 414–422 (2015)
16. Mu, J.B., et al.: ACS Appl. Mater. Interfaces **3**, 590–596 (2011)
17. Liu, X., Wang, K., Fan, X., Xiong, Z., Tong, X., Dong, B.: A Route Planning method of UAV based on multi-aircraft cooperation. In: Zhao, X., Yang, S., Wang, X., Li, J. (eds). Web Information Systems and Applications. WISA 2022. Lecture Notes in Computer Science, vol. 13579. Springer, Cham (2022). https://doi.org/10.1007/978-3-031-20309-1_38
18. Jie, Y., Jiang, Q., Zhang, Y., Wang, N., Cao, X.: Nano energy **27**, 554–560 (2016)

Brain-Machine Based Rehabilitation Motor Interface and Design Evaluation for Stroke Patients

Jiayi Cai[1]([✉]) [iD] and Jialiang Cai[2] [iD]

[1] Faculty of Humanities and Arts, Macau University of Science and Technology, Macau, China
Cai849027924@qq.com
[2] Macao Institute of Materials Science and Engineering, Macau University of Science and Technology, Macau, China

Abstract. Based on the brain-computer interface is an important way of existing and future medical rehabilitation medicine, based on post-stroke motor rehabilitation plays a very important role, researchers through the study and clinical brain-computer interface experiments and evaluation of the way, can improve 18% the efficiency of rehabilitation. The intention of the movement is converted into electrical signals that the machine can recognize and interact with the body to achieve active control of the patient. A technology that helps patients in the rehabilitation period to regain their motor functions and self-care abilities. In this paper, a brain-machine interface-based upper limb motor rehabilitation training system is designed, which uses brain-machine fusion as the rehabilitation motor therapy model and combines the muscle strength and movement morphology characteristics of the upper limb on the hemiplegic side of the patient. This study is based on the first clinical trial application of this system, and the focus was validated and improved in all six patients who were followed up. This study was designed from the perspective of combining a brain-machine interface system and upper limb rehabilitation assistive robot system, and combined with the human-computer interaction platform to apply this wearable upper limb movement prosthesis for the disabled in clinical trials for rehabilitation training and assistive therapy, providing effective guidance and assistance for patients with such needs. The research results of this paper have been patented and granted.

Keywords: Brain-machine interface · Upper extremity rehabilitation motor interface · Design evaluation · Rehabilitation system development

1 Introduction

Brain-computer interface technology is usually based on stroke patients to provide another direct and recoverable modality [1, 2]. Also, noninvasive therapies have achieved significant results and feedback in rehabilitation [3–5].

In this paper, a brain-computer interface-based upper limb motor rehabilitation training system is designed, which uses brain-computer integration as the rehabilitation motor

L. Yuan et al. (Eds.): WISA 2023, LNCS 14094, pp. 625–635, 2023.
https://doi.org/10.1007/978-981-99-6222-8_52

therapy model, combines the muscle strength and movement morphology characteristics of the upper limb on the hemiplegic side of the patient, and uses MATLAB for brain-computer interface software programming and simulation; it also uses neural feedback technology and visual feedback technology as well as an optimized training method based on improved particle swarm through MATLAB for human-computer interaction design and simulation. Also to provide the accuracy of decoding [6, 7].

This study is based on the data map of the upper extremity rehabilitation movement interface and design assessment of brain-machine stroke patients.

This study is based on the first clinical trial application of this system and the focus was validated and improved in all six patients who were followed up. Brain-machine integration of rehabilitation movement therapy is aimed at the recovery of patient's motor functions, combining patients' subjective feelings about the treatment and the guidance of rehabilitation therapists to improve patients' quality of life and their ability to live through motor function recovery, and then achieve the purpose of returning to society.

The study improved the results and enhanced the quality of life of the patients based on the three participation benchmark maps of educational essentialism, namely, cognitive participation: the transmission of knowledge content; behavioral participation: the regulation of volitional behavior; and emotional participation: the comprehension of life connotation three integration models [8].

This paper reviews the brain-machine stroke patient upper limb rehabilitation motor interface and design assessment and analysis data map, the brain-machine stroke patient upper limb rehabilitation motor interface and data processing map, and the brain-machine stroke patient upper limb rehabilitation test assessment and analysis and game development data map, focusing on the development of a game and human interaction system, updating and studying the traditional closed-loop DBS system [9, 10], and based on the brain-machine interface technology architecture is re-researched and redefined. And possible future development trends are pointed out.

2 Materials and Methods

In this study, our researchers based on the brain-machine stroke patients' upper limb rehabilitation movement interface and design assessment and analysis data, the brain-machine stroke patients' upper limb rehabilitation movement interface and design assessment and analysis data map, the brain-machine stroke patients' upper limb rehabilitation movement interface and data processing map and the brain-machine stroke patients' upper limb rehabilitation test assessment and analysis and game development data map. A paradigm, decoder (ABSD), and software development were developed [11], and the incremental signals were translated into 110 ms and transmitted to the brainwave control system [12].

The aim was to enhance and improve clinical rehabilitation outcomes in stroke patients. Five types of research methods were used: questionnaire, field study, design evaluation, system programming, and ergonomics. Brain-machine integration of rehabilitation exercise therapy is aimed at the recovery of motor function of patients, combined with the patient's subjective perception of the treatment and the guidance of the rehabilitation therapist, and the effector can trigger signals and assist in control through the patient's residual volitional brain function with the assistance of EEG [13, 14], this study improves the quality of life of patients through motor function recovery and improves their ability to live, and then achieve the goal of returning to society.

Tests were conducted based on a $40M^2$ indoor environmental space with separate calibration of the starting and ending points. To study the kinetic theory and practice of neurons [15], the study integrates that information in rate encoding is decoded in terms of the number of spikes per time versus the average firing rate [16]. Although rate encoding is relatively stable on average, precise timing with high-frequency firing rates can be more accurate than using rate results [17, 18]. Tests were performed based on a $40 M^2$ indoor ambient space. The operation process is mainly the following six points respectively: destination selection; path planning; optimal path; path tracking; reference speed command; motion control and actual speed command, based on the obstacle map we researchers gave set vector map; automatic wheelchair positioning, and emergency obstacle avoidance.

Brain-computer interface technology is an important branch in the field of rehabilitation medicine and plays a very important role in the rehabilitation training of motor dysfunction after stroke. Brain-machine interface technology realizes active patient control by monitoring the movement intention of stroke patients in real-time, converting the movement intention into electrical signals that can be recognized by the machine, and interacting with the body with signals.

The brain-machine interface is a technology that uses the patient's movement intention as the core and rehabilitation assessment and feedback as the means to help patients recover their motor function and self-care ability during rehabilitation. Currently, the brain-computer interface research field is commonly used in motor imagination research based on EEG signal analysis, upper limb rehabilitation motor control based on combined EEG-EMG signal detection, and intelligent prosthesis design based on virtual reality technology and user models. Typical applications in the field of brain-machine interface research are rehabilitation robots, including upper limb rehabilitation robots, lower limb rehabilitation robots, and neuromodulation upper limb rehabilitation robots [19–21].

3 Analytical Data of Upper Extremity Rehabilitation Motor Interface and Design Assessment for Stroke Patients

This study analyzes the data map based on the brain-machine stroke patients' upper limb rehabilitation movement interface and design evaluation. Brain-machine interface technology is an important branch in the field of rehabilitation medicine and plays a very important role in the rehabilitation training of motor dysfunction after stroke. Brain-machine interface technology enables active patient control by real-time monitoring of stroke patients' movement intentions, converting the movement intentions into electrical signals that can be recognized by the machine, and interact with the body with signals, making it the most commonly operated acquisition method for BMI [22].

Brain-computer interface is a technology that uses the patient's movement intention as the core and rehabilitation assessment and feedback as a means to help patients recover motor function and self-care ability during rehabilitation. Currently, the brain-computer interface research field is commonly used for motor imagery research based on EEG signal analysis, upper limb rehabilitation motion control based on joint EEG-EMG signal detection, and intelligent prosthesis design based on virtual reality technology and user models.

This is shown in Fig. 1 below. The research medium is based on two major board diagrams of learner and brain science, and four major communication restrictions are classified as physiological restrictions; spatial restrictions; expression restrictions, and consciousness restrictions, based on BCI neural signal decoding and BCI neural signal encoding, which can be applied to scenarios such as neuroimaging, neurostimulation (transcranial stimulation TMS), and AI algorithms, respectively (1A). Brain-machine integration of rehabilitation movement therapy is aimed at the recovery of patient's motor functions, combined with patients' subjective feelings about the treatment and the guidance of rehabilitation therapists, to improve patient's quality of life and their ability to live through motor function recovery, and then achieve the purpose of returning to society.

The study is based on the baseline diagram of three types of participation based on the essential theory of education: cognitive participation: the transmission of knowledge content; behavioral participation: the regulation of volitional behavior; and emotional participation: the comprehension of life connotations (1B). In this context, this paper takes a patient with hemiplegia after a lower extremity stroke as the research object to assess and rehabilitate the motor function of the upper extremity of the affected side based on cognitive assessment. The brain-computer interface deep learning framework map is based on three behavioral criterion maps for knowledge content; operational execution and social support, and the learning dimensions are cognitive brains; behavioral brains, and emotional brains (1C). The study is based on simplistic planned paths and actual situational paths as the criterion maps, with the starting and ending points labeled respectively.

Tests were conducted based on an indoor environmental space of $40M^2$. Firstly, after action recognition, gait information extraction, and pre-processing, the patient then performs lower limb joint spatial 3D position estimation to obtain joint angle information and finger end distance information, forearm rotation speed information, and uses genetic algorithm and fuzzy neural network-based methods for prediction model training; secondly, it uses neural feedback based technology and visual feedback technology and improved particle swarm algorithm based optimization The training parameters are optimized by using neural feedback technique and visual feedback technique and improved particle swarm algorithm; finally, the rehabilitation kinematic analysis, brain-machine interface design, and simulation (1D) are realized by MATLAB programming simulation. The study is based on the complex planning path and the actual situation path as a criterion map with the starting point and the end point marked respectively. Tests were conducted based on a $40M^2$ indoor environmental space (1E).

In this context, this paper takes the lower limb post-stroke hemiplegic patient as the research object, based on the brain-computer interface cognitive assessment for the motor function assessment and rehabilitation training map of the affected lower limb. The operation process is mainly the following six points respectively: selection of destination; path planning; optimal path; path tracking; reference speed command;

motion control and actual speed command, based on the obstacle map we researchers gave set the vector map; wheelchair self-positioning and emergency obstacle avoidance. The five system components are a webcam; LIDAR; encoder; ultrasonic series with an automatic navigation system (1F).

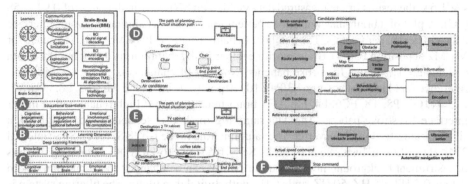

Fig. 1. The plot of data is based on the evaluation of the brain-machine stroke patients' upper extremity rehabilitation motor interface and design. (A) Study medium based on two panels of learner and brain science diagrams. (B) Study of three engagement benchmark maps based on educational essentialism. (C) Brain-computer interfaces deep learning framework diagram based on three behavioral criteria diagrams. (D) The path is based on simplistic planning with the actual situation path as a criterion map. (E) Paths are based on complexification planning with actual situation paths as criterion graphs. (F) Motor function assessment and rehabilitation training diagram for the affected lower limb based on brain-computer interface cognitive assessment.

4   Motor Interface and Data Processing for Upper Limb Rehabilitation of Brain-Machine Stroke Patients

Stroke is one of the major diseases threatening human health, featuring high incidence, high disability, and high mortality, and causing great harm to human health and socio-economic development. Based on the brain-machine stroke patients' upper limb rehabilitation exercise interface and design assessment chart, the purpose of stroke patients' rehabilitation exercise is to improve motor function, promote joint rehabilitation, and reduce the occurrence of disability.

This is shown in Fig. 2 below. The brain-machine-based upper extremity rehabilitation exercise interface and design assessment diagram for stroke patients (2A). Based on evoked potential (SSVEPS) is the most used system in bmi system studies [23, 24]. The maximum induced current and current electric field maps of the brain for the superimposed state were approximately equal to 43.2 μA/m2 and 90 μV/m, respectively, and were also within the same data range for five separate measurements (2B).

Brain-computer interface technology, as a new type of human-computer interaction, has been widely used in upper limb rehabilitation training after stroke, but there are limitations in its practical application due to the lack of systematic research on the working principle and design theory of brain-computer interface and the shortage of technical personnel in rehabilitation institutions. In this chapter, based on lower limb stroke patients as the research object, an in-depth study is conducted on the basis of brain-machine interface research theories and clinical applications, combined with patients' subjective feelings, motor function assessment, and rehabilitation training.

Based on four Volts form assessment charts with time standards set at 0.4 s; 0.8 s; 1.2 s & 1.6 s, the results of the measured data yielded an SNR within the range of 0,93–1.58 (2C). A plot of a brain-machine stroke patient upper limb rehabilitation motor interface data and design parameters. The number of channels is 14 (plus CMS/DRL reference for M1/M2 at position P3/P4). Brain-machine upper limb rehabilitation motor interface data and design parameters for stroke patients' Number of channels is 14 (plus CMS/DRL reference for M1/M2 at position P3/P4); 10–20 positions including AF3, F7, F3, FC5, T7, P7, O1, O2, P8, T8, FC6, F4, F8, AF4. brain-machine upper limb rehabilitation motor interface for stroke patients The data sampling method is sequential sampling with a single analog-to-digital converter. The sampling rate range is 128 SPS/256 SPS (internal 2048 Hz). EEG analysis rate is 14-bit 1 LSB = 0.51 uV (16-bit ADC, discarding two bits of instrument background noise), and the setting can be changed to 16-bit. Bandwidth data is 0.2 HZ–45 HZ, 50 HZ, and 60 HZ with digital trap filters. The filtering system is a built-in digital 5th-order Sinc filter. Dynamic fusion (reference input) 8400 uV (PP). The coupling method is AC coupling.

Brain-machine stroke patient upper limb rehabilitation motor interface data and design parameters connectivity proprietary 2.4 GHz wireless, BLE, and USB (diffusers only). Product battery life up to 9 hours (LIPO) Impedance measurement using the patented system for real-time contact quality. Motion sensor number ICM-20948. accelerometer 3-axis +/−4 g. magnetometer 3-axis +/− 4900uTesla. motion sampling 32/64 Hz. motion resolution 14/16 bit (user-defined).

Brain-machine stroke patient upper limb rehabilitation motion interface data with the quantitative output of design parameter channels. The sensor material is ag/agCI+blanket+saline (2D). Brain-machine stroke patient upper limb rehabilitation movement interface and data processing map, firstly, after action recognition, gait information extraction and pre-processing, then upper limb joint spatial 3D position estimation, joint angle information and finger end distance information, forearm rotation speed information, and prediction model training using genetic algorithm and fuzzy neural network based method; secondly, using neurofeedback technology based Secondly, the training parameters were optimized using neurofeedback and visual feedback techniques and an optimized training method based on improved particle swarm algorithm; finally, the rehabilitation kinematic analysis and brain-machine interface design and simulation were realized through MATLAB programming simulation (2E).

5   Brain-Machine Stroke Patient Rehabilitation Test Assessment Analysis and Game Development Data

Using the brain-machine stroke patient upper extremity recovery movement interface and design evaluation as a background, the investigators prepared game evaluation charts under six successful cases based on six software combinations. The control commands of the external electrical signal excitation system were determined by neural activity anticipation and feedback.

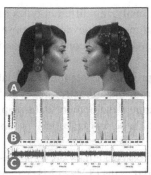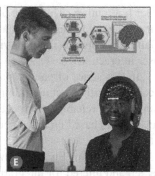

Fig. 2. Upper extremity rehabilitation motor interface and design assessment diagram for brain-machine stroke patients. (A) Evaluation diagram of the upper extremity rehabilitation movement interface and design for brain-machine stroke patients. (B) A plot of maximum induced brain current and current electric field in the superimposed state. (C) Assessment map based on four Volts forms. (D) A plot of upper limb rehabilitation motor interface data and design parameters for brain-machine stroke patients. (E) Diagram of the upper extremity rehabilitation motor interface and data processing for brain-machine stroke patients.

The software used to develop the study were: the first software was Unity (hyper-reality tool) Unreal Engine is a real-time game engine and 3D creation tool, the second software was Autodesk Maya (3D modeling and rendering software) Autodesk's VR software animation development tool, and the third software was A-Frame (supported VR headsets almost all) built an excellent open source web framework for VR experiences, the fourth software is OpenVR (motion tracking functionality) an SDK and application programming interface (API), the fifth software is SteamVR (Valve Index, HTC Vive and Oculus) VR enthusiasts can later use the SteamVR system, and the sixth software is VRTK (outstanding design patterns) toolkit for creating VR applications. Reusable scripts.

Basic VR rhythm training digital counterpart of the operation game, which has a strong sense of humor and a strong curiosity to explore new things in VR. And can be one of the preferred games for the evaluation of motor interface and design of upper limb rehabilitation according to brain-computer stroke patients.

The utility model is better to enable the controller to be more intuitive in the game to get the feeling when playing. VR is fast becoming a favorite medium for many people in the future. Due to the scarcity of game development for existing studies, our researchers decided to study the game topic and developed six of the preferred games based on the Brain-Computer Stroke Patient Upper Extremity Rehabilitation Motor Interface and Design Evaluation standalone games.

This study will focus on the following aspects: 1. use Unity3D virtual technology to implement the game; 2. customize the game content according to the user's needs for virtual reality environment; 3. use virtual reality technology to complete the project development, including virtual world creation and interactive operation; 4. use UG simulation and related algorithms to Realize the simulation of virtual reality scenes; 5. Create with the help of UG models. These projects will be helpful for early treatment after brain injury. The game development is carried out by analyzing different types of virtual reality technologies and some other related applications to achieve the purpose. Game development includes virtual scene creation, virtual environment construction and user interaction.

In this series of games, players need to choose the appropriate content and format according to their actual situation. The study was based on brain-machine stroke patients' upper limb rehabilitation test assessment analysis and game development data map, researchers based on a deep neural network decoder for stable performance in 17 years [19, 25].

As shown in Fig. 3 below, the brain-machine device based signals are first collected and filtered for integration as well as band-pass filtered maps (8 Hz–12 Hz), then transcoded and feature integration extracted by GPS for w-matrix, researchers customize 90 Trial data and these data can be obtained based on motor imagery data learning, and finally the collected data vectors are fed to the trained STM classifier The final acquired data vector was input to the trained STM classifier to output the corresponding indicators and control signals, with a single output signal of 200 MS.

Our researchers conducted signal acquisition and study of the left and right brain based on an adaptive approach to improve the accuracy of signal acquisition, and invited a total of six subjects, respectively, with the first three data ranges for subjects Aw (Train/Test: 50/243) and Ay (Train/Test: 32/ 251) for subsequent monitoring data integration (3A),Machine reading comprehension has been widely used in various fields such as question answering systems and intelligent engineering [26].

And brain-machine device based signal collection and filter integration and band-pass filter maps (8 Hz–12 Hz) were invited for a total of six subjects, respectively, with the last three data ranging from subjects Aw (Train/Test: 52/246) and Ay (Train/Test: 27/289) (3B).

The brain-machine interface design and simulation can be implemented using a variety of techniques, and the MATLAB language is used in the Matlab environment to model the motion intent algorithm and the motion trajectory to achieve the simulation graph, and the simulation results are analyzed to summarize the problems and propose improvement measures to finally make the system run stably and reliably. MATLAB is a computer simulation, scientific computing, and other fields that can be used Multifunctional platform.

The first type of brain-machine stroke patient design assessment method is based on three types of measurements of students' classroom activities in an episodic behavioral manner: cognitive level diagnosis, knowledge mastery, and skill level diagnosis, respectively. Feedback on the learning outcomes of students' real-time classroom interaction with the brain-machine and Pad interaction. The second type of brain-machine stroke patient design assessment approach, based on the teacher and student classroom activities implicitly state three types of measures, respectively: based on BCI cognitive characteristics measures, learning style identification, and teaching style identification. The activity profiles are neural brain activity profile, individual brain area activity profile, and group brain activity profile. The application scenarios are neuroimaging, neurostimulation (transcranial stimulation TMS), AI algorithm, etc. The learning style is based on intelligent support for pre-course learning and intelligent support for gram-weight teaching.

Pedagogical style identification and intelligent teaching system integrate four categories, which are: complex learning ability and measurement analysis, intelligent analysis of teachers' teaching behavior, scientific evaluation of classroom teaching effect, and intelligent learning support for students after class. The terminal port and the starting port of the above study are the upper limb rehabilitation movement interface device for stroke patients and the upper limb rehabilitation movement transmission terminal diagram for stroke patients (3C).

In this paper, the human-computer interaction effect of human upper limb movement is simulated by establishing the motion intention model and motion trajectory model respectively based on MATLAB language to verify the adopted method and implementation effect. The first type of brain-machine stroke patient design evaluation method is based on a brain-computer interface intelligent teaching application game model diagram (3D). The second type of brain-machine stroke patient design assessment approach is based on a brain-machine interface intelligent teaching application game model diagram (3E).

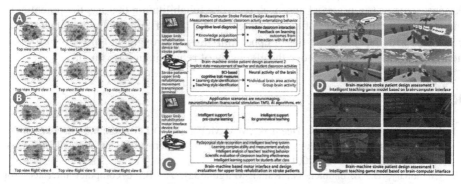

Fig. 3. The plot of data from brain-machine-based upper extremity rehabilitation test assessment and game development for stroke patients. (A) Brain-machine device signals are collected and filtered with band-pass filtering. (B) Brain-machine device-based signal collection and filtering integration and band-pass filtering plots. (C) Simulation diagram of motion intention algorithm modeling and motion trajectory implementation using MATLAB language in Matlab environment. (D) Diagram of the first type of brain-machine stroke patient design assessment approach based on the brain-computer interface for intelligent teaching application game model. (E) Diagram of the second type of brain-machine stroke patient design assessment approach based on the brain-computer interface for intelligent teaching and learning applications.

6   Conclusion and Future Work

This study is based on the fact that sports rehabilitation after a stroke plays a very important role. Through research and clinical brain-computer interface experiments and evaluations, researchers can increase recovery efficiency by 18%. This process is usually completed after the rehabilitation physician evaluates the patient and proposes a corresponding training program. Five types of research methods are adopted, namely: questionnaire survey, field investigation, design evaluation, system programming, and ergonomics. In this paper, a brain-computer interface-based upper limb exercise reha- bilitation training system is designed. The system uses brain-computer fusion as the rehabilitation exercise treatment mode and combines the characteristics of the upper limb muscle strength and movement morphology of the patient with hemiplegia. This study is based on the first clinical trial application of the system, focusing on the verifica- tion and improvement of the 6 follow-up patients. The disadvantage is that the number of patients followed up in this study is not enough, and the disease index based on different patients also lack of research and analysis. More comprehensive evaluation metrics and experimental settings are lacking to better evaluate the performance of the proposed method. This study is based on the key points of brain-computer stroke reha- bilitation exercise interface and design evaluation for patients with upper limbs. Future work will involve more patient recruitment and more sufficient data experimental sample collection, to realize and verify patients' active control. The technical and experimen- tal conclusions of helping rehabilitation patients recover motor function and self-care ability. Overall, this article provides a study of the development and evaluation of a non-invasive brain-computer interface system for motor rehabilitation in stroke patients. Research suggests a practical approach to stroke rehabilitation that has the potential to improve the quality of life of stroke patients. The use of brain-computer integration, combined with the characteristics of upper limb muscle strength and movement form of hemiplegic patients, can improve the efficiency of rehabilitation treatment.

References

1. Wolpaw, J.R., Birbaumer, N., Heetderks, W.J., McFarland, D.J., Peckham, P.H., et al.: Brain-computer interface technology: a review of the first international meeting. IEEE Trans. Rehabil. Eng. **8**, 164–73 (2000)
2. Chaudhary, U., Birbaumer, N., Ramos-Murguialday, A.: Brain–computer interfaces for communication and rehabilitation. Nat. Rev. Neurol. **12**, 513–25 (2016)
3. Ang, K.K., et al.: Clinical study of neurorehabilitation in stroke using EEG based motor imagery brain-computer interface with robotic feedback. In: 2010 Annual International Conference of the IEEE Engineering in Medicine and Biology. IEEE, pp. 5549–5552 (2010)
4. Pichiorri, F., et al.: Brain-computer interface boosts motor imagery practice during stroke recovery. Ann. Neurol. **77**(5), 851–865 (2015)
5. Ramos-Murguialday, A.: Brain-machine interface in chronic stroke rehabilitation: a controlled study. Ann. Neurol. **74**(1), 100–108 (2013)
6. Kiguchi, K., Hayashi, Y.: A study of EMG and EEG during perception-assist with an upper-limb power-assist robot. In: IEEE International Conference on Robotics and Automation, pp. 2711–2716 (2012)
7. Riccio, A., et al.: Hybrid P300- based brain-computer interface to improve usability for people with severe motor disability: electromyographic signals for error correction during a spelling task. Arch. Phys. Med. Rehabil. **96**(3), S54–S61 (2015)
8. Prioria, A., Foffani, G., Rossi, L., Marceglia, S.: Adaptive deep brain stimulation (aDBS) controlled by local field potential oscillations. Exp. Neurol. **245**, 77–86 (2013)
9. Zhou, A., Johnson, B.C., Muller, R.: Toward true closed-loop neuromodulation: artifact-free recording during stimulation. Curr. Opin. Neurobiol. **50**, 119–127 (2018)

10. Weiss, J.M., Flesher, S.N., Franklin, R., Collinger, J. L., Gaunt, R.A.: Artifact-free recordings in human bidirectional brain–computer interfaces. J. Neural Eng.**16**(1), 016002 (2018)
11. Eliseyev, A., Auboiroux, V., Costecalde, T., et al.: Recursive exponentially weighted n-way partial least squares regression with recursive-validation of hyper-parameters in brain-computer interface applications. Sci. Rep. **7**(1), 16281 (2017)
12. Rose, J., Gamble, J.G.: Human Walking. ed. Philadelphia, PA. Lippincott Williams & Wilkins (2006)
13. Milekovic, T., Sarma, A.A., Bacher, D., et al.: Stable long-term BCI-enabled communication in ALS and locked-in syndrome using LFP signals. J. Neurophysiol. **120**(1), 343–360 (2018Jul 1). https://doi.org/10.1152/jn.00493.2017
14. Hochberg, L.R., Bacher, D., Jarosiewicz, B., et al.: Reach and grasp by people with tetraplegia using a neurally controlled robotic arm. Nature **485**(7398), 372 (2012)
15. Harth, E., Csermely, T., Beek, B., Lindsay, R.: Brain functions and neural dynamics. J. Theor. Biol. **26**, 93–120 (1970)
16. Gerstner, W., Kistler, W.M., Naud, R. Paninski, L.: Neuronal dynamics: from single neurons to networks and models of cognition. Cambridge University Press (2014)
17. Dayan, P. & Abbott, L. F. Theoretical neuroscience: computational and mathematical modeling of neural systems (Computational Neuroscience Series, 2001).
18. Gollisch, T., Meister, M.: Rapid neural coding in the retina with relative spike latencies. Science **319**, 1108–1111 (2008)
19. Perdikis, S., Tonin, L., Saeedi, S., Schneider, C., Millán, J.D.R.: The Cybathlon BCI race: successful longitudinal mutual learning with two tetraplegic users. PLOS Biol. **16**, e2003787 (2018)
20. Danig, S., Orsborn, A.L., Moorman, H.G., Carmena, J.M.: Design and analysis of closed-loop decoder adaptation algorithms for brain-machine interfaces. Neural Comput. **25**, 1693–731 (2013)
21. Orsborn, A.L., Moorman, H.G., Overduin, S.A., Shanechi, M.M., Dimitrov, D.F., Carmena, J.M.: Closed loop decoder adaptation shapes neural plasticity for skillful neuroprosthetic control. Neuron **82**, 1380–93 (2014)
22. Weiskopf, N., Mathiak, K., Bock, W., Scharnowski, F., Veit, R., et al.: Principles of a brain-computer interface (BCI) based on real-time functional magnetic resonance imaging (fMRI). IEEE Trans. Biomed. Eng. **51**, 966–70 (2004)
23. Middendorf, M., McMillan, G., Calhoun, G., Jones, K.S.: Brain-computer interfaces based on the steady-state visual-evoked response. IEEE Trans. Rehabil. Eng. **8**, 211–4 (2000)
24. Wang, Y., Zhang, Z., Gao, X., Gao, S.: Lead selection for SSVEP-based brain-computer interface. In: Proceedings of the 26th Annual International Conference of the IEEE Engineering in Medicine and Biology Society, vol. 4, pp. 4507–4510. Piscataway, NJ: IEEE (2005)
25. Krauledat, M., Tangermann, M., Blankertz, B., Müller, K.R.: Towards zero training for brain-computer interfacing. PLOS ONE **3**, e2967 (2008)
26. Gao, F., Yang, Z., Gu, J., Cheng, J.: Machine reading comprehension based on hybrid attention and controlled generation. In: Zhao, X., Yang, S., Wang, X., Li, J. (eds). Web Information Systems and Applications. WISA 2022. Lecture Notes in Computer Science, vol. 13579. Springer, Cham (2022). https://doi.org/10.1007/978-3-031-20309-1_30

Author Index

© The Editor(s) (if applicable) and The Author(s), under exclusive license
to Springer Nature Singapore Pte Ltd. 2023
L. Yuan et al. (Eds.): WISA 2023, LNCS 14094, pp. 637–639, 2023.
https://doi.org/10.1007/978-981-99-6222-8

Printed in the United States
by Baker & Taylor Publisher Services